CANCER WARS

Cancer Wars

How Politics Shapes
What We Know and Don't Know
About Cancer

ROBERT N. PROCTOR

BasicBooks
A Division of HarperCollins*Publishers*

Designed by Ellen Levine

Library of Congress Cataloging-in-Publication Data
Proctor, Robert N. 1954–
   Cancer wars : how politics shapes what we know and don't know
about cancer / Robert N. Proctor.
      p.   cm.
   Includes bibliographical references and index.
   ISBN 0–465–02756–3
   1. Cancer—Political aspects—United States. 2. Cancer—Social
aspects—United States. 3. Carcinogenesis.   I. Title.
RC276.P76 1995
616.99'4—dc20   94–38792
                                                              CIP

95 96 97 98 ❖/HC 9 8 7 6 5 4 3 2 1

For Geoffrey and Jonathan

Contents

Acknowledgments

FINANCIAL SUPPORT for writing this book has come from several different quarters. I'd like to thank my colleagues at the Davis Center for Historical Studies, Princeton University, and the Ethical, Legal, and Social Implications division of the National Center for Human Genome Research for support for the genetics aspects of this study. I would also like to thank Gary Gallagher, head of Penn State University's Department of History, who generously granted me leave to research and write the book.

Those who granted me interviews include: Bruce N. Ames, Victor E. Archer, Donald S. Coffey, Leonard Cole, Devra Lee Davis, Edith Efron, Samuel S. Epstein, Lois Gold, Joseph Fraumeni, Michael Gough, Naomi Harley, John Higginson, Klaus Hueper, Jan Jerabek, Alfred Knudson, Arnold Levine, Jiri Majer, Thomas F. Mancuso, Allan Mazur, John Mulvihill, Anthony V. Nero, William Nicholson, David Ozonoff, David Rall, Vladimír Řeřicha, Arthur Rose, Marvin A. Schneiderman, Donald R. Shopland, and Arthur C. Upton. Readers who commented on one or another portion of the manuscript include, apart from people already mentioned, James R. Adler, John C. Bailar, III, Barry I. Castleman, Robert Cook-Deegan, Alan Derickson, Merril Eisenbud, Már Jónsson, Nancy Krieger, Matthias Leitner, Arno Motulsky, David Pimentel, Susan Rabiner, Londa Schiebinger, Geoffrey Sea, and Bert Vogelstein.

I would also like to give a special thanks to my prodigious research assistant, Keith Barbera, follower of leads par excellence, without whom many of the questions in this book would never have been raised. I dedicate this book to my children, Geoffrey and Jonathan.

CANCER WARS

Introduction

What Do We Know?

Nearly every medical man is assailed at frequent intervals
with the question, Why has cancer increased so much?
—Arthur Newsholme, 1899

For every Ph.D. there's an equal and opposite Ph.D.
—Gibson's Law

MORE THAN A THOUSAND people die of cancer every day in the United States. For every American alive today, one in three will contract the disease and one in five will die from it. Cancer is the plague of the twentieth century, second only to heart disease as a cause of death in the United States and most other First World nations. While most other diseases are on the decline, cancer is on the rise. The trend has been building for some time: Roswell Park, a New York physician, noted as early as 1899 that cancer was "now the only disease which is steadily upon the increase."[1] That year, cancer claimed about 30,000 U.S. lives. In 1994, cancer will kill more than fifteen times that many: 538,000, according to American Cancer Society projections.[2] If present trends continue, cancer will become the First World's leading cause of death sometime in the twenty-first century. It has already reached that status in Japan.[3] This year alone, cancer will kill nearly twice as many Americans as were killed in all of World War II.[4]

RADICAL FACTS

The tragedy is magnified by the fact that the causes of cancer are largely known—and have been for quite some time. Cancer is caused by chemicals in the air we breathe, the water we drink, and the food we eat. Cancer is caused by bad habits, bad working conditions, bad government, and bad luck—including the luck of your genetic draw and the culture into which you're born.*

---

*Cancer is primarily a disease of the elderly, which is why death rates are generally presented as "age-adjusted" figures. Florida, for example, has a much higher annual cancer death rate than Alaska (250 per 100,000 versus 80 per 100,000) due to the fact that Floridians are, on average, significantly older than Alaskans. Adjusting for age, one finds that Floridians actually have a slightly lower cancer mortality rate than Alaskans (163 per 100,000 compared with 184 per 100,000). See the American Cancer Society's *Cancer Facts & Figures—1992* (New York, 1992), p. 7. The failure to age-adjust was usually the single greatest flaw of early cancer registries—the otherwise impressive registry of *fin de siècle* Hungary, for example. See Juraj Körbler, "Die grosse ungarische Krebsstatistik aus der Zeit der Jahrhundertwende," *Krebsarzt* 14 (1959): 224–27.

Much of what is known about the disease stems from its distinctive political geography. The World Health Organization estimates that about one-half of all cancers
occur in the most industrialized one-fifth of the world's population.[5] Even industrialized nations, though, suffer dramatically different kinds of cancer. The leading cancer killer in the United States (for both males and, since the 1980s, females) is lung
cancer; in Japan, it is stomach cancer. Mouth cancer is nearly ten times as common
among French males as among Israeli males; women in England and Wales are more
than five times as likely to die of breast cancer as women in Japan. Global comparisons show that cancer incidence rates may vary dramatically from one nation to another.[6] African-Americans in Atlanta are 28 times more likely than Japanese in
Osaka to suffer cancer of the prostate.[7] Czechoslovakia has the highest colon cancer
death rate on record (41 per 100,000), but its breast cancer rate is only about one-
third that of the United States.[8] Queensland, Australia, is the skin cancer capital of
the world, an unhappy consequence of the migration of light-skinned Britons to a region of intense and unrelenting sun.

Cancer rates can also vary substantially within a given nation. In certain parts of
China, esophageal cancer is 600 times more common than in other parts of that
country. Cancer atlases published by the National Cancer Institute indicate that industrialized states such as New Jersey and New York have higher rates of cancer
than rural states such as Kansas or Arizona.[9] Nonsmoking Iowans have one of the
lowest recorded rates of lung cancer mortality. Americans who live in the Sunbelt
states are more likely to develop skin cancer than Americans who live in the North.
Blacks suffer significantly higher rates of death from nearly every form of cancer
than do whites (skin cancer being the only notable exception), though studies have
shown that it is poverty rather than race that is the root cause of the difference.[10]
Samuel Broder, director of the National Cancer Institute, put it bluntly: "Poverty is a
carcinogen."[11] And cancer may in turn produce poverty: Daniel M. Hayes at the University of Southern California has found that people who survive childhood cancer
tend to earn less money, have more difficulty getting life insurance, and be less
likely to marry than people who never had the disease.[12] Cancer, even if cured, casts
a long shadow.

Occupation, diet, and lifestyle are linked with the disease. Asbestos workers suffer cancers of the lungs and pleural lining; dye workers suffer cancers of the bladder
and stomach. Liver cancer can be caused by exposure to vinyl chloride, lung cancer
by work with arsenic, uranium, or beryllium. Bowel cancer is rare in countries where
meat is not a regular part of the diet and where large amounts of fiber are consumed.
(This is true even after adjusting for the fact that people tend to die younger in Third
World nations, leaving fewer people to reach the ages at which colon cancer is common.)[13] Mormons, whose religious beliefs forbid them to smoke or to drink alcohol,
coffee, or tea, have 20 percent lower cancer death rates than non-Mormons. In
China, esophageal cancer has been traced to diets that include particular kinds of
moldy bread.[14] In France, esophageal cancer has been linked to the consumption of
home-brewed alcoholic beverages (maps of alcoholism and esophageal cancer rates
are almost perfect matches).[15] The Japanese vulnerability to stomach cancer probably derives from their consumption of large amounts of pickled, burned, and high-

salt foods such as seaweed and shellfish, and other foods cooked with soybean pastes and sauces.[16] Cancer of the mouth is common among the peoples of India who chew betel nuts and tobacco leaves. Liver cancer is especially common among the Bantu (in Africa) and in Guam, where foods are often contaminated with afla-toxin, a potent fungal toxin. In China, correlations have been found between women's risk for lung cancer and the cooking fuel they use; how frequently they cook with oil and even the kinds of oil they use have also been implicated (canola [rapeseed] oil vapors appear to be particularly hazardous).[17] In the United States and elsewhere, correlations have been found (and disputed) between fat consumption and breast cancer. Pesticides and PCBs (polychlorinated biphenyls), for example, collect in fatty tissues, and this may explain the link. Fat alone, though, may be the more important offender.[18]

Sexual practices are also involved. Cervical cancer is rare in nuns, but common in prostitutes. A sexually transmissible virus may be the culprit, but perhaps some other environmental agent is at work. Age of childbirth is tied to risk of breast cancer: hav-ing children early tends to lessen a woman's risk, and having children later in life—or not having any children—seems to heighten the risk. Nuns, though they have low rates of cervical cancer, have high rates of breast cancer. Physical exercise seems to help prevent breast cancer: Susan Love, author of *Dr. Susan Love's Breast Book,* sug-gests that support for women's athletics may be one way to diminish the disease.[19]

We also know there is no greater cause of cancer than tobacco. C. Everett Koop, the U.S. surgeon general under Ronald Reagan, estimated in the mid-1980s that cig-arettes alone accounted for between 300,000 and 400,000 American deaths yearly—an annual death toll six or seven times higher than the number of American lives lost in all twelve years of the Vietnam War. Oxford's Richard Doll and Richard Peto argue that tobacco is the single most important carcinogen in the modern world: "In-deed, were it not for the effects of tobacco, total U.S. death rates would be decreas-ing substantially."[20] Two-pack-a-day smokers increase their risk of lung cancer by a factor of twenty and shorten their life expectancy by about eight years. And cigarette smoking has been found to endanger not just the smoker but anyone in the vicinity: the Environmental Protection Agency estimates that more than 3,000 nonsmoking Americans are killed each year by "secondhand" or "environmental" tobacco smoke.[21] A 1991 report in the official journal of the American Heart Association places the figure as high as 50,000.[22]

Cancer, in other words, is not a constant of the human condition but a product of the substances to which we are exposed at home or at work and the lifestyles we lead. Cancer is also a historical disease, insofar as patterns of incidence have changed over time. In 1900, lung cancer was a very rare disease; today, it is the sec-ond largest cause of death in industrial nations.[23] In 1900, more women than men died of cancer (18,000 female deaths compared with 11,500 male deaths in the United States); today, it is men who outnumber women, mainly due to smoking. There are cancers from which only men suffer and cancers from which only women suffer: about 43 percent of all cancers among American women occur in the breast, uterus, or ovary; nearly a third of all cancers diagnosed in men (excluding skin can-cer) are cancers of the prostate.[24] Cancer once seemed to affect European Americans to a greater extent than African-Americans,[25] but today the pattern is reversed. Who gets cancer, where, and how are cultural artifacts. Professional chemists suffer high rates of cancer in consequence of their distinctive occupational hazards.[26] American women suffer more than men from skin cancer on their lower legs, because the transparent nylons introduced in the 1940s exposed women's legs more to ultraviolet radiation from the sun (cotton stockings, woolens, and long skirts shielded previous generations).[27] Males suffer more from skin cancer on the trunk of the body. Skin cancer rates overall are up, thanks to sunbathing fashions (and the decline in hat wearing) begun in the 1960s. Ozone depletion is likely to increase rates, as the natural protective ultraviolet screen of the upper atmosphere begins to diminish with the release of chlorofluorocarbons.

Such knowledge as we have of causes, however, has done surprisingly little to aid us in our search for solutions. Notwithstanding repeated and glowing pronouncements from the American Cancer Society, cancer treatment has made little progress since Richard Nixon declared war on the disease in 1971. Five-year survival rates for the majority of cancers (lung, colon, breast, and stomach, for example) remain essentially what they were back then—even though more than $25 billion has been spent on research by the National Cancer Institute. A 1993 study released by the institute showed that while people diagnosed with cancer do generally live longer—53 percent now live for five years or more, compared with 49 percent in the mid-1970s—part, perhaps even all, of even this meager improvement is due not to improved treatment but to improved diagnostic techniques that make earlier diagnosis possible.[28]

Even where survival rates have improved, this may or may not have to do with the kind of treatment cancer patients receive. Chemotherapy has dramatically increased the life expectancy of children diagnosed with leukemia, and several other rare, once fatal cancers. Hodgkin's disease and testicular cancer are now quite treatable, even curable.[29] Malignant melanoma, a skin cancer that strikes more than 30,000 Americans every year, is nearly 100 percent curable if detected early, and many lymphomas respond well to chemotherapy. But the outlook is less rosy for the really big killers. Ninety-five percent of all lung cancers still prove fatal, regardless of how they are treated. A 1984 report in the *New England Journal of Medicine* reported that people with colon cancer—America's second largest cancer killer (after lung)—who receive chemotherapy alongside standard surgical removal of tumors do not live any longer than people who receive no chemotherapy. The study also found that chemotherapy administered over a standard seventy-week period to adults with colon cancer increased their risk of leukemia.[30] Similar doubts have been raised about the use of radiation and surgery—the other two legs of the traditional therapeutic triad.

Due to these gloomy findings (see the accompanying graphs), it has become hard to deny that the war against cancer is being lost. This was the conclusion reached in an article published in a 1986 issue of the *New England Journal of Medicine*.[31] Stanford University president Donald Kennedy expressed a somewhat stronger view when he called America's cancer campaign "a medical Vietnam." James Watson, co-discoverer of the DNA double helix and one of the nation's most widely respected scientists, called it "a bunch of shit."[32]

Cancer Death Rates by Site, Females, United States, 1930–1990

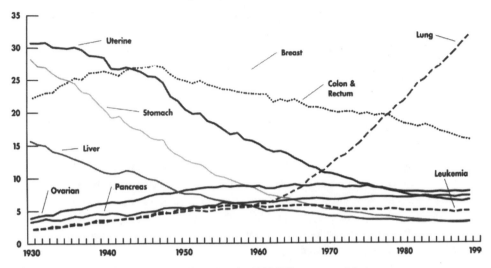

Rates are per 100,000 and are age-adjusted to the 1970 U.S. census population.

Source: American Cancer Society, *Cancer Facts & Figures—1994* (New York: American Cancer Society, 1994), p. 4.

Cancer Death Rates by Site, Males, United States, 1930–1990

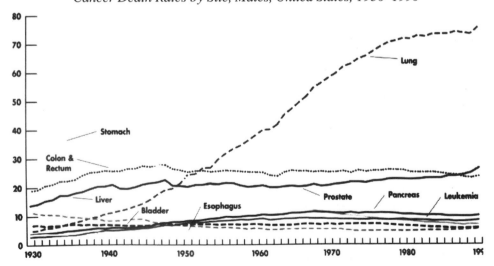

Rates are per 100,000 and are age-adjusted to the 1970 U.S. census population.

Source: American Cancer Society, *Cancer Facts & Figures—1994* (New York: American Cancer Society, 1994), p. 4.

What is going on here? How can one of the largest medical efforts in history, with some of the best and brightest brains behind it, have proved so futile? Why, if cancer is so obviously a product of things like smoking, radon, and the industrial way of life, has so little been done to solve the problem at its roots? Is it true that, as Samuel Epstein of the University of Illinois Medical Center put it: "While much is known about the science of cancer, its prevention depends largely, if not exclusively, on political action"? What, if anything, has changed since the medical historian Henry Sigerist, over sixty years ago, characterized the history of cancer research as "a dry history of errors and of many disappointments"?[33]

SEAS OF CONTROVERSY

The purpose of this book is to explore the political history of cancer, focusing especially on recent debates over its cause. I say *political history* because the question of what causes cancer—and how and to what extent—remains a politically charged one. Environmentalists argue that industrial pollutants are a major cause of the disease; industry trade groups defend the innocence of specific substances with concepts like "acceptable risk" and "genetic susceptibility." There is controversy over the relative significance of lifestyle and occupation in the genesis of cancer; there is controversy over the role of stress, personality factors,[34] and the "natural carcinogens" in things like beer, barbecue, and bruised broccoli. No one really knows how serious a hazard is posed by radon gas seeping into basements or by electromagnetic fields from high-voltage wires and household appliances.[35] Do cellular phones cause brain cancer?[36] What about quartz dust and the natural asbestos (serpentine) on many unpaved California roads: can these cause lung cancer? What about the low-level ionizing radiation one receives while flying in airplanes, especially during sunspot activity, or from mammograms or dental X rays? Does having an abortion increase one's likelihood of getting breast cancer, as pro-life advocates have argued?[37] Does breast-feeding lower the risk?[38]

I say no one knows, but it would be more precise to say that there are islands of agreement separated by deep seas of controversy. Evidence accumulates that chlorinated water causes bladder and rectal cancer, but the Chlorine Institute, a trade association, disputes that claim.[39] Hair colorings are suspected of causing non-Hodgkin's lymphoma, a cancer of the immune system, but that too is disputed. Bruce Ames of Berkeley has generated enormous controversy with his contention that peanut butter and spinach contain natural carcinogens, and that drinking a glass of wine is thousands of times more hazardous than drinking pesticide-contaminated well water. Doubts have been raised about whether the threat of dioxin has been exaggerated,[40] whether all or only one kind of asbestos (amphibole) is dangerous, whether industrial life is as perilous as we are often led to believe.

These debates are often fierce, as competing interest groups jostle one another to exculpate or blame a particular chemical, dietary, or lifestyle nemesis. Tempers run high, because much is at stake. Environmental groups depend to a certain extent on public fears to solicit funds, and manufacturers stand much to gain or lose from whether and to what extent dioxin, vinyl chloride, or pesticide residues are judged public health hazards. Cancer research is big business, and the cancer research establishment has tools that cut in certain directions rather than others. The interests at stake, in other words, are complex and sometimes shifting. Even industry interests are not what they used to be. Billion-dollar industries have sprung up to "abate" or "remediate" environmental hazards such as asbestos, lead, and radon, and much of the industry of the twenty-first century will probably be devoted to cleaning up the industry of the twentieth.[41] The commercialization of environmental troubles has led to novel conflicts of interest,[42] along with unprecedented stakes in the exaggeration or diminution (often by fiat) of environmental hazards. The net result is that experts commonly quarrel over the gravity of specific hazards. Even when experts do agree, opinions often differ about what is to be done, when and where, by and for whom, to what end, and at what cost.

My primary focus is on the *causes* of cancer rather than, say, the efficacy of treatment, because while the rough ingredients are known, their relative proportions (how much each causes) are not—at least not with the kind of consensus we would like. Indeed, the question of the relative contribution to human cancer of environmental pollution, smoking, and natural carcinogens such as radon and plant toxins is among the most controversial in modern science. Strange as it may seem, there is no broad or well-defined consensus to the most important questions: Are overall cancer rates on the rise or the decline? Why do rates for some cancers appear to have risen and others to have fallen? How should the hazards of specific carcinogens be assessed? Is it better to utilize animal studies, bacterial bioassays, or longer-term epidemiological studies? What are the relative risks of polluted water, radon gas in homes, and the natural toxins in peanut butter? Is it dangerous to use a cellular phone or to live near high-voltage wires? How significant is it that Brazil nuts are the most radioactive of all natural foods, that talcum powder contains asbestos, or that antique orange Fiesta dishware emits radon? Does undisturbed asbestos in schools really constitute a major health hazard? Are there "thresholds" of exposure to radiation or chemical toxins below which a given carcinogen is safe? How much do individuals vary in their susceptibility to specific carcinogens—and what should we do with our knowledge of that variability? What are the most pressing research needs? Is it better to fund basic research into mechanisms (for example, molecular genetics), basic research into causes (for example, epidemiology), or applied research into cures (for example, chemotherapy)? When do we know enough to act, and when does the quest for knowledge impede our ability to act? And what kinds of questions never get asked?

These are troubling questions, coming as they do in the midst of growing accusations that environmentalists have exaggerated the risks posed by toxic contaminants. Was the widely publicized Alar scare of 1989 a prudent response to a real threat or hysterical media hype fostered by power-hungry environmentalists? Are worries over asbestos the product of what its manufacturers call "fiber phobia"? Was the 1982 evacuation of the dioxin-infested town of Times Beach, Missouri, a well-advised flight from the deadliest of all modern toxins or an ill-advised cave-in to apocalyptic prophets of doom willing to sacrifice scientific substance to an environmentalist agenda?

Scientists are bitterly divided over how to answer these and many similar questions. Anyone who reads the popular or scientific press will have noticed that the past dozen years or so have seen a massive and concerted effort to reevaluate the threat posed by environmental pollutants, though it is not yet clear whether and to what extent that effort will succeed. A 1991 review of cancer causation in *Science* magazine discussed the cancer-causing potential of gallstones, head trauma, and stomach bacteria, while cautioning that "the widespread public perception that environmental pollution is a major cancer hazard" is incorrect.[43] Such assertions—and there are many—have only recently begun to hit the political fan. The Delaney Clause, the 1958 amendment to the Food and Drug Act barring the addition of artificial carcinogens to foods, will almost surely fall—its detractors note that almost everything is carcinogenic to some degree—but what will it bring down with it?

GENERAL PRINCIPLES

Tracing the outlines of these controversies is a primary goal of this book, but I also want to say something about the practice of modern science. The framework within which I am pursuing these questions is what I call the "political philosophy of science," the central questions of which are the following:

1) Why do we know what we know, and why don't we know what we don't know?* If the politics of science consists (among other things) in the structure of research priorities, then it is important to understand what gets studied and why, but also what does *not* get studied and why not. One has, in other words, to study the social construction of ignorance. The persistence of controversy is often not a natural consequence of imperfect knowledge but a political consequence of conflicting interests and structural apathies. Controversy can be engineered; ignorance and uncertainty can be manufactured, maintained, and disseminated. (As one tobacco company privately put it: "Doubt is our product.") Uncertainty is a central feature of modern knowledge about cancer.

2) Who gains from knowledge (or ignorance!) of a particular sort and who loses? Scientists and science mongers often speak royally about how "we" know this or that—but who is that *we* and why does it know what it knows? What, for example, is the connection (if any) between the fact that 90 percent of cancer researchers are males and that men predominate as subjects in clinical trials for cancer drugs?[44] Who does science and who gets science done to them? What or whom is knowledge for, and what or whom is it against?[45]

---

*Historians and philosophers of science have tended to treat ignorance as an ever-expanding vacuum into which knowledge is sucked—or even, as Johannes Kepler once put it, as the mother who must die for science to be born. Ignorance, though, is more complex than this. It has a distinct and changing political geography that is often an excellent indicator of the politics of knowledge. We need a political *agnatology* to complement our political epistemologies.

3) How might knowledge be different, and how *should* it be different? What are the virtues of looking at ultimate rather than proximate causes, for example, or of seeking prevention rather than cure? What are the social responsibilities of the cancer theorist or, for that matter, of the science theorist?

Questions such as these are self-consciously political and ethical; they take us beyond the merry-go-round of realism versus relativism that plagues so much of recent science studies, perhaps even beyond the postmodern accounts of science that rework the primacy of theory over experiment or experiment over theory without challenging the direction of research itself.[46] These questions allow us to take seriously the problem of the social responsibility of science, but they also require that we entertain a broader notion of science than is common coin among positivists, postpositivists (notably Thomas Kuhn), and some of the more timorous social constructivists. Science has a face, a house, and a price: it is important to ask who is doing science, in what institutional context, and at what cost. Understanding such things can give us insight into why scientific tools are sharp for certain kinds of problems and dull for others. It might also help us see why the war against cancer is going so badly.

PLAN OF THE BOOK

Chapter 1 begins with a look at early conceptions of trends and causes. Cancer is an ancient disease, but little is known about rates prior to the nineteenth century. Scholars today debate how much of the recent increase is due to diagnostic improvements and better publicity, but readers may be surprised to find that the debates over whether cancer is on the rise—complete with arguments over diagnostic artifacts—date from the middle of the nineteenth century. Cancer was dubbed a "disease of civilization" more than 150 years ago, and its apparent statistical increase has been disputed ever since.

Chapter 2 examines the life and science of two of the foremost twentieth-century advocates of the environmentalist view of cancer: the little-known Wilhelm Hueper and his most influential disciple, Rachel Carson. Hueper was head of the Environmental Cancer Section of the National Cancer Institute from 1948 to 1964 and was celebrated by Rachel Carson in *Silent Spring* as "the father of environmental carcinogenesis." Hueper is interesting for a number of reasons, not the least of which is his suppression by the Atomic Energy Commission (AEC) in the 1950s, following his efforts to document lung cancers among the uranium ore miners of the Colorado Plateau. Hueper later complained that censorship of his and other studies had delayed measures to remedy the situation, leaving countless men to die who otherwise could have been saved. The same was true, as we shall see, in other countries, like East Germany and Czechoslovakia, which had their own Huepers and many more victims kept in the dark. We also look at how Hueper's ideas about environmental carcinogenesis informed Rachel Carson's views on human health. Carson is best known for her concerns about wildlife killed by pesticides, but one chapter in *Silent*

Spring explains the growth of human cancer as a consequence of unrestricted pesticide spraying—an idea that remains as controversial today as it was in 1962.

In chapter 3 I trace the history and impact of the announcement in September 1978 by Joseph Califano, secretary of Health, Education, and Welfare, that in the coming decades as many as 20 to 40 percent of all cancers could be caused by exposures to only six industrial pollutants commonly found in the workplace. Samuel Epstein buttressed these claims that same year in his provocative book *The Politics of Cancer;* few books have ignited such a firestorm of controversy. I explore the response of industry scientists to these claims and the influential critiques of Richard Doll and Richard Peto, authors of *The Causes of Cancer* (1981), written as part of a congressional Office of Technology Assessment project to elucidate what causes cancer and to what extent. My goal here is to broaden the usual notion of cancer causation, to challenge the neat separation of causes into categories: smoking, diet, occupation, and so forth. Cancer, like any other disease, has broader social causes, and recognition of these broader social causes can help us plan appropriate avenues of prevention.

Chapter 4 explores the political chill that swept over environmental and occupational health agencies in 1980 with the election of Ronald Reagan as the nation's fortieth president. The numbers here tell much of the story. Agencies such as the Occupational Safety and Health Administration, the Environmental Protection Agency, and the Consumer Product Safety Commission saw their budgets slashed and their department heads replaced by pro-industry appointees. Cost-benefit analysis and risk assessment came of age as the new administration championed the view that public health and safety are either not worth the cost or are better served by deregulation. I explore the ideological effluent of the Reagan revolution and its life-and-death consequences, also some of the rhetorical strategies used by architects of environmental deregulation.

In chapter 5 I examine the rather shady influence of industrial research bodies and trade associations on cancer research and policy. The Tobacco Institute and Council for Tobacco Research are well known, but there are hundreds of lesser known examples of this growing trend to wield science as an instrument of public relations. The Asbestos Information Association warns against "fiber phobia"; the Institute of Shortening and Edible Oils funds research seeking to demonstrate the food value of tropical oils and other fats. The American Meat Institute discredits charges that meat causes breast and colon cancer; the National Coffee Association disputes evidence that caffeine causes cancer of the pancreas. The Calorie Control Council has managed to keep saccharin on the market ever since it was shown to cause cancer in rats; the Chlorine Institute has helped to rescue the reputation of dioxin.

This chapter explores how these trade associations have come to recognize the value of what *Time* magazine has called "Hertz rent-a-scientists." Such associations make it their business to exploit, disseminate, and, if need be, produce uncertainty in the area of environmental carcinogenesis. One consequence is what I call the MacNeil-Lehrerization of science: the media display of equal and opposing views on questions of carcinogenic hazard, the triumph of what PR men call Gibson's Law—"For every Ph.D. there's an equal and opposite Ph.D." A related phenomenon is the "smokescreen effect"—the effort to tie up scientific traffic with true but trivial work, to draw attention away from what is really going on. I argue that more attention needs to be paid to the nature of bias in such research, and to the more general question of how science is used for political and ideological purposes—how light is used to cast confusing shadows.

Chapter 6 looks at recent debates surrounding the notion, popularized by Bruce Ames of Berkeley, that "natural carcinogens" in foods pose a far greater human health hazard than industrial pollutants or pesticides. The thesis has been elaborated in several lengthy articles and letters in *Science* magazine, and has been endorsed by several leading scholars in the American Association for the Advancement of Science, notably Philip Abelson and Daniel Koshland. Ames is not easily dismissed: he is, as he likes to point out, the twenty-third most cited scientist in the English-speaking world. He was once better known for his environmental activism: he crafted the science that led to the ban of Tris, the carcinogenic pajama flame retardant, in 1977; he was also one of the first to discover carcinogens in hair dyes. He is the inventor of the bacterial bioassay (the "Ames test") used in laboratories throughout the world to determine carcinogenic hazards.

Ames's argument, in short, is that plants have developed natural chemical defenses against many insect and fungal pests, and that human consumption of such "natural pesticides" poses a far greater cancer hazard than industrially derived toxins. His point is not to introduce new fears but to assuage old ones: Ames argues that public fears of pesticides and food additives are baseless, and that DDT and dieldrin would actually, if relegalized, reduce cancer by making cancer-fighting fruits and vegetables cheaper. The critical response, as one might imagine, has been immense, and in some cases vicious.

Chapter 7 is an exploration of the political history of what might, at first, appear a purely technical issue: the shape of dose-response curves. Central here are the accusations that the animal studies that form the stock-in-trade of carcinogenic bioassays are fundamentally flawed. Bruce Ames is again at the center of controversy. In the early 1970s he helped establish the popular notion that a single molecule of a carcinogen could cause cancer; today he is one of its strongest critics. Ames now argues that animal studies cannot provide an accurate basis from which to judge human risks. High-dose exposures may themselves trigger cancer by stimulating cellular replication, which in turn amplifies endogenous mutations. If cellular proliferation is as important in the generation of cancer as damage to the genetic material, then there is good reason to believe that animal studies exaggerate the cancer risks of low-level exposures to carcinogens.

The question of the magnitude of the hazard posed by low-level carcinogens remains hotly debated. Important to recognize in the debate, I shall argue, are different conceptions of the body and different conceptions of regulatory prudence. Environmentalists tend to champion a "body victimology"—regarding the body as passive, helpless in the face of carcinogenic insults. Threshold advocates, by contrast, tend to express a "body machismo"—putting a great deal of emphasis on the fact of DNA repair, cellular shedding, and other means by which the body responds to toxins. Environmentalists tend to see nature and the body as weak, and the human capacity to right wrongs as strong. Anti-environmentalists tend to see nature and the body as strong and therefore able to withstand a certain level of toxic insult.

Chapter 8 looks at the history of radiation cancer—from the first death, in 1904 in Thomas Edison's laboratory, to the scandalous censorship at home and abroad of scientists who sought to warn us in the 1950s and 1960s of the hazards of uranium mining. I discuss the tragedy of the radium-dial painters of the 1930s and the controversies surrounding the cancer hazard posed by the atomic bombs dropped on Hiroshima and Nagasaki. I also look at the famous uranium mines of Schneeberg and Joachimsthal, in the mountains separating Germany and Czechoslovakia, where the lung cancer hazard of mining was first discovered. I traveled to Czechoslovakia in the summer of 1992 to examine archives pertaining to the 1879 discovery of the so-called *Schneeberger Krankheit* (uranium mine-induced lung cancer); what I found was that more people have died from the disease since the 1950s than in all of previous history. Czech atomic authorities provided me with details on the uranium miners and the rates of lung cancer, and even on the total quantity of uranium produced—but there was more to the story: the shocking revelation that "uranium mine concentration camps" had been established in the late 1940s and early 1950s to force tens of thousands of political prisoners to work the mines.

Chapter 9 zeroes in on one of the hottest recent radiation controversies: whether and to what extent household radon poses a lung cancer hazard. The puzzling question here is why it was not until the mid-1980s that a widespread public health threat was broadly recognized. The first scientific article postulating a radon hazard in household cellars appeared in 1931 in Germany, and scientists were well aware of the radon hazard in uranium mines by the 1940s. I investigate the idea that concerns about radon followed closely on the heels of a broader interest in "indoor air pollution," stimulated primarily by fears that by insulating our homes to conserve energy, we were also trapping inside a number of dangerous gases.

Here again we have a case of the social construction of ignorance, a case where it is the *absence* of a discovery (the dog that did not bark, as both Hegel and Holmes once had it) that is the most poignant fact in need of explanation. Social forces can leave scientific gaps; historians of science can profitably study those gaps—the rich history of scientific nonevents. Prior to the 1980s, household radon appears to have fallen into such an ideological gap. Environmentalists concerned about radiation tended to focus on the hazards of military, industrial, or medical exposures (especially from atmospheric testing, atomic power, and X-ray diagnostics); health physicists trained in Atomic Energy Commission traditions had their own distinctive myopias, insofar as they were foxes guarding the henhouse. Neither group was prepared to appreciate the possibility of a household hazard. The absence of well-defined interest groups allowed the oversight, though once it was discovered, the nuclear industry was able to make political hay out of the issue. Nuclear advocates argued that the weather-tightening of homes to conserve energy was actually *increasing* our radioactive exposures by a far larger measure than the emissions of nuclear power plants and nuclear waste. The specter of household radon allowed the industry to argue that nuclear energy could actually lower our risks of radioactive exposures.

Chapter 10 examines the history of the idea of differential susceptibility to cancer. The question of predisposition is likely to become more and more "loaded" as new genetic technologies allow us to detect—far in advance of symptoms, often in utero—a broad range of what have come to be called genetic lesions. Cancer predisposition genes are prominent among the oracular goals of the new genomics, and there are widespread hopes (amid occasional fears) that markers for susceptibility even to very common cancers will soon be discovered. Such a prospect, of course, brings with it a host of troubling ethical prospects. As genetic predispositions are identified for more common cancers, public demand for testing may increase by leaps and bounds, proving once again that invention is the mother of necessity (Thorstein Veblen's perceptive caution). People designated as "carriers" may find themselves faced with new forms of stigma and discrimination, and there is the prospect of genetic screening in the workplace and racial stereotyping. Policy implications are likely also to emerge in the area of environmental regulations: if the (mis-?) conception grows that "genetics is more important than environment" in the onset of certain diseases, lawmakers may find themselves less willing to enact pollution prevention legislation. There is a growing tendency to argue that certain individuals may actually be invulnerable to certain kinds of cancer: tobacco lobbyists are masters at this art, having long sought to convince the public that smoking causes cancer only in those persons for whom there is already a genetic predisposition—a rhetorical ploy that loses none of its force by virtue of being a tautology.

A final chapter on "How Can We Win the War?" looks at the broader concept of cause, the question of trends again, and some of the professional and political obstacles to effective prevention. This chapter also explores an insufficiently appreciated conflict of interest stemming from the fact that there are two very different senses of what it means to be "conservative" when it comes to cancer. On the one hand, scientists qua scientists often recognize that it is important not to overstate the gravity of a particular hazard: estimates of risk are therefore commonly expressed conservatively, as minimum values. Environmental and public health agencies, by contrast, have a responsibility to protect the public, and "conservatism" for them can mean something quite different. Environmental conservatism means erring on the side of human health and safety, restricting the public's exposure to a toxic substance, for example, even in the face of uncertainty about its danger. These two conceptions are often at loggerheads. Appreciating how this tension is resolved (or left unresolved) can help us understand the enduring volatility of both the science and the politics of cancer.

This chapter also reasserts the importance of understanding knowledge of cancer in terms of the social construction of ignorance. Ignorance is surely one of the most prominent features of knowledge about cancer in the late twentieth century. That ignorance is not just a natural consequence of the ever shifting boundary between the known and the unknown but a political consequence of decisions concerning how to approach (or neglect) what could and should be done to eliminate the disease. Ignorance is socially constructed by outright censorship (admittedly rare), by failures to fund, by the absence or neglect of interested parties, and by efforts to jam the scientific airwaves with noise. Science, public policy, and public opinion are all affected.

My main concern throughout is that inadequate attention has been given to where and how one should look and act to discover causes and organize prevention. The concept of causation is part of the problem: much of the politics of cancer research lies in how far down in the chain of causation one is willing and able to look. Does one stop (or start) with genetics, or immunology, or epidemiology? When is a nation's health care or environmental policy (or the absence thereof) the cause of cancer? Can cancer be caused by elections, or recessions, or ad-induced habits and fashions? The question of where the "real cause" lies is politicized, with the Left favoring ultimate (or social) causes and the Right favoring more proximate causes or mechanisms. The emphasis on immediate mechanisms rather than social causes means that some of the more practical questions of cancer research—like how to get people to stop smoking or to remediate the radon in their workplaces—are rarely the objects of well-funded scientific attention.

Two caveats: I do not claim to have treated all of the contested causes of cancer. Viral oncology, for example, is a well-established field that I almost totally ignore in these pages.[47] Viruses have been implicated in some rare leukemias, in nasopharyngeal tumors in China (the Epstein-Barr virus), in pediatric lymphoma in Africa, and in liver cancer in Africa (hepatitis B). Viruses may be important in the progression of cervical irritation to cancer, and the HIV virus can leave one vulnerable to cancers typical of the AIDS epidemic (notably Kaposi's sarcoma, but also certain kinds of invasive cervical cancer). Viruses have proved useful in research into the mechanisms of carcinogenesis (the first in vitro cancer tissue cultures, for example, used the Rous chicken sarcoma), and viral research laid the groundwork for the discovery of oncogenes.[48] Contrary to long-held expectations, however, viral infection appears to play a role in only a tiny minority of cancers. Even considering the impact of AIDS, viruses probably account for no more than 1 percent of all cancers.[49] Granting even that viruses may play a heretofore undiscovered role in other cancers, it is probably fair to say that undue attention has been given to this particular collection of agents. The same could be said for infectious agents more generally.

Second, I have slighted recent debates concerning the role of diet in cancer. I deal at various points with some of the elements of this issue: the controversy over the relative import of lifestyle and occupational factors, for example, and Bruce Ames's thesis that foods contain natural carcinogens. I do not look, however, at the history of debates over the perplexing role of dietary fat and dietary fiber. As is increasingly becoming apparent, prevention means not just reducing one's exposure to carcinogens but consuming an adequate supply of "anticarcinogens," especially the vitamins and other antioxidants found in green and yellow vegetables. Such areas are not without controversy, as indicated by a recent *New England Journal of Medicine* study suggesting that vitamin supplements may actually increase certain kinds of cancer.[50] I have not explored these important themes only because they have been admirably treated by other scholars.[51]

Much of what follows is not happy reading. A colleague once characterized my work as treating "the pathology of science," and I must confess that I do share a certain sympathy with Hueper's characterization of his method as "one piece of dirt leading to another." The story of cancer and cancer research is not an upbeat one, and the stables of the cancer research establishment are not what one would call clean. Understanding requires that we look at the seamy side of science, not just at its polished carapace. We are rightfully suspicious of fairy tales that recall the triumph of medical reason over popular superstition, but I don't take that as grounds for any kind of pessimism. There are good reasons to believe that straightforward public health measures could greatly lessen our cancer load, and that is reason, I believe, for optimistic activism.

Optimistic activism, though, requires a solid understanding of the politics of science: how priorities and practices are shaped by power formations, ideological gaps, interests and apathies, government and industrial support (or lack thereof), disciplinary dogmas, and professional or institutional parochialisms. Recent debates over what causes cancer bring these concerns into focus. The terrain is among the most highly contested in all of modern science. My hope is that by examining how and why such questions have become the objects of controversy, we may learn more about how politics shapes science—and perhaps even more about what really causes cancer and what to do about it.

Chapter 1

A Disease of Civilization?

With regard to cancer, it is not only necessary to observe the effects of climate and local situation, but to extend our views to different employments, as those in various metals and manufactures; in mines and collieries; in the army and navy; in those who lead sedentary or active lives; in the married or single; in the different sexes, and many other circumstances. Should it be proved that women are more subject to cancer than men, we may then inquire whether married women are more liable to have the uterus *or breasts affected; those who have had children or not; those who have suckled, or those who did not; and the same observations may be made of the single.*

—Society for Investigating the Nature and
Cure of Cancer, Edinburgh, 1802

CANCER IS AN ANCIENT DISEASE. Evidence of tumors has been found in dinosaur bones more than a hundred million years old and in the human tissues of Egyptian mummies dating back four or five millennia.[1] Healers commented on the presence of tumors as early as 1500 B.C.: the famous Papyrus Ebers contains a section on malignant and nonmalignant tumors, and there are several other early discussions of this sort.[2] Hippocrates of Cos (circa 460–377 B.C.), who came up with the modern term for cancer (*carcinoma,* from *karkinos,* meaning "crab," referring most likely to the crablike appearance of a heavily vascularized breast tumor), described what are clearly cancers of the lip, skin, rectum, stomach, uterus, and female breast; he also distinguished ulcers (*carcinoma apertus*) from internal tumors (*carcinoma occlusus*). Hippocrates held that cancer, like all diseases, was an imbalance of bodily "humors" brought on by improper diet and exercise, or the vagaries of climate, age, and season; cancer was an excess of "black bile" (melancholia). Galen, in the second century A.D., followed Hippocrates in his belief that cancer was at root a humoral imbalance; his advice, offered by European and Middle Eastern physicians for 1,500 years, was to follow a strict dietary regimen to restore the body's balance—avoiding walnuts, for example, and other foods that might provoke the "melancholic" humor.

It is often hard to tell, though, whether an ancient malady should be classed as cancer in the modern sense. Today we distinguish sharply between a *cancer* and a *canker,* for example, but prior to the advent of microscopic diagnostics, that distinction was routinely blurred. The *Oxford English Dictionary* defines cancer in the modern sense as "a malignant growth or tumour in different parts of the body that tends to spread indefinitely and to reproduce itself. . . . it eats away or corrodes the part in which it is situated, and generally ends in death." A canker, by contrast, is "an eating, spreading sore or ulcer: a gangrene." The terms were often interchanged, revealing a basic ignorance of the cellular processes that distinguish a cancer in the modern sense from a benign tumor or ulceration. We can look back and try to guess, from symptoms and other clues, how a particular *cancre,* scirrhus (tumor), or ulceration would be classified today, but this is obviously an effort fraught with difficulties. The "cancerated" breast and nose described in the seventeenth century might well have been cancer in our sense, but a "cancerated hand" caused by banging on a nail was almost surely not.

THE QUESTION OF TRENDS

In light of these and other difficulties,[3] it is not hard to understand why we have little sense of what cancer rates were prior to the mid-nineteenth century. We do know—with varying degrees of certainty—that some very famous people died of what we would now probably recognize as cancer: Atossa (daughter of Cyrus), Caesar Augustus, Nero, Genghis Khan, Attila the Hun, Charles the First, Anne of Austria, and Frederick the Great. David Hume is said by some to have died of cancer, as is Napoleon Bonaparte and Ludwig van Beethoven.[4] Art historians have speculated that the woman portrayed in Michelangelo's "La Notte" may have suffered from breast cancer, and that the same fate may have befallen Rembrandt's mistress, the model portrayed in "Bathsheba at Her Toilet."[5] Maximinus I (A.D. 173–238), emperor of Rome, is sometimes said to have suffered from a pituitary tumor, and at least one scholar has speculated, judging solely and rather imaginatively from a bulge in his breastplate, that Tutankhamen, king of the eighteenth dynasty of Egypt, may have suffered from an adrenal cancer.[6]

We do not know, though—and presumably have no way of finding out—how *often* people died of cancer in comparison with other diseases.[7] (We cannot yet tell such things from bones alone.) European governments began to record causes of death, but statistics as a formal body of inquiry did not exist until the first part of the nineteenth century.[8] The few early records we have are of little use in illuminating cancer incidence: the remarkably well kept register of Market Deeping in Lincolnshire, for example, lists the causes of death for 387 persons between 1711 and 1723, but records only two deaths from cancer. Was the disease indeed that rare, or did cancers of the internal organs simply go undiagnosed?[9] The same problem attaches to the records compiled in the *Collection of the Yearly Bills of Mortality from 1657 to 1758,* published for London in 1759. About a quarter of 1 percent of the 2 million deaths recorded in this register are listed as "cancer," though the term used at this point included "fistula" and gangrene. And many cancers of internal organs, again, were no doubt not recorded.[10]

The first quasi-reliable health statistics for England and Wales date from 1837, when laws were passed to mandate the registration of all deaths. The registry published for the year ending in June 1838 shows an overall cancer death rate of 166 per million per year, a figure that increased more than fivefold by 1905.[11] What, though, are we to make of such figures? Decline of mortality from other causes was no doubt partly responsible for the increase, as was the overall rise in longevity. The quality of reporting for many of these years was also poor: many deaths are listed as due to either "indefinite causes" or "old age," and disease reporting was anything but uniform. British and American government registries didn't distinguish different kinds of cancer until the early years of the twentieth century; German national statistics didn't even list cancer as a cause of death until after 1905.[12]

Cancer was also often difficult to diagnose prior to the era of medical microscopy, cellular pathology, and X-ray photography. Compound microscopes were invented in the seventeenth century, but cancers were not identified by this means until the 1830s and pathologists were generally not well trained in the art until the second half of the century.[13] Lung cancer was notoriously difficult to distinguish from other respiratory ailments prior to the invention of X-ray photography in the 1890s;[14] imperfect access to hospitals must have further depressed these early figures. Even today, there are those who claim that cancer statistics in the second half of the twentieth century are distorted by ever-improving diagnostics, increased access to medical care, and media attention to the cancers of national celebrities—Betty Ford's breast cancer, for example, or Ronald Reagan's colon polyps, both of which stepped up rates of diagnosis in the general population. As *Time* magazine puts it, "history teaches that there is no greater publicity for any disease than to afflict a part of the presidential anatomy."[15]

The question of whether cancer is on the rise was already debated in the first half of the nineteenth century.[16] Stanislas Tanchou (1791–1850), a Parisian physician, was one of the first to attempt a proof using statistics. In a 1843 memoir addressed to the illustrious French Academy of Sciences, Tanchou suggested that cancer was "like insanity, much more common in civilized nations." Tanchou took great pains to document, based on mortality records collected by the Département de la Seine, that cancer deaths had increased in Paris from 595 in 1830 to 779 in 1840; he also showed that cancer rates were higher in Paris than in the suburbs. Cancer was more common among Christians than among Moslems, among the French than among the English.[17] Cancer was more common among domestic than among wild animals and, in Egypt, was more often found among Turkish women than among Fellah women. Cancer was "practically unknown" in America and Africa.[18]

Domenico Rigoni-Stern (1810–55) was another who argued that cancer was more common in cities. In 1842, in a remarkable paper based on his analysis of Veronese death registries from 1760 to 1839, the Italian surgeon concluded that cancer was more common in towns than in the countryside, and that it increased with age. He showed that women died from the disease more often than men (about eight times more often, according to his figures), and that unmarried women, especially nuns, were particularly vulnerable to cancer of the breast. Nuns, indeed, were about five times more likely to die of cancer than other women. Cancer of the face was about as common among men as among women, indicating that cancers in different parts of the body must have different causes.[19] Rigoni-Stern backed up all these conclusions with "statistical facts," though he recognized that there were systematic biases in how such facts were gathered and analyzed. He later criticized Tanchou for generalizing about cancer trends on the basis of only ten years of observation.[20] He was unsure why cancer struck women more often than men, but wondered whether it might have something to do with the "natural functions" of the uterus and the breast or perhaps with the susceptibility of these organs to "mechanical damage." He speculated that the high rate among nuns might have something to do with their consumption of fish or their long and repeated fasts, or perhaps with their practice of compressing their breasts during prayer or their custom of wearing tight-fitting habits.

John Le Conte (1818–91) was apparently the first American to argue that there had been a "remarkable increase of mortality from cancerous affections" in England and in Wales. He first published such a claim in 1842, making him—along with Tanchou and Rigoni-Stern—one of at least three scholars to publish such a thesis independently within the space of about a year.[21] In 1846, in an article for the *Southern Medical and Surgical Journal,* the young physician cited Tanchou's observations that cancer was predominantly a disease of old age rather than of youth, of cities rather than of rural districts. He conceded that it would be "premature" to attempt to express this quantitatively, suggesting instead that intelligent discussion of such questions "must be postponed until some zealous investigator of vital statistics arises, who has the leisure and the courage to properly analyze the vast accumulation of valuable facts which are entombed in the mortuary registers of the last forty years." He also seems to have kept a sense of humor about the matter: his 1846 endorsement of the Tanchou Law (that cancer increases in proportion to the intensity of human civilization) is tempered by his observation that it was "doubtless in no small degree flattering to the national vanity of the French *savant,* that the average mortality from cancer at Paris during 11 years is about 0.80 per 1,000 living annually, while it is only 0.20 per 1,000 at London!!! Estimating the intensity of civilization by these data, it clearly follows that Paris is 4 times more civilized than London!!"[22]

Le Conte defended Tanchouism for nearly half a century: as chair of chemistry at the College of Physicians and Surgeons in New York, as the first president of the University of California (in 1869), and as an expert consultant for the California State Board of Health.[23] In 1888, he speculated that increased longevity, the more thorough certification of death, and the more common use of microscopic diagnostics must account for part of the difference.[24] As early as 1842 he argued that the increase might be due to "the wretched condition of the immense population employed in the manufacturing and mining districts of England"; he was also already aware, though, that part of the increase was "only *apparent,* arising from more careful registration, and from improvements in pathology and diagnosis."[25]

Testimonials to the rarity of cancer among uncivilized peoples proliferate around the turn of the century. Missionaries reported that cancer was rare or absent among the Inuit of the Arctic, the Hunza of the Himalayas, and other "primitive" peoples.[26] The missionary Livingstone French Jones stated in his 1914 *Study of the Thlingets of Alaska* that "tumors, cancer and toothache" were unknown to the people of the Pacific Northwest until contact with the white man. Albert Schweitzer reported that when he arrived in the West African nation of Gabon in 1913, he was "astonished to

encounter no case of cancer," attributing this to their salt-free diet.[27] Vilhjalmur Stefansson, the Arctic explorer, reviewed dozens of claims by medical missionaries,
whalers, frontier physicians, and others with firsthand experience that cancer was
rare or absent among the natives of the Pacific Northwest and the Arctic, as among
primitive peoples more generally. Cancer, according to Stefansson, was known only
in those few places where European diets and other habits had been adopted—especially smoking and drinking. It was not until 1933, he claimed, that a cancer was
positively identified among native northern Alaskans.[28]

Statistical analyses continued to indicate an increase among "civilized" populations. In 1899, for example, Roswell Park, a professor of surgery at the University of
Buffalo Medical Department, asserted that "cancer is now the only disease which is
steadily upon the increase." Park supported this statement with statistics from England and Wales indicating that cancer death rates had grown fivefold from 1840 to
1896. In New York State, too, cancer had increased faster than one would expect
from population growth and aging alone: Park argued that the increase was "certainly not due to improvements in methods of diagnosis," since the previous era's
misdiagnoses were as likely to exaggerate as to underestimate rates. Park issued a
"startling prophecy": if things were to continue on their present course, it would be
only a decade before cancer deaths would outstrip deaths from consumption, smallpox, and typhoid fever combined.[29] Arthur Newsholme, an examiner in preventive
medicine at Oxford, pointed out that same year that cancer was already killing more
Englishmen than any other disease, with the exception of bronchitis, pneumonia, and
"phthisis" (tuberculosis).[30]

The fact of increase all along, though, was controversial. In 1844 Rigoni-Stern had
argued that at least part of the increase might be due to how the data were gathered.[31]
Walter H. Walshe, professor of pathology at University College in London and author
of the most comprehensive mid-nineteenth-century treatise on cancer, agreed, as did
several others.[32] George King, secretary of England's Institute of Actuaries, and
Arthur Newsholme, author of a highly regarded treatise on vital statistics, pointed out
in 1893 that in Frankfurt, Germany, where cancer deaths had long been classified by
site of origin, there had been no increase in cancer between 1860 and 1889 in those
parts of the body in which the disease is most easily detected (breast, tongue, and
uterus, for example), the implication being that previously "inaccessible" cancers
(lung and liver, for example) were simply being better diagnosed now than in former
times.[33] August Hirsch, in the 1883 edition of his *Handbook of Geographical and
Historical Pathology,* used these same data to argue that the apparent increase in cancer was "merely an illusion due to improved and more comprehensive collection of
the statistics of mortality." The distinguished Berlin medical professor concluded that,
despite testimonials to the rarity of cancer in many parts of the world, there was "no
support" for the idea of cancer as a disease of civilization. The problem, as he saw it,
was that the number of cases appeared to be independent of population density; there
was, furthermore, he argued, little evidence that the disease was more common in
large towns than in small. In Massachusetts in 1850, for example, cancer appeared to
be more than twice as common in small towns and rural districts as in Boston.[34]

In 1899, Newsholme extended the skeptical argument to England. Many of the people formerly classified as dying from "old age" or "indefinite causes" were now being correctly classed as cancer victims, and there were enough such cases to account for most of the observed increase. English deaths from "indefinite causes" declined from 143,000 in 1867 to only 69,000 in 1895: How many of those in the earlier years had actually been cancer? Insurance records showed a slower rate of increase than the public registry; this was significant, he suggested, given that insurance companies would presumably be especially keen to determine actual causes of death. He also compared rates of increase between men and women, believing (wrongly) that a differential increase would be evidence of diagnostic bias. Cancer registries indicated a faster increase among men than among women, and he used this to argue that the increase was only apparent, since the cancers suffered by men were traditionally more difficult to diagnose. Interestingly, he does not seem to have considered the possibility that men could have been exposed to more rapidly increasing levels of carcinogens—through smoking or workplace hazards, to name just two possibilities. Nor did he consider the fact that "the insured" might have experienced a slower rate of increase due to the fact that, being fairly well off, they were less likely to have been exposed to certain kinds of carcinogens than were members of the lower classes.[35]

It is difficult to sort out who was right on the question of long-term cancer trends. Time seems to have favored the argument for increase. A 1906 report by the U.S. Census asserted that there was a "steady increase in the death rates due to cancer," and a 1909 report noted that cancer was the eighth leading cause of death in the country, accounting for more than 4 percent of all deaths. W. Roger Williams devoted several pages of his 1908 *Natural History of Cancer* to a critique of Newsholme's thesis, pointing out that though improved diagnosis had indeed led to additions in the registry, it had also led to subtractions, since several ailments once classed as cancer (notably lupus, fibroid tumors, and polyps) were now no longer so classed.[36] Frederick L. Hoffman, chief statistician for the Prudential Insurance Company, extended Williams's critique in his 1915 book, *Mortality from Cancer Throughout the World*. Hoffman pointed to several flaws in King and Newsholme's analysis (skin cancer, for example, had been classed as "inaccessible"), and refigured the Frankfurt data to show an overall increase of cancer for the period 1906 to 1913. Diagnostic improvements alone, in his view, could not account for the increase: the low rate of breast cancer among Japanese women, for example, was proof that a nation's low cancer incidence was not necessarily a sign of its medical incompetence. His conclusion, buttressed by some 600 pages of charts and statistical tables, was that cancer mortality was increasing "at a more or less alarming rate throughout the entire civilized world."[37]

Cancer by this time, according to contemporary statistics, was killing 10 percent of American men and 19 percent of American women above the age of forty-five. Worldwide, according to Hoffman, cancer was killing roughly half a million people per year, and in "civilized" nations such as the United States, the rate of increase was on the order of 2.5 percent per year.[38] Cancer surpassed tuberculosis as a reported cause of American deaths in 1924, and by 1934 was the nation's number-two killer. Only heart disease was (and still is) more deadly.

The apparent increase in cancer was already used, in the 1920s, to try to bring the U.S. government into the struggle against cancer. In a famous 1928 speech before Congress, Senator Matthew Neely of West Virginia waxed apocalyptic about "Cancer—Humanity's Greatest Scourge":

I propose to speak of a monster that is more insatiable than the guillotine; more destructive to life and health than the mightiest army that ever marched to battle; more terrifying than any scourge that has ever threatened the existence of the human race. The monster of which I speak . . . has fed and feasted and fattened . . . on the flesh and blood and brains and bones of men and children in every land. The sighs and sobs and shrieks that it has exhorted from perishing humanity would, if they were tangible things, make a mountain. The tears that it has wrung from weeping women's eyes would make an ocean. The blood that it has shed would redden every wave that rolls on every sea. The name of this loathsome, deadly, and insatiate monster is "cancer."[39]

Waving a copy of Hoffman's book, Neely warned that this monstrous plague, if unchecked, could "in a few centuries depopulate the earth." The government was spending $10 million a year to eradicate the corn borer and $5 million to study tuberculosis, yet nothing was being done for this worst of all scourges, a disease that inflicted suffering "greater than any ever devised by American Indians," more intolerable than anything ever imposed by "the fanatical fiends of the Dark Ages." Neely's agitation led the Senate to pass its first cancer bill, dedicating $50,000 to the cause, but the bill later died in a House committee. The National Cancer Institute was finally established in 1937, but as late as 1938 the total budget for the Public Health Service was only $2.8 million, a paltry sum compared with the more than $26 million allocated (just to take one example) to the Department of Agriculture.

Critics all along, albeit with diminished success, were continuing to argue that much and perhaps even all of the observed increase was due to improvements in diagnostic techniques and opportunities. Charles P. Childe, the British author of the 1906 book *Control of a Scourge,* cautioned that "the verdict as to the increase of cancer in modern times must be the cautious Scotch one, 'not proven.'" Walter Willcox, a professor of economics and statistics at Cornell, updated King and Newsholme's data in 1917, arguing that "improvements in diagnosis and changes in age composition" explained "more than half and perhaps all of the apparent increase in cancer mortality."[40] Louis I. Dublin, a statistician at the Metropolitan Life Insurance Company, as late as 1937 argued that since almost all of the observed increases were in the more inaccessible parts of the body, improvements in diagnostic methods, coupled with new opportunities for detection—the increasing frequency of hospitalization, for example—were responsible for much of the observed increase. The cancer situation in the United States was "far from alarming"; the answer to the question "Is Cancer Mortality Increasing?" was therefore "probably not."[41]

One puzzling aspect of this question of trends is that, until sometime in the twentieth century, cancer was generally looked upon as a female disease. It had been observed in the ancient world to be especially common among women, and nineteenth-century statistical inquiries confirmed the imbalance. Rigoni-Stern and Tanchou had made note of this, and British medical statistics for the period 1851 to 1860 showed that women were more than twice as likely to die of cancer as men.[42] In data gathered between 1853 and 1856 at London's Middlesex Hospital, the first English hospital with a separate cancer ward, Septimus W. Sibley showed that more than 80 percent of all cancer patients were female; breast cancer alone accounted for more than a third of the 520 total cases.[43]

How can we account for the fact that, prior to the twentieth century, cancer was far more likely to strike women than men? Female longevity cannot be the answer, since before about 1900 women lived only about two years longer than men—a difference that even John Le Conte recognized as "wholly insufficient to account for the vast disparity" in the cancer mortality of the two sexes. Le Conte recognized that cancer generally strikes women at an earlier age than men, but this again simply raised the question, Why? Le Conte speculated that the distinction might have something to do with the "greater feebleness" of the female constitution,[44] and Rigoni-Stern speculated that cancer might be "naturally" higher among women because of some particular function fulfilled by the uterus and female breast. The twentieth-century version of this argument suggests that female hormones are inherently carcinogenic.[45] That is one reason people today worry about the possible carcinogenic effects of birth control pills, synthetic hormones added to meat, and hormones in drugs used to treat menopause, depression, and heart disease. All are thought to add to the body's natural carcinogenic burden.[46] Evidence for the hormonal theory came from the observation that early pregnancy helps to prevent breast cancer, though childlessness works to increase a woman's risk. (Breast cancer is still more common among nuns than among mothers of comparable age.) The hormonal theory would also explain at least part of the rise in breast cancer incidence in recent years: the earlier onset of menses—probably due to better diet—may have contributed to the rise, and the general trend toward having children later in life also seems to be playing a part. It is not yet known whether the pressure to abandon breast-feeding in the 1950s and '60s could have contributed to cancer incidence; these things are difficult to sort out from other factors working at this time, like the increasing use of pesticides.

Part of the explanation may lie in the fact that, until some time in the twentieth century, women's cancers were more easily diagnosed than men's. King and Newsholme argued in 1893 that breast and uterine cancers were "accessible cancers" and were therefore overdiagnosed in comparison with cancers that were more likely to strike men.[47] Louis Dublin's 1937 review of cancer mortality among the policyholders at the Metropolitan Life Insurance Company showed that between 1911 and 1935, more than twice as many women died of cancer as men (175,350 versus 86,696). By the end of this period, however, rates for men and women were nearly the same. Dublin attributed this shift to the fact that "a high proportion, probably 80 percent, of male cancers are inaccessible, as compared with about 50 percent among females." Previous statistics had therefore probably missed many cancers of men. Dublin also noted that the equalization of male and female cancer rates occurred earlier in Britain, perhaps because the National Insurance Act of 1911 offered diagnostic facilities and hospitalization to a much broader segment of the population than was available in the United States.[48]

One could probably argue, though, that the higher death rate for women was real, and that this was a consequence of the fact that women's reproductive organs are particularly sensitive to environmental trauma. The environmental trauma theory was advanced as early as 1904, when Wilhelm Weinberg, a prominent geneticist and racial hygienist in Stuttgart, showed that uterine cancer was more common among the poorer women of that city.[49] Female cancer mortality was still higher than men's as late as 1950; Harold Dorn of the National Cancer Institute pointed out that year that the greater susceptibility of the female breast and genital organs was responsible for the higher female rate. Dorn speculated that "infections such as cervicitis" and "lacerations and tears during child birth" might increase the likelihood of cervical cancer, but he also admitted there was little evidence on which to base such claims.[50] More recent studies have shown that early sexual intercourse and multiple sex partners are associated with high rates of cervical and uterine cancer; the root cause here may be some kind of chemical irritation or a sexually transmissible virus.

The hormonal and environmental theories are both supported by the fact that women suffer a major proportion of their cancers in their reproductive organs. Recent investigations suggest that a high-fat diet may trigger the production of the hormones suspected of giving rise to breast cancer; the role of pesticides and other carcinogens remains unknown. Devra Lee Davis, now a senior adviser at the Department of Health and Human Services, suggests that fat-soluble pesticides and several common plastics may mimic or amplify the physiological effects of estrogen. Experiments are under way to test this "xenoestrogen" hypothesis, and the preliminary results are contradictory.[51] One thing is fairly certain: the twentieth-century shift from females to males as the main sufferers of cancer is primarily due to the fact that men have been smokers much more often than women have.[52] Occupational exposures may also have elevated male rates. By the 1960s, men were dying of cancer at significantly greater rates than women—lung cancer being the main cause of the difference. But the nineteenth-century pattern of female predominance may be restored as women's smoking patterns come to resemble those of men.

Skeptics such as Dublin notwithstanding, the dominant view by the 1930s was that overall cancer rates were indeed on the increase. Population-based cancer registries made it harder and harder to dispute the rise in cancer's mortality and incidence rates.[53] Diagnostic improvements and better access to health care accounted for part of the increase, but surely not all. People were living longer, but even age-adjusted rates were climbing.

One of the things that clinched the belief in an increase was better evidence that cancer rates varied dramatically from place to place. In the 1940s, National Cancer Institute surveys showed that age-adjusted rates for specific cancers varied widely in different states of the union. Death from stomach cancer, for example, was about two and a half times as common in Minnesota and Arizona as in Georgia, Alabama, and South Carolina. Breast cancer death rates were four times higher in New Jersey than in New Mexico. (New Jersey was dubbed a "cancer belt" even before World War I, presaging its classification in the mid-1970s as "cancer alley.") Lung cancer death rates ranged from 2 per 100,000 in Utah and Nevada to more than three times this high in Rhode Island. Death rates for all cancers combined were highest in New York and Rhode Island (145 and 138 per 100,000) and lowest in Arkansas and New Mexico (62 and 71 per 100,000). These, again, were age-adjusted rates, meaning that the uneven distribution of the cancer-prone elderly could not be the root cause of the disparity.[54] (Differences in reporting, combined with differences in medical care, might account for at least some of the variations.)

More dramatic even than state-by-state variations were those that began to be discovered among different peoples of the world, and among peoples with distinctive religious or dietary practices. John Higginson, a physician from the University of Kansas Medical Center, showed that cancer was rare among the Bantu of Johannesburg, claiming the lives of only 1 in 100, compared with 1 in 7 in the United States. The higher incidence of death from other causes explained part, but not all, of the difference.[55] Ernst L. Wynder and colleagues showed that American Seventh-Day Adventists, whose members generally neither smoke nor drink alcohol, suffer exceptionally low rates of both cancer and coronary heart disease when compared with other Americans.[56] A 1950 Oxford symposium on "The Geographical Pathology and Demography of Cancer" showed that stomach cancer, while common in industrial nations, was rare in Java and in Africa. Cancer of the pancreas was rare in wealthy nations but common in Uganda; stomach cancer was much more common among Swedish men than among American women—and so forth.[57] World Health Organization statistics for 1952 showed that overall cancer rates could vary by as much as twenty-fold from one nation to another.[58] Later studies showed that esophageal cancer was 300 times more common in Iran than in Nigeria, and that stomach cancer death rates were 25 times higher in Japan than in Uganda. Uterine cancer was rare in Israel but extremely common in Colombia. Ugandans, Nigerians, and South African blacks had the world's lowest incidence of cancers of the lung, stomach, bowel, and uterus, while Singapore residents had the lowest rates of cancer of the pancreas, bladder, and breast. Liver cancer rates were highest in Mozambique and among South African blacks and were lowest among U.S. populations—both black and white.[59]

The significance of these findings was that, barring a purely racial explanation, it was *the environments* of these peoples that gave them a small or big chance of getting cancer. Higginson concluded in 1960 that if the majority of cancers were environmentally induced, then the majority were presumably preventable. He advised further study of "the most primitive members of the human race, i.e., the Australian aboriginees and the bush peoples of Southern Africa," to determine how low the cancer rate might be in a state of nature, free of exposure to human carcinogens. This would define the "basic" cancer rate below which it would be difficult to go, given the spontaneous cancers resulting from "chance mutation, background radiation," and so forth. The lowest rates presumably defined an unavoidable genetic contribution.[60]

In the postwar, antiracist climate of the 1960s, Higginson's suggestion that most cancers were "theoretically preventable" gained a wide audience.[61] Migration studies confirmed that people tend to die of the same kinds of cancer as their neighbors and that "race" was not a very good predictor of cancer rates (see chapter 10). In the 1970s, with the environmental movement at its peak, Higginson was forced to remind his readers that "environmental" did not mean "industrial" (see chapter 3), but most experts by this time conceded that more than half of all cancers could be prevented.

THE QUESTION OF CAUSES

Cancer is a historical disease, in at least two separate senses. On the one hand, views of what causes cancer and how best to treat it have changed dramatically over time. Cancer is also historical, however, in the sense that the actual causes involved have changed over time. Both of these aspects are important for our story.

Prior to the eighteenth century, as already noted, European diseases were most commonly regarded as the expression of an imbalance in bodily humors. Tumors were classed by their macroscopic features—ulcerating, hard, soft, of this or that shape, and so forth. Cancer was conceived as a disorder of the entire bodily system, not of any one given part, and no clear distinction was drawn between malignant tumors and ulcerating inflammations. Diseases were best known by their symptoms, and in this respect cancer was not unlike several other diseases leading to a wasting away—like tuberculosis or consumption, as it was popularly known up until the bacteriologic revolution and the "pasteurization of France" at the end of the nineteenth century.[62]

In the sixteenth and seventeenth centuries, suspicions began to grow that diseases might have specific, material, or local causes. Paracelsus replaced the idea of a balance of humors by the idea of a balance of chemicals, tracing cancer to an excess of alchemical "arsenic." Nicolaas Tulp and Zacutus Lusitanus suggested that cancer might be contagious; Tulp, the demonstrator in Rembrandt's *The Anatomy Lesson,* actually claimed to have been "infected" while operating on a breast tumor. George Bell, an English surgeon, proposed that breast cancer might be caused by "a languid circulation" occasioned by an arterial blockage, by a sudden or intense cold, or by anger, fear, or anxiety acting to impede the circulation, especially in "the delicate frame of the female sex."[63] Others proposed the irritating effects of acids or the traumatic effects of bumps and bruises: in 1676, for example, a prominent English surgeon named Richard Wiseman suggested that two of his patients had developed cancer at the site of a previous contusion. Wiseman also believed that improper diet could provoke the development of a cancer, especially meat and drinks containing "great acrimony."[64] John Hunter in the eighteenth century maintained that cancer was caused by a coagulation and fermentation of lymph; the lymphatic system had been discovered in the previous century, and Hunter speculated (correctly) that it was through the lymphatic channels that cancer spreads to distant parts of the body. Specific causes were invoked that are still acknowledged today: John Hill warned in 1761 that "the immoderate use of snuff" could cause cancers of the nose, and Samuel T. von Soemmerring noted in 1795 that pipe smokers show an unusually high incidence of lip cancer.[65] Efforts were made at this time to transfer cancer from one person to another and from humans to dogs, though neither was very successful. In 1775, Bernard Peyrilhe won a prize from the Académie des Sciences of Lyon for his account of the experimental transfer of cancer from a human patient to a dog. The experiments were never carried very far, according to the medical historian Henry Sigerist, for "the dog barked so terribly that it was killed by the doctor's landlady."[66]

The most famous early indication of a physical cause for cancer is Percivall Pott's 1775 discovery that English chimney sweeps were suffering from cancers of the scrotum ("soot wart") caused by exposures to chimney soot and tar.[67] Chimney sweeping had become customary to guard against fires from clogged chimneys: "climbing boys" from the age of four or five were employed to shimmy up stacks to scrape out the soot from incompletely burned coal.[68] Pott, a London physician, decried the singularly hard fate of these boys, "treated with great brutality, and almost starved with cold and hunger." The British parliament eventually moved to limit child labor of this sort, though it was ultimately the falling price of soot that slowed the disease. (Chimney sweeps sold the soot they gathered to gardeners to kill pests; the sifting and sorting of the soot could be as dangerous as the original sweeping.) The latency period between exposure and development of the disease was first recognized in this context: Astley Cooper, a London surgeon, observed that men formerly employed as chimney sweeps could develop the "soot wart" years after having quit this line of work, and T. B. Curling, another surgeon, described the case of a man who developed scrotal cancer nineteen years after quitting work as a sweep.[69] Soot was eventually recognized as causing other kinds of cancer too. Pott's own son-in-law, James Earle, documented the case of a forty-nine-year-old gardener who developed skin cancer on the back of his left hand, five years after spreading soot in his garden to kill slugs. The gardener had carried the soot in a pot hung over his hand, inadvertently allowing soot to be rubbed into his wrist.[70] As late as 1900, the gardener's amputated hand and arm were still on display at the Museum of St. Bartholomew's Hospital in London.[71]

Percivall Pott is widely cited as the first to document an occupational cancer, but the honor could probably just as well go to Bernardino Ramazzini, the Italian physician whose 1713 *Diseases of Workers* is generally recognized as the most comprehensive early treatise on occupational health.[72] Ramazzini, whom we know Pott to have read—he is indeed Pott's only cited authority—provides an excellent corrective for anyone with a romantic longing for preindustrial forms of labor. His catalog of occupational illnesses is sobering. Miners were dying young and often palsied, toothless, and short of breath. Gilders were becoming deaf, dumb, and dazed by their work with mercury. Potters were being poisoned by the deadly fumes of metals (such as lead) used to calcine and glaze their vessels, while glassworkers' eyes were shriveling up from heat, their lungs debilitated by pleurisy, asthma, consumption, and chronic cough. Painters suffered blackened teeth and melancholia from handling red and white lead, and bakers were becoming hoarse from breathing in flour dust and bowlegged or even lame from straddling a stool to knead dough. Laundresses were becoming sick from the noxious fumes from soiled clothing, and Jews were falling ill from their unenviable work of gathering rags. These and dozens of other, equally pitiable trades formed the basis for Ramazzini's lament that "many an artisan has looked at his craft as a means to support life and raise a family, but all he got from it is some deadly disease, with the result that he has departed this life cursing the craft to which he has applied himself."[73]

Ramazzini, interestingly, mentions only one kind of cancer among the various tradespeople he discusses. In the middle of his discussion of the diseases of wet-nurses, he notes that nuns suffer a particularly high incidence of breast cancer due to their abstinence from sexual intercourse. He was struck by the apparent fact that "the breasts suffer for the derangements of the womb" and speculated that between these "two sources of desire" was some "mysterious sympathy" that had so far escaped the attention of scholars.[74] Today we recognize his observation of a higher incidence of breast cancer among nuns as correct—though he wrongly attributed the excess to their practice of celibacy, rather than to the failure to become pregnant and bear children.

In the nineteenth century, with fears beginning to grow that cancer was on the rise, wide nets were cast in the search for causes. Urbanization, dietary insufficiency, dietary excess, sedentary lifestyle, affluence, shifts in climate, physical trauma, and reversal of sex roles all came under scrutiny. Changing morals were sometimes blamed, as were sexual abstinence, sexual promiscuity, the failure to breast-feed,[75] and the misfortunes of hereditary constitution. Psychological theories abounded, with stress, depression, and sedentary reflection all coming under suspicion. In 1854 a French physician, Jean-Zulema Amussat, read a paper to the Parisian Academy of Medicine suggesting that "grief," along with "all the physical and moral disturbances which are its result," was "the most common cause of cancer." Gerard von Schmitt, the physician who treated Alexandre Dumas for cancer, suggested in 1871 that "deep and sedentary study and pursuits, the feverish and anxious agitation of public life, the cares of ambition, frequent paroxysms of rage," and "violent grief" were the principal causes of cancer; and Sir James Paget, a well-known London surgeon and pathologist, agreed that "deep anxiety, deferred hope, and disappointment" were often "quickly followed by the growth or increase of cancer."[76] An 1883 review in the *British Medical Journal* traced the increase to the prevalence of "nervous tension" due to increased luxury—also to the excessively refined "delicacy" to which civilization had led us.[77]

Mechanistic theories were equally popular, stressing the action of chemicals (such as soot, following Pott), chronic irritation (Rudolf Virchow's theory), parasitic agents (especially popular toward the end of the century), misplaced embryonic cells (Julius Cohnheim's theory), or abnormal conditions during regeneration of injured tissues.[78] Cancer was imagined to grow from the disturbed tissues around a healing scar, or from the after-effects of a blow or other injury. Cancer was said to stem from alcoholic intemperance, overindulgence in spicy foods, and diverse "digestive derangements."[79] Cancer was also said to strike individuals with a particular "constitutional predisposition": Friedrich W. Beneke of Germany claimed that the disease was especially common in people with large hearts, small lungs, capacious intestines, and abundant body fat.[80] Arsenic-contaminated water was another theory, leading its advocate, a certain William Lambe, to recommend distilled water as a cancer cure. Iron-contaminated water was yet another suspect, prompting protests in the early part of the century that cancer was being caused by the shift from using hollowed-out elm trunks to iron pipes to deliver London's drinking water.[81] Contradictions are easy to find in nineteenth-century discussions: cancer is traced to hard water and then to soft, to wealth and to penury, to too much protein and too little, and so on. Some said that infection was key, others that heredity was the crucial factor. Some blamed meat, others blamed fish, still others the potato or tomato. The presumption is often that one single cause might explain all cancers, a presumption reflected in—and perhaps required by—the failure of most nineteenth-century mortality registries to distinguish among different kinds of cancer.

I noted earlier that theories of cancer change over time, and that cancer itself appears with uneven frequency at different times and in different places. Prior to the nineteenth century, most investigators concerned themselves primarily with cancers of a preindustrial era: Pott's chimney sweeps were injured by the soot from coal- and wood-burning fireplaces; clerical celibacy was nothing new; and tobacco had been enjoyed in Europe for nearly two hundred years before it was recognized as a carcinogen. Many of these early causal links were new not to human experience but to medicine. Most were simply newly discovered causes—traceable, as we have seen, either to the growing power of medical diagnostics or the growing interest of activist physicians and governments in matters of public health.

The same is true for several subsequent landmarks in environmental carcinogenesis. In 1879, for example, two German physicians correctly identified the "mountain sickness" suffered by Saxony's silver and uranium miners as lung cancer. What was new was the diagnosis, not the disease: nearly three hundred years had elapsed since Paracelsus and Agricola had first described the ailment.[82] Arsenic was another case in which a cancer link was discovered after centuries of exposure. Arsenic had long been used in pharmacopoeia, though it was not until the 1870s that people treated with potassium arsenite for psoriasis were found to have a higher than average risk of skin cancer.[83] (In the fifteenth century Paracelsus had attributed *all* cancer to an excess of arsenic, though this particular substance had for him a broader alchemical meaning.) What was new, again, was the diagnosis, not the disease. The same was true in 1906, when James Hyde showed that skin cancer was especially common among men whose employment took them out into the sun: farmers, boatmen, draymen, hackmen, gardeners, nurserymen, vine growers, lumbermen, raftsmen, pilots, fishermen, railway employees, and sailors, for example. In the American Midwest, Hyde found that 90 percent of those who died from skin cancer were agricultural laborers.[84] People had presumably been dying from sun-induced skin cancer at least since hairless humans evolved from the australopithecines; what was new, again, was the diagnosis ("country cancer"), not the disease.

The same is true if we look at how novel kinds of cancer came under medical scrutiny. British imperial expansion allowed Western medical expertise to notice types of cancers entirely foreign to European experience—especially in those parts of the world where European interests were at stake. In the second half of the nineteenth century, for example, a peculiar form of cancer was discovered in Kashmir, in the Himalayas. This is a cold part of the world, and to warm themselves the local shepherds carried close to their bodies a small earthen pot into which were placed red hot coals. The pot—known locally as a *kangri*—was suspended from around the neck when the person was standing and set between the thighs when sitting. In 1866, a British physician at the Medical Mission Dispensary in Srinagar noticed that the people who carried these pots commonly suffered from skin cancer of the lower abdomen and inner sides of the thighs. The author of this report recorded 30 separate cases among the inhabitants of the valley, and by the 1920s more than 2,000 cases of the now notorious "kangri cancer" had been recorded.[85] In Egypt, too, imperial physicians noticed an association between bladder cancer and infestation by the parasitic blood worms (*Bilharzia haematobia*) responsible for the disease known as schistosomiasis or bilharzia. Reginald Harrison of the Liverpool Royal Infirmary in 1889 showed that Egyptians afflicted with the parasite suffered extraordinarily high rates of bladder cancer.[86] International medical researchers were alerted to both kangri and bilharzia cancers as a consequence of the work of British physicians in the colonies. The same is true of betel chewing and cancer of the mouth in India, first linked by British physicians in the early years of this century.[87]

In other cases, however, the cancers themselves were new—or at least new in their placement or frequency or manner of presentation. Cancers of this sort began to appear in the nineteenth century, often in the wake of changes brought about by the Industrial Revolution. New forms of power, new construction materials, new chemical processes, and new social habits all changed the extent to which people were exposed to environmental carcinogens. Smelting, for example, was a very old art, but the engines of the new age required a rapidly growing supply of metals, exposing ever more workers to the irritating fumes of carcinogenic oxides and salts. In 1822 John A. Paris, a fellow of the Royal Society of London and senior physician to the Westminster Hospital, reported that skin cancer was common among the Welsh and Cornish workers exposed to arsenic fumes in copper smelters and tin foundries. Animals in the region were also affected, as testified by a contemporary account reporting that "horses and cows commonly lose their hoofs, and the latter are often seen in the neighboring pastures crawling on their knees and not infrequently suffering from a cancerous affection of their rumps."[88]

Petrochemical and coal-tar technology was another development that changed the nature and scope of human cancer. Prior to the 1850s, machines were lubricated almost exclusively with animal oils. With the growing use of heavy, power-driven machinery, new kinds of lubricants were needed. Techniques were developed to "crack" and distill oil, coal, and shale to produce not just lubricants but fuels, dyes, solvents, and fertilizers. Prolonged contact with novel chemical substances produced hitherto unknown kinds of cancers. In the second half of the century, physicians began to see high rates of cancers among Scottish workers in the shale distillation industry, German miners in the coal industry, and workers handling asphalt, tar, pitch, and petroleum fractions.[89] Several cases of aplastic anemia caused by exposure to coal tar–derived benzene were reported in 1897.[90]

Synthetic dyes became a particularly grievous source of cancers. Coal-tar dyes had been discovered in the 1850s, and the subsequent shift from natural to synthetic dyestuffs exposed tens of thousands of workers to new carcinogens. In 1895, a German physician published the first evidence of bladder cancer among the manufacturers of aniline dyes in Germany.[91] In 1912, a Swiss physician showed that German dye workers were thirty-three times more likely than the general population to develop bladder tumors.[92] Dye-related bladder cancer was discovered in Switzerland in 1905, in England in 1918, in the Soviet Union in 1926, in the United States in 1934, in Italy in 1936, and in Japan in 1940—in each case, roughly ten to twenty years after the introduction of synthetic dye manufacturing into that nation. Wilhelm Hueper would use this sequence of events to prove that there was normally a period of latency between chemical exposure and the onset of cancer.[93]

As smokestacks and chimneys fouled ever larger areas of European cities, fears began to grow that the air itself could cause cancer. An English surgeon by the name of Henry T. Butlin speculated in 1892 that if soot could cause skin cancer, as was clear from Pott's work, might it not also cause cancers of the internal organs? Butlin forecast a dismal fate for humans living in an increasingly polluted world:

We who live in large cities swallow and inhale soot every day in greater or less quantity. We accept the position, grumblingly no doubt, still we accept it; we know that our great smoke fogs make many people ill, and that they kill a certain number with acute disease. But it is possible that we owe far more than this to the influence of *floating soot* and that a part of the increase in the occurrence of that awful disease, cancer, of which the national statistics tell so striking a tale, is due to the daily contact of soot with areas of the lining of the mucous membrane, or to the entrance of soot into some one or other of the internal organs, in which the conditions are favorable to its action.[94]

Butlin was probably the first to identify air pollution as a possible cause of the apparent upsurge in cancer.

People who worried about cancer as a disease of civilization, however, usually worried about more than dirty air and the hazards of work. One common idea in the nineteenth century was that the increasing consumption of meat might be to blame for the rise of cancer.[95] Aristide A. Verneuil, a French physician, in the 1870s reported that he and other hospital surgeons in Paris were seeing many more cases of cancer of the tongue than in former years; he attributed this to the fact that people were eating much more meat than previously, an idea that was endorsed, understandably, by vegetarians of his day.[96] W. Roger Williams, a surgeon at London's Middlesex Hospital, suggested in 1896 that the "gluttonous consumption of meat, which is such a characteristic feature of the age," was likely to prove a major cause of cancer.[97] Carnivorous critics threw cold water on this idea by pointing out that, in India, cancer appeared to be as prevalent among the Hindus ("to whom the fleshpot is an abomination") as among cow eaters.[98] A British Medical Association survey of 194 breast cancer victims found that "123 were moderate feeders, 59 small feeders, and 12 large feeders," and that most were moderate consumers of meat.[99] Butlin thought that altitude and geological factors were much more important, so he investigated whether cancer victims lived in towns or rural areas, on hills or in valleys, near streams or on dry fields, on sandy soil or clay, and so forth.[100]

Some of these theories sound rather fabulous today. In the 1880s and '90s, for example, consumption of tomatoes was regularly offered as a cause of cancer, to the point that "whenever the question of cancer is touched upon in the lay press, this tomato theory is sure to be brought forward with a gravity worthy of a better cause." The origins of this belief, according to one contemporary skeptic, may have derived from "some fancied resemblance between the interior of the vegetable and a fungoid growth." The editors of the popular American medical magazine *The Practitioner* ridiculed the "tomato theory" and suggested that the increase in cancer was simply "the necessary penalty of sanitary progress": more people were living to a ripe old age, and more people were therefore liable to succumb to cancer.[101]

Equally bizarre—at least to modern ears—was the theory popular between 1908 and 1910 that drinking water from trout-fed streams was a major cause of cancer. President William Howard Taft called upon Congress to allocate $50,000 to study the problem, following the advice of Dr. H. R. Gaylord, director of the New York State Institute for the Study of Malignant Disease. Gaylord had reason to believe that cancer was "most prevalent in the well-wooded, well-watered, and mountainous regions" or in poorly drained areas with alluvial soil. It was astonishing, at least to him and his supporters, how closely the distribution of cancer followed the distribution of trout: "A map of one might well be taken as a map of the other." The *New York Times* devoted eight columns to the idea, endorsing President Taft's proposal as "the greatest stroke so far toward the conquest of the dread disease." Despite reports of 100 million U.S. trout with cancer and Gaylord's insistence upon a human health hazard, this notion didn't last, and Taft's call for funding died in Congress's Committee on Fisheries.[102]

Dietary theories proliferated in the early years of the twentieth century. Alcohol consumption was linked to cancer of the mouth and esophagus, and cancer was imagined to be, like scurvy, a vitamin-deficiency disease. Frederick Hoffman of the Prudential Insurance Company endorsed the dietary theory in his 1915 book, *Mortality from Cancer Throughout the World,* though he rejected the idea of vitamin deficiency in favor of cancers being a product of "excessive nutrition," especially the consumption of fatty, sugary foods high in carbohydrates and proteins. His 1937 *Diet and Cancer* further tied the growth of cancer to dietary excess, suggesting that the increased consumption of canned and preserved foods might play a role, along with the growing fondness for white bread, meats, sweets, and other foods leading to constipation, ulcers, and obesity. Hoffman even suggested that the vulnerability of women to cancer might have something to do with their purported tendency to overeat. Hoffman's calls for far-reaching changes in the dietary habits of the American population are prescient in the light of recent efforts to establish dietary links to cancer.[103]

Experimental work also began to confirm the role of diet. Carlo Moreschi, an Italian physician working in Germany, had shown in 1909 that sarcoma tumors transplanted into mice that had been fed a low-calorie diet grew more slowly than in mice given all the food they wanted.[104] Peyton Rous of the Rockefeller Institute later showed that by underfeeding rats he could slow the development of transplanted tumors; he also showed that exercise could have similar effects.[105] In 1940 Albert Tannenbaum showed that mice fed a low-calorie diet were less likely to develop both spontaneous and chemically induced tumors, and within a few years half a dozen studies had confirmed his observations.[106] M. G. Mulinos and Roswell K. Boutwell suggested that this might be due to the fact that overfed laboratory animals produce an excess of cancer-causing hormones.

The idea of cancer as a "disease of civilization" was a diverse cluster of ideas with diverse meanings. For some, cancer was the necessary price of civilization, an ineluctable accompaniment of our material progress and industrial vitality. This is what Le Conte had in mind when he stated that cancer was on the rise as a consequence of the "prolongation of human life incident to the progress of civilization."[107] Charles H. Moore similarly argued that the increase in cancer might be ascribed "to corn-laws and good living, to the discoveries of gold, to the good government which has reared to adult life and to old age a larger proportion than heretofore of the entire population."[108] James Sawyer, a physician in Birmingham, concluded about this time that the "enormously increased consumption of meat by the great masses" was responsible for the rise of cancer, and that "steam appears to have brought us cheap food, and cheap food has multiplied our cases of cancer by two."[109] Progress, for this group, was the overarching cause of cancer.

For another group, cancer was the canary in the mine, the index of our collapsing morals and degenerate civilization. Some of those who embraced this idea did not even believe that "nurture" was more important than "nature" in the genesis of cancer: John Cope, the misogynist pseudonymous author of a 1932 book called *Cancer: Civilization: Degeneration,* argued that "heredity" was five to ten times as important as "environment" in the genesis of cancer. Cope's was a moral rhetoric: cancer was a disease of civilization, but then civilization was itself a disease, a cancer on the march, threatening the racial stock of the nation. Cancer was a symptom of a culture in collapse, an expression of degenerate manners and morals.[110] Similar views were popular on the Continent, especially in Nazi Germany, where cancer as a "disease of civilization" harmonized with the curious blend of romanticism and modernism that guided fascist philosophy.[111]

Civilization was thus a broader concept even than its successor notion, *the environment.* Civilization might entail pollution, dietary excess, physical sloth, and various forms of immorality and gender-role inversions, but it could also include the overcrowding that attends urbanization (consistent with an infectious etiology), the diverse stresses of industrial life, the increased life span offered by modern medicine (cancer as a natural consequence of aging), and the increasing detection of cancer through improved diagnostics and methods of registration. Each of these factors figured, at one time or another, in the early idea of cancer as a disease of civilization, an idea that mixed a kind of Rousseauian (and Jeffersonian) romance for the countryside with a celebration of the healthy habits of "primitive" peoples.

Cancer eventually came to be seen as a product of the environment rather than of civilization, though the important reasons for this shift were probably due more to successful rhetoric than to new evidence. The 1950s was not a period of impressive research into environmental carcinogenesis, though some efforts were made, as we shall see, to sort out the offending agents. The list of potential culprits by the middle of this century was long: ubiquitous pollution, improper diet, delayed childbirth, radiation, tobacco, various and sundry germs, psychological distress, and a host of other factors. The particular agents stressed often reflected fashions of the time: germ theories were popular when Pasteur and Koch convinced scholars that infectious agents were the root causes of disease; racial theories were popular at the height of the eugenics movement (see chapter 10); occupational theories became fashionable when labor's voice was strong; and genetic and biochemical approaches would predominate as the research establishment secured its role in the war.

By the 1960s, as we shall see, it would be common to argue that the majority of cancers were environmentally caused and therefore potentially preventable. Debates would continue to center on which particular aspects of the environment were to blame—and even on whether in fact the incidence of certain kinds of cancer was on the rise. The question of causation became increasingly politicized as ever more powerful interest groups staked out claims, developed interests, extended their powers. As is often the case in science, what one sees has a lot to do with where one is standing.

Chapter 2

The Environmentalist Thesis

A grim specter has crept upon us almost unnoticed.
— Rachel Carson, 1962

FACED WITH CONFLICTING theories and hard-to-interpret statistics, researchers developed experimental methods in the early years of the twentieth century to decipher the causes of cancer. Transplantation research had already shown that once a normal cell turns cancerous, it will always remain so.[1] In 1908, Jean Clunet of Paris showed that the "radiologic imprudence" of his colleagues in radiology could be replicated in the laboratory setting; in 1910 he published his work on the artificial induction of cancer in laboratory animals. In 1915 Katsusaburo Yamagiwa and Koichi Ichikawa at the University of Tokyo produced coal-tar cancers in laboratory animals, and in 1916 experiments were published showing that female mice whose ovaries had been removed developed fewer than expected breast cancers.[2] In 1922 Ernest Kennaway at the Cancer Hospital in London began an effort to identify the chemical agents active in the Japanese coal-tar experiments, and by the early 1930s he had identified a previously unknown hydrocarbon—3,4-benzpyrene—as the primary culprit.[3] George M. Findlay in 1928 showed that ultraviolet radiation from a mercury vapor lamp could induce skin cancer in mice, confirming two centuries of speculation and numerous epidemiological studies suggesting that sunlight could cause cancer.[4]

Evidence was also continuing to mount that workplace exposures were adding to the cancer burden. The first person ever known to have died from an X ray–induced cancer succumbed in 1904, and the tragedy of the radium-dial painters of the 1920s provided a dramatic example of the dangers of radioactivity (see chapter 8). In 1919 physicians in England and Wales were required to register every case of occupational cancer that came to their attention; this resulted in both a jump in the number of cancers registered and the identification of new kinds of workplace hazards.[5] In 1922 British physicians found that cotton mule spinners were suffering high rates of scrotal cancer from exposure to lubricating oils; a 1932 exhibit at the New York Academy of Medicine presented mule-spinner's cancer as "the most frequent type of occupational cancer known."[6] Medical attention focused more and more on the problems of industrial cancer: a 1926 bibliography listed 290 scientific papers on occupational cancer, and a 1933 supplement on tar cancers added another 391.[7]

The social forces driving such concerns were several. Governments were worried about the public health of their citizens, and labor activists were troubled by the lack of attention to workers' health in the industrial context. Private insurance examiners were aware that knowledge of cancer hazards could be useful in their line of work: Frederick Hoffman of the Prudential Insurance Company pointed this out in 1915, when he published evidence that certain occupations were more "cancerigenic" than others.[8] Hoffman was convinced that cancer was more common among the middle and upper classes—the people most likely to be able to afford insurance.[9] Corporations were worried about lawsuits from employees injured by their work, and this was yet another stimulus for the focus on environmental causes of cancer.

WILHELM C. HUEPER: A HERO HOUNDED

It was this latter, corporate context that inspired the century's most profound and colorful critic of occupational cancer hazards. Wilhelm C. Hueper (1894–1978) is largely forgotten today, but for three or four decades in the middle of this century he was the most powerful American champion of the view that increased exposure to industrial chemicals was producing unprecedented levels of cancer. He was one of the first to study radiation-induced leukemia,[10] and the first American to document a lung cancer hazard in the chromium industry. He directed the National Cancer Institute's Environmental Cancer Section from 1948 to 1964, and he provided decisive testimony at the congressional hearings that led to the 1958 Delaney Amendment of the Food and Drug Act, barring synthetic carcinogens in foods.* Rachel Carson praised his 1942 *Occupational Tumors and Allied Diseases* as the "classic monograph" on the topic; a whole chapter of her *Silent Spring* is devoted to environmental cancer, drawing heavily from Hueper's work. Cancer theorists often cite a 1964 statement by the World Health Organization that the majority of human cancer is traceable to environmental exposures and therefore potentially preventable,[11] yet the role played by the father of occupational carcinogenesis in establishing that consensus is often not appreciated.

Hueper, as one can easily imagine, was not an uncontroversial figure. The Du Pont Company accused him of being a Nazi in the 1940s and a Communist in the 1950s—he was neither. He was a student of industrial duplicity as well as of chemical carcinogenicity, who once described his method as "one piece of dirt leading to another." Historians of medicine have paid little attention to him even though, according to an archivist at the National Library of Medicine, his papers deposited

---

*Hueper in his 1952 testimony suggested that the dyes used in many foods, the arsenic used in hair dyes, the estrogens in certain facial creams and skin lotions, and the lampblack used in eyebrow pencils might cause cancer. When asked whether studies had been done to determine whether dyes shown to cause cancer in animals could also cause cancer in humans, Hueper deadpanned: "I don't think I would volunteer for that experiment." See U.S. Congress, House Committee Hearings, 82nd Cong., Senate Lib. 1362 (1952); also the House Committee Hearings, 85th Cong., Senate Lib. 1666 (1958): 369–372. According to his son, the Hueper family never mixed yellow food coloring into their margarine, as was common practice in the 1930s and '40s (interview with Klaus Hueper, June 20, 1994).

there are the single most widely used collection in the manuscript division of that library. Two groups apparently find him fascinating: tobacco industry attorneys, who like his denunciations of "the tobacco theory of cancer,"[12] and asbestos industry attorneys chastened by his early advertisement of the asbestos–lung cancer hazard.[13] Hueper is as interesting for his myopia as for his perspicacity, as shown by his failure to appreciate the magnitude of the tobacco hazard (he himself smoked a pipe until 1938). He is also interesting because his efforts to track down industrial carcinogens were blocked at many points by industrial and governmental authorities who were less eager than he to find out what causes cancer.[14]

Hueper emigrated to the United States from Germany in 1923, in the wake of war, revolution, and economic collapse. The Kiel-trained physician had seen his own medical practice fail, and for a time had worked at a steel mill in the Ruhr valley, where he probably saw firsthand some of the hazards faced by industrial workers. Three years after coming to the United States, he published his first of more than a hundred articles linking the growing incidence of cancer to exposure to synthetic chemicals. (He had earlier worked on noncancer hazards, including an analysis of the pathological effects of cosmetic paraffin breast implants.)[15] The problem for many theorists, then as now, was how to explain the apparent growth of lung cancer rates in Europe. The leading theories focused on the role of tobacco, infectious agents, and artifacts of diagnosis. Smoking had become popular during World War I, when cigarettes were distributed with rations, which led to suggestions that tobacco might be causing the increase. Others suspected a delayed effect of the influenza epidemic of 1919, and a few diehards scoffed at the phenomenon as an artifact of diagnostic imperfections. Delayed effects of inhaling mustard gas was another popular theory at the time.

Hueper argued that none of these was right: the increase in smoking was too recent to explain the observed rise in cancer rates, and there was little historical evidence correlating other recorded flu epidemics with increases in cancer. The more likely cause, in his view, was the inhalation of coal-tar fumes and automobile exhaust, along with asphalt dust from roads and soot-laden smoke from furnaces and chimneys.[16] Hueper's persistent questioning of the cigarette theory of cancer, even late into the 1950s and '60s, would eventually cost him a measure of scientific support.[17]

In the spring of 1930, Hueper moved from Chicago to Philadelphia to assume a position as chief pathologist at the University of Pennsylvania's Cancer Research Laboratory and director of pathology at the American Oncologic Hospital. It was here that a chance personal contact gained him entrée to the industrial setting that would allow him to make his most lasting scientific discovery. Mr. Irenée du Pont, the philanthropist and industrial magnate, was the primary funder of the laboratory at the University of Pennsylvania. Hueper came to know Mr. du Pont because his superior at the laboratory, Ellice McDonald, was also the industrial magnate's personal physician.

Hueper was interested in industrial hygiene and asked, soon after his arrival in Philadelphia, to be taken on a tour of the giant Du Pont Dye Works at Deepwater, New Jersey. On his tour, Hueper was told that in the process of manufacturing dyes, certain aromatic amines were used that he knew, from his familiarity with German medical literature, could cause cancer of the bladder. Hueper wrote to Mr. du Pont to inform him of the hazard, and six months later the company's medical director appeared at his lab to report that twenty-three cases of bladder cancer had been found among the aromatic amine workers at several of the company's plants. Du Pont had begun production of these dyes in 1917, when war conditions had made it difficult to obtain German supplies. The first Du Pont worker was diagnosed with the disease in 1929.[18] It made sense that, with a ten- to fifteen-year latency period, it would be then that cancers were beginning to appear. As Hueper would later tell the story, "a new industrial cancer hazard of occupational origin had arrived on schedule in the United States."[19]

Hueper worked at the University of Pennsylvania until the summer of 1934, when he learned that his contract was not going to be renewed. Facing unemployment, he traveled to Europe with his wife "to test our luck in Germany, where openings had become available because of the Hitler turmoil." (Hitler had seized power in January 1933, and Jews had been driven out of most university jobs by the end of the year.) Hueper was by no means a Nazi or an anti-Semite,* but he did consider himself a German patriot and had fought faithfully for the Fatherland in World War I. (His autobiography begins with a long and rather tedious account of his exploits—including delivering poison-gas canisters to the front.) Despite inquiries all over Germany, however, he was unable to secure a position. He returned to the United States, where he worked briefly at Philadelphia's University Hospital and later at a hospital in Uniontown, Pennsylvania, a coal-mining city southeast of Pittsburgh. Finally, in November 1934, he was offered and accepted a position as pathologist at Du Pont's new Haskell Laboratory of Industrial Toxicology in Wilmington, Delaware. Hueper was already known to the company for having urged—in the summer of 1930—the creation of such an institute to investigate the bladder cancer hazard (Mr. du Pont had replied that this was not feasible during the depression).[20] By the time of Hueper's appointment, dozens of additional cases had turned up at the company, and workers were beginning to protest the medical department's requirement that all of the thousand or so vulnerable workers undergo regular and painful cystoscopic exams.[21]

In his three years at Haskell, Hueper performed animal studies on a broad range of Du Pont products, including seed grain vermicides (dimethyl and diethyl mercury compounds), carbon disulfide, ethylene glycol and related solvents, refrigerant

---

*There is no evidence of anti-Semitism or pro-Nazi sentiments in any of Hueper's writings or correspondence. In 1936, though, he did allow a speech of his on the racial specificity of cancer to be translated into German and published in a leading German racial journal (edited by the notorious racist Hans F. K. Günther). Hueper here argued that different races suffer different rates of cancer and other diseases, but he also noted that it was difficult to separate out the contributions of heritable racial traits (*ererbten Rasseanlagen*) and the more plastic contributions of habit, custom, and hygiene. He did warn, though, that it was dangerous for a race with relatively few genetic defects to mix with races with a greater incidence of diseased traits; see his "Krankheit und Rasse." Hueper had joined the notorious Freikorps shortly after World War I: the image of the young physician rounding up Communists with a swastika painted on his helmet is not a very pleasant one.

gases such as Freon, and Teflon coatings for kitchen utensils. Prior to the Japanese attack on Pearl Harbor, Du Pont had made large deliveries of explosives to the Japanese (who were then fighting in China); Hueper was asked to investigate why a number of young workers had died from acute heart failure after handling "safety dynamite" containing ethylene glycol dinitrate. Hueper found that many of the workers showed evidence of chronic nitrite poisoning after inhaling vapors from the treated dynamite. German dynamite makers were invited to the United States to discuss this new occupational disease, and even though the Europeans' findings made their way into print, Du Pont officials—according to Hueper—barred publication of the company's own findings.[22]

Hueper was disappointed from the outset by his lack of scientific freedom at the laboratory. Research projects were determined not by the scientific staff but by the various departments of the Du Pont Company. Hueper and his colleagues were originally able to publish their work freely, but eventually even this was disallowed, allegedly because other companies might thereby gain free medical information. Hueper was also bothered by the fact that the laboratory was not permitted to carry out long-term studies (more than a year or two) that might reveal delayed biological effects of exposures—notably, cancers. The only exception was a study in which he showed that betanaphthylamine given orally to dogs for a period of three years could cause bladder cancer—a contribution that earned him international acclaim and corporate notoriety.[23] Hueper concluded from this experience that the carcinogenic effects of many substances were probably being overlooked, and that industrial laboratories were "unsuitable media to be entrusted with safeguarding the health of their employees and of the general population."[24]

Industrial secrecy was a perennial problem: even as chief pathologist of the Du Pont lab, Hueper was often barred from learning about the nature of the chemicals to which workers were exposed. One day, when he noticed that his experimental animals were suffering from what appeared to be heavy metal poisoning, he asked a company chemist whether this might be the case. The colleague confirmed that they had been adding lead to several batches of their "synthetic silk" (nylon). When Hueper asked his research chief whether he could include this information in his report, Hueper was accused of having discovered one of the company's closely guarded manufacturing secrets. On another occasion, Hueper asked to visit the company's famous Chambers Works in Salem County, New Jersey, to determine how extensively workers were being exposed to betanaphthylamine vapors and dust. Du Pont's Chambers Works was already notorious as the site where, in the 1920s, hundreds of workers had been injured and several killed in the manufacture of tetraethyl lead as an antiknock compound for gasoline. (The site had acquired the nickname House of Butterflies, in reference to the fact that many of the lead-poisoned workers developed insect hallucinations.)[25] He was taken to the plant floor with several colleagues and was surprised to find that the plant was very clean. He mentioned this to a foreman, who retorted: "Doctor, you should have seen it last night; we worked all night to clean it up for you." Hueper realized that what he had seen was a well-staged performance; he therefore asked to see the benzidine operation in a separate building, for which management had not been prepared. There he found the powdery white benzidine dust everywhere: "on the road, the loading platform, the window sills, on the floor, etc." He wrote a memo to Mr. du Pont describing his experience, but he never received an answer and was never again allowed to visit the Chambers Works.[26]

In November 1937 Hueper was fired from his job at Du Pont, effective at the end of January. The excuse given was economic, but Hueper had also run into trouble when he protested the efforts of two of his superiors—the company's medical director and the head of the Haskell Lab—to claim credit for his experimental demonstration of the naphthylamine bladder cancer hazard. Hueper was informed at the time of his dismissal that he was contractually obliged never to publish any of his Du Pont work without the company's consent, but he refused to comply and, as a result, was hounded by company officials for the rest of his life. When he accepted an invitation by the International Union Against Cancer to speak on his experimental production of bladder cancer in dogs, for example, the company's medical director informed him that presentation of such a paper would result in a lawsuit against him for violating his contract. Financially unable to challenge the Du Pont demand, he was forced to decline the invitation.[27] Over the next twenty years, Hueper would have to overcome many similar obstacles in his efforts to inform the public and the medical profession of occupational hazards.[28]

In the summer of 1938, Hueper began work on his 896-page magnum opus, published in 1942 as *Occupational Tumors and Allied Diseases*. The book presents the first major review of world literature on occupational causes of cancer, but Hueper for a time had difficulty finding a publisher. Yale University Press offered to publish the book, but only on the condition that the manuscript undergo "a drastic revision" (Hueper's words) by two scientists appointed by the press to become co-authors. Hueper refused. C. C. Thomas of Springfield, Illinois, finally accepted the manuscript, but only after Hueper's new employer, the pharmaceutical firm William R. Warner, agreed to furnish $3,000 to offset printing costs. Hueper had worked for Warner since the summer of 1938, and had been given freedom to work on scientific projects of his own choosing. Much of his time was spent preparing his manuscript, but he also conducted animal studies to determine the carcinogenic effects of sunlight, hormones, coal tar, arsenic, asbestos, ionizing radiation, and blood plasma substitutes made from liquid plastics.

Hueper began his book by suggesting that "the gigantic growth of modern industry" had generated a multitude of chronic and insidious diseases never before observed. The chemical industry had showered us with "new synthetic substances in never-ending number," including "dyes, mordants, explosives, plastics, fertilizers, insecticides, fungicides, solvents, rubber, resins, lacquers, pigments, paints, finishes, textile fibers, fuel and lubricants for motors and machines, refrigerants, building materials, radioactive substances, food components, drugs, toilet articles, pharmaceuticals, household supplies, and innumerable other articles." The discharge of wastes from chemical factories, paper plants, gas works, rayon plants, and tanneries had fouled our water, while carbon monoxide, sulfur dioxide, and arsenious oxide had fouled our air. Metallurgical industries were exposing people to lead, mercury, and chromium, while chemical insecticides had introduced lead, arsenic, copper, and nicotine into foods. New mining technologies had introduced novel threats into the workplace (pneumatic tools, for example, raising unprecedented levels of dust), as had new medical diagnostic tools (roentgen, radium, and ultraviolet rays) and new chemicals for coloring and preserving foods. Workers exposed to aromatic compounds alone included "aniline makers, artificial-leather makers, calico printers, camphor makers, coal-tar workers, compositors, compounders (rubber), dye makers, dyers, explosive workers, feather workers, germicide makers, lithographers, millinery workers," and several dozen other classes of workers. All were living within what Hueper called the "new artificial environment"; all were exposed to one degree or another to cancer-causing substances.[29]

Hueper then suggested that while acute poisonings were surely a problem, longer-term chronic effects—such as cancer—were a far greater cause for concern. (Trench warfare had earlier provided Hueper with a kind of metaphor for chronic health hazards: in his autobiography, he notes that by contrast with conventional warfare, where the primary danger is from "acute and severe losses," the most serious threat in trench warfare is from "low-level mental and physical erosion.")[30] Recognition of chronic hazards had been made difficult by the fact that most occupational cancers do not show symptoms for months or even years after exposure. Physicians were inadequately trained to recognize such hazards, and managers were hesitant to entrust this information to outsiders. Physicians might not inquire at all into the cause of a particular disorder, especially if it could be caused by a wide range of agents.

Hueper went on to note that most occupational cancers were, at least in principle, preventable, and that the key in each case was to reengineer the production process. The hand cancers formerly suffered by radiologists had been largely stopped by the introduction of safety procedures and protective clothing; the closing of the copper-smelting works and tin-burning houses of Cornwall and Wales had ended the rash of cancers historically associated with that region. Hueper also predicted that the epidemic of arsenical keratoses and epitheliomas in towns such as Reichenstein, Germany, would probably end, since the water had been cleaned of arsenic. Hueper pointed to a number of cases in which differential cancer rates could be explained by differences in manufacturing methods. German mule spinners did not suffer from cancers of the scrotum, unlike their English counterparts, because of differences in how the thread was produced: German thread was thicker and so the spindles in German cotton mills revolved more slowly, throwing off less oil (the lubricant used may also have contained less of the carcinogenic shale oils). An earlier shift from mule spinning to the less hazardous ring spinning prevented an equivalent epidemic in the United States; a more thorough treatment of the lubricating oils with sulfuric acid apparently further reduced the cancer hazard.

Scrotal cancer appears to have become common earlier in England than in Germany, largely because the shift from wood burning to coal took place earlier in England than on the Continent. English rates may also have been higher because the fireplaces there burned at low temperatures with a relatively weak draft, encouraging the accumulation of tarry soot. Making matters even worse, English flues were often tortuously narrow—sometimes only seven inches in diameter—combining both horizontal and vertical segments, requiring one to crawl through the chimney to clean it. (As discussed earlier, children—dubbed "climbing boys"—were employed for this task, sometimes stolen for the purpose or sold by their parents for a small fee.)[31] German chimneys, by contrast, were generally straight and therefore easier and safer to clean.

Cancer, in other words, could be prevented, but Hueper was not optimistic about society's willingness to take the necessary steps. Chimney sweeps had been found to suffer from high rates of scrotal cancer as early as 1775, but the first English laws designed to bar children (under age eight!) from engaging in labor of this sort did not pass until 1788. The minimum age for apprenticeship was raised to ten in 1834 and to sixteen in 1842, but the new laws were apparently resisted, evaded, and frequently broken. An epidemiological survey of 1861–90 showed that chimney sweeps still had far and away the highest cancer mortality of any occupation,[32] and even as late as 1912, they were more than a hundred times more likely to die of scrotal cancer than the general population of England and Wales.[33] Similar results, only slightly improved, were found in a survey of 1931–35.

Other occupations showed equally appalling neglect: Hueper denounced the "utter disregard" for health and safety among the dusters and sprayers of arsenic insecticides and the irresponsible release of twenty-six tons of arsenic daily from the smelters of the Anaconda copper mines of Montana. Substances such as these did not easily disappear from the environment, and future generations would have to reckon with the 80 million pounds of lead and other arsenates estimated to have been applied as insecticides on American farms. He also lamented the fact that as late as 1937, only two states—Colorado and Michigan—had laws fixing an upper limit for arsenic contamination of foods. The only long-term solution, he suggested, was to replace arsenic insecticides by agents that spontaneously degrade into harmless substances.[34]

Hueper's insistence that synthetic chemicals are important causes of cancer was a heterodoxy for which he was ostracized by the cancer establishment and lionized by its critics (Rachel Carson praised the book as "the Bible" of environmental carcinogenesis). Hueper's book was almost totally neglected when it first appeared, though. Not a single cancer journal published a review, and there was only one medical review of any note, in the *Archives of Pathology*.[35] One explanation for the neglect may lie in the fact that Hueper's book appeared in the spring of 1942, just months after the United States had declared war on Japan. Occupational carcinogenesis was hardly a priority in a society mobilizing for war: physicians were more concerned about keeping people at work than about the long-term effects of industrial pollutants. And Hueper's being a German émigré with a pronounced accent certainly did not help his case (he spoke German at home his entire life); nor did the fact that, only a few years previously, in the early years of Hitler's Third Reich, he had returned to Germany to seek employment.

Some people did, of course, read and appreciate the book—most notably, Morris Fishbein, editor of the *Journal of the American Medical Association*. Fishbein apparently so liked the work that he invited Hueper to address the issue of occupational and environmental carcinogenesis on the editorial page of the journal. Several of

Hueper's editorials did appear—unsigned, as was the practice—until 1949, when Fishbein was replaced as editor.[36]

In 1948, Hueper's pioneering efforts in the area of environmental carcinogenesis won him a post as founding director of the Environmental Cancer Section of the National Cancer Institute (NCI), a position he held until his retirement in 1964. The NCI had been founded in 1937 by a special act of Congress; it was part of the National Institutes of Health (NIH) in Bethesda, Maryland, the research arm of the Public Health Service. The NCI had greater autonomy than other branches of NIH (until 1944 it was the only branch authorized to provide grants to outside investigators) but, as we shall see, it was not entirely free of political pressures. Medical research funding was growing rapidly in this period: the NIH budget, for example, rose from only $180,000 in 1945 to more than $46 million in 1950.[37] Hueper's appointment came in the middle of this period of rapid, almost euphoric, growth, fueled by the optimistic faith that scientific research was the key to social progress.

Paul Starr's history of *The Social Transformation of American Medicine* states rather categorically that scientists at this time "enjoyed autonomy within the constraints of professional competition." This was indeed the position articulated by the official in charge of the NIH Division of Research Grants—Hueper's superior—who, in 1951, wrote:

> The investigator works on problems of his own choosing and is not obliged to adhere to a preconceived plan. He is free to publish as he sees fit and to change his research without clearance if he finds new and more promising leads. He has almost complete budget freedom as long as he uses the funds for research purposes and expends them in accordance with local institutional rules.[38]

The NCI may well have enjoyed greater autonomy than other branches of the NIH, but for Hueper personally, there were many strings attached to freedom. Shortly after joining the NCI, he received a letter from the Federal Loyalty Commission informing him that he was under investigation for disloyalty to the United States. G. H. Gehrmann, Du Pont's medical director, had denounced him as a Nazi, a trumped-up charge that was dismissed when Hueper was able to show that it came from a former colleague jealous of the acclaim Hueper had won for his work. Gehrmann later wrote to the director of the NCI accusing his foreign-born colleague of harboring communistic tendencies, though this was shown to have no more substance than the Nazi charge.

Political pressure was also felt from the Public Health Service's own Division of Industrial Hygiene (DIH), which was wary of Hueper's untiring quest to document chemical and nuclear hazards. On one occasion, the DIH director forced Hueper to remove his name from a study identifying lung cancer hazards in the chromate industry; industry officials with influence at the DIH—notably A. J. Lanza—pressured the PHS officials (via Surgeon General Leonard A. Scheele) to tone down or withdraw the paper, which eventually appeared under the sole authorship of Thomas F. Mancuso, director of the division of industrial hygiene at the Ohio Health Department.[39] On another occasion Hueper learned, to his astonishment, that carbon copies of his papers, submitted to the editorial board of the NCI for clearance, were being routinely submitted to the Du Pont company for commentary and appraisal. Hueper found this out by chance when, in 1960, a former member of the company's Haskell Laboratory confided that he was familiar with his work, having been asked to provide prepublication critiques. Hueper later said he had been startled by this "objectionable collusion" between the Public Health Service and "an industrial private party which had been engaged for years in my persecution." He never managed to find out who had been passing along his papers.[40]

The most consequential act of censorship against Hueper was a decision to bar him from investigating the lung cancers among the uranium ore miners and millers of the Colorado Plateau. Hueper had begun this project in April 1948, well aware of the fact that European uranium miners had long suffered from very high rates of lung cancer. His efforts in this area were stymied, however, by the fact that all NIH papers dealing with the health effects of radiation had to be cleared by the Atomic Energy Commission (AEC). When Hueper sought to present evidence of the hazard at a 1952 meeting of the Colorado State Medical Society, the AEC's director of biology and medicine, Shields Warren, ordered the NCI to instruct Hueper to delete all references in his paper to the hazards of uranium mining. Hueper initially refused, on the grounds that he had not joined the institute to become a "scientific liar." Pressed with the demand, though, he felt he had no alternative but to withdraw from the conference. He did send a copy of his paper to the society's president, explaining his decision. When word got around that Hueper was not silently accepting his censorship, Warren again wrote to the director of the NCI, this time asking for Hueper's dismissal. Hueper continued on, but was barred from all epidemiological work on occupational cancer—on orders from the surgeon general. Hueper was henceforth allowed to do only experimental work on animals (an interesting commentary on the politics of experimentation and the form of scientific censorship) and was prohibited from further investigations into "the causation of cancer in man related to environmental exposure to carcinogenic chemical, physical, and parasitic agents."[41] As Victor Archer recalls the incident, Hueper was essentially barred from research travel west of the Mississippi.[42] (Archer was in a position to know: he was the one who was eventually allowed to undertake an investigation of uranium mining epidemiology for the Public Health Service—see chapter 8.) The director of the NIH, William H. Sebrell, also canceled Hueper's promotion to a higher grade. Hueper later complained that government censorship of his and other studies on the health effects of uranium mining had delayed measures to remedy the situation, leaving countless men to die who could have been saved.[43]

Hueper's career at the National Cancer Institute was understandably a bitter one. Nowhere was this more starkly revealed than in his 1959 memorandum to the institute's director, chronicling what he called the "dark" and "somewhat shady" events in the history of his Environmental Cancer Section. In over twenty painful pages, he recited a long list of political impediments to his work. Pressure had been exerted on him (by G. Burroughs Mider, the NCI's chief of research) to restrict his speaking engagements at medical schools and schools of public health, his travel to industrial sites, and his serving as an expert witness at compensation trials; he had also been ordered to delete the names of manufacturers from papers submitted for publication. The approval of scientific papers of his had been held up—in one instance, until he resigned as a consultant to the Department of Labor on matters of occupational carcinogenesis. He had canceled a contract with Paul Hoeber, Inc., to publish a second edition of his *Occupational Tumors,* fearing that the book would be "mutilated and perverted" by the institute's editorial board and the "politically spirited" information officers at the surgeon general's office. He complained that such restrictions obstructed his research and also worked to deprive "occupational cancer victims, their widows and orphans" of the legal evidence they might need to file for compensation.[44]

Hueper deplored the use of political convenience to sacrifice the reputation of "a valuable scientist who had the courage of reporting embarrassing facts"—even when it worked, as it sometimes did, in his favor. In one such case, in late 1957, a paper of his on the cancer hazards of pesticides and food additives was refused clearance by the NCI editorial board on the nebulous grounds that it was "unsuitable." One year later, the secretary of Health, Education and Welfare barred the sale of all cranberry products shortly before Thanksgiving because they were contaminated with the herbicide aminothiazole. Hueper resubmitted the paper, and this time it was rapidly approved; the NCI apparently recognized its political value given the turn of events, and allowed it to pass. Hueper deplored such machinations and advised "a more realistic, far-sighted, impartial and competent approach" to the problems of environmental cancer. He also cautioned, rather ominously (if naively), that NIH officials should not expect help from him if it should ever come to light that they had helped to suppress important public health information: he specifically mentioned William Sebrell's collaboration in halting his study of the miners of the Colorado Plateau.[45]

When Hueper finally retired in 1964, he was disappointed with what he had achieved at the NCI; as he later wrote, he had "not succeeded in laying a foundation . . . on which others might erect a solid building." His Environmental Cancer Section was abolished on the day of his retirement, and the library he had worked hard to assemble was dispersed.[46]

The Public Health Service may have misgauged his talents, but elsewhere he was not without supporters. In 1959 he won the Anne Frankel Rosenthal Memorial Award for Cancer Research, presented to him by the American Association for the Advancement of Science. In 1962 he was the recipient of the United Nations Award for Outstanding Research in the Cause and Control of Cancer (shared with L. M. Shabad of Moscow's Cancer Research Center); the World Health Organization presented the award at a plenary session of the United Nations—though he was disappointed that his PHS superiors never sent him a note of congratulation. This was followed by the Humanitarian Award of the National Health Federation (in 1965) and the Gold Medal of the Cancer Prevention Center in Rome (1968). In 1975 he won the First Annual Award of the Society for Occupational and Environmental Health (presented to him by Samuel Epstein). Finally, in 1978, more than a dozen years after his retirement and only months before his death, the National Institutes of Health honored him with its Director's Award on the recommendation of Arthur C. Upton, the Carter administration's newly appointed director of the National Cancer

Institute. The NIH award cited his work on radiation-induced leukemia, his exposé of the cancer hazard posed by water-soluble chromate chemicals, and his classic book on occupational tumors. The April 1979 issue of the *Journal of the National Cancer Institute* was dedicated to the memory of Hueper, who died of a heart attack at age eighty-four on December 28, 1978.[47]

Hueper lived long enough to see a dramatic rebirth of interest in environmental carcinogenesis, but this never stilled his bitterness toward those who resented his one-man cancer crusade. In 1976 he predicted a steady increase in cancer as a result of the nation's ever growing arsenal of environmental and occupational carcinogens—"biological death bombs . . . which may prove to be, in the long run, as dangerous to the existence of mankind as the arsenal of atom bombs prepared for future action."[48] Blame for this condition he laid at the doors of both industry and government. Industrial managers had devised a comprehensive set of strategies to escape moral and legal obligations for their actions, including suppressing or delaying publication of evidence of carcinogenicity; "manufacture of statistically negative evidence by packing the studied population with a majority of individuals having no exposure and thereby diluting any positive evidence to the vanishing point"; employing workers for too short a time to cover the latent period; and locating "flexible experts" to testify that animal evidence of carcinogenicity is insignificant. Industrial health authorities had been known to ignore or reclassify diseases to avoid responsibility: Hueper cited one case in which an Ohio health commissioner had faithfully reported all cases of scrotal cancers among wax workers in oil refineries—until the mid-1930s. Hueper visited one such refinery in the early 1950s and found that, while such cancers had not actually disappeared, they were now classified as "venereal ulcers," which were not compensable. It took another ten years for industrial hygienists to publicize the problem, by which time technological changes had rendered many of the wax-pressing operations obsolete.[49]

Hueper had little tolerance for such activities. In an unpublished interview shortly before his death, the aging pathologist railed against the manufacturers of carcinogens: "We have to put those guys who think they are cleverer than we are, they are smarter, they can cheat us, in jail. Not money, just jail. For instance, I would put the president of a company which produces cancer-producing material . . . for ten years as an orderly in a hospital for incurable cancer patients. Then he will know what he has done." Public health authorities, though, were also culpable because they had allowed themselves to be captured by industrial interests. Valuable work was done, but often not until political pressure had been brought to bear from groups outside the medical establishment.

Hueper was always fond of pointing out areas in need of investigation, and in 1976 he produced one last list. There had been inadequate study of the hazards of living in homes built on the radioactive slag of phosphate mines and of the risks from chromium fumes in foundries and metal shops. More needed to be known about the effects of acid mists in chrome-plating operations and chromite ore dust in brick manufacture. Studies were needed of migrant laborers handling pesticides and of the nickel refiners of West Virginia. More needed to be known about whether the hormones given to women as contraceptives or during menopause increase the risk of breast or uterine cancer, and whether carcinogens in the "maternal organism" may be transported into the fetus to cause cancer. Hueper complained that public health authorities had generally been slow to study these and many other problems. Evidence of a bladder cancer risk among chemical workers handling aromatic amines (in rubber factories, for example) had been around for decades, but where were the studies of the effects of these substances on dyers, painters, or printers using colored inks? Why was the human liver cancer hazard of vinyl chloride discovered only in the 1970s, when evidence of a danger to rodents had been known for almost twenty years? Where were the studies exploring whether mechanics and service station attendants working in grease pits show increased cancers of the skin or other organs? Why had there never been long-term studies of the "considerable number of young soldiers" fed irradiated foods in the 1950s?[50]

Hueper's campaign to track down industrial carcinogens earned him a reputation as something of a maverick in the health profession. Harold Stewart, one of his closest friends at the NCI, described him as someone with a unique ability "to shock and alarm." His dry and rather Germanic prose suggests that he favored scholarly precision over readability, but he was not above appealing to popular sentiment. On one occasion, Hueper raised eyebrows at a National Research Council meeting when an industry representative stormed in waving a copy of the *Police Gazette* in which an article by Hueper appeared, "wedged in among the pictures of scantily attired bathing beauties." The fact that a prominent cancer researcher would publish in such a journal ("found in barber shops, cheap saloons, pool halls, and like places") appeared to his rather staid colleagues "like waving a red flag in the face of a bull."[51]

Hueper's main problem, though, was that he was out of step with the political currents of his time. Environmental carcinogenesis was generally regarded with suspicion in the 1950s and early 1960s. In his emphasis on chemical irritants, Hueper could even be regarded as old-fashioned: his 1942 treatise stressed the importance of generalized trauma or chronic irritation as a cause of cancer, a viewpoint that flourished from the middle of the nineteenth century to the early decades of the twentieth. This view came under criticism in the 1950s, as the emphasis shifted to nonchemical carcinogens—especially viruses and genetic factors—as keys to the disease. Conventional wisdom held that chemicals worked at most to trigger cancer viruses already present in cells, or perhaps to stimulate cancer genes or misplaced embryonic cells. Hueper denounced the "purely assumptive character" of such allegations, the net effect of which was, in his view, to diminish the role of chemicals, radiation, and other irritants. He never did believe that viruses or genetic factors play an important role in carcinogenesis.[52] Interestingly, the older focus on irritants has been revived to a certain extent in recent years, both in theories focusing on the role of stress and in Bruce Ames's thesis that cancer is produced by a process akin to chronic wounding.

I don't mean to leave the impression that Hueper was alone in his fight against occupational and environmental cancer hazards. Industrial hygiene was a well-established field by the 1930s, and labor activists had fought for health reforms for decades.[53] Prior to the 1920s and 1930s, however, workplace exposures were usually so high that it was normally acute, rather than chronic, poisons that were of greatest concern. Cancer attracted little attention. Neither of the first two English-language treatises on industrial hygiene had much to say about cancer: J. T. Arlidge's *Hygiene, Diseases and Mortality of Occupations* (1892) mentions cancer only in the context of chimney sweeps, and Thomas Oliver's *Dangerous Trades* (1902) is equally brief on the topic. Even Alice Hamilton's influential *Industrial Poisons in the United States* (1929) makes surprisingly little reference to cancer.[54] The more central concerns were accidents and poisonings—as in the 1880s, when several Du Pont workers were killed by breathing nitric acid fumes used in the manufacture of dynamite, and several others were injured when nitroglycerin was absorbed through their skin. Corporations eventually responded to such incidents by conducting medical research and attempting to establish safe practices—but again, little attention was given to cancer. Cancers almost always appear long after the exposures that cause them, making it difficult to recognize a link. The larger problem, though, was that in the postwar conservatism of the 1950s, Hueper's pro-labor and, as it was perceived, anti-industry emphasis on cleaning up the environment and the workplace found little support.

RACHEL CARSON: PROTECTING THE WEB OF LIFE

Wilhelm Hueper was never what one could call a folk hero, but one of his best-known admirers was. Rachel L. Carson (1907–64) was an extraordinarily talented writer, whose passionate and simple prose brought her a large and sympathetic audience. Her prizewinning essays appeared in *The New Yorker,* the *Yale Review,* and the *Atlantic;* her 1951 book, *The Sea Around Us,* remained on the *New York Times* bestseller list for eighty-six weeks and was translated into thirty-three foreign languages. Like Hueper, she was a governmental employee—first a writer for and later editor-in-chief of all Fish and Wildlife Service publications—but she never had to endure the censorship suffered by Hueper. Her prescient and, some say, exaggerated warning of the dangers of pesticides soon became the most widely cited document of the environmental movement; six years after its publication, a British expert on pesticides proposed that the history of pest control be divided into two periods: BC and AC—"Before Carson" and "After Carson."[55]

Rachel Carson's 1962 *Silent Spring* is best known for its worries about birds and butterflies; less well known are her cautions against the human health hazards of agricultural chemicals and industrial effluents. The chapter on "Elixirs of Death," for example, begins by noting: "For the first time in the history of the world, every human being is now subjected to contact with dangerous chemicals, from the moment of conception until death." Pesticides were the most prominent source of that danger. Carson noted that the synthetic pesticides discovered during World War II were found during a search for chemical warfare agents—agents coincidentally found to be lethal (also) to insects. In light of this historical association, and especially the fact that humans share many biological processes with insects, it should have come as no surprise that pesticides (more properly *biocides,* in her view) could be harmful to humans as well as to insects, plants, and rodents. In every case, their potency depended upon their ability to block and/or alter biological activity—activity that was often common to both animal friends and animal foes.[56]

Carson's critique was broader, of course, than the problem of human cancer. Her primary focus was on the health of the ecosystem as a whole, of which humans are just one part. The crux of her argument was that since pesticides kill "good" as well as "bad" insects, petrochemical pest control can worsen the very condition it is designed to improve—by wiping out the predators of pests along with the pests (causing the latter to "flareback"). Insecticides almost invariably foster the selection of resistant strains, requiring farmers to apply ever greater quantities of poison to achieve the desired effect. Pesticides can also kill valuable soil microbes, with long-run consequences that are difficult to predict. The larger lesson was that a fragile "web of life" connected everything to everything else, and that in our rush to poison a few insect enemies we are poisoning the earth itself.

Carson did not see this larger ecological context as something separate from the question of human health and well-being. The ecological "web of life" embraced humans no less than plants or other animals. Harm to the one is often harm to the other. Insecticides sprayed on lakes to remove gnats will end up in plankton, fish, and humans. Carson warned that pesticides could become more and more concentrated as they move through the food chain: unlike earlier mineral-based pesticides, organic pesticides tend to concentrate in fatty tissues, where they lie dormant, "like a slumbering volcano," erupting to injure their hosts when that fat is consumed. DDT spraying at Clear Lake, California, for example, produced concentrations of .02 parts per million in water, 5 parts per million in plankton, and 1,600 parts per million in the fatty tissues of birds. Carson pointed out that DDT had become so common in the human diet that it was hard to find anyone free of the substance. Americans averaged 5 to 7 parts per million in fatty tissues, and even Eskimos in the Arctic carried residues.[57] Pesticides were being found not just in the eggshells of eagles but in mothers' milk and infant tissues. Carson cautioned against the new use of "systemic" insecticides—insecticides that render the very tissue of a plant or blood of an animal poisonous to insect pests. She wondered whether the time might come when humans would take pills to poison their blood against mosquitoes.

Silent Spring also warned about the synergistic effects of multiple exposures—to substances that, even if relatively harmless in isolation, could become lethal in combination. Synergistic effects could occur if, for example, one of the substances destroyed a protective detoxifying liver enzyme. Interactive effects such as these could pose a danger to the farmer who sprays with one poison one week and another the next, or even to the consumer who ingests two or more different pesticide residues in a salad.[58] Exposure to multiple pesticides was more likely the rule than the exception: it was meaningless to talk about the safety of this or that contaminant in isolation, because we are continually ingesting pesticides in complex and unique combinations with unknown or even unknowable consequences.

Though her primary emphasis was on acute poisonings, Carson also considered the case of chronic disease, most notably in her chapter with the ominous title "One in Every Four." Here she pointed out that cancer had grown from a minor to a major cause of American deaths, accounting for only 4 percent in 1900 but 15 percent by

1958 (today's figure is more than 20 percent). A relative rarity in previous centuries, cancer had become the foremost cause of death by disease among American school-children and one in four adults would eventually contract the disease. Evidence was still sketchy concerning which among the various pesticides were most hazardous, but five or six were clearly carcinogenic, notably the arsenic-based herbicides but also several of the newer petrochemical pesticides such as DDT. (Wilhelm Hueper was Carson's source for the view that DDT was a carcinogen.) Given the long latency period of most cancers, it was still too early to judge the overall impact of these exposures. DDT, for example, was not even recognized as an insecticide until 1939; its widespread use dated from 1942 for military purposes and 1945 for civilian use. Only time would tell what the long-term effects might be. For shorter-term cancers, the early results were already disturbing: from 1950 to 1960, death rates from blood and lymph malignancies had grown from 11.1 per 100,000 to 14.1 per 100,000. Leukemia rates appeared to be rising at about 4 or 5 percent per year, and not just in the United States.[59]

Carson postulated that chemical carcinogens might act in a similar manner as radioactive agents, by mutating the genetic material. She proposed that certain substances may also work in an indirect manner, as when the balance of sex hormones is disturbed. Toxic injuries to the liver, typical of pesticide damage, may promote this kind of "indirect carcinogenesis" by allowing excess production of estrogen. Cancer may be caused by vitamin deficiencies, or even by "the complementary action of two chemicals, one of which sensitizes the cell or tissue so that it may later, under the action of another or promoting agent, develop true malignancy." Detergents, for example, may irritate the lining of the digestive tract, making the tissues more vulnerable to carcinogenic hazards. Cancers such as leukemia may begin through a two-step process: "the malignant change being initiated by X-radiation, the promoting action being supplied by a chemical, as, for example, urethane."[60]

Whatever the mechanism, it was hard to deny the danger. Dutch studies had shown that cities receiving their drinking water from rivers had higher cancer rates than those drawing their water from wells. Arsenic-fouled runoff waters from mining waste dumps had been linked to cancer, and one could expect similar dangers from the arsenic-based insecticides sprayed on tobacco and other crops. Carson noted that in the spring of 1961 an epidemic—actually an *epizootic*—of liver cancer had appeared among the rainbow trout in U.S. fisheries; in some areas, practically all the fish over the age of three were affected. Hueper at the NCI had arranged a collaborative project with the Fish and Wildlife Service to report all fish with tumors, in order to warn of possible human health hazards from polluted waters. Carson seconded Hueper's warning that "the danger of cancer hazards from the consumption of contaminated drinking water will grow considerably within the foreseeable future."[61] She also cited his comparison of the present cancer epidemic with the situation at the end of the nineteenth century, when common killers such as cholera were recognized to originate from fouled water. The key in both cases would be prevention rather than cure. The task should not even be as hard as in the nineteenth century, since it is we who have fouled our nest: "man has put the vast majority of carcinogens into the environment, and he can, if he wishes, eliminate many of them." We are living in a "sea of carcinogens" and it is up to us to clean up that sea.

Rachel Carson's *Silent Spring* has become a classic of environmental literature. The book has sold more than a million copies and has never gone out of print; at least twenty-two foreign-language translations have appeared. A recent survey of 250 environmental activists and environmental organizations ranked it second only to Aldo Leopold's *Sand County Almanac* in terms of numbers of people influenced by an environmental text.[62] As one can well imagine, the immediate response to it was polarized. Agribusiness and pesticide interests were predictably incensed. *Chemical World News* branded the book "science fiction, to be read in the same way that the TV program 'Twilight Zone' is to be watched." An official of the Canadian Department of Agriculture dismissed it as "based on emotion rather than sound scientific logic"; *Chemical and Engineering News* produced a snide review headlined "Silence, Miss Carson."[63] Several other attacks were equally ad hominem (and sexist): at a meeting convened shortly after the publication of the book, a member of the U.S. Federal Pest Control Review Board queried: "I thought she was a spinster. What's she so worried about genetics for?" F. A. Soraci, director of New Jersey's Department of Agriculture, suggested that Carson's book was typical of that "vociferous, misinformed group of nature-balancing, organic-gardening, bird-loving unreasonable citizenry that has not been convinced of the important place of agricultural chemicals in our economy." Thomas H. Jukes of American Cyanamid called the book a "hoax"; P. Rothberg of Montrose Chemical (a manufacturer of DDT) called her "a fanatic defender of the cult of a balance of nature."[64]

Several chemical firms took more direct action. Even before the book appeared, the Velsicol Chemical Corporation of Chicago wrote a five-page letter to Houghton Mifflin seeking to persuade the publisher to reconsider its plans to publish the book, especially in light of its "inaccurate and disparaging" remarks about chlordane and heptachlor, pesticides manufactured solely by Velsicol. The letter also intimated that Carson was allied with "sinister parties" working to undermine the chemical industry and free enterprise more generally "so that our supply of food will be reduced to east-curtain parity." The National Agricultural Chemicals Association expanded its public relations department, and the Manufacturing Chemists' Association mounted a campaign (including 100,000 mailings) to stress the "positive side" of pesticides. *Monsanto Magazine* parodied the book's first chapter with a fable of its own ("The Desolate Year") warning of the famine and destitution that would surely accompany a world without pesticides. The Nutrition Foundation, representing fifty-four food and chemical manufacturers, put together a "fact kit" to refute Carson; the package was mailed to nurses, librarians, agricultural experiment stations, the entire membership of the American Public Health Association, and many other groups. This steady barrage of propaganda yielded fruit: in November 1962, the *AMA News* advised its readers to contact chemical trade associations to find out how to reassure their patients' fear of pesticides. Carson was stunned to hear that the nation's leading physicians' association was looking to industry for answers: as she told a reporter, "I can't believe that the AMA seriously believes that an industry with $300 million a year in pesticide sales at stake is an objective source of data on health hazards."[65]

Opposition to the book was more than matched, though, by the enormous wave of support from readers—many of them quite prominent (attention was first drawn to the book when a condensed version was serialized in *The New Yorker* in June 1962). Scientists who immediately endorsed the book included Loren C. Eiseley, Julian Huxley, Hermann J. Muller, and many lesser lights. Paul H. Mueller, the Swiss chemist who developed DDT as a pesticide, suggested on the *New York Times* editorial page that Carson should be awarded the Nobel Prize for Peace. Favorable reviews poured in from dozens of popular and scientific magazines, including *Scientific American, Saturday Review,* and *Pacific Discovery.* Supreme Court Justice William O. Douglas wrote a sympathetic article on Carson for the Book-of-the-Month-Club's newsletter, and Congressman John V. Lindsay entered the book's concluding paragraphs into the *Congressional Record.* President John F. Kennedy cited it at a White House press conference on August 29, 1962, in response to a question about whether the administration was doing anything about the pesticide problem. Shortly thereafter, he asked his President's Science Advisory Committee to look into the question. The committee's final report reflected many of Carson's concerns (she had been solicited for advice), much to the dismay of the Agriculture Department and the Public Health Service.

Rachel Carson was ill in the final months of writing *Silent Spring.* She had a growth removed from her breast in March 1960, and though she asked whether the tissue removed was cancerous, her physician lied and told her it was benign. Six months later, with a different physician, she found out the truth.[66] (Cancer patients at this time—particularly women—were commonly not told the nature of their illness; this practice continues today in many parts of the world, for example, Japan and France.) Carson's cancer had apparently already metastasized by the time of diagnosis. Tumors eventually formed in her cervical vertebrae, causing numbness in her hands. One can only wonder how this must have affected someone who made her living almost entirely through writing. One also wonders what else Carson would have written had cancer not frozen her pen. She confessed shortly before her death that she had several more books planned. We know of three: a book of nature study for children ("Help Your Child to Wonder"), a plea for the conservation of unspoiled shorelines, and an ambitious "man and nature book" on what it means to live in a world where natural forces can be changed and perhaps ruined by human agency. She confided to a friend that, if she could live to be ninety, she would surely have even more to say.[67]

Carson died on April 14, 1964, at the age of fifty-six, amid ever-growing acclaim for her ideas. By the end of 1962 more than forty state legislatures had introduced bills to regulate pesticide use, and Kansas and Iowa were enacting legislation requiring licenses for professional pesticide applicators. Loopholes in the Federal Insecticide, Fungicide and Rodenticide Act were being closed that had allowed pesticides to be kept on the market long after they were found to be hazardous. At her funeral in Washington's National Cathedral, Stewart L. Udall, Secretary of the Interior, helped carry her casket, along with Senator Abraham Ribicoff of Connecticut, another longtime supporter. On the floor of the U.S. Senate, Ribicoff paid tribute to "this gentle lady who aroused people everywhere to be concerned with one of the most significant problems of midtwentieth–century life—man's contamination of his environment."[68]

*   *   *

It may be too much (and perhaps too little) to say that Hueper and Carson did for industrial disease what Pasteur and Koch had done for infectious disease. Hueper was more of a synthesizer, Carson more of a popularizer. The comparison is interesting, though, insofar as in both cases, a new class of disease-causing agents was identified that changed how we think about illness and health.

Effaced in any such comparisons, though, is the fact that Hueper and Carson were activist scientists, concerned not just to identify new kinds of hazards but to question the cultural authority of science as it was then practiced. This is more true for Carson than for Hueper: Hueper was worried about "shoddy science," science twisted by industry interests, but Carson was worried about the broader relation between humans and nature. Her notion of a "web of life" diverged radically from the instrumental rationality of most of her contemporaries; her challenge to how we understand science and nature may ultimately outlast whatever specific attention was drawn to pesticides and other chemical hazards. Neither Carson nor Hueper, after all, was without particular myopias—some of them rather glaring. Both, for example, were slow to appreciate the greatest cancer hazard of the century: tobacco smoke.[69]

Hueper and Carson were both aware that cleaning up industrial carcinogens could prove much more difficult, in certain respects, than guarding against biological germs. Germs, after all, were almost never the direct and intentional product of industry. (Feces-contaminated water was the most common cause of the devastating cholera and typhoid epidemics of the nineteenth century.) No one felt about diphtheria or TB the way Monsanto felt about pesticides, Johns-Manville felt about asbestos, or the Atomic Energy Commission felt about radiation. Industrial hazards, unlike infectious germs, could also be seen as the unfortunate but inevitable consequence of economic progress—a "cost" to be balanced against the "benefits" of modern life. There was therefore more to overcome in convincing the scientific public that industrial carcinogenesis was something to be taken seriously. When such a recognition was finally institutionalized, with the founding of the Occupational Safety and Health Administration and the Environmental Protection Agency in 1970, Hueper and Carson could be seen as having helped to clear the path.

Chapter 3

The Percentages Game

*I have a suspicion that we're losing the war on cancer
because of mistaken priorities and misallocation of funds.*
—Senator George S. McGovern, 1978

STRANGE AS IT MAY SEEM in our own environmentally conscious era, the 1970s were actually far more energetic in terms of environmental activism and legislation. The first Earth Day in 1970 was followed by the passage of the Environmental Protection Act and the Clean Air Act—two of the greatest triumphs of American environmental history. The Occupational Safety and Health Act of 1970 required an employer to provide a work site "free from recognized hazards that are causing or are likely to cause death or serious physical harm to his employees"; the Clean Air Act imposed strict and unprecedented standards for urban air quality, enforced by threats to withhold federal funds if standards were not met. DDT was banned in 1972, and many other carcinogens followed in its path. President Richard Nixon has the distinguished honor of having signed into law the most potent environmental legislation of the century.

In the middle of that decade of environmental activism, a flood of criticisms of medicine and medical priorities began to emerge. Critics included physicians and folk healers, feminists and Marxists, nature lovers and cost-conscious health administrators. Thomas McKeown published two books in 1976 arguing that, compared to improvements in sanitation and nutrition, medicine had played only a minor role in the historical increase in longevity. In *Illness as Metaphor,* Susan Sontag condemned the "conventions of concealment" that allowed physicians to treat patients without telling them they had cancer. Peter Chowka pointed out in the *East-West Journal* that the National Cancer Institute's obsession with viruses and chemotherapy had deflected attention from more promising inquiries into diet. Others argued that the costs and benefits of cancer treatment were not as advertised: the libertarian Aaron Wildavsky suggested that the marginal benefits of health care dollars were essentially zero, and Ivan Illich, the radical Catholic cleric, maintained that medicine itself was contributing to sickness through a process of *iatrogenesis,* or "doctor-induced disease."[1]

As part of this upsurge, the "cancer establishment" came under attack from both the Right and the Left. On the Right, paranoid conservatives argued that Rockefeller philanthropic interests and the federal research establishment had conspired to suppress alternative therapies. G. Edward Griffin of the John Birch Society warned of a conspiracy by government and business to conceal evidence of the value of laetrile (Vitamin B_{17}) as a cheap and effective anticancer agent. How else could one explain the fact that the American Cancer Society owned half the patent rights to 5-fluorouracil, a leading chemotherapeutic agent manufactured by Hoffmann-La Roche, a co-conspirator "firmly within the IG-Rockefeller orbit"? What else could one expect from the American Medical Association, half of whose income came from advertising, rendering it "a captive arm and beholden to the pharmaceutical industry"? Griffin endorsed the suggestion of Milton Friedman, the Chicago economist, that the FDA should be abolished, and that people should be free to choose the kind of medical treatment they want.[2]

Critics on the Left, by contrast, argued that capitalist production was tending to reduce workers to "inputs" that, from the point of view of production, were expendable. Vicente Navarro, editor of the *International Journal of Health Services,* suggested that considering the class character of health administrators, it was hardly surprising that the problems of industrial workers were downplayed. David Ozonoff of Boston University's School of Public Health suggested that while large amounts of money had been spent on basic cell biology and even more on chemotherapies, little had gone into identifying the causes of cancer: the chemical and physical pollutants in the home, community, and workplace. Instead, undue attention had been given to one particular carcinogen: cigarette smoke. This was consistent with the more general tendency to "blame the victim" by implying that "we get the deaths we deserve as a result of our intemperate personal habits."[3]

Common to many critiques at this time was the idea that most cancers are of environmental origins and therefore preventable. Hints of such an idea go back a very long way; its specifically modern formulation dates from the 1950s, when John Higginson suggested that 70 to 80 percent of all human cancers are environmental and therefore avoidable. The World Health Organization (WHO) in 1964 endorsed his thesis that "the majority of human cancer is potentially preventable."[4] As director of the WHO's International Agency for Research on Cancer (IARC), Higginson championed this point of view in a much-cited article published in 1969, and John Cairns in 1975 echoed the view that "almost all cancers appear to be caused by exposure to factors in the environment."[5] The point of such claims was to inveigh against the more traditional view that cancer was simply a product of fate, bad luck, genetic infirmity, or aging. The practical implication was clear: if cancer was a result of things like personal habits, environmental pollution, and occupational poisons, then those should be the targets of cancer policy. Critics used this to argue that the "wrong war" was being waged against cancer, that far too much emphasis was being devoted to the elusive search for cures and not enough to prevention.[6] Senator George McGovern of South Dakota, the one-time presidential candidate, castigated federal cancer officials for ignoring diet and other preventive possibilities: a 1978 hearing of his Select Committee on Nutrition asked, "The War on Cancer: Is It a Multi-Billion Dollar Medical Failure?"[7]

As we shall see, though, scholars would use the term "environmental factors" broadly, leading to confusion as to how one should go about the business of prevention.[8] It was never easy to determine, for example, exactly what proportion of cancers were attributable to industrial effluents and what portion were to be assigned to other environmental factors such as smoking, alcohol, dietary fat, and viral infection. Where one stood on such matters often had as much to do with one's politics as with anything else. For environmental activists, blame was clearly to be laid at industry's door. Thus Samuel Epstein in 1974 wrote that "there is now growing recognition that the majority of human cancers are due to chemical carcinogens in the environment."[9] Brochures for the Rachel Carson Trust, an antipesticide group headed by Epstein, suggested in 1975 that "As much as 60–90 percent of cancer cases in the world today are caused by environmental and occupational factors. . . Many leading authorities believe that 90 percent of these environmentally induced cases are due to chemical carcinogens, most of which are man-made."[10] The growth of synthetic organic chemical production since World War II was often used to explain the apparent growth of cancer and the likelihood of its continued increase: Ralph Nader in 1976 prophecized before the U.S. Congress that we were about to enter the "carcinogenic century."[11]

Support for such a thesis came from the fact that many—and perhaps even most—of the reliably identified carcinogens were of industrial or occupational origins: chimney soot (1775), smelting fumes (1820), paraffin (1875), the air inside uranium mines (1876), the lubricating oils used by mule spinners (1887), aniline dyes (1895), X rays (1908 for sarcoma and 1930 for leukemia), coal tar (1915), radium-dial paint (1920s), nickel ores (1932), asbestos (1930s), betanaphthylamine (1930s), and so forth. The only notable exceptions were certain rare and inconsequential viruses, the parasitic worm bilharzia (traced to sewage-contaminated waters of the lower Nile), sunlight, certain hormones, and, of course, tobacco. The order in which these discoveries were made is suggestive: Edith Efron, the neoconservative author of *The Apocalyptics,* claims that if natural carcinogens had been discovered before industrial carcinogens, the regulatory zeal over synthetic chemicals might have been tempered.[12]

Apart from mavericks such as Hueper, however, the industrial emphasis had always been—until the 1970s—a minority position within the cancer research establishment. Michael Shimkin, the editor of *Cancer Research,* in 1969 characterized as "benzpyromaniacs" those who believed that polycyclic hydrocarbon pollutants would prove to be the primary cause of human cancer.[13] John Higginson, author of the "70 to 80 percent environmental" formulation, that same year pointed out that he had never said that industrial pollution was a major cause of cancer: "the environment" included much more than industry—notably diet, sexual behavior, hormonal influences, social pressures, and so on. In a 1979 interview, Higginson explained that:

> Environment is what surrounds people and impinges on them. The air you breathe, the culture you live in, the agricultural habits of your community, the social cultural habits, the social pressures, the physical chemicals with which you come in contact, the diet, and so on. A lot of confusion has arisen in later days because most people have not gone back to the early literature, but have used the word environment purely to mean chemicals.[14]

Higginson confessed that he was all for cleaning up the air and trout streams, but he also cautioned that we shouldn't use the wrong argument for doing it. Cancer should not be made the "whipping boy for every environmental evil," especially if in so doing we ignore more important "lifestyle" carcinogens like cigarettes.[15]

The magnitude of the cancer hazard from pollution remains unknown. What I want to explore here is how, in the late 1970s, powerful environmental voices emerged to argue that industry is responsible for a sizable fraction of human cancers. Nowhere was this view more powerfully set forth than in Samuel S. Epstein's 1978 *Politics of Cancer.*

SAMUEL S. EPSTEIN AND THE POLITICS OF CANCER

Samuel S. Epstein, born in Great Britain in 1926, earned diplomas in tropical medicine and pathology at London University before obtaining his medical degree in 1958. He moved permanently to the United States in 1961 after obtaining a position as chief of the laboratories of carcinogenesis and toxicology at the Children's Cancer Research Foundation in Boston. There he distinguished himself as an analyst of the health effects of tetraethyl lead, maleic hydrazide (a herbicide), and various petrochemical contaminants in air and water. He published papers on the synergistic toxic effects of Freons and piperonyl butoxide and tried to develop a practical screening test for chemical carcinogens, exploiting the same suspected links between mutagenesis and carcinogenesis that would become the basis for the Ames test (the most widely used bacterial bioassay) a few years later. Epstein at this time was concerned about the long-term genetic effects of chemicals on the human germ plasm. He became secretary of the Environmental Mutagen Society in the late 1960s, believing that the germ-line genetic effects of environmental chemicals could be as serious as the shorter-term cancer they caused.[16]

Much of this work combined Epstein's political acumen and his scientific expertise: he cautioned against food irradiation and against the continued use of cyclamates, for example. In 1970, he and William Lijinsky of the University of Nebraska's Eppley Institute for Research in Cancer published a widely cited review of the carcinogenic action of nitrosamines produced in the acid environment of the stomach from consuming the nitrates (and hence nitrites) commonly used as preservatives in meat and fish. Their conclusion was that "reduction of human exposure to nitrites and certain secondary amines, particularly in foods, may result in a decrease in the incidence of human cancer."[17] Epstein was among those convinced that "most human cancers are probably caused by chemical carcinogens in the environment" and therefore preventable.[18] His expertise in this area led him to testify for a broad range of governmental committees, and in 1971 he drafted a proposal for the Toxic Substances Control Act, a later version of which was passed in 1976.

The central thesis of his *Politics of Cancer* is that industrial pollutants are responsible for a sizable proportion of all cancers. Epstein sought to prove that cancer rates had been on the rise for several decades following the rapid growth of steel, petrochemical, and pharmaceutical production after World War II. He predicted that cancer rates would continue to grow as the exposures in the 1960s and '70s took their toll in the 1980s and '90s. Overall cancer rates were increasing at about 2 percent per year, but some were growing faster: cancer of the uterus, for example, was growing at about 10 percent per year. Food additives were another major problem: Epstein pointed out that Americans ingest about nine pounds of chemical additives per year, and that children consume a quarter-pound of coal-tar dyes—many of which are known carcinogens—added to food for purely cosmetic reasons. Occupational exposures were yet another major cause of cancer: Epstein estimated that 50,000 U.S. deaths were caused annually by asbestos alone.

Epstein also pointed out that increasing cancer rates were not simply due to the fact that people were living longer. Even after adjusting for age, the average incidence of cancer had increased more than 10 percent in the preceding four decades. Thus "a fifty-year-old man today is more likely to die of cancer than was a fifty-year-old man in 1950"—8 percent more likely, in fact (27 percent compared with 19 percent).[19] The likelihood of an individual contracting cancer of the esophagus, colon, rectum, pancreas, lung, breast, ovary, prostate, or bladder increased between 1937 and 1969, even taking into account the increased life span over this period. Every year, five times as many Americans were dying from cancer as had been killed in the Vietnam and Korean wars combined.

Warning of an "epidemic of cancer," however, was only part of his argument. Epstein also charged that industrial interests had conspired to deflect, distort, or even destroy evidence of the carcinogenicity of specific compounds. Industry had adopted a "complex and effective set of strategies to block regulation of hazardous products and processes," including minimizing hazards, maximizing the supposed costs of compliance, and exaggerating the unique efficacy of particular products. Industry strategies included diverting attention (by insisting on ever greater precision or long-term human studies); propagandizing the public (through media blitzes of the type waged by the Calorie Control Council to defend saccharin); blaming the victim (by pointing to genetic susceptibilities or to bad habits such as smoking); controlling information (that provided to government agencies, for example); influencing policy (by assisting in the writing of toxic legislation, for example); and exhausting agencies (by repeatedly challenging the assumptions underlying risk-assessment procedures). If these failed, Epstein suggested, corporations could always pull up stakes and move either to southern states, with their traditionally lax occupational and environmental standards, or abroad, where they might set up shop with virtually no environmental restrictions.[20]

Epstein charged that industry manipulation, suppression, and outright destruction of cancer data were commonplace, especially when profitable products or processes were involved. In the early 1970s, for example, European plastic manufacturers found that vinyl chloride could cause cancer; the information was shared with the U.S. Manufacturing Chemists' Association, and both groups agreed to withhold publication of the findings. The results were not released until the widely publicized deaths of several polyvinyl chloride workers from a rare liver cancer. FDA evidence that the insecticides aldrin and dieldrin (used primarily on corn) were carcinogenic date from 1962 but their manufacturer, the Shell Chemical Company, managed to stave off a ban on agricultural uses until 1975 (household use against termites was not, however, banned). Chlordane and heptachlor, two closely related organochlorine pesticides, were ordered phased out in the late 1970s, but only after a decade-long struggle following the FDA's discovery of a cancer hazard in mice. Reviewing such cases, Epstein asserted that industry had been aided "by the laxness of some regulatory agencies whose authority and function are often subverted by political pressures." It was improper for industries to be the main supplier of cancer information to Congress, a practice he likened to "asking the Mafia to give the police a list of people who should be imprisoned."[21]

Federal cancer agencies were also culpable, according to Epstein, having been dissuaded from responding appropriately to the cancer threat by virtue of their close ties to industry. The National Cancer Advisory Board, created by the National Cancer Act of 1971, was headed by Benno Schmidt, a New York investment banker with ties to oil, steel, and chemical industries through J. H. Whitney and Co., of which he was managing partner. Pharmaceutical interests were strongly represented on the board, but at no time had labor or a public interest group been represented.[22] These and other conflicts of interest led the nation's leading cancer agencies to focus excessively on basic research into the *mechanisms* of carcinogenesis and insufficiently on research into what *causes* cancer in the first place. Most important, federal cancer bodies had failed adequately to support efforts to prevent the disease. A 1974 index of current NCI grants showed that only 1 out of a total of 307 pages dealt with epidemiological studies. That year NCI spent $134 million on "cancer cause and prevention," but more than half that amount was devoted to exploration of viral causes. Not one of the twenty-three members of the National Cancer Advisory Board had significant professional experience in either preventive medicine or epidemiology, and only one had experience in chemical carcinogenesis. There was also the rather curious fact that the U.S. government was spending $50 million per year on tobacco subsidies while also claiming to be fighting cancer.[23] Epstein would later sum up the situation with a hyperbolic complaint about "Billions for Cures, Barely a Cent to Prevent."[24]

In the final chapter of *Politics of Cancer,* called "What You Can Do to Prevent Cancer," Epstein noted that there were many practical steps that could be taken to reduce the risk. Quitting smoking was at the top of his list, followed by reducing one's alcohol consumption and eating intelligently (including low-fat and fiber-rich foods and avoiding junk foods with purely cosmetic additives such as nitrites, colorings, and oil-derived protein flavorings). Exploiting the authority of his medical degree, he cautioned menopausal women against taking estrogens unless the symptoms were crippling and advised against the use of drugs such as Flagyl to treat vaginal infections, griseofulvin for athlete's foot and ringworm, and Lindane shampoos for head lice. Given the uncertainties in the safety of the pill, he recommended consideration of other forms of contraception; he also advised against using cosmetics sold with warning labels indicating an uncertain cancer hazard. Unnecessary X rays should be avoided "like the plague," whether offered by doctors, dentists, or chiropractors. Genital hygiene should be practiced, especially by uncircumcised males, and everyone—particularly people with fair skin—should avoid prolonged exposure to the sun and to tanning lamps. He cautioned against the use of hair sprays with vinyl chloride propellants and hair dyes containing carcinogens. Parents should not allow their children to sleep in pajamas treated with Tris, the cancer-causing flame retardant, and everyone should avoid close contact with paint and varnish removers containing carbon tetrachloride, trichlorethylene, perchloroethylene, or benzene.

Less conventional were his suggestions to weigh the health impacts of where we live and work, how we play, and how we are treated by physicians. Epstein suggested that industrial workers should try, wherever possible, to work for large, well-organized plants with a reliable and informed union leadership. Housing should be chosen as far as possible from chemical or asbestos plants, refineries, highways and expressways, hazardous waste–disposal sites, and downtown areas with air pollution. He pointed out that arts and crafts and other hobbies may involve hazardous materials—such as asbestos (in plasterboard and spackle, for example), heavy metals (in many paints), and dangerous solvents (benzene, for example). Parents should be aware that their children may face hazards at school—in junk foods or laboratory and shop chemicals, for example. Finally, cancer prevention should be supplemented by Pap smears, periodic pelvic exams for women over fifty, routine proctoscopy for men over forty, breast self-examination from adolescence on, and laryngoscopy for smokers and drinkers to detect cancers as early as possible. Specialized cancer centers should be consulted for rare kinds of cancer, and legal remedies should be sought if one suspects that an employer's negligence may have caused one's cancer.[25]

Epstein discussed these and other precautions and remedies at length, but his primary focus was always on preventing cancer before it arises. Ultimately, he argued, triumph over cancer can come only through a political process. Since neither industry nor government had the political will or power to campaign effectively against carcinogens, it was up to public interest groups and labor unions to lead the fight. "More science" was not what was needed: Epstein ended his book with a plea that "while much is known about the science of cancer, its prevention depends largely, if not exclusively, on political action."

Epstein's indictment of industry malfeasance, research impotence, and regulatory incompetence generated enormous controversy. Mickey Friedman of the *San Francisco Examiner* called the book a "bombshell";[26] Congressman Andrew Maguire of New Jersey lauded it as one of the most important books in a decade. Carl M. Shy of the Institute for Environmental Studies at the University of North Carolina stated that the book could very well have the impact of *Silent Spring* in arousing the public to demand removal of carcinogens from food, drugs, and the workplace. Television networks arranged interviews: in November 1978 Epstein appeared on the *Today* show and *Donahue;* Tom Brokaw of NBC began planning a show featuring Epstein, prompting chemical industry authorities to demand equal time. (NBC complied, though Brokaw was clearly more sympathetic to Epstein.) The book was serialized in newspapers, sold to the Sierra Club (where it became one of their most successful publications), and translated into Japanese.[27]

As one can imagine, political sympathies had much to do with how one responded to the book. Julian DeVries, medical editor of the conservative *Arizona Republic,* lambasted Epstein for espousing "the Sierra Club line"; DeVries endorsed the view of Warren Winkelstein, dean and professor of epidemiology at Berkeley, that the public's fear of environmental cancer was "a federally fostered form of hysteria which has little basis in scientific fact." Michael Halberstam in the *New York Magazine* complained that while the case histories were moving, the book was flawed by the zeal with which its author approached the problem. Epstein's unrelenting vilification of the toxic environment had produced a "free-floating paranoia among many people about everything they eat or breathe and, in others, a sense of hopeless resignation."[28] Labor, left-leaning, and popular health publications were predictably far more supportive. The United Automobile Workers' *Washington Report* declared that the book could "save human lives"; A. F. Grospiron, president of the Oil, Chemical, and Atomic Workers Union, wrote that it laid the foundation "for political action by trade unionists and public interest groups." William Tucker, a contributing editor of *Harper's,* suggested that the book was "probably going to serve as a Bible for the forthcoming effort by the President's Council on Environmental Quality to drive all carcinogenic materials from the marketplace."[29]

Most of those who raised technical objections were worried that Epstein had exaggerated the magnitude of environmental hazards. Halberstam criticized him for minimizing the dangers of smoking and denying that there might be thresholds of exposure below which a particular carcinogen is harmless: "We are almost all exposed to three of the classic carcinogens—sunlight, soot, and sex—and we do not all develop cancer of the skin, scrotum, or cervix." Tucker noted that Epstein failed to mention that natural foods such as spinach and other common leafy vegetables contain nitrates, the precursors of cancer-causing nitrosamines; Epstein had also failed to acknowledge that the body has some natural defenses against cancer, even against genetic damage caused by carcinogens. Tucker questioned Epstein's "leap of faith" that allowed him to say that the majority of cancers were caused by industry, and worried that the book would invite the "tragedy" that we will "scrub the environment clean of carcinogenic materials, only to find that cancer—as it has always been—is still with us."[30]

Epstein's most persistent critic was Richard Peto, reader in epidemiology at Oxford and soon-to-be co-author of a landmark report for the United States Office of Technology Assessment on *The Causes of Cancer.* Peto responded to Epstein's thesis with an article in *Nature* that began by suggesting that cancer research was in danger of becoming polarized between environmentalists and industrialists. The Oxford epidemiologist was by no means eager to spare industry blame for cancer, but for him it was the tobacco industry that deserved the lion's share of that blame. (His mentor, Richard Doll, also of Oxford, had been one of the first to demonstrate, in the 1950s, a clear and irrefutable link between smoking and lung cancer.) Cigarettes, Peto argued, account for the majority of lung cancer deaths in the United States—a point steadfastly denied by the tobacco industry.[31] Peto also challenged Epstein's view that overall cancer rates were on the rise. While U.S. mortality rates from several cancers were up (pancreas, melanoma, and certain lymphomas), deaths from other kinds (stomach, cervix) were down. Peto claimed that apart from the large increase in lung cancers due to smoking, there had been no noticeable increase in cancer incidence or mortality.

Above all, Peto argued that Epstein's "environmentalist zeal" had distorted his scientific judgment. In an early draft of his *Nature* review, circulated in the fall of 1979, Peto accused Epstein of forgery, prompting both the journal's editors and Epstein to hire lawyers to explore the possibility of a libel lawsuit. Peto charged Epstein with having deliberately omitted data from one of his tables that would have weakened his conclusion that saccharin may be carcinogenic; Epstein responded that he had simply cited an abbreviated source and that his general point was valid in any event. Peto accused him of having presented his facts in a misleading or unbalanced manner, as when he spoke of a "medical military complex" or the "plastic age which symbolizes how the value of our lifestyle has been degraded." Peto objected to Epstein's claim that cancer cost Americans $25 billion a year and countered by suggesting, bluntly and perhaps callously, that in a world without cancer the cost of supporting the increased numbers of people living long past retirement would probably be more than the $25 billion saved by eliminating cancer. Peto expressed his opposition to further regulatory control of industrial carcinogens since there was "both qualitative and vast quantitative uncertainty about the health benefits of restriction."[32]

Epstein responded to Peto's charges with a letter in *Nature,* defending the extrapolation of saccharin rodent data to humans and pointing out that high-dose animal studies could as easily underestimate as overestimate the carcinogenicity of a particular substance (see also chapter 7). Epstein pointed out that Peto's correlation of breast cancer and dietary fat did not necessarily prove that the one was caused by the other. Tobacco could also not be regarded as the sole cause of the rise in lung cancer. Epstein pointed out that Marvin A. Schneiderman, associate director for science policy at the National Cancer Institute, had drawn attention to the fact that "big and frightening" increases in lung cancers had been observed even among nonsmokers, and that these could not be attributed to personal lifestyle.[33] (No mention was made, however, of the possibility that secondhand tobacco smoke might be at least partly responsible.) Epstein continued his counterattack in an article on the "Fallacies of Lifestyle Cancer Theories" in the same issue of *Nature,* suggesting that Peto's emphasis on lifestyle played right into industry's hands and would probably be used to thwart efforts to regulate environmental carcinogens. Epstein admitted that smoking was the single most important cause of lung cancer, but he also pointed out that 20,000 people died from lung cancer every year who never smoked.[34] Blue-collar workers did tend to smoke more than those in professional and managerial classes, but this alone could hardly explain the fact that in certain industries (asbestos and steel manufacturing, for example), lung cancer rates were as much as ten times higher than in the general population. Tobacco, Epstein concluded, was being used as a "smokescreen" to deflect attention from occupational hazards.[35]

The bitterness of the Epstein-Peto debate may be partly explained by the fact that, in both England and the United States, *The Politics of Cancer* was a big hit with labor organizations. In Britain, the health and safety officer of the General and Municipal Workers' Union (GMWU), David Gee, invited Epstein to speak on the politics of cancer. In June 1979, Epstein lectured at the Trades Union Council Centenary

Institute of Occupational Health, co-sponsored by the GMWU and the Association of Scientific, Technical and Managerial Staffs (ASTMS), two of the most powerful unions in the country. All ten regional health and safety officers of the unions were invited, as were representatives from several other unions. Epstein repeated his claim that cancer research bodies had failed to address the question of causes, and that industry interests had conspired to prevent the problem of cancer from being tackled at its roots.

The British media's coverage of Epstein's visit was divisive and inflammatory, especially by English standards. *Time Out,* the popular London entertainment magazine, noted that one in four of its readers would contract cancer, that treatment had improved little in thirty years, and that the disease was preventable insofar as it was caused by unhealthful habits and toxic exposures. The magazine endorsed Epstein's thesis that the persistence of cancer was due not to a lack of scientific information but to a lack of political will.[36] The *New Scientist* reported favorably on the suggestion that the chemical industry had orchestrated a massive conspiracy to prevent regulation, and that industry had managed to maintain "a stranglehold on the whole process of generating and assessing scientific data."

Other British observers were much less sympathetic. Several warned against the mixing of science and politics, implying that Epstein and his ilk had politicized an otherwise pure and decent area of science. Ruth Hall, London diarist for the *New Statesman,* bristled that the very idea of a "politics" of cancer was "almost as ludicrous as the *canard* that if you vote Labour (or is it Tory?) your balls will drop off." (Percivall Pott might have had something to say about this.) Hall found the idea that industrial carcinogens could be responsible for a sizable portion of cancers "simplistic to the point of lunacy," noting that "cancer has always been with us," as evidenced by a reference in Oswaldus Gabelhouer's 1599 *Book of Physic* to a "canker" of the breast. The author of the *Time Out* report, John Mathews, responded in a letter to the *New Statesman* that, though it was true that carcinogenic mechanisms were poorly understood, this was "as true as it is irrelevant." One may correctly identify a cause without knowing all the details of a mechanism. Mathews expressed the matter through an analogy: "If I attack Ms. Hall with a club, will she ignore my molestations on the grounds that she cannot give a molecular explanation of how my blow knocks her to the ground?" David Basnett, general secretary of the GMWU, wrote that Hall's spirited debunkery was now "framed and in use on our shop steward training courses as an example of getting the maximum of false logic, irrelevancies, naivete, and straw men in the smallest number of words."[37]

Favorably impressed with Epstein's thesis, several of Britain's leading unions took steps to curb carcinogens in the workplace.[38] The GMWU drafted a Preliminary Cancer Prevention Program in the fall of 1979, deploring the lack of research and action on prevention and calling for chemical labeling and monitoring of workplace hazards. The document appeared to endorse the Epstein-Califano thesis that occupational exposures were a primary source of cancers. The union sent a copy of the draft to Peto for comment, and the Oxford epidemiologist responded by warning of the danger of "overkill interest" in the Epstein approach—that if thousands of substances were labeled carcinogenic, this might detract attention from more genuine hazards such as asbestos and smoking. Peto reminded the union that for every 1,000 workers, "1 will be murdered, 6 will die of traffic accidents, 150 will be killed before their time by smoking and probably at most only about a dozen . . . will die from all forms of occupational exposure to carcinogens." He also stated that while such estimates could not be defended against hostile criticism, at least they could not be proven demonstrably wrong "unlike the larger [Califano-OSHA] estimates which were generated in America last year."[39]

JOSEPH CALIFANO'S
20 TO 40 PERCENT

In the United States, the Epstein-Peto controversy would not have been so heated had the basic outlines of Epstein's approach not been endorsed by several of the nation's most powerful medical, environmental, and occupational health organizations. On September 11, 1978, about the same time that Epstein's book appeared, nine distinguished scientists from the National Cancer Institute, the National Institute of Environmental Health Sciences, and the National Institute for Occupational Safety and Health produced a draft summary of what may well be the most radically environmentalist U.S. government document ever written. The report, released in full four days later, predicted that over the next three decades the proportion of cancer mortality attributable to asbestos and five other carcinogens commonly found in the workplace (benzene, arsenic, chromium, nickel oxides, and petroleum fractions) could rise to as much as 40 percent. Asbestos alone was going to account for 13 to 18 percent. These figures did not even include the effects of radiation or other industrial pollutants, nor did they include "nonoccupational" effects of pollutants on the community at large.[40]

This "Estimates Paper"[41] is remarkable not only because it was radical (and was soon denounced as such) but also because it was endorsed by a cabinet-level officer of the Carter administration. Joseph Califano, secretary of Health, Education, and Welfare (HEW), announced the main conclusions of the report in a rousing speech before the AFL-CIO National Conference on Occupational Safety and Health in Washington. Califano pointed out that preventive medicine had almost always been more successful than curative medicine, as illustrated by the conquest of the infectious scourges of the last century. Preventive public health measures had led to a dramatic change in the diseases from which modern Americans suffer: far more people die today from chronic than from infectious diseases, especially heart disease, cancer, and accidents. Policy makers had not yet recognized this shift, however, and continued to focus "almost obsessively" on treatment rather than prevention. Only 4 percent of the nation's $48 billion federal health care budget went to prevention, despite the fact that 25 million workdays were lost each year because of occupational injuries and disease. Something was clearly wrong with the nation's priorities. Califano ended his speech on a moral tone: American workers were willing to sweat on the job, but "we should not ask them to bleed." Organized labor was urged to continue to "preach the Gospel of prevention and protection."[42]

The paper was not just about numbers, however. It also argued that there were pitfalls inherent in any effort to measure occupational carcinogenesis, given the failure to collect data, ambiguities resulting from the phenomenon of latency, and the difficult issue of synergy. The paper stressed that since most cancers have multiple causes, it was often difficult or even impossible to measure the individual contribution of any given agent.

The political response was immediate. In October 1978, shortly before adjournment, Congress amended the National Cancer Act by requiring that the HEW issue an annual report listing known or suspected carcinogens and the nature and extent of human exposure. In November, Califano announced the establishment of a National Toxicology Program with an annual budget of $41 million to improve research and detection of toxic substances. The Estimates Paper was used by OSHA (Occupational Safety and Health Administration) in preparing its cancer policy; it also became the basis for a 1980 report on "Toxic Chemicals and Public Protection" prepared by President Carter's Council on Environmental Quality. The Atlanta-based Center for Disease Control had released a similar report of its own a few weeks earlier, confirming the prognosis that the United States was on the verge of "an epidemic of occupational disease" and recommending the development of a national occupational surveillance capability parallel to that already in place for infectious diseases.[43]

Critics jumped on the paper, arguing that it had grossly exaggerated the risks of occupational exposures. Central to many objections was the study's methodology. Its authors had relied on a 1974 study (the National Occupational Hazard Survey, or NOHS) commissioned by the National Institute for Occupational Safety and Health (NIOSH) to determine the extent to which workers were exposed to specific carcinogens. The government investigators extrapolated from these samplings and concluded that 38.2 million employees had nearly 4.38 billion exposures, or an average of 115 exposures per worker. The authors of the paper multiplied the numbers exposed by cancer risk ratios associated with those exposures, and from this calculated the number of cancers one could expect.[44]

The problem with this procedure, as numerous critics pointed out, was that it failed to take into account the fact that exposures had declined in recent decades. Risk ratios were derived from epidemiological studies from the 1930s and 1940s, when exposures had been much higher. As *Science* magazine reporter Thomas H. Maugh II put it, these were "clearly inflated" figures:

In each case, the investigators have taken the highest risk ratio available—ratios obtained for workers exposed to massive concentrations of the carcinogens—and multiplied that by the total number of workers who might have been exposed to the carcinogen, even though most or all of the workers have never been exposed to the concentrations upon which the risk ratios are based. In a simple analogy, one might find that the risk of the driver dying in an automobile crash is one in ten if the automobile is consistently driven at speeds in excess of 120 miles per hour, and that there are currently 100 million automobiles on American highways. Using the logic of the government report, one would conclude that there will be 10 million excess deaths as a result of driving at high speeds.[45]

The asbestos estimate incited the most controversy. The Estimates Paper projected that 13 to 18 percent of *all* U.S. cancer mortality over the next several decades would be due to asbestos, based on extrapolations from studies of American shipyard workers. (Asbestos was widely used to insulate engine-room boilers and steampipes on American ships; other workers were heavily exposed because the ships were very poorly ventilated.) The document stated that 8 million to 11 million American workers had been exposed to asbestos since the beginning of World War II, approximately 4 million of them heavily exposed. Among those in this latter group, the number expected to die of cancer was estimated at between 35 and 44 percent (20 to 25 percent from cancer of the lung, 7 to 10 percent from mesothelioma, and 8 to 9 percent from gastrointestinal cancer). This meant that the total number of cancers attributable to asbestos would be somewhere in the neighborhood of 2 million over the next thirty to thirty-five years, or roughly 67,000 per year.

On December 9, 1978, an editorial in *Lancet,* Britain's leading medical journal, criticized the asbestos estimate for relying on extrapolations from studies of shipyard workers without considering the degree to which other workers had been exposed. The editorial disparaged the entire report as "insubstantial—remarkably so."[46] Industry also threw its hat into the ring. The Asbestos Information Association, a trade group, charged in a letter to Califano that the cancer death estimates were "grossly overestimated." Ralph L. Harding, Jr., president of the Society of the Plastics Industry, dismissed the document as "ludicrous to the point of absurdity" and "an embarrassment to medical science." Harding was especially bothered by the Estimates Paper's calculation that vinyl chloride was responsible for 1,940 excess cancers annually; the plastic society's president claimed that there had been only 23 confirmed fatalities from the substance in the last sixteen years.[47]

The most sustained industry challenge came from the American Industrial Health Council (AIHC), an industry body formed in 1977 to counter OSHA's plan to strengthen the regulation of occupational carcinogens. The AIHC produced an anonymous report denouncing the conclusions of the study as little more than "a figment of the collective imaginations of the government investigators" and calling the HEW predictions "scare tactics" designed to buttress support for new guidelines on carcinogenic exposures. The AIHC noted that if the numbers exposed to asbestos were really as given in the Estimates Paper, the epidemic should already be evident in death rates from mesothelioma, a disease found almost exclusively among people exposed to asbestos. In fact, data from the National Cancer Institute showed that deaths in the United States from mesothelioma were not in the range of 7,000 to 10,000 per year predicted by the Estimates Paper, but were fairly stable at about 1,000 per year. Industry accounted for at most 4 to 5 percent of all cancers, and the more important causes were to be found in diet, lifestyle, and natural agents such as cosmic radiation and sunshine.[48]

Seeking external confirmation of its critique, the AIHC hired Dr. Reuel Stallones, dean of the University of Texas School of Public Health, and Thomas Downs from the same institution, to prepare a review of the NCI document. The team reworked the NCI data and found, to the AIHC's dismay, that it was "reasonable" to conclude that the proportion of total cancer attributable to occupation was on the order of 20 percent. The review concluded that this estimate might be "in the lower mid-range of likelihood," but that the full range was most likely "from 10% to 33%, or perhaps higher if we had fuller information on some other potentially carcinogenic substances." Stallones and Downs concluded that while the NCI might have overstated the dimensions of the asbestos-cancer problem, it was hard to avoid the conclusion that "any reasonable projection, whether higher or lower than the one presented, is horrifying, and fully supports the conclusion that this experience constitutes a public health catastrophe, and that the official response to it is fully justified."[49] Stallones and Downs's conclusions were an embarrassment to their employers, and were promptly ignored. Subsequent AIHC literature cited the more congenial estimate by Oxford University's Richard Doll and Richard Peto that occupational exposures account for no more than 1 to 5 percent of all cancers.[50]

One ironic consequence of the attention given the 1978 Estimates Paper was to elevate the status of previous estimates of occupational hazards, especially those predicting less dire consequences from workplace exposures. John Higginson in 1969, for example, had estimated that 1 percent of mouth cancers, 1 to 2 percent of lung cancers, and 10 percent of all bladder cancers were occupational in origin; in 1976 he and Calum Muir had conjectured that 1 to 3 percent of all cancers were probably caused by workplace exposures.[51] Ernst L. Wynder and Gio B. Gori produced a figure of 1 to 5 percent in a 1977 editorial for the *Journal of the National Cancer Institute,* and Richard Doll that same year estimated that occupational cancers "cannot have been responsible for more than a very small proportion" of all cancers.[52] Philip Cole, an epidemiologist at Harvard's School of Public Health, estimated that occupational exposures accounted for about 15 percent of all cancer in men and 5 percent of all cancer in women.[53] The Estimates Paper reviewed these and similar calculations, concluding that all were "speculative," most were out of date, and many were "seriously incomplete or deficient." Part of the problem was that the question of percentages had never been the object of serious and systematic investigation. That was still the case in 1980 when Vincent T. DeVita, director of the National Cancer Institute, suggested that "when the dust settles, industrial or chemical exposure will account for only 7 percent of all cancers."[54]

If these early estimates of the contribution of workplace exposures to the cancer burden were speculative, they were nonetheless quickly adopted as authoritative, especially by those eager to trash the Califano report. Indeed, by the end of the 1970s the 1 to 5 percent figure had achieved a kind of canonical status among conservative Anglo-American scientists. Philip Abelson, for example, in an October 1979 editorial in *Science* railed against the "cancer stampede" that had followed in the wake of Califano's "alarmist" estimates of the previous year. Abelson complained that people's natural tendency to become "spooked" by cancer had been exploited by opportunists; contradicting Califano's estimates, he cited Higginson's claims that though 80 to 90 percent of all cancers were environmental, only 1 to 5 percent were due to occupational exposures. Abelson noted that the paper had not been published in a peer-reviewed journal and that its overall effect was to detract attention from what, in his opinion, was probably "the best hope for reducing cancer incidence—careful study of foods and effects of cooking."[55]

Protests did continue to be voiced that the 1 to 5 percent figure was not something that could be assumed without discussion. It was surely too low, some argued, given that it was calculated from estimates pertaining to known carcinogens, ignoring the much larger number of chemicals not yet tested.[56] Samuel Epstein questioned the means by which such numbers were derived: "The reasoning went as follows. Everybody knows that diet causes 35 percent of cancer. Everybody knows that smoking causes 40 percent. Everybody knows that sunlight causes this and medicine that. That all adds up to 96 percent. Gee whiz, what are we going to do with the other 4 percent? Let's make that occupation." He noted that he examined "30 papers from 1978 on this 4 percent. And A refers to B as the basis. B refers to C, C refers to D, D refers to E, and E refers back to A as the authority for it. There is no evidence whatsoever for the 4 to 6 percent."[57] In each case, Epstein suggested, the role of occupation was calculated as a residue, following estimation of the purportedly better-established figures for diet, smoking, alcohol, and so forth. Ignorance of the magnitude of occupational hazards was used to derive their insignificance.

The more common view, however, was that the Estimates Paper had exaggerated the significance of occupational exposures. The most consequential expression of this view came in 1981, when Richard Doll and Richard Peto published *The Causes of Cancer,* summarizing the results of an inquiry performed at the request of Congress's Office of Technology Assessment to quantify the proportion of cancers due to known risk factors.[58] Doll and Peto concluded that the majority of U.S. cancer deaths (roughly 65 percent, but with a large margin of error) were caused by diet and smoking. Tobacco was by far the largest reliably known cause, accounting for about 30 percent of all U.S. cancer deaths. Diet was estimated to account for roughly 10 to 70 percent; alcohol accounted for another 3 percent, occupation 4 percent, and industrial pollution another 2 percent (see table).

This was a far cry from the numbers in the Estimates Paper. Doll and Peto attacked the government document for assuming that exposures were always of the

Doll and Peto's Estimates of What Causes Cancer

Factors	Estimated Percentage of All Cancer Deaths
Diet	35
Tobacco	30
Infection	10 (?)
Reproductive and sexual behavior	7
Occupation	4
Alcohol	3
Geophysical factors	3
Pollution	2
Medicines and medical procedures	1
Food additives	< 1

Source: Adapted from Richard Doll and Richard Peto, *Causes of Cancer* (New York: Oxford University Press, 1981), p. 1256.

same degree and duration, an error that led to risk estimates more than ten times too large. The OSHA/Califano estimates were "so grossly in error that no arguments based even loosely on them should be taken seriously."[59]

In the dozen or so years since its publication, Doll and Peto's *Causes of Cancer* has been widely praised. It was endorsed by the Office of Technology Assessment, the NCI, and countless government and academic treatises. Popular magazines embraced both the emphasis on diet and smoking and the devaluation of the industrial threat.[60] The *New York Times* credited the book with bringing about "a less alarming view of the danger from carcinogenic pollutants" within the cancer establishment,[61] though one should not forget that the timing of the report did much to help it. Ronald Reagan's election in November 1980 prepared the ground for a more dismissive approach toward occupational disease; even Michael Gough, the OTA administrator who hired Doll and Peto to write the report, recognized that its dismissal of environmental pollution as a major cause of cancer "fit into the antiregulatory bent of the new administration."[62] In subsequent years, the report would often be invoked to dismiss calls for regulation of environmental pollution—a consequence its authors may or may not have anticipated.[63]

Critics responded to the book by pointing to weaknesses in the analysis. Samuel Epstein and Devra Lee Davis charged that Doll and Peto had omitted blacks and the elderly from their analysis—the very groups for whom cancer rates were highest and most rapidly increasing.[64] Davis concludes that there remains a great deal of uncertainty about the magnitude of the hazard posed by environmental and occupational pollutants, and that the best available figures provide little grounds for complacency.

THE MYSTIFICATION OF NUMBERS

Most of the authors of the 1978 Estimates Paper today concede that their estimates of the role of occupational carcinogens were too high.[65] David Rall says he wishes they had left out the words "at least" in the statement that "at least 20 %" of all cancers could be attributable to occupational factors. Marvin Schneiderman says they should have called it not "Estimates of the Fraction" but rather "Estimates of the Potential for Prevention." Schneiderman defends the original figures on the grounds that, as a measure of the possibility of prevention, the numbers still work. His argument is subtle, however, and rests on a set of philosophical assumptions that sees causality in somewhat different terms than is commonly imagined.

Part of what was in question in the 1978 Estimates Paper was what it means to attribute percentages to specific causal factors. The paper suggested that most cancers have multiple causes acting in combination to produce higher rates of cancer than the sum of the individual agents acting alone. Working with asbestos increases your odds of getting lung cancer by a factor of 5, and smoking by a factor of 10, but smoking in combination with asbestos work appears to increase your odds by a factor of about 50.[66] The relation is not additive but multiplicative. A reduction in one of several carcinogens could therefore, presumably, result in a disproportionate lowering of cancer. In the example just cited, either stopping smoking or reducing asbestos exposure could dramatically lower lung cancer rates among asbestos workers who smoke. The same is true of many other carcinogenic exposures. The primary purpose of the 1978 Estimates Paper, according to Schneiderman, one of its principal authors, was not to diminish the significance of nonoccupational causes but rather to gauge how much cancer could be reduced by cleaning up the workplace.

Schneiderman's qualification is intriguing, because if you want to know the potential impact of a specific intervention on cancer incidence (regulation of workplace exposures, for example), then your estimate of the total contribution of diverse carcinogens can exceed 100 percent by an arbitrarily high amount.[67] Imagine for the sake of argument that 60 percent of all automobile fatalities are caused by excessive drinking. Reduction of excessive drinking would therefore offer a sizable potential for reducing traffic fatalities. But it might also be the case—without contradiction— that 60 percent of all automobile fatalities could be prevented by installing air bags. Absence of air bags and excessive drinking, therefore, in this hypothetical example, are *each* responsible for 60 percent of all traffic fatalities. Curbing drinking *and* installing air bags would presumably (though not necessarily) reduce traffic fatalities by more than 60 percent—though obviously not by more than 100 percent. The point is simply that contributing causes need not add up to 100 percent; they may in fact go arbitrarily high.[68]

What this also means is that one cannot calculate the impact of a specific carcinogen as a residue, derived by adding up other "known" factors and then subtracting from 100 percent. It might well be true that only 4 percent of all cancers could be prevented by cleaning up the industrial environment, but this cannot be derived from estimates suggesting that diet causes 35 percent of all cancers, smoking 30 percent, sunlight 3 percent, and so on. It is altogether possible—indeed, it is conceptually essential—for a broad-minded list of contributing causes to exceed 100 percent by an indeterminate margin. That, apparently, was the presumption of the authors of the Estimates Paper, though it has been widely misunderstood. Appreciating this can help us understand how environmentalists and molecular geneticists (for example) can each claim to be dealing with causes that account for 90 percent of all cancers. Cancer can be understood as a consequence of defects in genetic regulation, but also as a result of environmental insults. There are proximate and ultimate causes of cancer, biological and social causes, and a focus on the one need not preclude a focus on the other.[69]

However one resolves the question of the magnitude of lifestyle versus occupational factors in the genesis of cancer, the figure of 1 to 5 percent has often been used to argue that workplace hazards are trivial, contributing only a "minuscule" addition to the overall cancer burden. Ernst Wynder of the American Health Foundation cited the figure during an attack on Epstein and Califano at the 1978 annual meeting of the American Public Health Association in Los Angeles;[70] the figure was cited repeatedly in response to Epstein's London lecture in 1979. (D. G. Harnden of the Birmingham University Medical School suggested that most scientists felt that a figure of 5 percent was "fairly generous.")[71] A 1979 British trade union document endorsing the Califano estimate prompted that country's Chemical Industries Association to counter that occupational risks were likely to be "no higher than one per cent of the total, and probably closer to zero";[72] Elizabeth Whelan had used almost identical language the year before.[73] The percentages involved became a matter of dispute at countless public forums: in 1981, for example, at public hearings on what to do about pollution in the Niagara River, Geraldine Cox of the Chemical Manufacturers Association (CMA) cited Doll and Peto to argue that toxic pollutants account for only a tiny proportion of all cancers. People do die of cancer, Cox reassured those assembled, but the "naked truth" was that "people have to die *eventually* of something."[74]

Ignored in most such analyses is that if the cancer burden from environmental and occupational pollution is low, one reason must be that decades of environmental and labor struggles have kept it from going any higher. The United States's environmental protection and occupational health and safety legislation are among the strongest in the world—despite persistent efforts to weaken them (see chapter 4). Also ignored is the fact that, even if one accepts the figure of 1 to 5 percent, this still means that five thousand to twenty thousand U.S. cancer deaths are caused by workplace exposures every year—hardly a trivial amount by any measure, especially if you are one of the victims. William J. Nicholson of the Mount Sinai School of Medicine notes that even if only 3 percent of all cancer deaths are occupationally derived, the figure for blue-collar workers will probably be in the neighborhood of 25 percent.[75] To those who suffer the lion's share of such cancers—industrial workers—it is little consolation that the general population is only trivially affected.

Even granting the game of percentages, though, may be granting too much. The question of greatest human interest, after all, is not, "What percentage of cancer should be assigned to factor X?" but rather, "What percentage of cancer might be reduced by changing factor X?" Diet, for example, might be responsible for 35 percent of all cancers, but this does not necessarily mean that 35 percent of human cancer might be prevented by changing dietary practices. It might be the case, for example, that 80 or 90 percent of the dietary contribution is unavoidable, and that changes in food consumption patterns might reduce human cancer incidence by only, say, 10 or 20 percent. The question of policy interest, again, is where one should intervene in the process of carcinogenesis, and that is not necessarily the same as the question of percentages as most commonly formulated. A particular occupational cancer might be entirely avoidable (by eliminating exposure to a known carcinogen) whereas a much lower fraction of a dietary cancer might be avoidable (one cannot survive on a no-fat diet). Synthetic carcinogens have received greater attention than natural carcinogens partly because it is generally thought—sometimes wrongly, as in the case of radon (see chapter 9)—that the former are more dispensable or easier to control than the latter. It is possible to live without dioxin, but it is very hard to live without food.

There are also deeper conceptual problems, stemming from the fact that it is not often easy to distinguish environmental, lifestyle, and genetic causes of cancer. Are the lung cancers suffered by Asian women using canola (rapeseed) oil to stir-fry better understood as dietary or occupational? What about the cancers suffered by thousands of British coal-tar workers in the first half of the twentieth century? Japanese workers in the very same industry were spared by virtue of their more rigorous practice of personal hygiene, and the same was apparently true for chimney sweeps on the Continent.[76] Does this exonerate the British disease as a nonindustrial phenomenon?

And what about tobacco? Tobacco is usually classed as a lifestyle carcinogen, as there is presumably a choice involved in smoking. But is it easier to quit smoking than to quit your job? Why does the tobacco industry get classed as a purveyor of lifestyle cancer, while Dow and Du Pont are saddled with causing occupational cancer? Thirty percent of all cancer may be due to smoking, but what is smoking due to? The poor smoke more than the rich, and high school dropouts are three or four times more likely to smoke than high school seniors.[77] Why not say that poverty or lack of education is a primary cause of lung cancer? Why not say that cancer is caused by cigarette advertising, tobacco subsidizing, obsessive fears of weight gain, and youthful illusions of immortality? (Smoking used to be more common among the wealthy—the inversion probably says something about who is more receptive to health education.) Smoking is more than a lifestyle; it's also a consequence of policy: witness the long-standing U.S. policy of supporting tobacco farmers and efforts that continue even today to encourage the export of cigarettes into Third World nations. The failure to control the world's most deadly carcinogen will one day be looked upon as the biggest mistake of twentieth-century health policy.

Lifestyle, diet, occupation, genetics, infection: the boundaries between these categories are routinely blurred. Cancers are misfits in the world that scientists want to see neatly divided up. The search for percentages to be accorded each of these "factors" belies the extent to which occupation is shaped by culture, culture by skills and technologies, technologies by lifestyles, lifestyles by infectious possibilities, infectious agents by economic structures—one could go on and on. Environment, occupation, and lifestyle are twisted together with custom, habit, economic history, and social policy in ways that make neat separations into percentages much more difficult than is commonly imagined. Dietary cancers are usually classed as "lifestyle," but what about the limited choices available to consumers in consequence of the structure of agribusiness? Fat consumption, linked to colon cancer, usually figures as "diet" (= lifestyle), but what about the fact that people in high-stress, low-earning jobs are more likely to eat fatty, high-salt, carcinogenic foods? What about the colon cancers caused by exposure to chemical toxins working in harmony with predisposing genes? The bilharzia bladder cancers of the Nile Valley and Brazil usually figure in the "infection" columns of causal tables, but what about the fact that the 200 million people worldwide who suffer from the parasite suffer mainly because they bathe or wash in sewage-contaminated water?

How, in other words, in a discourse dominated by concepts of genetics, lifestyle, and occupation, does one classify the cancers due to medical neglect, environmental injustice, media-induced fashion, or industrial malfeasance? If poverty is what distinguishes African-American from European-American cancer rates, why do the American Cancer Society and National Cancer Institute classify by race rather than by income? How does one rest with the classification of cosmic radiation as "natural," when no effort is made to alert airline passengers (especially pregnant women) to a possible radiation hazard during sunspot activity?[78] The presumption is often that diet is freely chosen, but why not attribute at least part of the blame to transport-based agriculture (which encourages the use of long-shelf-life tropical oils) or incomprehensible food labels? And if the U.S. Army's $210 million strategic breast initiative somehow comes up with a miraculous cure or even moderately successful treatment, how should we regard the previous neglect of this disease? How would we classify the cancers caused by the medical establishment's failure to address what, for most of human history, has been the single most deadly form of cancer?

My point is not to challenge the idea that 30 percent of all cancers are caused by cigarettes, or that diet plays a major role in cancer. I do want to stress, though, that the causal agencies most commonly invoked are too often defined as physical substances without social context: tobacco, asbestos, dioxin, and so forth. Insufficient attention is given to causal factors that transcend the physical: elections, advertisements, policies, and so forth. One could argue that some fraction of African-Americans' cancer may be due to being black in a racist society, or to the unequal distribution of wealth, or to the failure of politicians to enact reforms, and the like. Is more known about what causes cancer in men than about what causes it in women? In cultures with a history of selective inattention to female illness (except in matters of reproductive health—hence the strong historical focus on the role of hormones in female cancers), some fraction of women's cancer could presumably be blamed on sexism. Women have historically had higher rates than men. What was the cause of that inequality prior to the advent of smoking? Why has ignorance, like knowledge and like cancer, been so unevenly distributed? I shall return to these questions in the conclusion of this book.

In the 1990s, controversy continues to swirl around the magnitude of environmental hazards and the question of long-term trends. Devra Lee Davis points out that many forms of cancer—including breast, colorectal, brain, non-Hodgkin's lymphoma, testicular, and kidney—are still on the rise. Not all of this can be due to smoking, she points out, because heart disease—a major cause of which is smoking—has been declining for several decades. It is unlikely that all of the reported increases in cancer can be due to improvements in diagnosis and increased access to care, because recorded deaths from "unspecified" causes are also on the rise.[79] In fact, there is reason to believe that at least some of these increases are due to increased chemical and radiation exposures. Childhood cancers have been linked to pesticide exposures, and there is evidence to suggest that farmers suffer more than others from certain kinds of cancers, due to agricultural chemical contacts.[80] It is not yet clear why brain cancers are on the increase—the answer may have to do with exposures to home termite treatments or low-energy radiation from household appliances and high-voltage wires. No one really knows.

As we shall see, the picture has become complicated by mounting evidence that industry is not the only—and perhaps not even the major—culprit. Bruce Ames has claimed that natural carcinogens in foods are far more dangerous than industrial carcinogens, and several newly discovered "indoor pollutants" have been factored into the debate. Much is at stake, and science is in the middle of the struggle. To understand this, we must return to the early 1980s, when political events transformed the landscape of the

Great Cancer Wars. The election of Ronald Reagan to office in the fall of 1980 shifted the balance away from the Epstein-Califano consensus and placed it firmly in the industry camp. Suspicious of environmentalist claims, Reagan officials would begin dismantling consumer safety, occupational health, and air- and water-quality controls, with consequences that are difficult to fathom, even today.

Chapter 4

The Reagan Effect

A society that forgets the benefits of its reforms becomes vulnerable to the callousness of special interests who prey on its amnesia.

—Ralph Nader, 1986

IF THE BEGINNING of the 1970s was a time of environmental optimism, the end of the decade was a time of environmental collapse—at least from the point of view of regulatory agencies. In December 1979 *Fortune* magazine declared that it was "roundup time for the runaway regulators," that "all of a sudden, just about every Senator and Congressman wants to get into the act of curbing the regulatory agencies." OSHA was discarding "hundreds of detailed safety specifications" and had begun experimenting with labor-management committees to monitor safety in the workplace. The EPA was beginning to explore previously unpalatable ideas such as emissions charges, marketable rights to pollute, pollution offsets, and the "bubble concept"—a pollution control experiment by which overall plant emissions would be regulated by whatever means a company wanted to use.[1] Companies were also strengthening their protests that many environmental regulations were overly complex and unnecessarily burdensome. OSHA regulations were singled out as exorbitant, but the EPA had also required substantial environmental investments, especially from the chemical industry. Companies were beginning to complain that the costs of complying with federal regulations, on the order of $50 billion to $150 billion per year, were rivaling the amount spent on plant and equipment.[2]

The election of Ronald Reagan in November 1980 sent a chill through environmental organizations that would be felt for nearly a decade. Federal agencies, particularly OSHA and the EPA, were the hardest hit. The liberal and, in many cases, aggressively environmentalist leaders appointed by President Carter were replaced by people more committed to "regulatory relief" than to environmental well-being. Marvin Schneiderman summed up many people's fears in 1981 when he stated that "regulatory decisions in this country for the next several years may be as likely to derive from ideology as from science."[3] Ellen Silbergeld, chief toxicologist at the Environmental Defense Fund, charged that "creationism" had taken over toxicology, referring to the new administration's questioning of the relevance of animal studies for assessing human cancer hazards.

The environmental malice of Reagan and his administrators is notorious. Here was a man who mocked forest preservation with his maxims that "trees are the biggest polluters" and that "if you've seen one redwood, you've seen them all."[4] Those who lived through this time may recall such slogans, but it is also important to appreciate the reasons behind the steps taken to weaken occupational and environmental health. We therefore have to examine the ideological backlash that spirited away the liberal science and policy of the 1970s, and how very different views concerning what causes cancer came to the fore.

ENVIRONMENTAL HEALTH DISMANTLED

Carter administration health authorities had made it clear that, when it came to national health and safety policies, the emphasis would be on public health rather than on high-priced, high-tech medicine. In the area of cancer policy, this meant a shift from curative to preventive strategies to conquer the disease. The Carter-era appointments to health and environmental agencies reflected this new emphasis: Eula Bingham, appointed in 1977 to head OSHA, was a zoologist with activist credentials in public health; Donald Kennedy, tapped to head the FDA, was also a zoologist and the first nonphysician to hold this position in more than a decade. Bingham supervised a radical reorientation of OSHA and urged the writing of the 1978 Estimates Paper; Kennedy supported the FDA's controversial ban on saccharin (identified as a carcinogen in animal tests) and joined with Bingham to coordinate the attack on toxic pollutants. Arthur Upton, the new director of the National Cancer Institute, had studied the health effects of radiation from the atomic bombs dropped over Hiroshima and Nagasaki; he also shared with Bingham and Califano a concern to put prevention near the top of health care priorities. Upton distinguished himself as the first NCI chief to win the approval of Sidney M. Wolfe of the Ralph Nader–affiliated Health Research Group ("he's our candidate") and became the first NCI director to display an open admiration for Wilhelm Hueper. The temper of the administration was signaled by Labor Secretary Ray Marshall's notification that OSHA would henceforth be going after "whales rather than minnows."[5]

For the Reagan administration, by contrast, the regulations enforced by agencies such as the EPA and OSHA were seen as impediments to the progress of business enterprise. Reagan won the election on a probusiness, antigovernment platform, promising regulatory relief and an end to the high interest rates and double-digit inflation that had plagued the final Carter years. Reagan was "prolife" but, as several of his critics pointed out, his administrators sometimes appeared indifferent to saving postbirth lives when it came to things like highway safety, toxic-waste protection, and consumer product safety.[6] He was a Sagebrush Rebel, sympathetic to the Western cattle and sheep ranchers who wanted to graze their herds free of government restrictions—even if this meant, as critics warned, turning federal lands into dustbowls and seas of mud.

Reagan's environmental myopia became a focus of attention during his 1980 bid for the presidency, when Senator Edward Kennedy mocked the former governor's assertion that 80 percent of all air pollution was caused by trees. Reagan denied ever having made the claim (he claimed only to have asserted that "growing and decaying vegetation" was responsible for "93 percent of the oxides of nitrogen," apparently confusing the nitrous oxides produced by plants with the nitrogen dioxide produced by smokestacks), but he never retreated from his stance that nature was far more culpable than industry in the genesis of pollution. On October 7, 1980, in a speech to a steel industry group, he announced that he had twice flown over the recently erupted Mount Saint Helens, confirming his suspicion that "that one little mountain out there in these past several months has probably released more sulphur dioxide into the atmosphere of the world than has been released in the last ten years of automobile driving or things of that kind that people are so concerned about." (In fact, as subsequent commentators pointed out, Mount Saint Helens, even at its peak, was producing only about 2,000 tons of sulfur dioxide a day, compared to the 81,000 tons produced daily by automobiles.) Shortly thereafter, Reagan's aides issued a statement that since air pollution had been "substantially controlled," further clean air regulations were unnecessary. The Reagan camp was embarrassed when, that same day, a record-breaking smog began to build up over Los Angeles, delaying a homecoming rally for the candidate. The Hollywood-Burbank Airport was forced to shut down, and Reagan's plane had to be rerouted to the Los Angeles International Airport.[7]

Reagan began dismantling occupational, environmental, and consumer product safety agencies soon after taking office. Many of his appointments came from industries with little interest in—or even open hostility toward—environmental protection. Thorne G. Auchter, OSHA's new administrator, was a thirty-five-year-old millionaire operator of a Florida construction business repeatedly cited by OSHA for health and safety violations; his most recent political experience had been as director of special events for Reagan's Florida campaign.[8] One of his first acts in office was to destroy or withdraw more than half a million dollars' worth of training materials prepared by his predecessor: one such pamphlet was on the hazards of cotton dust, depicting a victim of brown lung on its cover. Auchter ordered 100,000 copies destroyed on the grounds that it was "offensive" and "obviously favorable" toward labor (he also objected to the fact that separate versions had been prepared for managers and blue-collar workers).[9] Auchter later supervised a shift from an emphasis on engineering controls (such as vacuum hoods) to personal protective devices (such as respirators)—a policy that drew praise from the Chemical Manufacturers Association but sharp criticism from labor, which saw it as part of an effort to force individual workers to shoulder the responsibility for worksite safety.[10] In November 1982 Anthony Robbins, president of the American Public Health Association and former head of NIOSH, charged Auchter with violating the OSHA charter and listed twenty-five reasons he should resign. Auchter finally did resign in 1984 to take a job as president of B. B. Andersen, Inc.; Congress and the FBI launched an investigation into conflict-of-interest charges when it turned out that Auchter had forgiven an Andersen-owned company of $12,600 in OSHA penalties and fines in 1981.[11]

Auchter's immediate boss, Secretary of Labor Raymond J. Donovan, also set out to reverse the agency's activism. Shortly after taking office, he canceled a chemical labeling proposal (the "right-to-know" rule) that had taken five years to construct.

Critics accused him of acting in response to a letter from the Manufacturing Chemists Association asking him to pull it. Like Auchter, Donovan was familiar with OSHA from the other side, as it were. Prior to his government post, Donovan had been executive vice president of a New Jersey construction company that had been cited by OSHA for safety violations 135 times in six years. Fifty-eight of these were "serious," meaning that, according to OSHA standards, the violations could have resulted in death or severe injury. Auchter's company, by contrast, had run up only 48 violations.[12]

EPA appointments were similarly designed to rein in environmental activism. Anne M. Gorsuch, the new EPA chief, had actually been involved in legal action against the agency (her former job was as an attorney for the Mountain Bell Telephone Company and a Colorado State legislator); she had also defended the right of sheep and cattle ranchers to poison coyotes using Compound 1080, a substance banned by Nixon's EPA because it harmed other wildlife. Many of her appointees were equally ill disposed toward environmental activism. Kathleen Bennett, the agency's new air pollution watchdog, had lobbied for Crown Zellerbach, the paper manufacturer. John Daniel, the agency's new chief of staff, had pressed the flesh for Johns-Manville, the asbestos manufacturer. The new head of enforcement was Robert Perry, a former Exxon trial lawyer. Thornton Field, responsible for hazardous waste and later for enforcement, had worked for Adolph Coors, the conservative anti-union brewer. John A. Todhunter, the assistant administrator for pesticides and toxic substances who showed up for work one day wearing a "Born to Deregulate" T-shirt, had worked as a consultant for the industry-financed American Council on Science and Health.[13]

One of the most notorious EPA maladapts was Rita M. Lavelle, an Edwin Meese protégée appointed by Gorsuch in the fall of 1981 to head the EPA's $1.6 billion "Superfund" program to clean up the nation's worst hazardous-waste dumps. In a 1982 interview for *The Nation,* Lavelle argued that corporate leaders were not to blame for the nation's festering toxic-waste sites: "You're basically dealing with illiterate workers who don't care to read. It's not usually the president of a company who says, 'Go hide that drum.' It's usually a worker who doesn't care, who violates company policies." A year after Lavelle took office, only 4 of the 160 Superfund sites had been cleaned and only one lawsuit had been filed. The EPA had spent no money on cleaning up the three sites owned or created by subsidiaries of Aerojet-General, Lavelle's former employer. Representative James Florio of New Jersey, the primary author of the Superfund legislation, complained that little had been done to enforce it; an EPA hazardous-site control officer, Hugh Kaufman, pointed out that there was now "more red tape in Superfund than in any program in the agency's history." Lavelle responded to Kaufman's criticisms by having him tailed by EPA inspectors.[14]

Joseph A. Coors, the radical anti-environmentalist beer magnate, was another major actor in the Reagan revolution. In 1977, Coors had led a group of wealthy ranchers, oil men, and mining magnates to found the Mountain States Legal Foundation (MSLF), a body particularly adept at challenging federal legislation for wilderness protection and clean air and water. As a Reagan confidant and member of his informal Kitchen Cabinet, Coors assisted the president in making appointments; in fact, so many Coors associates got jobs in the administration that they earned the nickname: the Colorado Mafia. One was James Watt, a born-again Christian and president of the MSLF, named Secretary of the Interior shortly after Reagan assumed office. Watt did for the Interior Department what Gorsuch did for the EPA, slashing the land-acquisition budget for parks, for example, and hiring an Indiana legislator known for his opposition to limits on strip mining to head the Office of Surface Mining. Robert Burford, a Colorado rancher and mining engineer with grazing rights on 32,000 acres of Bureau of Land Management land, was another captain promoted from this Colorado outfit. Burford was named to head the Bureau, where he successfully crippled its efforts to prevent overgrazing (he had been a leader of the Sagebrush rebellion). Burford was close personally to Anne Gorsuch, another Coors confidant; the two were married in February 1983.[15]

Reagan's key instrument in this process of environmental retreat was the budget. In his first year of office, the former governor proposed a 60 percent reduction in the EPA's real spending and a 40 percent cut in the agency's staff.[16] Congress disputed some points, but Reagan got much of what he wanted. OSHA, NIOSH, and the Consumer Product Safety Commission faced similar cuts. In the 1982 budget, nuclear research and development (R & D) got a 7 percent boost, while solar R & D was cut by 41 percent, geothermal energy by 63 percent, and state and local conservation by 48 percent. Preventive health was cut by 16 percent and health planning by 50 percent.[17] William Drayton, a former EPA official from the Carter years, complained that Reagan was effectively repealing U.S. environmental legislation "through the personnel and budgetary back doors."[18] Russell Train, the conservative Republican EPA chief under presidents Nixon and Ford, stated that it was hard to look at Gorsuch's managerial modus operandi without concluding that her intention was to destroy the agency.[19]

The beneficiary of what became known as "death by a thousand cuts" was not the national debt, which rose dramatically in every year of Reagan's two terms, but rather national defense. Secret military funding skyrocketed, and projects like Star Wars (an antimissile system) and Milstar (a supersecret military satellite communications system) devoured billions of dollars every year. By the end of the 1980s, more than 70 percent of all federal research dollars was being funneled into military projects, and military R & D exceeded the nation's environmental R & D by a factor of about a hundred. As late as 1991, Ted Weiss, a New York City Democrat, could still complain that only $67 million had been budgeted for breast cancer research, while forty-five times that amount was being spent on the B-2 bomber—a program that had recently been canceled.[20]

The consequence of such priorities for occupational and environmental policy was profound. OSHA was already understaffed by some accounts—having only enough compliance officers to visit about 1 percent of U.S. workplaces every year. (In 1977, under Carter, the agency had reduced its number of annual inspections by one-third in response to complaints that many of the requests were frivolous.)[21] The new administration further reduced the agency's enforcement capacity, diminishing the average number of inspections from 5,680 per month at the end of the Carter period to 4,700 a month in Reagan's first year in office. From 1980 to 1985, OSHA staffing declined by nearly a third, from 3,015 to 2,176.[22] Follow-up inspections dropped by more than half, and the number of major citations dropped by nearly 30 percent.[23] Auchter defended this latter move on the grounds that in the vast majority of cases, companies cited on a first visit cleaned up their act by the time a second inspection took place. Critics pointed out that this was probably because they had known there would be a follow-up—a burden from which they now were generally relieved. By the end of the Reagan era, critics could complain that the U.S. government had six times as many fish and game inspectors as job and safety inspectors.[24]

Another consequence—one that didn't get a lot of press coverage—was an increase in the censorship of environmental officials. In 1982, when Jim Sibbison, an EPA public information officer, drafted a news release pointing out that the pesticide dibromochloropropane was suspected of causing cancer and that several workers had become sterile after handling it, Gorsuch's office deleted the reference to cancer and replaced "sterility" with "adverse health effects." References to cancer and to hazards to pregnant women were subsequently eliminated from an EPA news release about radiation protection. Sibbison learned that some topics were virtually taboo: "After a while, I simply stopped mentioning cancer, birth defects and damage to genes. As a colleague of mine said, 'The administrator's office will take the words out anyway.'"[25]

The EPA's friendly relations with industry meant that companies with a stake in environmental investigations had new opportunities to influence agency reports. According to Anthony Roisman, who resigned in 1982 as chief of the Justice Department's Hazardous Wastes section, "It was a well-known fact that if a company didn't like what it was getting in negotiations with agency lawyers, it could go to the assistant administrator, the deputy administrator, or to Burford's office. They could just go all the way up the agency ladder to get what they wanted." In 1981, for example, the EPA's regional administrator for Chicago, Valdas Adamkus, asked his staff to prepare a report on pesticide and herbicide contamination of the Great Lakes, especially near Saginaw Bay, into which a Dow Chemical plant based in Midland, Michigan, had been discharging wastes. Dow officials contacted John Hernandez, deputy administrator at the EPA, who provided the company with a copy of the report, not yet released to the public. Dow's vice president for health and environmental sciences contacted EPA's Chicago office requesting changes in the report's title, lower estimates of the risks from dioxin, and deletion of certain references to miscarriages, reduced fertility, and Agent Orange, the defoliant manufactured by Dow. The EPA's Washington staff began to urge revisions of the document, resulting in the deletion of six lines asserting, among other things, that the Dow plant was "the major source, if not the only source," of dioxin in the Saginaw Bay area. Dow officials later dismissed the incident as a routine exercise in peer review, though Public Citizen, the consumer group, charged that Dow's concern might have had something to do with the fact that a group of Vietnam veterans had sued the company for withholding information showing that dioxin could be "exceptionally toxic" to humans.[26]

At OSHA, Reagan officials went so far as to crack down on unauthorized correspondence with other health professionals. In 1981, for example, the OSHA scientist

Peter Infante almost lost his job when he wrote to the World Health Organization using OSHA letterhead that formaldehyde caused cancer in laboratory animals. Attorneys for the Formaldehyde Institute, an industry group, found out about the letter and wrote to Auchter, asking whether it would be possible to better "control" the behavior of bureaucrats like Infante. OSHA complied by notifying Infante that similar behavior in the future could cost him his job. Albert Gore, then a young member of Congress (D-Tenn.), heard about this apparent act of censorship and called Auchter before a congressional subcommittee, where the future vice president denounced the administration's action as "a blatant effort to rid the government of a competent scientist." Gore warned that OSHA was sending a message that "those who try to do their jobs and protect the health of the American people will instead lose their jobs to protect industry profits."[27]

Dramatic cutbacks in funding meant dramatic cutbacks in environmental enforcement: in his first year in office, Reagan reduced by about 70 percent the number of cases against polluters referred to the Justice Department. By the fall of 1982, the EPA had filed only two suits against chemical-waste dumpers. The Reagan agency refused to regulate formaldehyde, despite evidence from its own scientific advisory panel that it was a carcinogen. Pesticide regulations were loosened, and a Carter-era recommendation to ban ethylene dibromide went unheeded. The Reagan administration stalled action on acid rain and raised allowable levels of sulphur dioxide emissions.[28] Implementation of new water-quality standards was delayed until 1983, returning standard setting to the states and to industry. The Clean Air Act was weakened, allowing higher levels of pollution from automobiles and electrical utilities.[29] The Council on Environmental Quality, established by Congress in 1970 to advise the president on environmental matters, was cut to about one-third its former size and its staff was reshuffled. In March 1983 the new head of the group, A. Alan Hill, admitted that in two years in office he had never discussed environmental affairs with President Reagan.[30]

In such a climate, it is not surprising that carcinogenic science and policy suffered. At the Food and Drug Administration, rules for food additives were revised to allow carcinogenic contaminants in foods and cosmetics, so long as they were introduced unintentionally as a side effect of the manufacturing process. At the EPA, regulations were revised to allow liquid hazardous wastes to be dumped, as had once been the practice, into landfills. OSHA officials stalled "right-to-know" legislation that would have forced employers to identify hazardous substances to exposed workers; critics lamented that in the first three years of Republican rule, the nation's foremost occupational health and safety watchdog had failed to issue standards for a single new chemical.

Similar obstacles were placed in the path of the National Institute for Occupational Safety and Health, the research wing of OSHA and the Mine Safety and Health Administration. Anthony Robbins, director since the fall of 1978, was dismissed after being attacked as a "social activist and self-appointed guardian of the working class" by the U.S. Chamber of Commerce.[31] The institute was relocated from Rockville, Maryland, to Atlanta, purportedly to improve administrative efficiency but probably also—as recognized by labor officials and the institute's own scientific staff—as part of an effort to weaken its access to Washington. (The Atlanta-based Centers for Disease Control, the parent organization of NIOSH, has traditionally been more conservative than the Labor Department's OSHA or MSHA.)[32] Funding for NIOSH was slashed from its Carter-era peak of $83 million to $57.5 million in 1983, making it the smallest member of the National Institutes of Health. Specific programs were blocked on White House orders: this was the case with a $1.3 million NIOSH program to notify some 200,000 workers of their risk from previously undisclosed carcinogens in their place of work. The point of the notification was to allow those exposed to seek medical advice, in the hope that early detection might improve their chances of successful treatment. The White House intervened, ostensibly on the grounds that it might lead to lawsuits against the corporations responsible for exposing the workers.[33]

Amid cuts in environmental protection and occupation health and safety, the administration launched a campaign to convince the American public that "the environment" was not to blame for the majority of cancers. On March 7, 1984, Margaret Heckler, secretary of Health and Human Services, announced a $700,000 Cancer Prevention Awareness Campaign to tell "the simple truth that cancer is usually caused by the way we live."[34] Donald Hodel, Reagan's energy chief, expressed a similar philosophy when he stated that the way to deal with ozone depletion and the consequent threat of skin cancer was to apply stronger suntan lotion.[35] Such attitudes led a House Energy and Commerce Committee to suggest that the Reagan administration was "soft on cancer." A 1984 congressional study group went further, branding the administration "a public health hazard."[36]

COSTS AND BENEFITS

The Colorado Mafia was brought down not so much by its disregard for environmental health as by its cavalier attitude toward the law. When Gorsuch refused to comply with a congressional request for documents showing how Lavelle had run the Superfund, she was cited for contempt of Congress, the first such citation of a cabinet-level official in U.S. history. Rita Lavelle herself was fired for improprieties in her administration of the fund, and eventually went to jail for lying to Congress. Several of her colleagues and co-workers were either fired or investigated, following charges of conflict of interest or other improprieties. John Todhunter resigned in March 1983, after reports began to surface that he had ordered his staff to alter EPA documents to make dioxin and formaldehyde appear less dangerous.[37] Gorsuch resigned that same month, with agency morale and congressional relations at an all-time low.

Seeking to save face, Reagan appointed William Ruckelshaus, the agency's first chief under Nixon (and a man with a reputation as Mr. Clean), to head the beleaguered agency. Ruckelshaus replaced the Gorsuch-Watt "meat-axe" approach with a softer touch, combining managerial efficiency with diplomatic tact, though some of the same conflicts of interest were still apparent. Ruckelshaus was a former vice president of the Weyerhaeuser Company, best known among environmentalists for its rating as one of the five worst polluters in the country. Weyerhaeuser had sprayed millions of acres of the Pacific Northwest with 2,4,5-T, a defoliant implicated in birth defects and several other problems. Ruckelshaus removed himself from deliberations on the safety of the herbicide, but environmentalists still wondered whether his appointees would feel secure in dealing aggressively with their boss's former employer.[38]

Two of the most powerful weapons in the new EPA's political arsenal—especially under Ruckelshaus—were cost-benefit analysis and risk assessment.[39] Ruckelshaus announced the new approach in a June 1983 speech to the National Academy of Sciences, arguing that risk should now be regarded as an unavoidable fact of life: "Life now takes place in a mine field of risks. No more can we tell the public, 'You are home free with an adequate margin of safety.'" Risk assessors would determine the magnitude of a given risk, while risk managers—presumably the experts at the EPA—would determine whether that risk was acceptable.[40]

Cost-benefit analysis was not, of course, entirely new: President Ford had required that regulatory agencies prepare an "inflation impact statement" for regulatory proposals, and under Carter, too, there were requirements that measures pass a "cost-effectiveness" test.[41] A 1976 EPA document asserted that it was reasonable to spend $200,000 to $500,000 per "health effect" (meaning death) avoided in the nuclear industry.[42] President Nixon had favored similar ideas. A 1973 study by his Scientific Advisory Committee argued that "the rigid stipulations of the Delaney clause [barring synthetic carcinogens from foods] place the administrator in a very difficult interpretive position. He is not allowed, for example, to weigh any known benefits to human health, no matter how large, against the possible risks of cancer production, no matter how small."[43]

Under the Reagan administration, cost-benefit analysis became a central feature of environmental and health legislation.[44] Cost-benefit analysis and cost-effective regulation had been mainstays of the American Enterprise Institute (AEI), the conservative think tank from which the new president drew many of his ideas for regulatory reform.[45] Several of the administration's earliest initiatives showed evidence of AEI input: Reagan's famous Executive Order 12291, issued on February 17, 1981, required that all regulations meet a test showing that "the potential benefits to society for the regulation outweigh the potential costs." It also gave unprecedented power to the Office of Management and Budget (OMB) to approve or reject regulatory measures. (James Watt admitted shortly after taking office that the budget was going to be used as "the excuse to make major policy decisions.") The OMB, headed by David Stockman, was empowered to direct agency officials to drop, delay, or reconsider agencies' recommendations, a procedure branded unconstitutional by a Congressional Research Service study.[46]

Recognition of a need for cost-benefit analysis grew partly from the fact that, prior to the 1980s, a number of environmental regulations required a "zero tolerance" standard for carcinogenic exposures. The 1958 Delaney Amendment to the Pure Food and Drug Act, for example, barred any detectable trace of chemical carcinogens from food. Limits for environmental and occupational exposures were often established without regard for cost of implementation. No distinction was made between measures that might cost a million or a billion dollars per death averted.

Critics of the new procedures, however, tended to point out that cost-benefit tests of specific regulatory measures usually placed inordinate emphasis on the "cost to industry" side of the equation—since short-term financial costs to industry are almost always easier to calculate than long-term health benefits to society as a whole.[47] Ignoring such considerations, Murray Weidenbaum of the American Enterprise Institute labeled cost-benefit analysis a "neutral" concept, which a writer for the liberal magazine *The Nation* characterized as "absurd."[48] Weidenbaum had co-authored a widely cited 1978 estimate that federal regulations would cost American business $103 billion in 1979; critics charged that by misrepresenting costs and ignoring benefits (especially the intangibles of improved health, but also financial savings from redesigned manufacturing processes), Weidenbaum's calculation was "grossly exaggerated" and "ideological arithmetic."[49] Mark Green of Public Citizen's Congress Watch, a consumer advocacy group, suggested that cost-benefit tests were "about as neutral as voter-literacy tests in the Old South" and that "the abolition of slavery or child-labor laws" would never have passed a cost-benefit test.[50]

The problem, again, was that while one could hardly object in principle to the effort to balance costs against benefits, the net effect was almost invariably to stymie health and environmental regulations. Efforts to have air bags installed in all new cars were blocked: Drew Lewis, the new Secretary of Transportation, proposed a delay for Standard 208, the "passive restraint rule" requiring air bags or automatic seat belts in all vehicles by 1983; within eight months of Reagan's inauguration, the standard was a dead letter. In 1982, the Center for Auto Safety estimated that the Reagan ruling could cost as many as nine thousand American lives per year.[51] The OMB watered down the Transportation Department's recommendations for minimum insurance carried by truckers delivering hazardous materials (gasoline, for instance); the OMB-approved standard was only one-tenth what the department had called for.[52]

Exposure standards for a whole string of carcinogens were also relaxed. In December 1987, an EPA study reduced its estimate of the cancer-causing potential of dioxin by a factor of sixteen. Four months later Lee Thomas, Ruckelshaus's successor, proposed a standard for environmental benzene so lax that David Doniger, a senior attorney with the Natural Resources Defense Council (NRDC), called it "the most shocking public health decision to come from the EPA during the Reagan era." (The NRDC later sued the EPA for negligence in regulating benzene.) The EPA's Risk Assessment Council about this time proposed a downward revision of the cancer risks of arsenic in drinking water, based partly on the fact that the skin cancers arsenic caused had been responding increasingly well to treatment. The logic behind such a move, as explained by Doniger, was apparently that "it is okay for polluters to make you sick, so long as doctors know how to make you well again."[53]

Under the cost-benefit approach, the value of a particular regulatory action is generally calculated in terms of the financial costs involved per human death (or illness or birth defect) averted. Regulatory measures that save relatively few lives per dollar spent are not cost-effective. In 1989 and 1990, for example, the EPA implemented a series of regulations to reduce benzene emissions from gasoline terminals, chemical vents, waste operations, and the like. Business advocates soon pointed out that the cost of such regulations was often extremely high per death averted. Albert L. Nichols showed in an article in *Regulation,* a conservative publication, that the cost ranged from $3.8 million per life saved (for benzene storage vessels) to nearly $3 billion per life saved (for gasoline bulk terminals). Nichols pointed out that this was a lot to pay to save a single life—much more, in fact, he suggested, than even exposed workers would be willing to accept, judging from the trade-offs they had been found to make between wages and risk of death.[54]

Supporters of the cost-benefit approach point out that it allows one to compare the social value and burden of qualitatively different kinds of regulatory actions— mandatory seat belts and sulphur dioxide emissions, for example. Such an approach presumably gives one a rational basis from which to choose what to regulate. William Ruckelshaus made this argument in his 1983 speech before the National Academy of Science. New tools of risk assessment would allow agency officials to compare the different degrees of harm caused by different toxic agents; one could then plug this into the cost-benefit equation to see whether the costs (risks) are worth the benefit. Such an approach was necessary, the new EPA chief claimed, given that nothing in life is risk-free. We regularly face risks from "hundreds, perhaps thousands of substances" and must learn to decide which of those risks are worth the costs and which are not.[55]

Critics of the approach sometimes point to the inhumanity or inequity of the process. Daniel Becker, an attorney at Environmental Action, protests that risk assessment distills people's lives into abstract numbers, putting "the government imprimatur of 'acceptable' on a death rate due to pollution." Barry Commoner suggests that risk assessment represents "a return to a medieval approach to disease in which illness—and death itself—was regarded as a debit on life endured as payment for original sin." Pollution, for the risk assessors, has become "the inevitable price to be paid for the material benefits of modern technology." Commoner and others worry that the science involved is not yet reliable enough to provide a sound base for policy, given how little is known about exposure routes (pharmacokinetics) or synergistic effects. And since risk assessment is generally done by those with a financial stake in the outcome, it is difficult to control for bias. The process tends to remove the decision-making process from the people "at risk" and place it into the hands of experts with the mathematical, chemical, and medical skills required to generate such figures. As one environmental analyst put it, risk assessment provides "an ability to cloak political or value decisions in the false objectivity of science."[56]

One of the problems with cost-benefit analysis is that it is often unclear how to define costs and benefits. Take the example of the Reagan EPA's efforts to relax auto-emissions standards. The auto industry in 1981 supported such a bill to save $80 to $350 on a new car selling for $10,000. Consumer safety advocates argued that the medical bills caused by the retreat could cost at least this much. James Ridgeway, writing in *The Nation,* forecast in 1981 that such compromises in air-quality enforcement would result in "a rise in diseases related to environmental pollution;

there will be more Love Canals and increased danger from acid rain."* California
Governor Jerry Brown during his 1980 presidential campaign pointed with disgust
to the fact that a worker making $10,000 a year would contribute only $30,000 to the
GNP in three years, whereas if environmental pollution gave that person cancer, the
medical costs alone would add $100,000 to the GNP in just one year. Brown sug-
gested that that was the kind of economic growth Reagan would probably get as a result of his attack on environmental regulations.[57]

It is always difficult to gauge the long-term financial impact of a given pollution-
control technology. In some cases, pollution prevention actually leads to industry
savings. In the dry-cleaning industry, for example, worries about the carcinogenic
effects of the most commonly used dry-cleaning fluid (perchloroethylene, or "perc")
led to improvements in equipment design that not only reduced exposures but saved
on input costs. By the beginning of the 1980s, under pressure from OSHA and the
Consumer Product Safety Commission, most commercial dry cleaners had shifted
over to a new machine design that combined the washing and drying cycles, allowing very little "perc" to escape. The new machines were 30 to 60 percent more expensive, but they saved a great deal on dry-cleaning solvent—not to mention saving
workers' lives. The story was roughly similar for the case of vinyl chloride (vc).
Manufacturers had argued that the stiff exposure standards mandated by OSHA following the discovery of liver cancers among vc workers would cripple the industry.
In fact, the price of the plastic did not increase because two simple methods were devised that allowed the industry to recapture the escaping gas at the same time it reduced workplace exposures. *Chemical Week*, a leading trade journal, celebrated the
profitable new development with headlines announcing "PVC Rolls Out of Jeop-
ardy, into Jubilation."[58]

Trade associations like the Chemical Manufacturers Association and the American Industrial Health Council lament the increasing costs of environmental regulations, but environmentalists have countered that these are often exaggerated. Robert
Harris of the Environmental Defense Fund estimated in 1981 that environmental
pollution control had added only about 0.2 percent per year to the total cost of U.S.
manufacturing over the period 1975 to 1980. Environmental and occupational pollution control had added only 0.1 percent to the inflation rate—and had created
400,000 new jobs. Benefits to the public from improved air quality were meantime
on the order of $21 billion per year: $17 billion from the avoidance of pollution-

---

*In Love Canal, in upstate New York, the Hooker Chemical Company dumped 21,000 tons of chemical
waste from 1947 to 1952, without the knowledge of local residents or town officials. In 1953 the company covered the site with dirt and sold it to the Niagara school board for one dollar, on the condition
that the company would not be held liable for any future injuries caused on the property. A school and
housing development were constructed on the site, though families who moved to the area soon noticed
a disturbingly high incidence of birth defects, miscarriages, and other health problems. In 1976, local
reporters from the town of Niagara broke the story, and the national press picked up on it soon after. Two
hundred and forty families were ordered evacuated by New York State authorities after health officials
found extensive contamination of homes near the canal. Love Canal became the first toxic waste dump
to gain national attention, and was a key event in the establishment of the EPA's Superfund program. See
Epstein et al., *Hazardous Waste,* pp. 89–132; Adeline Levine, *Love Canal: Science, Politics, and Rights*
(Lexington, Mass., 1982); and Beverly Paigen, "Controversy at Love Canal," *Hastings Center Report,*
March 1982, pp. 29–37.

related deaths and illnesses and $2 billion from things like cleaning-cost reductions, corrosion prevention, and improved agricultural production.[59]

Figures such as these have been the topic of much debate, though again one can argue that the focus on dollar values biases the argument from the outset. There are usually benefits from pollution prevention that are difficult or impossible to quantify: How does one put a value on the aesthetics of urban visibility or the absence of foul-smelling fumes? What is the cost of the personal or familial pain caused by cancer? Since cancers accrue mainly to the elderly, is the value of a life lost to cancer lower than, say, a life lost to an accident or to AIDS? How does one measure the cost of species extinctions or of habitat degradation?[60] Long-term "costs" extending over decades or even centuries are difficult to measure, and it is callous at best to cost out the pain of lost life or family members' grief. There is also the distributional problem: those who suffer the costs of pollution are often not the ones who reap the benefits. Discussions of the costs of regulation too often fail to ask: Costs to whom? Benefits for whom?[61]

IDEOLOGICAL EFFLUENT

The Reagan effect is not to be measured just by numbers. It also sustained an ideological attack on the "apocalyptic" science and policy of the 1970s. Environmentalists were labeled a diseased other: Luddite naysayers, "toxic terrorists," "chic-apocalyptic neoprimitives,"[62] zealous malcontents, a "penthouse proletariat" pursuing chemicals in the environment with the fanaticism of a witch hunt. The literature of this sort is vast and spans a longer period of time than the eight years of Reagan rule.[63] Here I would like to focus on four of its more articulate and influential advocates: Edith Efron, Elizabeth Whelan, Aaron Wildavsky, and Dixy Lee Ray.

Edith Efron versus the Apocalyptics

Edith Efron's 1984 book, *The Apocalyptics: Cancer and the Big Lie—How Environmental Politics Controls What We Know About Cancer,* is probably the most forceful and competent example of this genre. More than six years in the making, with over five hundred pages of carefully argued text, the book chronicles "one of the most astonishing scientific scandals of our time: the ideological corruption of cancer in the United States." Much of the book is rather straightforward intellectual history, describing, among other things the ambiguities of animal testing; the controversies surrounding the Estimates Paper; the discovery of natural carcinogens; Ames's thesis that "carcinogens are mutagens"; the multihit theory of cancer; the concept of thresholds; and several other elements of modern cancer theory. Sandwiched around these histories is a radical and unrelenting critique of environmental science and politics. Efron (1922–), a Rochester-based freelance writer for *TV Guide, Look, Life,* and other popular magazines, argues that environmental extremists have insinuated themselves into the federal establishment, twisting the truth, distorting science, and needlessly scaring the public.[64]

The central misrepresentation is what she calls the Garden of Eden theory, which claims that cancer is "a man-made byproduct of rampant industrialism and that nature is pristine and pure." The two central dogmas of this orthodoxy, maintained by conspiratorial scholars and compliant media, are the scam of animal testing and the fallacy that there are no thresholds below which exposure to carcinogen is not dangerous. Animal studies are illegitimately used to infer hazards in humans; the linearity of dose-response is used to buttress a puritanical and outdated "one-molecule" theory of carcinogenesis. Efron believes that it is moral fervor rather than dispassionate reason that has propped up these two dogmas. Scientists know that industrial chemicals are trivial causes of cancer, but they hold their tongues to preserve their jobs—many of which were created, she says, by fear of chemicals.

Efron ridicules the "cancer 'prevention' establishment" for assuming, without evidence, that industry is responsible for a sizable fraction of human cancers. Samuel Epstein is a major target, but so is Rachel Carson, Barry Commoner, Ralph Nader, Wilhelm Hueper, the leaders of the Carter-era OSHA and NCI, and the authors of the 1978 Estimates Paper. All, in her account, pay insufficient attention to natural carcinogens and the possibility of thresholds; all have sacrificed their scientific integrity to further their regulatory goals. Efron clearly believes that little can be done by governments to prevent cancer: the implication is that since "nature" is a major cause of cancer, regulators should simply relax, repent, and retreat. Low-level carcinogens should not be feared, since there is uncertainty regarding the magnitude of those hazards. She contrasts "basic science" and "regulatory science," the former having understanding as its goal, the latter being science "severed from the quest for understanding." Environmentalists, in her view, fall almost without exception into this latter camp. The typical regulatory scientist is "an intellectual policeman whose judgments . . . are backed up by the guns of the state."[65]

The problem, as Efron sees it, is that scientists have abandoned the norms of science (disinterestedness, organized skepticism, and so forth) put forward by Robert K. Merton in the 1940s to distinguish American science from its fascist counterparts. Scientists no longer think for themselves, but bow their heads to the environmental elite. In an apparent reference to George Orwell, or perhaps to Aldous Huxley's *Brave New World,* Efron writes that we now (in 1984) live in a Biologist State run by malevolent, pseudoscientific cardsharp "cancer preventionists" peddling myths derived from the "fairy tale" Rachel Carson told the American public, a tale "she had learned from an old man at the NCI" (Hueper).[66]

The success of Efron's book (it sold about 30,000 copies, received some 250 reviews, and was translated into German and Japanese) must be understood against the backdrop of ascendant environmentalism of the late 1970s. Ironically, the appeal of *The Apocalyptics* is similar in certain ways to the sweeping simplicity of Epstein's *Politics of Cancer:* both are products of cancer's cold war years; both assume that powerful, ill-willed men in high places are making us sick (with cancer, according to Epstein; with fear of cancer, according to Efron). Both recognize that politics saturates the pursuit of research in these areas.

The shortcoming of *The Apocalyptics,* however (as of Epstein's book, arguably), is that the taint of politics is presumed to lie on one side only. Efron derides the "moral-political approach" of the environmentalists, likening their "cultural crime" to "a mitigated Lysenkoism," but she fails to recognize that science was politicized even prior to the emergence of "regulatory science" in the 1970s. There is no sense that the environmentalists of the 1970s were responding to real industrial abuses of the 1940s, '50s, and '60s, or that there is also a politics implicit in threshold-based industrial toxicology (see chapter 7). The net effect is to leave both the ideas she endorses and the ideas she opposes in a free-floating, ahistorical void. Ames's emphasis upon natural carcinogenesis is said to have fallen through "a sudden opening of the skies" in 1983; the apocalyptic paradigm is likewise said to have swept "through the brains of the press and the 'elite'" like a "typhoon"—without social cause or precedent.[67]

Efron's rhetoric of all-embracing conspiracy ("unknown men in the nation's bureaucracies have been arbitrarily deciding what is and is not scientific truth") makes one wonder whether the charge of politicizing science could be raised against the author herself—notwithstanding her pretense to "militant neutrality."[68] The money trail offers one clue: funding for her research came from the conservative Olin Foundation, Pepsi-Cola, William F. Buckley's Historical Research Foundation, and Richard Mellon Scaife, a right-wing financier and confidant of Reagan who, in the 1970s, had supported the professional careers of Edwin Meese and James Watt.[69] Efron's neoconservative credentials were already established in her previous book *The News Twisters* (1971), in which she seconded Spiro Agnew's charges that the broadcast news media was strongly biased in favor of the "Democratic-liberal-left axis of opinion."[70] The networks were "sanctioning, inflating, and sympathizing with the positions of the far left splinter of the spectrum," speaking out "for black militants and against the white middle class." Efron deplored the abandonment by network liberals of the norm of neutral objectivity, typified by David Brinkley's claim in 1968 that to be objective is to be "a vegetable."[71]

Efron is no vegetable. The book would perhaps be more convincing if it did not claim not to take sides, but its message is clear. *The Apocalyptics* is arguably the most emphatic book-length expression of the 1980s revolt against the cancer environmentalism of the previous decade—the "*Silent Spring* of the counter-revolution," as Bruce Ames put it.[72]

Elizabeth Whelan: Every Man for Himself

Elizabeth M. Whelan (1944–) is an equally distinguished spokesperson for the Reagan retreat, and one with a great deal more institutional and media clout than Edith Efron. A libertarian antismoking activist and Harvard-trained Ph.D. demographer, Whelan is co-founder, with Harvard's Nutrition Department chairman, Fredrick J. Stare,[73] of the American Council on Science and Health (ACSH), a kind of Ralph Nader consumer group in reverse, dedicated to showing that food additives are not as hazardous as the chemophobic "toxic terrorists" have led us to believe. Her philosophy, expressed in radio talk shows, countless ACSH brochures, and many of her more than twenty books, is basically an individualistic one: in the war against cancer, it's every man for himself. Like Efron, Whelan suggests that people need to stop proposing social solutions for what are essentially personal problems; people need to face up to the fact that it is our own actions that put us at risk. Her politics are almost uniformly pro-industry, the only exception being her repudiation of the tobacco trade, which she rightly vilifies as the nation's leading peddler of cancer.

Whelan is not as hostile as Efron to the possibility of prevention; indeed, many of her publications stress the perils of smoking, drinking, and a high-fat diet.[74] She is no less vehement, however, in her rejection of environmentalism, food faddism, and back-to-naturism. Like Efron, Whelan seeks to paint environmentalists as antipatriotic technophobes who would just as soon burn the flag as smash machines. Her 1975 book, *Panic in the Pantry,* co-authored with Stare, attacked the organic food fad, along with the foolhardy fear of pesticides and food additives. In a 1977 speech before a meeting of the National Soft Drink Association, Whelan argued that "junk food" should really be regarded as "fun food," assuring bottlers that "the availability of something like soft drinks to children does not pose any known health hazard that would harm them."[75] Her subsequent writings show her to be a visceral optimist and patriot, an unflinching champion of the view that "healthwise, Americans have never had it so good." Americans live longer, eat better, and enjoy other benefits as never before, and things like chlorinated water and pesticides are responsible: "Modern life may have its risks but it sure beats old-fashioned death. . . . Sometimes the good news is almost overwhelming."[76]

The good news is presented most forcefully in her 1985 *Toxic Terror,* advertised as "the truth about the cancer scare." The tone of the book is set in a foreword by Nobel Prize winner Norman Borlaug, best known as the brains behind the Green Revolution's high-yielding rice and wheat. Borlaug complains that it is a First World luxury to worry about the "carcinogen of the week" and the "toxic of the month." He asks the reader to keep in mind that "banning chemical fertilizers and pesticides will inevitably lead to deadly serious food shortages; it's as simple as that. Banning other industrial chemicals will likewise impoverish our nation. That is the price the toxic terrorists are asking us to pay."[77] The toxic terrorists, of course, are the environmentalists: toxins in the body politic. Whelan herself goes on to suggest that the media share part of the blame, insofar as they feature bad news over good, but so do the "comatose scientists" who fail to respond to media bias. Much of the material here is familiar: Epstein and the Estimates Paper, OSHA's excesses, the fallacies of extrapolating from animals to humans, and so forth. Her argument is that it is important to focus on big causes rather than small ones—"elephants rather than ants." Elephants for her include cigarettes (causing 500,000 U.S. deaths per year), alcohol (100,000 deaths), addictive drugs (35,000 deaths), and hazardous lifestyles (failure to use seat belts or smoke detectors, for example, causing 50,000 deaths per year). Ants include things like pesticides, food additives, and foam insulation—products that, by her account, do little or no harm to human health.[78]

Whelan is most famous today not for what she wrote but for being president of the American Council on Science and Health, a "non-profit, tax-exempt, nonpartisan public education group" based in Manhattan. Whelan co-founded the ACSH in 1978 to defend food producers against charges of fouling the American food system with harmful colorants, pesticides, sweeteners, and plasticizing agents.[79] Today the council presents itself as "a unique voice, backed by scientific evidence, defending the

achievements and benefits of responsible technology within America's free enterprise system." By contrast with consumer advocacy groups (Ralph Nader's Health
Research Group is its perennial nemesis), the council claims to provide a "balanced"
approach to public health issues. ACSH goals, apparently shared by the organization's two hundred scientific and policy advisers, include "communicating consumer
information about the alleged health effects of trace amounts of chemicals—whether
man-made or naturally occurring—emphasizing a frequently ignored truth, 'the dose
makes the poison'"; "assessing the risks to health and the environment of banning
pesticides, food additives, and other advances of modern technology in pursuit of
'zero risk'"; and "helping Americans distinguish between hypothetical and real
health hazards."[80] The rhetoric of reassurance is important here: in the fifteen years
since its founding, the ACSH has become the leading voice for food industry efforts
to establish that our food supply is safe.

Despite the protests of its introductory brochure ("IS ACSH AN INDUSTRY
FRONT? NO!"), it is not hard to detect a certain selectivity in the problems addressed by the council. The ACSH warns against "excessive sun exposure, alcohol
abuse, failure to wear seat belts or to use smoke detectors, unbalanced diets, and
promiscuous sexual behavior," but never is there a warning against anything recognizable as an industrial effluent, food additive, or agrichemical product. There is a
blissful optimism here, as when the brochure asserts that the number of deaths
caused by "trace levels of dioxin, PCBs, EDB, chlordane, heptachlor, DDT (and all
other pesticides), ionizing radiation from U.S. nuclear plants, lead in air and water
[and] food additives" is "unknown but negligible." Ignorance of harm is apparently
evidence of the absence of harm; what we don't know can't hurt us.

Consumer groups have challenged the ACSH claim to scientific neutrality. In the
fall of 1978, the National Nutritional Foods Association, a trade group, filed a $1.3
million damage suit against Whelan and Stare, alleging that the council had been
formed as a conspiracy to defame and disparage the health food industry (the suit
was eventually dismissed).[81] In 1982, Michael Jacobson's Center for Science in the
Public Interest prepared a book-length critique of the council, denouncing it as a
consumer fraud: "Through voodoo or alchemy, bodies of scientific knowledge are
transmogrified into industry-oriented position statements." Peter Harnik, author of
the study, denounced the ACSH as "an industry front" (the motive, perhaps, for the
council's capital-letter denial.)[82] Harnik pointed out that ACSH funding came from
many of the nation's major food and chemical producers;[83] with this kind of support,
Harnik asked, was it any wonder that the council opposed taking "fun foods" such as
candy and soda out of school cafeterias? Was it surprising that the council defended
food colorings against the charge that they may cause hyperactivity among children,
or that it lampooned critics of the herbicide 2,4,5-T?

Exculpating purported environmental hazards is Whelan's forte. ACSH newslet-
ters state that nuclear power is safe, that pesticides may actually help wildlife, and
that it is wrong for the media to focus on the "carcinogen of the week." The council
defends Alar against the NRDC, saccharin against Epstein, bovine somatotropin
against a broad spectrum of detractors, and New Jersey against the charge of being
cancer alley. Since the mid-1980s, the ACSH has also become a big promoter of

Bruce Ames's thesis that natural carcinogens are a far greater threat than synthetic chemicals produced by industry. A recent brochure follows Ames in asserting that "human dietary intake of nature's toxins is at least 10,000 times higher than the intake of man-made pesticides." It is therefore high time for Americans "to stop acting on the presumption that 'natural' is safe and 'man-made' is suspect." The brochure suggests that both can be toxic in excess, but that "neither man-made nor natural food chemicals are hazardous in the quantities we consume on a daily, monthly, or yearly basis."[84] Whelan's own enthusiasm for pesticides borders on the misanthropic: in a 1989 tongue-in-cheek speech announcing her candidacy for president (with Norman Borlaug as her running mate), the ACSH chief proclaimed her desire to form a cabinet-level position for national pest control directed not at insects but at environmentalists: "I'm talking about controlling human pests, the pests who continue to terrorize us about the quality of our environment. Human pests are damaging the quality of life in America. These people are not promoting health but their own political agenda."[85]

Aaron Wildavsky: Pollution as Moral Coercion

Aaron Wildavsky (1930–93) was a professor of political science at the University of California, Berkeley, and is another notable in this roster of environmental retreatists. Wildavsky's perspective differed somewhat from that of Efron or Whelan, insofar as he added what one might call an anthropological cast to the discussion. The perspective is one that attempts to defuse environmentalism by framing pollution as a "social construct" and environmental organizations as parochial, divisive, and ultimately impotent sects. The net effect is a kind of right-wing relativism, a complacent "cultural constructionist apology for industrial capitalism."[86]

Wildavsky, though, was not an insubstantial thinker. In a 1982 article in the libertarian *Cato Journal*, the Berkeley professor defended libertarianism as a form of government that does not attempt to guard its members against misfortune:

> Pollution is a problem for libertarians because some people exercise power over others without these others being able to fight back effectively. Pollution appears to violate the norm of reciprocity. Market mechanisms do not provide reciprocity because there is no way in which the costs imposed on the polluted can be transferred to the polluter. And the polluter lacks incentive to bear his part of the burden.

In fact, this is an illusory problem, since:

> Properly understood, I contend, libertarianism (unlike other modes of social organization) knows no pollution.... If there are no boundaries against transactions—competitive individualism tears down rather than builds up boundaries— there can be no pollution. For the libertarian there are no permanent uncompensated bads, because eventually markets will arise to internalize the external costs.[87]

The very idea of pollution, he maintained, is an antiestablishment construct used to attack market institutions: "saying 'the environment is dirty' is equivalent to charging that 'markets are dirty.'" The charge that environmental pollution is ubiquitous can therefore "be understood for what it is—a fundamental attack on libertarian culture." Wildavsky cites the work of Bernard Cohen of Pittsburgh showing that nuclear energy is safe, and suggests that this particular technology is popularly chosen for opprobrium not because it is especially dangerous ("quite the opposite") but rather because such an attack has a larger symbolic value. An attack on nuclear power is an attack on "bigness," "corporate capital," and indeed "the entire establishment." Pollution in sectarian hands (including most environmentalists and consumer advocates) is a means of moral coercion, "a political weapon against individualism and collectivism."[88]

The interesting question, then, for Wildavsky, became, "What kind of culture routinely invokes nature and the natural in defense of its way and in opposition to contrary cultures?" He distinguishes collectivists, individualists, and sectarians. Collectivists and capitalists are generally optimistic about technology, though for different reasons. Individualists believe that an infallible knowledge of the future is neither necessary nor desirable; collectivists believe either that they already know the future or that their successors eventually will. Sectarians, by contrast, are technological pessimists because they oppose the hierarchy and competition they see as integral to technological production. Sectarians have an "Armageddon complex" that leads them to fear immense future risks: oxygen depletion, population explosion, thermal inversion, and so forth. Sectarians are "perennial outsiders":

> They cannot rule, except for brief periods, because they cannot solve the problem of authority. They tend to split, either because they will not acknowledge leaders or the charismatics they prefer tend to drive out rivals. What they can do is criticize the establishment, whether it be collectivism or individualism or some combination thereof. Being small and critical, however, they are often on the defensive. Sectarianism needs a strategy that will keep its members together and will threaten the stronger establishment. With what? With doom. If machines are broken, they can be repaired. If a given technology proves defective, it can be replaced. But if the threat is to human life itself, or to the natural environment that sustains life, incremental adaptation is ruled out. Only radical change toward sectarianism will save the day, indeed, the planet.[89]

For Wildavsky, environmentalists are sectarians in this sense. The problem, though, is that not all—nor perhaps even most—of those who call themselves environmentalists subscribe to the antitechnology vision outlined here. George Bush, after all, claimed to be an environmentalist (he likes to hunt and fish), and Bruce Ames today maintains that "we are all environmentalists."[90] There is a great deal of diversity in what comes under this label: neo-Malthusians who worry about Third World population growth are at odds with civil libertarians who worry about the infringement of reproductive rights; deep ecological monkey-wrenchers have little in

common with the pragmatists at the Environmental Defense Fund. Most environmentalists are antinuclear, but not because they oppose technology in general: few
would oppose federal support for efforts to improve the efficiency of solar energy
panels, and few would oppose the aggressive development of pollution prevention
technologies.

Dixy Lee Ray: "Ms. Plutonium"

As former governor of the State of Washington and chairman of the Atomic Energy Commission under Nixon, Dixy Lee Ray* (1914–94) is better known than
Efron, author of a single cancer treatise, Whelan, head of an industry front ("NO!"),
or Wildavsky, author of essays directed primarily at an academic audience. Her 1990
book, Trashing the Planet, became a national best-seller, with Newsweek going so
far as to call it "the most absolutely necessary book of the year, perhaps the decade."
Talk-show host Rush H. Limbaugh III in his The Way Things Ought to Be suggested
that reading Ray's book was "the best way you can arm yourself against the junk scientists" who favor intrusive government in the name of "doomsday environmentalism." Limbaugh noted that Ray's book was "the most footnoted, documented book I
have ever read,"[91] which may say more about the talk-show host's reading habits
than about the bibliographic skills of the former governor.

Like Efron, Whelan, and Wildavsky, Ray opposed the hysterical, antiscience,
antitechnology, "doom-crying opponents of all progress" who yearn for a simpler
age. The zoologist-turned-publicist deplored the "sob-sister journalism" that treats
individual stories of childhood leukemia as if they were epidemic, and contrasts the
dominant "alarmist environmentalism" with the "calm reason" required for the solution of environmental problems. Ray traces modern environmentalism back to the
original saboteurs who threw their wooden shoes (sabots) into the machinery in their
"opposition to the introduction of anything new." Though most environmentalists
are "fine, decent citizens," a powerful and vocal minority are "determinedly leftist,
radical, and dedicated to blocking industrial progress and unraveling industrial society." Ray warns against the teaming up of "militant vegetarians with radical environmentalists" to terrorize the public in the name of animal rights and ecodefense; she
then lays blame for the annual starvation of "60 to 100 million people" on the shoul-
ders of environmentalists who, by limiting pesticide production, have deprived the
world's hungry of food.[92] Many of these themes are elaborated in her 1993 book, Environmental Overkill, dedicated to Petr Beckmann and Edward Teller, pronuclear activists and promulgators of yet another kind of apocalypse.[93]

Ray got her first political break in 1972, when Nixon named her to the five-
member board of the Atomic Energy Commission. Before that she had been an associate professor of zoology at the University of Washington, director of the Pacific

---

*Dixy Lee Ray was born Margaret Ray; she changed her name to honor her favorite part of the country
and her favorite Civil War general, Robert E. Lee, to whom she was distantly related. At the age of
twelve, she became the youngest girl ever to climb Mount Rainier, Washington's highest peak.

Science Center (a local museum), and a teacher in the Oakland public school district. In 1973, Nixon appointed Ray to head the AEC (on the recommendation of the outgoing head, James Schlesinger, who left to become director of the CIA), apparently hoping that a female naturalist might mollify critics of the beleaguered commission. Ray oversaw some changes at the agency, most notably the separation of the safety and development aspects of the commission's work.[94] Nuclear detractors pointed out that much of the cold war mentality of the commission remained intact: Ray's enthusiasm for all things nuclear earned her the title Ms. Plutonium from Ralph Nader ("Raydiation" was another occasional epithet); Ray, in her turn, branded the consumer advocate "an ignorant man" with "no credentials."[95]

Ray left the AEC in 1975 to become Assistant Secretary of State in the Bureau of Oceans, a position she left after six months to return to Washington State, where she ran for governor as a pronuclear, promilitary Democrat, backing the construction of a Trident submarine base on Puget Sound, unrestricted logging of old-growth forests, and expansion of the state's oil-transport business with Alaska. (She wanted supertankers to be allowed to carry Alaskan oil into Puget Sound.) Her strong-arm approach to government combined with her inability to compromise left her with little popular or media support (she once named the pigs on her farm after reporters she had grown to dislike), and she was unable to win the 1980 Democratic party primary for the governorship. She retired to her farm on Fox Island, where she grew vegetables and carved Northwest Coast art until her death in 1994.

Many of Ray's ideas repeat those I have already considered, but we should recall that the texts these four figures wrote are part of a much broader literature. Ben Bolch and Harold Lyons's 1993 *Apocalypse Not,* published by the Cato Institute, laments the "irrational fears of cancer" that have led to $1.4 trillion in EPA compliance costs and billions sometimes spent to prevent a single cancer. Michael Fumento's 1993 *Science Under Siege* decries the technophobic "cult of the natural" that has led us to blame industry, rather than nature, for our environmental ills. Andrea Arnold's 1990 *Fear of Food,* with an introduction by Senator Steve Symms of Idaho, denounces the "media mendacity" that led to the 1989 Alar scare, and M. Alice Ottoboni's 1991 *"The Dose Makes the Poison"* invokes the Paracelsian maxim to argue that a little bit of a carcinogen can't hurt you.[96] The D.C.-based Center for the Defense of Free Enterprise publishes a continuous stream of such works, as does the Cato Institute, the American Enterprise Institute, libertarian magazines such as *Reason* and *The Freeman,* and the organs associated with Reed Irvine's archconservative "Accuracy in Media."[97]

It would be wrong to exaggerate the unity of these various authors, but there are several common threads. All are neoconservatives of one stripe or another—and several call themselves libertarians. All lament a purported "chemophobia" gripping the nation; all worry that the liberal news media have exaggerated the hazards of industrial life. All wrap themselves in the rhetoric of liberty, science, and faithfulness to entrepreneurial individualism; all operate on the assumption that it is wrong for the state to adopt protections in the face of uncertainty. The presumption is that environmentalism is a Trojan horse to advance the cause of socialism and that "the only difference between the socialists and the environmentalists is the color of their flags."[98] All exploit the language of environmental degradation: environmentalists are a toxic contaminant in the body politic, toxic terrorists promoting despair rather than hope. These are images that resonate with an older, conservative rhetoric of termitelike subversion undermining the body politic—metaphors of the 1950s brandished occasionally even by environmentalists.*

Frequently encountered in such writings is a rhetoric of reassurance—often expressed in terms of the naturalness of purported hazards. The "amazing" truth, as Wildavsky says, is that "life is chemicals; nature is chemicals; people are chemicals."[99] Ray points out that "we live in a radioactive world. We always have and we always will";[100] "electromagnetic fields are part of our normal environment."[101] The point of such claims is usually to diminish the hazard: that which is natural and familiar is purportedly benign. An interesting variant on this argument asserts the opposite: for Efron, as for Ames, the naturalness of a hazard is no reason to doubt its capacity to cause harm. Nature's carcinogens are far more potent than "man-made" carcinogens; it is wrong to equate the natural and the beneficent.[102] What unites these apparently contrary arguments is a rhetoric of reassurance based on the potency of nature: the natural is that which we cannot avoid (and therefore shouldn't be worried about), but it is also that which exceeds industry in potency (industrial carcinogens should therefore not be of great concern). The rhetorical force is similar in the two cases: the point is to diminish fear, to undermine activism, to slow or redirect efforts to prevent environmental cancer.

The rhetoric of reassurance blends with a rhetoric of risk. Whelan recounts how a man died from drinking too much water; Ray tells how a nuclear power plant is "infinitely safer than eating," because three hundred people choke to death on food every year.[103] Risk is not something to be feared but to be embraced: risk is inevitable, necessary, even desirable insofar as it is "potential profit." Wildavsky writes that "hazard is a fact of life," that "the secret of safety lies in danger." Paul Johnson of the American Enterprise Institute, reviewing Efron, contends that "man" is "a risk-taking animal" and that "not to take risks is the biggest risk of all."[104] This is the entrepreneurial ethic writ large: there is no free lunch, no gain without pain. Effete intellectual environmentalists forget the inevitability of trade-offs, that modern conveniences simply cannot be had without a certain cost. The language is perhaps more readily familiar to the investment banker than to the industrial or domestic worker; there is no sense that illness caused by pollution, for example, is not a choice made in anticipation of profit, that there is something different about the risks assumed by portfolio managers and by children who inadvertently eat lead dust after touching the window sill. There is little sense that the people taking the risks may not be the ones harvesting the rewards.

---

*Rachel Carson can be seen as extending the metaphor of communist threat to internal bodily health: carcinogenic pesticides are an alien, insidious, often invisible force undermining the body; insecticides have an "immense power not merely to poison but to enter into the most vital processes of the body and change them in sinister and often deadly ways" (*Silent Spring*, p. 16).

The rhetoric of antiscience is important in this context. Science is presented as the natural enemy of environmentalism; genuine science is contrasted with disingenuous politics. Science is said to be "under siege" by a band of elitist, ecoemotional, doomsday-preaching saboteurs who yearn for a preindustrial age free of modern harms and inconveniences. Environmentalists are cast as otherworldly zealots: Efron talks of environmental "evangelists" and Wildavsky likens environmentalism to witchcraft, mysticism, and the fear of Satan.[105] Environmentalists are a privileged class: "they want a society in which the elite have organic strawberries and cream, and the rest of the people thank them each day for saving the tsetse fly and the precious mosquito."[106] The populist rhetoric of freedom joins with a rhetoric of technocracy, as when Ray contrasts her *Trashing the Planet* with Vice President Gore's *Earth in the Balance:* her book (unlike Gore's) was "grounded in science, not pseudo-science; it was not filled with hair-raising warnings of impending disaster; and it didn't earn half a million dollars for the author."[107]

This is a profoundly naive view of science, combining as it does a 1950s-style confidence in the technocratic future with a 1980s-style longing for an imaginary past. Ray thus asserts at one point that it is "ironic" that so many women are "anti-technology," given that technological change has been the key to women's liberation. She claims that the invention of the spinning wheel in medieval times was "the very first step taken to release women from true domestic bondage"; progress was not made again until "the industrial revolution brought three new technical inventions—the sewing machine, the typewriter, and the telephone, all of which opened new avenues for the employment of women." Never mind the political battles fought for universal suffrage or the fact that many so-called labor-saving devices actually ended up making "more work for mother."[108] Technology, in Ray's view, has always had a liberating effect: indeed, "electricity has done more to liberate women than all the speeches and protests and affirmative action programs that have often jolted our sensibilities."[109]

Elsewhere, though, Ray confesses, wistfully, that she herself yearns for a simpler and nobler time:

> There was a time in my youth when reliable experts were believed. It was a time when most people and most institutions were presumed to be well-meaning and honest until and unless proved otherwise. It was also a time of unprecedented increase in our knowledge about the world, in our belief in ourselves, and in our ability—through understanding and logic—to provide adequate solutions to technical problems.

The former governor laments the passing of this Norman Rockwellian "time of optimism and progress" (apparently the 1930s to the 1950s), a time when alarmists could be ignored and the fruits of science were appreciated by a grateful public. Progress has continued, she says, but we no longer welcome it: "We seem to have become a nation of easily frightened people—the healthiest hypochondriacs in the world!"[110] Ray's hope—shared by Efron, Whelan, and Wildavsky—is to restore the cultural authority of science. But they do not recognize that scientists are themselves divided on questions of environmental science and policy. Or that policy is often—and necessarily—made in the face of uncertainty, that experts can wear blinders, that struggles over how to deal with environmental hazards cannot be reduced to strife between all-knowing scientists and an ignorant public.

BODY COUNT

The Reagan legacy continued largely unchanged into George Bush's presidency. As vice president, Bush had headed up the Presidential Task Force on Regulatory Relief; as president, he assigned a similar task to his own second in command. Vice President J. Danforth Quayle soon earned a reputation as the most aggressive environmental deregulator since Ronald Reagan, using as his instrument in this effort the "Council on Competativeness" (*sic*), an executive branch group founded in June 1990 to continue the regulatory relief begun by the previous administration. In the two and a half years of its existence, the council "blocked a ban on the dangerous practice of burning lead-acid batteries, stalled a rule that protects workers from the hazards of formaldehyde, and delayed new standards for clinical laboratories."[111] Supporters argued that the council saved consumers billions of dollars by helping to speed the approval of drugs: a May 26, 1992, action by the council, for example, cleared the way for most genetically engineered foods to be marketed without extra safety tests. Representative Henry Waxman of California characterized the council as a "rogue agency" that "flouts the law," a "domestic version of the National Security Council during the Iran-contra scandals" (he later led a successful effort to cut off funding for the council). Even William Ruckelshaus, the former EPA administrator, admitted that the council had created an impression that "Republican administrations oppose environment, health, and safety regulations because of their ties to big business."[112]

President Bush announced in his January 28, 1992, State of the Union address that he was levying a ninety-day moratorium on all new occupational health and safety regulations in order to speed up economic growth. Among other things, the measure delayed methylene chloride restrictions and standards for cadmium and lead, indoor air quality in the workplace, ergonomic safety and health to reduce repetitive strain injuries, and legislation designed to protect workers from falls and respiratory ailments.[113] Environmental measures were said to be exerting an intolerable financial burden on industry: Senator Orrin Hatch of Utah on March 10, 1992, stated at hearings on the moratorium that 13 percent of U.S. business income was spent on complying with regulations. *Science* magazine's Philip Abelson supported the president in this regard: countless editorials in the nation's most prestigious general science journal supported the Reagan-Bush environmental retreat, leading Samuel Epstein to characterize the magazine's editorial page as "a bully pulpit to trivialize concerns on environmental pollution and occupational hazards."[114]

The Clinton-Gore administration promises to change much of this. Several prominent environmentalists have been appointed to key positions in occupational and environmental health, and wildlife management. George T. Frampton, Jr., president of the Wilderness Society from 1986 to 1993, was named assistant secretary of the Interior, responsible for national parks and fish and wildlife. Jim Baca, a former board member of the Society, was named head of the Bureau of Land Management. The DOE under Hazel O'Leary has begun to air some of its dirty laundry, including a far-reaching inquiry into the AEC's support for human radiation experiments. Labor Secretary Robert Reich has breathed new life into OSHA, increasing criminal penalties for "willful violations," extending OSHA coverage to workers in DOE nuclear plants, requiring employers with eleven or more full-time workers to establish safety and health committees, and so on.[115] Pesticide policy has begun to change: the new EPA has promised to implement a 1993 National Research Council recommendation to recognize children and infants as more susceptible than adults to harm from pesticides; the older policy had regarded adults, children, and infants as indistinguishable in this respect.[116] Barry Commoner has asserted that the Clinton-Gore EPA, headed by Carol Browner, is indistinguishable from the Reagan-Bush agency,[117] though this assessment is probably overly harsh. The EPA's 1995 budget included an 8 percent increase, bringing total funding for the agency up to $7.2 billion and prompting Friends of the Earth, an environmental group, to characterize the Clinton administration budget as an "earth budget."[118]

The National Cancer Institute periodically reminds us that cancer is largely a preventable disease. For the Reagan administration and its ideological allies, however, preventive medicine was a nonstarter. Environmental health measures were dismantled, and occupational safety supervision was dramatically diminished. Research funding even for "natural" and "lifestyle" carcinogens like radon and tobacco was gutted, despite the attention given by antienvironmental publicists to such factors.[119] The EPA's radon budget was axed in the first Reagan budget (Congress later forced a restoration), and in 1985 Reagan's ally in the Senate, Jesse Helms, gained Senate support for a measure to cut the federal excise tax on a pack of cigarettes from sixteen cents to eight cents. Food labeling laws were shelved and nutrition programs for low-income children were drastically cut (remember the reclassification of ketchup for school-lunch purposes as a vegetable?). Reagan administration officials hampered the distribution of Carter-era dietary guidelines recommending low-salt, low-sugar, and low-fat foods, prompting Michael Jacobson of the Center for Science in the Public Interest to accuse the Reagan USDA of having been captured by the meat industry.[120]

An interesting exercise would be to try to calculate how many lives will have been lost as a result of the Reagan-Bush administration's abrogation of federal regulatory responsibilities in the area of environmental health, consumer product safety (including tobacco), and occupational health and safety. It is impossible, of course, to conduct such an exercise with any pretense to precision. One would have to factor in things like the wealth and the poverty created by the administration—ambiguous elements that could easily push the figure higher or lower. Carter OSHA administrators in the late 1970s estimated that, if the agency's proposed toxic legislation were implemented immediately, the nation could see a 20 percent drop in cancer death rates over the following twenty years.[121] Assuming that the number of lives saved would have been averaged out over the entire twelve-year period of Republican rule, this would mean that Reagan-Bush policies may have cost the nation about 50,000 cancer deaths per year, or 600,000 total cancer deaths for the entire twelve-year period. Even if such a calculation were too high by a factor of ten or even a hundred, the result is still disturbing.

We do have some empirical knowledge of specific consequences, smaller in magnitude than cancer effects. In 1992 Patricia Buffler, dean of the University of California, Berkeley, School of Public Health, published a study showing that the Reagan administration's five-year delay in requiring warning labels on aspirin led to the needless deaths of 1,470 children from Reye's syndrome, a fatal complication suffered by children treated with aspirin for chicken pox or the flu.[122] What is true on a very small scale for a rare and preventable disease like Reye's syndrome is almost certainly true on a much larger scale for a disease like cancer. Ronald Reagan may have been the most potent new carcinogen of the 1980s.

Chapter 5

"Doubt Is Our Product": Trade Association Science

"Just the place for a Snark!" the Bellman cried
As he landed his crew with care;
Supporting each man on the top of the tide
By a finger entwined in his hair.

"Just the place for a Snark!" I have said it twice:
That alone should encourage the crew.

"Just the place for a Snark!" I have said it thrice:
What I tell you three times is true.
                    —Lewis Carroll, *Hunting of the Snark*

SCHOLARS HAVE GENERALLY ignored the extent to which trade associations seek to produce science to influence public opinion and governmental legislation.[1] Even Martin Gardner, the great chronicler of scientific "fads and fallacies" (the Flat Earth Society, Uri Geller's spoon bending, and so on), conspicuously ignores the "science" peddled by trade associations in his several books on such topics.[2] There are thousands of such associations in the United States, promoting everything from asbestos to zinc. The Beer Institute defends brewers against the charge that drinking causes crime or traffic accidents; the Lawn Institute (representing grass-seed producers and pesticide manufacturers) works to assure consumers that it is okay "to ChemLawn." The Asbestos Information Association cautions consumers against a hysterical "fiber phobia"; the Calorie Control Council defends artificial sweeteners such as cyclamate and saccharin against charges that they cause cancer.[3] In 1986, according to one estimate, trade associations and their member corporations spent nearly $2 billion on what has come to be known as *issues management, advocacy communication,* or *image advertising.*[4] Washington, D.C., alone is headquarters to some 1,700 trade associations, making trade association business the second-ranking private industry in the nation's capital, after tourism.[5]

Many of the trade associations formed in the past twenty years or so were established in response to consumer movements or environmental legislation. The Asbestos Information Association was formed in 1970 in anticipation of the establishment of OSHA; the American Industrial Health Council was formed in

1977 in response to OSHA's push to regulate generic workplace hazards. When scientific consensus began to shift in favor of the global warming hypothesis, a Global Climate Coalition produced experts pointing to ambiguities in the analysis.[6] When public opinion seemed to be moving against disposable diapers, a National Association of Diaper Services was formed to show that washables consume three times as much nonrenewable energy and generate ten times as much water pollution.[7] The Polystyrene Packaging Council was likewise formed (with initial funding of $14 million) in the wake of local government efforts to ban Styrofoam containers in fast-food restaurants: critics who suggested that plastics were overwhelming landfills could now be countered with statistics on how plastic occupies only a minuscule proportion of landfill mass (but not volume!).

Such institutes are almost always champions of science, calling for ever more elaborate studies of this or that substance. There is a complex bias in such research, though. The Tobacco Institute can support "good" medical research, but the hoped-for effect may be "goodwill" toward the industry. A chemical manufacturer may promote research into the hazards of a hydrocarbon, but the hoped-for effect may be to insinuate ambiguity and forestall regulation. A lead zinc association may call for greater precision in studies of how ingested metals affect intelligence, but the main goal may be to raise doubts about the health risk.

Science, in other words, serves several different functions in the industrial context. It is perhaps best known as an inspiration for technological innovation, but science is also an instrument of public relations. We are familiar with the science *of* advertising, but we must also acknowledge the phenomenon of science *as* advertising. Support for science can provide a company with the appearance of progressive responsibility or even patriotism. Support for science can also provide an effective means of delaying regulation or otherwise influencing policy. Trade associations have become expert in the discovery, manipulation, and manufacture of both knowledge and (most notably) ignorance; appreciating how this works can help us recognize that ignorance is more than an ever-shrinking residue, left behind in territory not yet claimed by the steady march of knowledge. Ignorance can be a managed product, just as light may be cast to produce the comfort of shadows.

SCIENCE AS PUBLIC RELATIONS

American corporations have long been adept in the area of public relations. In 1934, Du Pont formed its first public relations department after Senate investigations of its military contracting led critics to brand the firm a "merchant of death"; the American tobacco industry employed aggressive PR experts in the 1950s to counter consumer fears that cigarettes were causing cancer. The growth of environmental activism in the 1970s and 1980s has moved an ever broader array of industries to engage public relations expertise. In 1977, for example, Monsanto launched a $4.5 million magazine and television campaign to reassure consumers that "without chemicals, life itself would be impossible." Monsanto's Chemical Facts of Life campaign succeeded in boosting the industry's image, judging from opinion polls reporting a 6 percent increase in "positive attitudes" toward the industry over a period of about two years.[8] Du Pont shortly thereafter announced a $4 million commitment modeled on the Monsanto success, and the Chemical Manufacturers Association in 1980 began a $7.5 million campaign "to increase recognition that the chemical industry is committed to doing a responsible job to protect the public."[9]

Public relations firms are often hired to conduct such efforts. When the Bureau of Alcohol, Tobacco and Firearms proposed a ban on whiskey bottles made from vinyl chloride, polymer industry officials hired Arthur D. Little to produce a report predicting staggering economic losses and unemployment from such legislation.[10] When the nuclear reactor at Three Mile Island failed, its operator, Metropolitan Edison, expanded its public relations department to counter criticism of its handling of the incident. Union Carbide did likewise in the aftermath of the 1984 leak of methyl isocyanate that left more than 3,000 dead and 200,000 injured near the Indian city of Bhopal.[11] Corporations employ "crisis management" teams to neutralize or even profit from such mishaps, the goal being to make lemonade, as one analyst recently put it, "out of life's lemons."[12]

Such imaging can produce results. Hill and Knowlton, the largest PR firm in the world with more than a thousand clients worldwide, recently boasted that:

When all Chilean fruit shipments to the U.S. were embargoed, we succeeded in getting the ban lifted in five days. When the U.S apple industry was faced with a similar life-or-death fight for existence, we had a comprehensive communication plan in motion within two days. When a beverage was alleged to be contaminated, we kept it on the shelf and demonstrated it was safe.[13]

Hill and Knowlton was actually the first PR firm ever to establish an Environmental Health Unit; the move was made in anticipation of the establishment of OSHA and the EPA in 1970. The company had already helped the British asbestos industry polish its image (its London office had set up the Asbestos Information Council) and had been a key player in helping the Tobacco Institute get its message out.[14] By the end of the 1960s the firm was urging Johns-Manville, the asbestos giant, to establish a similar association: the net result was the creation of the Asbestos Information Association.[15] Hill and Knowlton later handled Du Pont's worries about ozone depletion; Ashland Oil's spill into the Monongahela River; the Mexico City earthquake (for the Mexican secretary of tourism); the fire at the DuPont Plaza Hotel in Puerto Rico; the Alar scare for the Washington Apple Commission and the International Apple Institute; and the Iraqi invasion of Kuwait (for a group calling itself Citizens for a Free Kuwait). Environmental PR, according to Hill and Knowlton, generally involves a kind of selective silence, half the time keeping your clients in the news and the other half keeping them out. Selective silence helps to keep up the company's own image: Hill and Knowlton's brochures fail to mention its management of the Tobacco Institute's PR in the 1950s or of Johns-Manville's image in the 1960s and '70s.

Trade associations, assisted by public relations talent, have come to recognize the value of science-as-public-relations. Drawing upon the expertise of "volunteers" from member companies, trade associations are able to combine research and advocacy in ways that insulate members from charges of individual bias or legal liability. Analysts of "strategic issues management" point out that trade associations "broaden the political base, spread costs, and lessen the political visibility" of individual companies.[16] Trade associations act as a "lightning rod on controversial issues, taking some of the heat off members."[17] There are financial benefits to the arrangement: according to a 1988 U.S. Supreme Court ruling, trade association expenses are tax-deductible if the association can prove that its policies are based on "objective, scientific research in setting industry standards."[18]

In consequence of these and other advantages, there is hardly an industrial product or process without a trade association council or coalition ready to come to its defense. The National Dairy Council promotes milk and cheese; the American Meat Institute promotes pork, poultry, and beef. The Northwest Forest Resource Council, representing logging interests, opposes species-preservation legislation in Oregon. There are firms that specialize in trade association management: the Kellen Company on Peachtree-Dunwoody Road in northeast Atlanta, for example, oversees the activities of some 20-odd associations, including the Calorie Control Council, the Horseradish Information Council, the Lignin Institute, the Infant Formula Council, and a dozen other food-related outfits.[19] The Society of the Plastics Industry, with a $50 million budget and a staff of 165, coordinates twenty-nine subsidiary "operating units," including the Vinyl Institute, the Degradable Plastics Council, the Plastic Drum Institute, and the Plastic Bottle Institute. The Synthetic Organic Chemical Manufacturers Association embraces twenty-one associations promoting specific substance interests; affiliates range from the Silicones Health Council to the Methyl Tertiary Butyl Ether Task Force.[20]

In an age of environmental activism, industrial promotion often means opposition to government regulation. When several states sought to require newspapers to print on recycled paper, the American Newspaper Publishers Association attacked such efforts as a violation of press freedom.[21] When the National Academy of Sciences reported that chemical agriculture had been a failure, the Washington-based Fertilizer Institute (representing the eight-billion-dollar industry) produced a carefully argued refutation of the academy's conclusions.[22] When Herbert Needleman produced his now-classic research linking lead exposures to children's diminished intelligence, the International Lead Zinc Research Organization disputed the claims.[23] When a University of California study showed that a thousand children died in consequence of the Reagan administration's five-year delay in requiring warning labels on aspirin, the Aspirin Foundation of America responded that the authors of the report had overstated their case.[24]

Environmental carcinogenesis is often what is at stake in such encounters. When OSHA moved to establish exposure limits for benzene, the American Petroleum Institute persuaded the U.S. Fifth Circuit Court of Appeals to overturn the limit. When the National Academy of Sciences published a review of evidence linking breast and colon cancer to the consumption of meat, the American Meat Institute organized a coalition of dairy, poultry, and livestock associations to discredit the charges.[25] When the *New England Journal of Medicine* published evidence that caffeine causes pancreatic cancer, the National Coffee Association responded with a point-by-point refutation.[26] When NCI scientists showed that coal tar dye–based cosmetics (especially hair dyes) contained carcinogens, the Cosmetic, Toiletry, and Fragrance Association, representing manufacturers with total annual sales of $18 billion, trotted out experts to counter the accusations.[27]

SMOKESCREEN: THE TOBACCO INDUSTRY

In the cancer field, the best-known efforts to produce and manage uncertainty are those of the tobacco industry. The incentives are strong, given that roughly a quarter of all adult Americans still smoke, earning billions in profits for the industry and billions for the government in the form of taxes. (Philip Morris made more money in 1992 than any other company in the United States: $4.9 billion.) The human costs are unrivaled: 150,000 Americans died of lung cancer in 1994; most epidemiologists figure that at least 80 percent of those, or 120,000, were caused by tobacco. Tobacco has been estimated to cause another 22,000 deaths from cancer elsewhere in the body, not to mention an additional 245,000 deaths from heart disease and stroke. The U.S. Centers for Disease Control in 1991 calculated that cigarettes kill 434,000 Americans every year, including several thousand victims of fires caused by cigarettes.[28] A 1993 estimate gave a lower figure of 419,000, the change being due to the slowdown in smoking since the 1960s.[29]

Evidence of cancer harm from tobacco goes back a long way. Pipe smoking was recognized as a cause of cancer of the lip in the eighteenth century, and by 1915 Frederick Hoffman could suggest that the relation of smoking to cancer of the mouth was "apparently so well established as not to admit of even a question of doubt."[30] The lung cancer connection was more difficult to prove, and it was not until the 1920s that German and American epidemiologists began to document a link.[31] Ernst Wynder and Evarts Graham in the United States published much stronger proof of a correlation in 1950; in England that same year, Richard Doll and A. Bradford Hill showed that people who smoked twenty-five or more cigarettes a day were about fifty times more likely than nonsmokers to die of lung cancer.[32] The American Cancer Society funded a major study beginning in 1951, relying on volunteers to follow the health of some 187,000 smokers. By the mid-1950s E. Cuyler Hammond, one of the project's principal investigators, had good evidence to show that two-pack-a-day smokers were dying about seven years earlier than nonsmokers.[33] Doll and Peto argued in 1981 that 30 percent of *all* U.S. cancer deaths were due to tobacco, and a 1985 study by R. T. Ravenholt of World Health Surveys in Seattle suggested that the figure might be as high as 40 percent.[34] The British Royal College of Physicians has characterized the annual death toll caused by cigarettes as "the present holocaust":[35] on a global scale, tobacco has probably killed more people than have died in every war ever fought. The figure could easily be in excess of 100 million.[36]

Following the widely publicized studies of the 1950s that linked smoking and cancer, industry officials launched efforts to "study" the problem. In 1954, the Tobacco Industry Research Committee (rechristened the Council for Tobacco Research in 1964) was established by tobacco manufacturers, growers, and warehousers to promote research by "independent scientists into tobacco use and health."[37] Full-page ads in 448 newspapers reaching 43 million Americans announced the creation of the council, along with its promise to support impartial research and to "let the results speak for themselves." The council's research budget grew from about $1 million per year in the early 1960s to $5 million per year in 1978 and nearly $20 million in 1994. Since the late 1950s, the council has spent more than $240 million on research, resulting in the publication of some 5,000 scientific papers. Grantees include more than a thousand scientists and three Nobel laureates; three members of the Council's Scientific Advisory Board are also members of the prestigious National Academy of Sciences.[38]

Tobacco industry research has long been ridiculed as biased, and it is not hard to see why. For nearly forty years, industry spokesmen have charged that the evidence linking cancer and cigarettes is "merely statistical," that it is "premature" to accept the "causal theory" linking smoking and cancer. The Tobacco Institute, formed in 1958 as an offshoot of the Tobacco Research Council, questions "conventional wisdom" by focusing attention on a series of apparent "mysteries," the rhetorical force of which is to cast doubts on the connection between smoking and disease. Why, the institute asks, does lung cancer rarely occur in both lungs simultaneously, when both are exposed equally to smoke? Why does lung cancer often occur "in the parts of lung *least* exposed to smoke"? Why is cancer of the trachea (windpipe) rare, even though this part of the body "is exposed to *all smoke* going into and out of the lungs"? Why do nonsmokers get a kind of lung cancer identical to that of smokers? Of course, there are good answers to most of these queries;* the point in each case, though, is to insinuate doubt, to reassure smokers, and to stave off regulation through a combination of wishful thinking, nonsequiturs, and faulty logic.

The Tobacco Institute has always drawn freely—and rather creatively—on a mixture of scientific fact and fiction to bolster its case for cigarette safety. In 1983, for example, during congressional debate of a bill requiring the rotation of warnings on cigarette packages, the institute produced a document asserting that the "vast majority" of smokers do not get lung cancer (true); that perinatal problems—low infant birth weight, for example—are correlated not with smoking but with factors such as "nutritional and economic status" (false); that smoking cannot explain the "ethnic and geographic patterns" of lung cancer mortality (false); that studies linking smoking and disease do not adequately control for variables such as "lifestyle, alcohol consumption, occupational and environmental exposures, genetics, aging, and the immune process" (generally false); that animal studies cannot provide a reliable measure of human risks (a half-truth); that some animal studies show that smokers actually live longer than nonsmokers (true but misleading); that lung cancer early in

*It is not surprising that lung cancer would rarely occur in both lungs simultaneously, since cancer has a long latency period and people do not usually live long enough to develop cancer in both lungs. Cancer of the trachea is rare because the epithelial tissues lining that organ do not trap smoke particles to the extent that the inner tissues of the lung do. Nonsmokers can get a kind of cancer identical to that of smokers because things other than tobacco smoke do cause cancer—though not nearly to the extent that tobacco does. And so forth.

the century was *under*diagnosed while recent lung cancers tend to be *over*diagnosed (true but beside the point); that many lung cancers spread to the lungs from other sites, "hiding the true incidence of cancers originating elsewhere in the body" (perhaps true); and that epidemiological studies do not adequately take into account the fact that smokers are a "self-selected group"—and therefore different from other people (perhaps true, perhaps not).[39]

This last point, about smokers being unique, deserves some elaboration. Tobacco Institute propaganda has long sought to portray smokers as either macho, virile, and rugged or slim, sexy, and liberated. Tobacco ads show smokers as nonconformist risk takers who scoff at the square advice of experts. A recent ad in *Time* magazine links the antismoking personality to classical music and the smoking personality to jazz.[40] This emphasis on the specialness of the individual smoker also explains why the industry has for so long interested itself in genetics. Clarence Cook Little, director of the Council for Tobacco Research until 1971, used to argue—echoing the statistician-geneticist Ronald A. Fisher—that there might be a gene or personality type that both leads one to smoke and predisposes one toward cancer.[41] Epidemiologists sometimes call this the "itch in the lung" hypothesis.

The legal and public relations value of such studies is obvious: if differential cancer susceptibilities could ever be established, one could plausibly argue that people who come down with the disease have at least partly their own heredity to blame. Perhaps many of those who smoke are invulnerable: after all, only about one in five smokers ever gets lung cancer—why doesn't everyone? Recent work at the Tobacco Institute has centered on efforts to dispute evidence of the hazards of "passive," "secondhand," "sidestream," or (as the industry prefers to describe it) "environmental tobacco" smoke. Strong evidence of the dangers of breathing other people's smoke emerged in 1980, when a study of 40,000 members of the Seventh-Day Adventist Church, whose adherents rarely smoke, showed that the nonsmokers among them had only half the risk of getting lung cancer as other nonsmokers. One year later, Tokyo's National Cancer Center Research Institute showed that the wives of smokers were twice as likely to contract lung cancer as the wives of nonsmokers.[42] The industry has responded that "poor ventilation" and what it calls "sick building syndrome" (including allergenic fungus, bacteria, and invisible gases and fumes—but not tobacco smoke) are the more common causes of indoor air pollution.[43] The industry has also responded by bullying investigators seeking evidence of tobacco hazards, accusing them of being not reputable scientists but "long-time, highly vocal antismoking activists."[44]

The question of environmental tobacco smoke has become a major concern of the tobacco lobby, primarily because it has emerged as one of the strongest reasons for courts to uphold the right of employers to prohibit smoking at the workplace and for governments to bar smoking in public buildings. In 1970 there were ten U.S. bills introduced by local governments to restrict smoking in public spaces; by 1980 there were nearly two-hundred (many modeled on Minnesota's Clean Indoor Air Act of 1975). The tobacco industry managed to defeat a 1978 California bill (Proposition 5) calling for restrictions on smoking in public spaces, but only after spending $6 million on ads to overcome three-to-one support for the bill. Over the next several years, bills limiting smoking were defeated in several other states; victory was bought with money channeled through industry-sponsored "citizens' councils" with names like Californians for Common Sense, Californians Against Regulatory Excess, Dade Voters for Free Choice, and Floridians Against Increased Regulation.[45]

With the nonsmokers' rights movement steadily growing, and with evidence on the dangers of secondhand smoke mounting, legal support for the industry's position has been eroding. In 1990, according to Tobacco Institute figures, 20 percent of all companies barred smoking on the job. The Pentagon announced a ban on smoking in all Department of Defense workplaces in March 1994, and it is probably only a matter of time before OSHA moves to regulate smoking on the job more generally. You can be sure, though, that creative ploys will be devised to dismiss the threat. In December 1991, for example, PBS-TV's *Frontline* featured an unruffled Philip Morris spokesman arguing that birdkeeping was six times more likely to cause cancer than sidestream smoke (birds apparently release cancer-causing mites when they fluff their feathers). In 1993, when the EPA released its report showing that secondhand smoke was responsible for 20 percent of all lung cancer deaths among nonsmokers (roughly 3,000 fatalities per year), the Tobacco Institute derided the study as "another step in a long process characterized by a preference for political correctness over sound science."[46]

The Tobacco Institute's chairman, George Allen, once claimed: "We are not on a crusade either for or against tobacco. . . . If we have a crusade, it is a crusade for research." Anyone with even a cursory familiarity with institute literature, however, will find such claims rather fabulous. Tobacco Institute propaganda suggests that scientists remain divided over the question of cigarettes and cancer, when in fact no stronger consensus exists in all of medicine. Questioned privately, even industry-supported scientists concede a hazard: in a recent survey of scientists receiving funding from the Council for Tobacco Research (CTR), more than 90 percent of the respondents admitted that cigarettes are addictive, that secondhand smoke is dangerous, and that "most deaths from lung cancer are caused by smoking."[47] Industry hypocrisy is also evident in the fact that the three major life insurance companies owned by tobacco companies charge smokers almost twice as much for life insurance as nonsmokers. Insurance companies apparently recognize what the tobacco industry won't: at any given age, smokers are about twice as likely to die as nonsmokers.[48]

The Tobacco Institute's elaborate efforts to deceive are, by now, notorious. More than twenty-five years ago, a *New York Times* editorial characterized the industry's challenge to the evidence of tobacco's dangers as having "the hollowness of a cough in a grave yard."[49] In the case of the CTR, however, the bias is often subtle, resting not typically in the falsification of data or even their misrepresentation but rather in the diversion of attention away from the *causes* of cancer and toward the *mechanisms* by which cancer arises. The overwhelming bulk of CTR support goes to genetics, biochemistry, or cell physiology—none of which tells you very much about the causes of cancer or what might be done to prevent it. A reader of CTR-funded research abstracts will find little to suggest that tobacco is the single largest cause of cancer in the United States, or that smoking kills even more people by heart attacks than by cancer. Epidemiology is rarely supported by the council.[50] The council has never funded research on how to prevent tobacco use or to stop smoking. A survey of principal investigators receiving CTR grants in 1989 showed that 80 percent of the respondents had never examined the health effects of smoking and were not doing so as part of their current research.[51]

None of this should come as a surprise to anyone who appreciates the political clout of the industry. Recall that the American Medical Association itself refused to concede a link between smoking and cancer throughout the 1950s: as James Patterson points out in his history of this period, the AMA apparently felt that to make such a concession would antagonize political allies supporting its fight against national health insurance. The nation's foremost professional medical association found itself allied with what critics called the Smoke Ring, the conservative protobacco political circle embracing southern Democrats and most Republicans.[52] The AMA's pension fund held 1.4 million dollars' worth of stock in R. J. Reynolds and Philip Morris until 1981, when pressure from some of its members compelled the association to divest. In the early 1990s, the AMA was still giving more money to congressmen who supported tobacco exports than to congressmen who opposed them.[53]

Governmental health institutions have not done all they could to help, either. The surgeon general's report of 1964 was important in alerting the public to the fact that cigarettes "contributed substantially" to lung cancer and heart disease,[54] but it was not accompanied by an effective antitobacco policy. Even the National Cancer Institute, formed in 1937 to coordinate the nation's anticancer effort, was surprisingly slow to take on the tobacco lobby. The NCI's annual plan for 1977–81 failed even to mention tobacco or cigarette smoking in its discussion of the origins and impact of cancer.[55] That oversight has since been corrected, but government intervention is still not what it could and should be. The surgeon general's Office on Smoking and Health in 1994 received only about $20 million per year (about double what it received, in real terms, when it was founded in 1966)—hardly the funding one would expect to battle the nation's number-one health hazard. As recently as 1991, the figure was only $3.5 million.

One interesting aspect of the tobacco hazard is that it has been used by different political groups for very different purposes. Some environmentalists have tended to downplay the hazard, on the grounds that excessive attention to smoking has drawn attention away from chemical and occupational hazards. Wilhelm Hueper made this argument,[56] as did Samuel Epstein. Susan Stranahan, in an otherwise informed essay in the *Philadelphia Inquirer* on industrial carcinogenesis, cited unnamed experts from the National Cancer Institute in support of her contention that "Although cigaret smoking is a better-known cause of cancer, chemicals far surpass it as a source of cancer."[57]

Tobacco industry advocates have been understandably receptive to this notion.[58] Joseph Califano's estimate that 20 to 40 percent of all cancers are occupation-related has been favorably cited by Tobacco Institute publications—a curious alliance, given that Califano himself was dismissed as secretary of Health, Education, and Welfare by President Carter in 1979, allegedly "acting on behest of tobacco industry friends" incensed by the health secretary's characterization of smoking as "slow motion suicide" and "Public Health Enemy Number One."[59] The CTR established a secret Special Projects Unit in the 1960s to conduct research favorable to the industry; Theodor D. Sterling and his associates at Simon Fraser University, for example, received $3.7 million in support of his research showing that occupational pollutants were more often important causes of cancer than smoking.[60] Hueper himself was reportedly offered a $250,000 job at the Tobacco Institute! He refused, explaining that he and his scientific talents were not for sale.[61]

The primary aim of tobacco industry science-as-propaganda is to produce a semblance of a need for "balance" in the "debate." Doubts are raised about established medical opinion. Scientific knowledge is by its very nature supposed to be provisional and therefore open to doubt; industry authorities have turned this self-professed tentativeness to their own advantage, arguing that ever more research is needed to clarify ambiguities and remove residual doubts. Of course, the industry call for ever more research is insatiable. It is insatiable because it serves an important social function: research buys time. The call for more research implies that scientific opinion is in flux, that important questions remain unresolved.

But tobacco industry officials do not just exploit ambiguity; they manufacture it. One reason tobacco industry spokespersons can point to a divergence of scientific opinion on the safety of environmental smoke is because the production of uncertainty is one of the primary goals of such institutes. This was privately admitted in an internal document produced by Brown and Williamson (a cigarette company) and subpoenaed by the Federal Trade Commission in 1969, which reads: "Doubt is our product since it is the best means of competing with the 'body of fact' that exists in the mind of the general public. It is also the means of establishing a controversy. If we are successful at establishing a controversy at the public level, then there is an opportunity to put across the real facts about smoking and health." Stanton A. Glantz of the University of California at San Francisco, a long-standing tobacco critic, has described this strategy as an attempt "to jam the scientific airwaves with noise."[62] The scientific search for clarity is twisted into a pseudoscientific celebration of doubt. H. Lee Sarokin, the federal judge dismissed from tobacco liability cases for his alleged anti-industry bias, recognized this when he characterized the industry as "the king of concealment and disinformation."[63]

ASBESTOS MISINFORMATION

The tobacco industry's efforts to manipulate scientific research and public opinion may well constitute "the longest-running misinformation campaign in U.S. business history,"[64] but there are other industries with equally impressive pedigrees of obfuscation. Asbestos is probably the second most fatal environmental carcinogen, after tobacco, and in defense of it, too, we see an extraordinary display of talent in the arts of constructing and disseminating ignorance.

Asbestos is a naturally occurring mineral with two remarkable properties: its silky, silica-rich fibers can be spun into a very fine thread from which can be woven fabrics that are virtually impervious to flame and fire (the Greek *asbestos* means inextinguishable, unquenchable). Charlemagne is said to have entertained guests on a tablecloth of asbestos that, after dinner, he would throw into the fire and later withdraw, clean and intact. Royal bodies were burned while wrapped in asbestos cloths, to prevent their ashes from scattering. By the Middle Ages, asbestos had already become known as the "magic mineral" and the "funeral dress of kings." A late-nineteenth-century commentator praised the mineral as "one of the most marvelous productions of inorganic nature," a "physical paradox" sharing properties of both the mineral and the vegetable kingdoms.[65]

Asbestos began to be used on an industrial scale in the nineteenth century, first as a packing material to tighten the seals of steam engine pistons and later as an insulating material to reduce heat loss from engines and steampipes. Asbestos insulation was especially important in steamships, because the retention of heat allowed them to conserve fuel and therefore travel further on a given coal capacity. Asbestos was also used in fireproof paints, roofing materials (the first formula for which was patented by Henry Ward Johns in 1868), cement, asphalt, wallboard, and, with the development of the automobile, brake-shoe linings and clutch facings. The explosive growth of steam-driven machinery led to a boom in the asbestos trade, and by the end of the nineteenth century the H. W. Johns Manufacturing Company (later bought out by the Manville Covering Company of Milwaukee to form the Johns-Manville Company) was the largest manufacturer in the world, with factories in New York, Chicago, and Philadelphia. Johns-Manville eventually acquired its own mines in Canada, expanding these to the point where the industrial giant was quarrying 40 percent of all the asbestos mined in that country.[66] Worldwide production continued to grow, from about 200,000 tons per year in 1920 to 40 million tons in 1970.

In today's climate of "fiber phobia," it is easy to forget that safety was one of the original reasons asbestos was pushed as a substitute for other kinds of building materials. Fires from cinders landing on wood shingle roofs claimed many lives in the early years of this century, and the switch to asbestos-impregnated substitutes was hailed as a victory for human health and safety. As an insulating material, too, asbestos was commonly believed to be a marked improvement over previous materials. In 1891, for example, it was pointed out that the mixture of cotton fibers and sodium silicate used for boiler insulation in the British Navy could give rise to attacks of respiratory irritation. Asbestos replaced cotton silicate, the presumption being that the new material would be less irritating.[67] Even today, there are those who suggest that asbestos cement piping is "the cheapest and most convenient method" of delivering water and disposing of sewage in developing nations; the lives thereby saved, according to this argument, outweigh the risks incurred by asbestos manufacture and transport.[68] Speculations have been put forward that the disaster of the space shuttle Challenger might have been averted if an asbestos-based putty had been used to seal joints in the booster rockets.[69]

As marvelous as the material is, it has also long been known to injure and kill those who handle it. Pliny the Elder in the first century A.D. recorded the fact that slaves weaving asbestos into cloth suffered from a sickness of the lungs, and many other commentators noted the corrosive "asthmas," "cancres," and "consumption" produced by working with the fiber. British health inspectors investigating the "dusty trades" in the nineteenth century noted the hazards of the material, and an 1898 British report on women's working conditions described the "insidious effects" of asbestos and other dusts in textile plants. That same year Henry Ward Johns, the entrepreneurial spirit behind the Johns-Manville Company, died from "scarring of the lungs"—probably caused by asbestos. In 1907, a British physician reported the first autopsy-confirmed death caused by asbestos: the man's lungs were heavily scarred; he had worked for fourteen years in the carding room of an asbestos factory and claimed to have lasted the longest among eleven people who had worked there since he began. All the others had died around the age of thirty.[70]

By 1918, the hazards of asbestos were sufficiently well known that insurance companies commonly refused to issue policies to workers who inhaled the fibers.[71] In 1924, a British physician reported that a young woman employed in a British asbestos factory had died from pulmonary fibrosis.[72] Three years later, in an effort to clarify her case, the term *asbestosis* was coined to designate the scarring of the lungs and impairment of breathing increasingly found among asbestos miners and others handling the fiber. (In the nineteenth century, dust-induced lung diseases had simply been lumped together under the rubric of "grinder's rot," "potter's rot," "miner's phthisis," and so forth.)

U.S. public concerns about dust diseases culminated in the 1930s, especially in the wake of the so-called Hawk's Nest incident at Gauley Bridge, West Virginia, where as many as a thousand workers (the exact number is not known; most were out-of-work black coal miners from the South) died after inhaling high levels of silica (quartz) dust* during the building of a tunnel for a Union Carbide hydroelectric power plant. As the historian Martin Cherniack has shown, the tragedy was primarily the result of the company's disregard for workers' health and safety. The Great Depression had made men desperate for work, and Union Carbide took advantage of this to circumvent accepted safety practices. The three thousand men who worked inside the tunnel were told to forgo wet drilling (using water to keep down dust) in order to speed up operations. Dry drilling meant that the miners were forced to breathe very high levels of dust. Many of those who succumbed to acute silicosis died after only a few weeks of drilling. Many of those who survived developed scarred lungs and a diminished resistance to pneumonia and tuberculosis. (Company physicians commonly told the men they had "tunnelitis" and prescribed a pseudomedication popularly known as "little black devils"—tablets of baking soda coated with sugar.) The case, celebrated in literature and the popular press as "America's worst industrial disaster," led to hundreds of millions of dollars in lawsuits and increased awareness of the hazards of industrial dust.[73]

Fearing the labor activism and public outrage already associated with silicosis, American asbestos manufacturers commissioned a study of the health effects of as-

---

*Silica dust is essentially powdered quartz: silicon dioxide. Asbestos is more properly defined in economic terms, as any mineral whose component particles have a length equal to or greater than three times their width. Silica and asbestos dust both cause fibrosis (scarring) of the lungs, though asbestos also causes cancer. It is not yet clear whether silica dust causes cancer; see the proceedings of the Second International Symposium on Silica, Silicosis and Cancer, published in the *Scandinavian Journal of Work, Environment and Health,* Suppl. 1 (1995).

bestos in the late 1920s. Johns-Manville and Raybestos Manhattan (the nation's leading manufacturer of brake-shoe pads and clutch facings) contracted with Anthony J. Lanza, assistant medical director of the Metropolitan Life Insurance Company, to investigate asbestos textile workers' health. By 1931, Lanza had drafted a report that must have shocked his patrons: 43 percent of workers with under five years' experience showed scarring of the lungs. For workers with over fifteen years' experience, the figure rose to 87 percent. Lanza did not publish his results until 1935, and only after Johns-Manville attorneys had carefully reviewed his manuscript, prompting its author to soften several of his conclusions. For example, Lanza included at the manufacturer's suggestion a statement that the health effects of asbestos were "milder" than those produced by silica dust. Silicosis was a hot political potato, and the industry was wary of anything that might link the two diseases. Lanza was well aware by this time that asbestotics were dying at a younger age than silicotics, but he apparently still felt justified in giving industry this "break."[74]

The second major U.S. asbestos study was a 1938 project commissioned by the North Carolina State Bureau of Health and the administrators of that state's Workmen's Compensation Act. The primary goal was to determine the maximum level of asbestos that could be inhaled without harm. Waldemar Dreessen, a U.S. Public Health Service physician, measured the levels of airborne dust in terms of millions of particles per cubic foot (mppcf), and multiplied this by the number of years of working experience to get an index of cumulative exposure (measured in terms of "particle-years") for each of the 541 workers at three asbestos textile plants. Workers were also X-rayed to determine which showed fibrosis of the lung. Dreessen found asbestosis in about half the workers exposed to more than 100 million particle-years and 20 percent of those exposed to 50 million to 99 million particle-years. None of the thirty-nine workers exposed to less than 2.5 mppcf for up to five years appeared to be affected. Nonetheless, Dreessen recommended a provisional safety "threshold concentration" of 5 mppcf, despite evidence presented in his own report that 5 of 192 workers exposed to 25 million particle-years (that is, for five years at his standard exposure) showed signs of asbestosis. As David Ozonoff has shown, this threshold value was settled on partly because it could be fairly easily achieved with contemporary dust-control technologies.[75]

Had Dreessen's admittedly high standard of 5 mppcf been rigorously adhered to, thousands of lives might have been saved. As it was, many asbestos workers ended up breathing much higher concentrations—often with no knowledge of the danger.[76] The most serious exposures were in the shipbuilding industry. Since the nineteenth century, asbestos had been used in ships to insulate bulkheads, boiler pipes, and (eventually) electrical lines. The 1921 Washington Treaty established limits on total naval tonnage among the Great Powers, one consequence of which was a shift from chrysotile to the lighter form of asbestos known as amosite. Amosite asbestos was more easily airborne and dispersed throughout a ship, posing a greater health hazard. The rapid expansion of the navy in World War II put a premium on production: shipyard workers' health was occasionally monitored, but turnover was great and little was done to document the long-term health effects of asbestos—or anything else, for that matter. Subsequent studies would confirm the suspicions of the 1930s that the fiber could do its deadly work long after one's exposures to it: once inhaled, many of the long, sharp, fibers remain permanently in the lungs, producing more and more scar tissue until death. The practice of setting asbestos-inhalation standards in the millions of fibers per cubic foot paid insufficient attention to this delayed, cumulative effect.

Too little attention was also paid to the fact that asbestos could cause not just scarring of the lungs but cancer. Medical evidence of a cancer risk among asbestos workers began to emerge in the 1930s.[77] In 1936 E. L. Middleton found three lung cancers among fifty-four cases of asbestosis reported to English inspectors, and in 1938 three German publications documented cancer among asbestos workers.[78] Wilhelm Hueper concluded in his 1942 *Occupational Tumors and Allied Diseases* that there was clear evidence of "excessive" lung cancers among victims of asbestosis, and in 1943 the German government recognized asbestos-induced lung cancer as a compensable occupational disease.[79]

U.S. industrial laboratories had begun to investigate the cancer hazard in 1936, pooling resources to support a series of studies conducted by Leroy Gardner and others at the Saranac Laboratory in upstate New York. Johns-Manville was the major funder. By 1938, two Saranac researchers were in a position to suggest that persons suffering from asbestosis might simply be especially susceptible to lung cancer.[80] Gardner himself, however, was more open-minded. In 1943, frustrated with his contractual obligation to clear his work with industry authorities prior to publication, he applied for funds from the National Advisory Cancer Council to begin a series of studies to determine the extent to which asbestos caused cancer. He had already been surprised to find, in a preliminary study, that eight of eleven mice developed lung cancer after breathing asbestos dust for fifteen to twenty-four months. The council turned down his request, however, and Gardner died shortly thereafter, distressed by his inability to publish freely. His successor at Saranac, Arthur J. Vorwald, prepared a report incorporating Gardner's data, but the sections linking asbestos to cancer were deleted from the published version at the request of the Asbestos Textile Institute, an industry association whose membership included most of the nation's leading asbestos manufacturers.[81]

In 1949, E. R. A. Merewether published a study for Britain's inspector of factories showing that, among 235 workers known to have died of asbestosis, more than 13 percent had lung cancer. *Scientific American* carried an article listing asbestos as a cause of cancer (citing Hueper), and a 1949 editorial in the *Journal of the American Medical Association* (unsigned but probably written by Hueper) endorsed this thesis. The Asbestos Textile Institute was understandably distressed, and at a March 7, 1956, meeting its board of governors entertained a proposal to mount a public relations campaign "to counteract the unfavorable publicity presently directed to the asbestos industries as a result of the work of Dr. Hueper." The institute had already rejected a plan by Gerrit Schepers of the Saranac Laboratory to examine tissue specimens from dead industry employees, justifying its decision on the grounds that such a study would imply that "a relationship existed between asbestosis and carcinogenic development, a condition which, to date, has not been established." Subsequent proposals to explore the relation between cancer and asbestos were refused on

the grounds that such studies "would stir up a hornet's nest and put the whole industry under suspicion."[82] Asbestos manufacturers had already privately conceded a
cancer hazard: in 1950, the medical director of Canadian Johns-Manville had written
to the president of the company cautioning that "there seems to be increasing proof
that asbestos fibers do cause lung cancer. . . . this whole subject could cause our
companies unlimited embarrassment and untold expense if labor leaders made use
. . . of the subject."[83]

Prompted by adverse media attention and the prospect of injury lawsuits, asbestos
industry associations began a series of further investigations to document the hazard,
and presumably also to document their concern to document that hazard. The most
important was a 1957 study of 6,000 Canadian chrysotile miners funded by the Quebec Asbestos Mining Association (QAMA) and the Industrial Hygiene Foundation.
What is interesting about this study is how its conclusions were manipulated to
slight the cancer hazard. Its author, Dr. Daniel Braun, medical director of the Industrial Hygiene Foundation, stated in the original report, printed for industry use but
not widely circulated (the cover is marked "confidential"), that 12 percent of those
suffering from asbestosis also showed signs of lung cancer—a very high rate. Braun
concluded that a miner who develops asbestosis "does have a greater likelihood of
developing cancer of the lung."[84] In an extraordinary act of (self-?) censorship, however, the concluding passages on asbestos and lung cancer were deleted from the
condensed version published in the American Medical Association's *Archives of Industrial Health* in 1958.[85] Johns-Manville's medical director, Kenneth W. Smith,
wrote to the general counsel of QAMA concerning the alteration:

> We have noted the deletion of all references to the association of asbestosis and
> lung cancer in this condensation. While we believe that this information is of
> great scientific value, we can understand the desire of Q.A.M.A to emphasize the
> exposure of the asbestos miner and not the cases of asbestosis. . . . It must be rec-
> ognized, however, that this report will be subjected to criticism when published
> because all other authors today correlate lung cancer and cases of asbestosis.[86]

Ivan Sabourin, chief counsel for Canadian Johns-Manville, later commented on
the "wonderful public relations value" of the document,[87] though over at the NCI
Wilhelm Hueper was less favorably impressed. He denounced the study as "statistical acrobatics" designed to obscure incriminating evidence by use of "a highly biased population group as a 'normal' standard." (Cancer rates among the largely rural
mining population had been compared with rates among the urban population,
whose overall cancer rates were one-third higher.) Hueper characterized the study as
"defective and biased," a "socially manufactured scientific merchandise of shoddy
quality."[88]

Hueper's opinion was confirmed by a growing number of epidemiological studies
nailing down the asbestos-cancer link. Richard Doll in 1955 showed that British asbestos workers were suffering high rates of lung cancer, despite efforts by that nation's largest manufacturer, Turner Brothers Asbestos Company, to suppress the
paper.[89] Finnish scientists in 1960 showed that a surprising fraction of farmers—not

miners—living near an asbestos mine in Finland had asbestos bodies in their lungs.[90] J. C. Wagner that same year found thirty-three cases of mesothelioma—a rare tumor of the lining of the chest and abdominal cavities now thought to be almost invariably associated with asbestos—in a crocidolite mining region of South Africa. All but one of the victims had been involved in either the mining or the transport of asbestos, or had some other form of exposure to the mineral.[91]

In 1963, Thomas Mancuso published the results of a long-term study in which he found that lung cancer, mesothelioma, and asbestosis rates were well above the national average among 1,495 American employees of a company manufacturing brake linings.[92] One year later, Irving Selikoff of Mount Sinai Hospital in Manhattan showed that, among 632 insulation workers exposed to asbestos dust for twenty years or longer, lung cancer death rates were about seven times higher than for unexposed populations of comparable age. Stomach, colon, and rectal cancer rates were also elevated, as were mesothelioma rates and, of course, death by asbestosis. Selikoff's study did not control for smoking, but even if one assumed that *all* of these men had smoked a pack a day, the lung cancer rates were still more than three times what one would expect for such a group.[93] Paul Brodeur later popularized Selikoff's work in a widely read series of articles for *The New Yorker;* Brodeur expressed the outrage of many other critics when he suggested that the industry had treated its workforce as "Expendable Americans."[94]

Victims turned to the courts for redress, and by the middle of the 1970s injury claims against asbestos manufacturers exceeded those for any previous industrial malfeasance. In 1974, former employees of a Tyler, Texas, asbestos textile plant filed a $100 million class-action suit against PPG Industries, the Asbestos Textile Institute, the Industrial Health Foundation, and other interested parties for failing to protect the plant's employees against asbestos dust.[95] In 1978, 5,000 California shipworkers filed a class-action suit against fifteen asbestos manufacturers, accusing them of conspiring to withhold and distort evidence of the hazard. That same year, HEW Secretary Joseph Califano warned that half of the 8 million to 10 million workers exposed to asbestos since the start of World War II could develop cancer or asbestosis. Califano instructed the surgeon general to send a "physician advisory letter" to each of the nation's 400,000 doctors warning them to be on the lookout for asbestos injuries; he also established an Asbestos Education Task Force to mount a public education program. Hundreds of television and radio stations broadcast a public service announcement (aired 60,000 times, according to one estimate) to alert the public to the hazard. That fall, 30 million Social Security beneficiaries looked in their mailboxes and found, along with their monthly checks, fliers with asbestos warnings.[96]

Outrage against asbestos peaked in the early 1980s when the Manville Corporation, inundated by lawsuits, filed for bankruptcy. In full-page ads taken out in the *New York Times,* the *Wall Street Journal,* and the *Washington Post,* the president of the company, John A. McKinney, explained the decision by noting that 16,500 lawsuits had already been filed against the company and that new claims were being entered at a rate of about 500 a month. The industry could not be regarded as negligent, the ads protested, since "not until 1964 was it known that excessive exposure to asbestos fiber released from asbestos-containing insulation products can sometimes cause certain lung diseases."[97] Public confidence in the industrial giant was shaken: its stock plummeted as trial lawyers geared up for action. A January 1985 poll by *Fortune* magazine ranked Manville last among 250 companies in terms of environmental responsibility, and third from last in terms of public confidence and admiration.[98]

The key question in most asbestos-injury compensation trials of the 1970s and '80s was whether asbestos manufacturers did indeed have and withhold from their employees, knowledge of the hazards of the mineral. In court, industry representatives repeatedly claimed that no one knew asbestos products were dangerous prior to 1964, when the New York Academy of Sciences organized an international meeting on the topic.[99] Industry attorneys relied heavily on the Braun-QAMA study for proof of both the safety of the fiber and the sincerity of the industry's claim to have honestly investigated potential hazards. Scholars working for the plaintiffs, though, were able to show that the danger was well known to industry officials prior to 1964. At the San Francisco hearings of a HEW congressional subcommittee on Compensation, Health, and Safety, papers were produced documenting industry knowledge of the hazard as early as the 1930s. Based on his reading of these so-called Asbestos Pentagon Papers, James Price, a South Carolina circuit court judge, charged Johns-Manville and another company, Raybestos-Manhattan, with downplaying or even suppressing publication of studies linking asbestos and cancer in order to forestall lawsuits from ailing employees.[100]

In subsequent proceedings, documents came to light showing that the industry had recognized the hazard, covered it up, and then sought to use the purported recency of its recognition to defend itself against injury claims.* As early as December 1934, for example, a Johns-Manville attorney had written to the company's general counsel, conceding with regard to injury claims that "one of our principal defenses . . . has been that the scientific and medical knowledge has been insufficient until a very recent period to place upon the owners of plants or factories the burden or duty of taking special precautions against the possible onset of the disease to their employees."[101] Industry attorneys repeated this same claim fifty years later, though by this time prosecutors were able to show clear and early evidence of the industry's knowledge of the hazard.

---

*David Ozonoff, a frequent expert witness for the plaintiff in asbestos trials, notes that company attorneys have tended to use a series of time-consuming retreats to defend their clients, typically running something like the following: "Asbestos doesn't hurt your health. OK, it does hurt your health, but it doesn't cause cancer. OK, asbestos can cause cancer, but not our kind of asbestos. OK, our kind of asbestos can cause cancer, but not the kind of cancer this person got. OK, our kind of asbestos can cause this kind of cancer, but not at the doses to which this person was exposed. OK, asbestos does cause this kind of cancer, and at this dosage, but this person got his disease from something else—like smoking. OK, he was exposed to our asbestos and it did cause his cancer, but we didn't know about the danger when we exposed him. OK, we knew about the danger when we exposed him, but the statute of limitations has run out. OK, the statute of limitations hasn't run out, but if we're guilty we'll go out of business and everyone will be worse off. OK, we'll agree to go out of business, but only if you let us keep part of our company intact, and only if you limit our liability for the harms we have caused" (personal communication, November 1992).

David Ozonoff of Boston University eventually catalogued more than 700 studies published before 1964 indicating a health hazard of the mineral.[102] Industry lawyers were also shown to have been something other than honest in their dealings with the public. In 1990, at the trial of *Richard Kulzer v. Pittsburgh Corning Corporation,* a confidential memo of January 12, 1979, to John A. McKinney, president of Johns-Manville, was introduced by the plaintiffs' attorneys, counterpoising the Johns-Manville (JM) position against the true facts, among them:

JM position: Prior to 1964, we lacked scientific knowledge about cancer hazards
               of asbestos.
FACT: In the mid-1950s, JM officials knew of scientific studies showing a
               relationship between cancer and asbestos.

JM position: Though sprayed asbestos inappropriate for use in schools, we see no
               evidence of danger.
FACT: If no problem, then why an inappropriate use?

JM position: There will be no future problems with workers.
FACT: Workers in Mexico are not being protected.

JM position: Since 1964, JM has "come clean" with employees about asbestos
               hazards.
FACT: A current 41-year employee of JM has called me to say that he has
               a spot on his lung . . . but no one at JM has ever communicated
               with him about it.

JM position: There was no cover up.
FACT: A current employee . . . says his father worked for JM for 35 years at
               Manville and was never told about dangers. Father was dismissed
               because of "miners' asthma" . . . he had to take JM to court to get
               compensation for an asbestos-related disease.

JM position: We have communicated openly with employees about illnesses.
FACT: . . . it was company practice into the early 70s not to tell a person
               about his illness.

These were extraordinary concessions, coming as they did from internal industry documents. The memo concluded that it was not hard to see "why we have members of congress calling us 'liars.'"[103]

In recent years, asbestos debates have taken a somewhat different tack, centering on the question of whether the general public is at risk from asbestos insulation in homes, schools, and commercial buildings. In the 1970s, Irving Selikoff showed that working in an asbestos factory for even one month was enough to raise one's likelihood of succumbing to lung cancer. A woman who pushed a tea cart through an asbestos plant for only an hour a week for one year contracted mesothelioma, as did a woman who worked in an office two floors below where asbestos cement panels were being sawed up.[104] If exposures as brief as these are hazardous, then it is likely that many people exposed outside the workplace may be at risk. This is worrisome, given that asbestos has been used in an estimated 3,000 products, including roof shingles, reinforced concrete, brake linings, clutch facings, floor tiles and linoleum, electric wire casing, ironing board covers, many kinds of fireproof cloth and clothing, and even in gas-mask and cigarette filters. (The much-celebrated "micronite filter" in Kent cigarettes, marketed by Lorillard in the 1950s as offering "the greatest health protection in cigarette history," contained crocidolite, a long-fiber form of the "amphibole" group of asbestos minerals notorious for its ability to cause cancer.)[105] Asbestos is naturally found in talcum powder, leading some physicians to advise against breathing in the dust or applying it near a baby's rectum.[106] Asbestos is used to filter beer, wine, liquor, and soft drinks. In the town of Asbestos, Quebec, home to one of the largest mines in that province, balls of asbestos fibers rolled down streets in the 1940s "like tumbleweed," and nearby homes were coated with the dust. The same was true in Manville, New Jersey, where for many years Johns-Manville had its largest plant and headquarters. Today, nearly every American city has asbestos in its ambient air.[107] Hairdressers have contracted asbestosis from washing and cutting the hair of asbestos workers; the same is true of the wives and children of workers, some of whom came into contact with the fiber from their husbands' or fathers' clothing.[108]

The Environmental Protection Agency barred the manufacture of most asbestos products in 1989,[109] but millions of tons of asbestos insulation, tiling, piping, and other products still remain in U.S. schools, homes, and other structures. Much of this has begun to crumble and disperse, giving rise to new concerns that the hazards of the fiber are going to postdate its manufacture by many decades. Epidemiological interest has begun to shift from the miners, manufacturers, and installers of asbestos to the so-called third wave of asbestos disease suffered by people involved in the repair, renovation, and demolition of buildings containing some 30 million tons of asbestos installed in the 1940s, 1950s, and 1960s. (Residents and neighbors of such operations are also, of course, vulnerable.)[110] In 1986 Ronald Reagan signed into law the Asbestos Hazard Emergency Response Act, requiring that every school in the nation be inspected for asbestos. Schools found to have hazards were required to begin cleanup operations by July 1989. The cost of the operation is staggering: asbestos sprayed on for 25 cents per square foot in the early 1960s may cost $25 to $50 per square foot to be removed in the 1990s. The EPA reckons total costs of the school cleanup alone at $3 billion, but the National School Boards Association puts the figure over twice as high. Cleanup of commercial buildings may be even more costly: the Port Authority of New York and New Jersey is expected to pay up to $1 billion just to clean up the World Trade Center and LaGuardia Airport; an abatement industry magazine has estimated that total U.S. abatement costs could exceed $2 trillion over the next three decades.[111]

In the midst of these worries, some have argued that asbestos removal is a foolish idea, and not just because of costs. Michael Fumento, best known for his book on

The Myth of Heterosexual AIDS, has pointed out that abatement workers operating under pre-July 1986 safety standards can expect to suffer 64 lung cancers per 1,000 workers, and that workers operating under the stricter standards imposed after this time can still expect to suffer 6 excess lung cancers per 1,000 men. This presumes, of course, that abatement is done according to OSHA regulations, though horror stories abound of immigrant workers knowing little or no English being hired to rip out asbestos with inadequate or no protection. Fumento argues that the only one really to profit from the process may be the abatement industry, which took in $2.7 billion in revenues in 1988, up more than tenfold from only five years earlier. Cashing in on the hysteria, these "parasites," as Fumento calls them, have helped to whip up a frenzy that he has dubbed "Abategate."[112] The EPA in 1990 conceded that undisturbed asbestos poses less of a danger than previously supposed; this slowed the abatement industry's expansion, though a 1991 report in the *Dallas Business Journal* suggested that the industry would eventually stabilize at about $4 billion to $5 billion worth of business per year.[113]

Given the stakes in the asbestos business, it is not surprising that some continue to argue that the hazard has been exaggerated. Central to most such claims is that not all forms of the fiber are equally dangerous. Asbestos, after all, is not a single mineral but a loose class of magnesium silicates, distinguished from other silicates (agate or jade, for example) by its fibrous presentation. Chrysotile, or "white asbestos," was the most commonly used form in the United States, accounting for about 95 percent of all production. Its fibers are curly and often bundled, in contrast to the straighter, more needlelike, amphibole forms such as crocidolite, often said to be the most hazardous. Other asbestiform minerals include the already mentioned amosite; tremolite; anthophyllite; actinolite; and serpentine, a gem form of the mineral used in the lapidary and jewelry trades.

This variety of mineral forms has been seized upon by industry apologists to argue that there are "safe" and "dangerous" forms of the fiber. This argument goes back at least to the early 1960s, when industry advocates tried to argue that chrysotile was safe, while Blue Cape asbestos from South Africa (crocidolite) was not.[114] Recent supporters of this view have argued that chrysotile fibers are more soluble than amphibole forms; chrysotile also tends to be more readily expelled from the body—probably because its fibers are curlier. This same line of reasoning suggests that the straighter amphibole fibers penetrate more deeply into the lung, where they do more damage.[115] OSHA today regulates all asbestos according to a single standard, but the American Conference of Governmental Industrial Hygienists recommends a standard that is two-and-a-half times less stringent for chrysotile than for crocidolite, and English laws also stipulate tougher exposure standards for amphibole forms.[116]

It is not my goal here to question whether some kinds of asbestos are in fact more dangerous than others. What I would simply like to point out is that the distinction has been used by asbestos manufacturers and their ideological allies to argue that the health risks posed by the fiber have been overstated. In the summer of 1989, the *New England Journal of Medicine* published an article postulating that many of the lung cancer deaths among American asbestos workers have been caused by crocidolite and tremolite contaminants in chrysotile asbestos; Philip Abelson of *Science* magazine endorsed this so-called "amphibole hypothesis" (also known as the "two-fiber hypothesis") and editorialized that the public had been led to believe, wrongly, that asbestos in homes and schools poses a major public health hazard.[117] Real estate and building supply interests launched a public relations campaign opposing the cleanup; the centerpiece of this campaign was again the hypothesis that there are safe and dangerous types of asbestos. The Safe Buildings Alliance, representing real estate and asbestos manufacturing interests (e.g., W. R. Grace and U.S. Gypsum), has asserted, for example, that you cannot get mesothelioma from chrysotile. There are scientists who support such a claim and scientists who dispute it—the difference often determined by whom one is working for.[118]

The question has polarized the U.S. scientific community to the point where entire conferences are held where advocates of one or the other position hold sway and the opposing side does not even attend. A December 1988 conference at Harvard University was dominated by those who believe that chrysotile poses relatively little risk to the general public and that abatement is therefore unwarranted; *Forbes* magazine called the conference "a veritable bibliography of research showing that the low levels of asbestos present in most buildings offer little, if any, risk."[119] By contrast, a June 1990 conference in New York, attended mostly by followers of the Selikoff school (the opposing camp declined the invitation), repudiated the "Harvard" conference and stressed that asbestos still remains a major hazard. The central issue in dispute is whether chrysotile is sufficiently different from amosite, tremolite, or crocidolite to warrant separate regulatory procedures. The debate has become heated, as evidenced by a 1991 *Science* report describing asbestos research as "a world riven by deep fissures and bitter disputes."[120] Here, as in other disputes over the hazards of economically important materials, there are different conceptions of danger. Archibald Cox, the former Watergate prosecutor and now chairman of the board of the industry-financed institute that administered a $4 million review for the EPA,[121] cautions that "there's a grave danger of overgeneralizing about asbestos as if all hazards are the same." The danger, in his view, is that unwarranted steps will be taken to limit use or handling of the material. The Selikoff group, by contrast, worries more about gambling with human health should Archibald Cox prove wrong. The Cox report plays down the health risks of chrysotile; Selikoff's group points to studies showing similar mesothelioma rates in people exposed to chrysotile and to amosite. The Cox report states that very short fibers "have much less carcinogenic activity than longer fibers"; Selikoff and his colleagues insist that short fibers may cause even more cancer than long fibers—simply because they are so numerous.[122]

While the debate rages on, asbestos remains a major cause of cancer in the United States, though estimates differ on the numbers involved. The 1978 Estimates Paper figured that "the expected average number of cancer deaths associated with asbestos per year" would be on the order of 67,000, but critics such as Richard Doll and Richard Peto argued that this was probably high by at least a factor of ten.[123] Irving Selikoff's Mt. Sinai group attributes 8,200 cancer deaths per year to the fiber, rising to about 9,700 per year by the end of the century and tapering to 3,000 in the year 2025.[124] Another group calculates a total of 131,000 asbestos cancer deaths over the period 1985 to 2009, and a third estimates 300,000 total U.S. asbestos deaths (not just from cancer) by the year 2000.[125] By the time the disease runs its course (assuming exposures drop to a negligible level by the end of this century), the magic mineral, funeral dress of kings, could easily claim more American lives than were lost in all of World War II (400,000). Shipyard workers suffered more than their share of misery: more lives were probably lost as a result of building ships than from fighting aboard them.

The history of asbestos is a history of scientific deception joining hands with industrial malfeasance. As Barry Castleman characterizes it, asbestos mining and manufacturing companies "profited from public ignorance about carcinogenicity of asbestos," delaying by decades the time when they would be forced to eliminate the mineral as a reinforcing agent.[126] For Ozonoff, the asbestos industry "entangled reputable scientists in a web of deceit and manipulation." Industrial hygienists were in a position to know and to act, but "did little or nothing in the face of ample warning."[127] Ozonoff blames the larger social and economic system that focused medical attention on the diseases of workers rather than on the workplace hazards that were causing those diseases. However one looks at it, industry officials clearly put the financial health of their clients ahead of the personal health of their workers, a decision whose tragic consequences will be felt for many decades.

PETROCHEMICAL PERSUASION

Environmentalists have long charged "industry" with misrepresenting toxic hazards, though industries are obviously not all of one voice on such matters. Chemical industry advocates tend to blame tobacco for the growth of cancer; tobacco advocates tend to blame genetics, asbestos, or chemicals. Asbestos sympathizers point out that "the lung scarring typical of asbestosis has 100 other causes, including many viruses," and that what looks like asbestos injury may really be the consequence of trace metals found with the fiber, or even of oils derived from the polyethylene bags in which the material is stored.[128] Petrochemical advocates are equally adept in this art of apologia, both because the hazards are so diverse and the industry is so large. Chemical manufacturing is the second largest industry in the United States; only agriculture is larger, and even agricultural fortunes are closely tied to petrochemistry. There are dozens of trade associations active in petrochemical risk assessment. Here I will look briefly at two of the most active: the Chemical Manufacturers Association and the American Industrial Health Council. Both have invested heavily in science as public relations; both exert a nontrivial influence on environmental policy.

The Chemical Manufacturers Association (CMA) is one of the oldest trade associations in North America. Founded in 1872 as the Manufacturing Chemists' Association, the group now represents 180 of the largest chemical companies in the United States.[129] It has a permanent staff of 165 and an annual operating budget of about $36 million—a figure that disguises the magnitude of the association's efforts, given that most of its expertise is provided by some 2,000 member-company volunteers. In 1990, member company volunteers held 1,800 meetings at CMA headquarters in

Washington, D.C.[130] The strength of the association derives largely from the strength of the industry itself: U.S. chemical sales in 1989 totaled $287 billion and chemical exports totaled $37 billion, or about 14 percent of all U.S. exports. The industry was immensely profitable in the late 1980s, recording annual trade surpluses of more than $16 billion at a time when U.S. manufacturing as a whole was suffering deficits of $73 billion.

The CMA is the most important promoter of the chemical industry in the United States. It coordinates emergency response to chemical spills and produces videos for the public on how to handle chemicals safely. It advocates a global ban on chemical weapons and cooperates with the Drug Enforcement Agency to curb the manufacture of illicit drugs. CMA staff work with state Chemical Industry Councils to defeat proposals viewed as injurious to the industry's interests: in 1990–91, for example, the organization helped defeat several "ill-conceived state environmental initiatives," including California's Proposition 128, "the most far-reaching environmental initiative ever to go before the public." Prior to the passage of the Clean Air Act amendments in November 1990, the CMA conducted a public relations campaign to weaken the bill, including dozens of press releases and position papers, seventy radio programs, thirty television releases, one hundred editorial board briefings, and several hundred interviews for reporters.

The expenses of such efforts are substantial—the CMA's Chemstar Fluorocarbon Panel alone had spent $24 million by 1990 when it concluded its activities—but then so, apparently, are the rewards. A 1981 CMA pamphlet reported that:

> Measured monetarily, CMA activities are estimated to be worth billions of dollars each year. More than $1 billion are saved annually by forestalling inequitable natural gas utility rates; $500 million are saved annually by reducing future increases in freight charges; more than $100 million are saved each year through the development of proper engineering standards. In a recent year CMA saved several billion dollars by deflecting inappropriate environmental and workplace laws, rules, and excessive taxes without endangering public health or employee safety.[131]

Such savings have continued into recent years. In 1990–91, the CMA's Chemstar panels on butadiene, ethylene oxide, and phosgene helped persuade the EPA to ease regulations on air emissions; that same year the Fluoralkenes Panel convinced the EPA to rescind a demand for additional safety tests on hexafluoropropene and vinyl fluoride, which would have cost over $2 million. By the middle of 1992, CMA member companies had spent $63 million on Chemstar activities since they began in 1972. The CMA distinguishes the "research" and "advocacy" components of this support (research for 1990–91 was $6 million, advocacy $3.5 million) but does not say exactly how the two are different.[132]

The CMA is the largest, but by no means the only, chemical trade association in the United States. Of particular relevance to cancer policy is the American Industrial Health Council (AIHC), a consortium of 130 of America's leading producers of steel, chemicals, pharmaceuticals, and textiles formed in the fall of 1977 at the request of Shell Oil executives "to assist the Occupational Safety and Health Administration in the development of a rational and practical regulation for control of exposure to carcinogens in the workplace."[133] The immediate cause for its formation was OSHA's announcement, in October of that year, of a plan to implement a comprehensive cancer policy involving new and far-reaching procedures to identify, classify, and regulate occupational carcinogens. OSHA was to be empowered to order sharply reduced exposure levels to chemical carcinogens, and companies were to be required to keep medical and exposure records on their employees for up to forty years.[134] Most controversial, the new procedures would also have allowed OSHA to regulate carcinogens generically—all brominated or halogenated hydrocarbons, for example—rather than on a case-by-case basis. OSHA administrators defended this as a practical alternative to the costly and time-consuming process of judging each chemical separately; the plan was endorsed by Anthony Mazzocchi of the Oil, Chemical, and Atomic Workers Union and by the United Steelworkers legislative director, John J. Sheehan. Secretary of Labor F. Ray Marshall called the new approach a major breakthrough, comparing the industry's preference for controlling carcinogenic substances on a case-by-case basis to "trying to put out a forest fire one tree at a time."[135]

The AIHC responded by arguing that the OSHA proposal would "freeze science" by ignoring new developments concerning the causes of cancer. The association was especially worried about costs: a 1978 report predicted initial implementation costs of between $6 billion and $88 billion, followed by annual costs as high as $36 billion. Smaller firms would be driven out of business and inflation would be ignited.[136] On December 14, 1979, the AIHC denounced the OSHA plan as "possibly the most costly act of regulation of producers and users of chemical products ever issued."

OSHA officially promulgated its Generic Carcinogen Policy in 1980, though the election of Ronald Reagan in the fall of that year put a halt to the effort. Administrative stays would continue to block the policy for the next twelve years.[137] The AIHC made no secret of its support for Reagan's policies: as the council's *Annual Report* for 1982 put it, "the climate of opinion in Washington is receptive." OSHA was reconsidering its entire cancer policy, "based largely on scientific and legal input offered by AIHC." AIHC input had also been important in the EPA's abandonment of the view that air pollution was a major cause of cancer; the AIHC boasted that "virtually all" of its objectives had been achieved through Reagan's regulatory reform.[138]

By its own account, the goal of the AIHC has always been to produce "sound science" that is value-free and faithful to the facts. There is no admission that politics enters into science in any major way—in the selection of research priorities or methodologies, for example, or in the rhetorical forms used to package conclusions. Politics for the council is a kind of unfortunate "residue" that enters science only in cases of uncertainty, where science cannot supply all the answers: "The politics of science arises from divergent interpretations in the face of residual uncertainty. It involves convincing others that one's interpretation or hypothesis is correct although adequate verification cannot be achieved."[139] The equation of politics and uncertainty is important for the AIHC. It obscures the fact that uncertainty itself may be one of the major goals of science-as-public-relations; it creates the image that when scientists differ, there must be room for reasonable doubt. It also opens up the possibility that uncertainty—or lack of consensus—can be engineered by casting about for "divergent interpretations."

The appeal to science is important for petrochemical public relations. The continual reference to science (the words *science* or *scientific* appear at least once in every paragraph of many AIHC publications) gives the council an aura of respectable neutrality, an image of standing above the petty interests of profit and personal gain. Research in this context has PR as well as legal value: the continued call for "more research" serves to buy time for industry—time to continue unregulated production or to plan counterattacks against regulatory moves. The impression produced is one of both a cautious conservatism and a healthy progressivism. The call for ever more research provides the council with an effective means to delay regulatory moves that might cut profits.[140]

RHETORICAL STRATEGIES OF AVOIDANCE

There are differences, of course, among the various associations I've described, but all share a desire to absolve their constituent client substances (tobacco, asbestos, petrochemicals) of blame for an alleged nuisance. Management of controversy is the name of the game in science as public relations, and this is achieved by a diversity of strategies and images—some implicit, some explicit.[141] In the area of environmental carcinogenesis, some of the more common, explicit arguments in trade association science are the following:

1) Chemicals are natural and therefore one shouldn't worry about them.
2) Environmentalists have bullied independent-minded scientists into adopting positions based not on sound science but on politically motivated hysteria.
3) Cancer is more the product of one's personal lifestyle (smoking, sexual behavior, indoor pollution, and so forth) than of industrial toxics or occupational exposures.
4) There are thresholds of exposure below which carcinogens are not dangerous. Everything causes cancer if you consume enough of it; low-level exposures either are completely safe or pose only negligible or acceptable risks.
5) The regulatory process (politics) must be separated from the scientific process (science). A sharp distinction must be drawn between risk assessment (science) and risk management (politics).
6) Fear of chemicals is often a greater hazard than the chemicals themselves.
7) The costs of chemicals (for example, cancer) must be balanced against their economic benefits.
8) Evidence of toxic hazards is ambiguous, inconclusive, or incomplete; we therefore need "more research" to clarify ambiguities, improve estimates of risk, elucidate mechanisms, and so forth.

Let us look in greater detail at how these arguments work.

1. *Chemicals are natural and therefore nothing to worry about.* A 1971 pamphlet published by the Manufacturing Chemists Association (forerunner of the CMA) celebrates the "fairyland" of modern food products and packaging made possible by chemical technologies; the association suggests that many people will be surprised to learn that "not only food but the elements that go into clothing and shelter, and even the earth itself and all its inhabitants, can be described in terms of chemicals." Fredrick J. Stare is cited cautioning consumers against letting "any food faddist or organic gardener tell you there is any difference between the vitamin C in an orange and that made in a chemical factory."[142] The chemicals-are-natural theme was the centerpiece of Monsanto's Chemical Facts of Life public relations campaign in the late 1970s and early 1980s; hundreds of Monsanto TV and magazine ads reminded consumers that chemicals were a fact of life; indeed, "without them, there would be no world." The AIHC similarly reminds us that "life is essentially a chemical process."[143]

This particular rhetorical strategy has gained force in the wake of Bruce Ames's suggestion that natural carcinogens in foods may cause far more cancer than pesticides (see chapter 6). It has also been used by many producers-under-siege to defend their products. Quebec's Natural Resources Minister Yves Berube has defended asbestos as "a natural product we have been breathing for thousands of years," a product to which we have developed "resistance" and which is therefore "no more harmful than any other product."[144] The Formaldehyde Institute claims that formaldehyde is "part of the normal life process," and the Asbestos Information Association boosts its namesake as "A Natural Substance for Modern Needs." The rhetorical force of such characterizations is to soothe; the argument usually boils down to the presumption that, as a Dow scientist once put it, "because dioxins are ubiquitous, we need not be concerned about them."[145] F. J. C. Roe, a consultant to the American Industrial Health Council, presumably intended a similar reassurance with his 1978 judgment that cancer "is probably one of nature's many ways of eliminating sexually effete individuals who would otherwise, in nature's view, compete for available food resources without advantage to the species as a whole."[146] Roe seems to have forgotten that the overwhelming majority of cancers affect people long past their reproductive years.

2. *Environmentalists have bullied independent-minded scientists into adopting positions based not on sound science but on politically motivated hysteria.* Trade associations commonly portray themselves as innocent victims of environmental scaremongering. The gist of this argument is that while industry has sought to achieve some sort of scientific base for its assessments, activists have politicized the process. The Tobacco Institute has played this card: a 1986 pamphlet on secondhand smoke decries "science at the mercy of politics," complaining that political pressures are preventing an open discussion of the hazard (or lack thereof). The pamphlet warns that "alarmingly, scientific integrity and academic freedom face a serious threat from political pressures being applied by government health officials and otherwise principled scientists."[147] Academic freedom is commonly invoked in this context, as is the First Amendment right to free speech. In the mid-1980s, the Fed-

eral Trade Commission was thwarted in its attempt to withdraw an R. J. Reynolds advocacy ad because the advertiser was judged to be entitled to "the full protection of the First Amendment, including the freedom to deceive."[148]

Other industries have assumed this Galileo-inspired posture of a victim bullied by powerful political forces. The rhetoric of "science versus politics" blends with more recent metaphors of political correctitude, as when Milton Green, Polaroid's associate director for chemical procurement for research, cautioned that the press's perennial focus on carcinogenic hazards threatens to undermine public confidence in science: "Scientific McCarthyism has become pervasive, and self-serving opportunists have helped foster hysteria about chemicals that, if unchecked, could destroy the already limited credibility of scientists." Green musters up the kind of old-fashioned scientific machismo one commonly finds among chemical industry managers, boasting that he himself was exposed—apparently without harm—to high levels of carcinogens in the 1940s, when "we just didn't know anything about the long-term dangers of exposure."[149]

3. *Cancer is more the product of one's personal lifestyle (smoking, sexual behavior, indoor pollution, and so forth) than of industrial toxics or occupational exposures.* Petrochemical trade associations have obvious reasons to make such claims: if petrochemicals are to blame for only a minuscule proportion of cancers, it is easier to argue that public health efforts must be directed toward behavioral modification rather than industrial regulation. Even a tobacco company, however, may find it useful to argue that cigarette smoking is more properly regarded as a matter of personal choice than of agricultural or advertising policy.

The problem with "lifestyle" theories, as Epstein and others have pointed out, is their presumption that cancer is, by and large, a matter of personal choice. The rhetoric of risk fits nicely here. Workers assume a "risk" when they take a job; society takes a "risk" when it indulges demands that producers satisfy consumers' wishes. The implication is that there are profits associated with risky behaviors. The problem begins as soon as we realize that many of the "risks" to which we are exposed are not of our own choosing. The Brazilian farmer who saws asbestos sheeting to make a roof can hardly be described as assuming a risk in pursuit of profit—especially if that risk is not well advertised or understood. The same is true, arguably, with addictive drugs like cigarettes or with pesticides, which are difficult to avoid in modern supermarket foods. The lifestyle emphasis shifts the burden of responsibility from society to the individual, "blaming the victim."[150]

4. *There are thresholds of exposure below which carcinogens are not dangerous.* This has long been a centerpiece of industrial argumentation and is fully discussed in chapter 7.

5. *The regulatory process (politics) must be sharply separated from the scientific process (science).* The AIHC suggests that the determination of carcinogenicity is "a scientific process, and not a regulatory one." The council has proposed the establishment of "a panel of qualified independent scientists" not subject to the influence of regulatory agencies to determine carcinogenic risks. Such a panel would avoid the mixing of "societal values" and "scientific judgments"—a confusion, the council argues, pervading present regulatory agencies' procedures. On the basis of decisions

by this neutral expert body, societal decisions might be made regarding "how much protection from the hazard the public will accept." The radical separation of science and politics allows that which can be classed as science to be regarded as apolitical, while barring that which is relegated to "the political" (for example, the public) from having a say in what kind of science should be done. The distinction allows trade associations to stand on the platform of science and rebuff charges that the platform is slanted; it has also been used to argue that risk assessments should not have "margins of safety" built into them. Trade associations generally favor scientific rather than public health conservatism (see conclusion).

6. *Fear of chemicals is often a greater hazard than the chemicals themselves.* *Chemophobia* and *cancerphobia* became common industry terms in the late 1970s, in response to environmentalist appeals and regulatory zeal.[151] Monsanto's president, John W. Hanley, thus warned that America was suffering from "an advanced case of chemophobia, an almost irrational fear of the products of chemistry."[152] Elizabeth Whelan has made chemo- and cancerphobia a centerpiece of her efforts to defend American food supplies as safe: in 1990, at the height of the Alar frenzy, the co-founder and president of the American Council on Science and Health suggested that "what we need is a national psychiatrist to figure out why consumers behave in such an irrational manner." (A psychiatrist apparently later called to tell her that people fear "the invisible.") Whelan pointed out that in a survey of American anxieties, fear of pesticides ranked third on the list—higher even than the fear of death (ranked seventh).[153]

Stephen Hilgartner, a sociologist at Columbia University, has noted that the emphasis on phobias in such literature implies that chemical fears are unfounded, that rational people should realize that the risks are vanishingly small.[154] The Asbestos Information Association thus warns against *fiber phobia,* the nuclear industry against *radiophobia,* Whelan against *nosophobia* ("morbid dread of illness"), and Accuracy in Media against *asbestophobia.*[155] Fear has become an object of scholarly study: in 1984, the U.S. Department of Energy paid $85,000 to Robert DuPont, a psychiatrist and president of the Phobia Society, to study how to overcome the public's "nuclear phobia."[156] Fear has been branded a mental illness: the Governmental Refuse Collection and Disposal Association has tried to medicalize public opposition to incinerators by denouncing the Nimby ("not in my backyard") mentality as a "recurring mental illness that continues to infect the public."[157] A 1992 article in *Psychological Reports* likens cancerphobia to fear of chemical AIDS, mass hysteria, "universal allergy," malingering, and phobias linked with needles, high places, airline travel, and insects. The authors concede that cancerphobia may be a response to "a specific event, such as a toxic exposure"; they even observe that cancerphobia may be "more prevalent in communities near chemical waste or toxic exposure sites." They then proceed to trivialize and medicalize that fear by characterizing it as "an Adjustment Disorder, with either Anxious Mood, Depressed Mood, Physical Complaints, Work Inhibition, or any combination of these." The authors suggest that fear of cancer "has been shown to reduce the tendency to process a persuasive message" (perhaps assurances that food additives and waste dumps are safe?). Fortunately for these authors, hope lies in the fact that cancerphobia "has been successfully treated with fluoxetine," a drug occasionally used in the treatment of other phobias.[158] The message, apparently, is that fear of chemicals can be treated by a properly administered medical regime of chemicals.

Fear has even been identified as a *cause* of cancer. In 1981, *Chemtech* carried an article on "Scientific McCarthyism" cautioning that since stress is a known cause of cancer (an unproved assertion), "it is not unreasonable to suppose that the present wave of cancerphobia might trigger more cases of cancer in susceptible individuals than actual exposure to the chemicals of which they are terrified."[159] A decade later, epidemiologists writing in the *American Journal of Public Health* reported that stress resulting from the near-meltdown of the Three Mile Island nuclear reactor may well have resulted in "a small wave of excess cancers" in the surrounding regions. Several newspapers jumped to the conclusion that "Stress, Not Radiation, Caused Cancers Near Three Mile Island."[160] Environmentalists sometimes claim that people are more likely to fear chemical corporations than to fear chemicals;[161] the corporate tendency to respond to public fears with public relations has no doubt tended to exacerbate these fears.

7. *The costs of chemicals (for example, cancer) must be balanced against their economic benefits.* In the 1970s, when federal occupational authorities moved to regulate vinyl chloride, industry spokesmen announced that the regulations would force plants to close, throwing 2 million people out of work and costing billions in lost production. In the 1980s, after Murray Weidenbaum calculated that the total annual cost of federal regulations would exceed $100 billion, trade associations cited these figures in defense of Reagan's policy of "regulatory relief." The force of many such arguments, again, is to suggest that regulatory costs often exceed regulatory benefits, the larger philosophical presumption being that regulation distorts the natural tendency of industry to regulate itself.

Political conservatives sometimes argue that environmental regulations can actually increase cancer risks. Warren Brookes in a 1990 article for *Forbes* suggests that there are many examples of environmental regulations "leading to results opposite to those intended." Alar, for example, by keeping apples longer on the tree, renders them less susceptible to leaf miners and obviates the need for harsher, cancer-causing insecticides. The banning of the fumigant ethylene dibromide in 1983 "probably raised cancer rates" by allowing the contamination of foods by powerful carcinogenic fungi. Brookes concludes that environmental regulations "can be more dangerous than the perceived original risk."[162] Bruce Ames has argued similarly that pesticide bans may increase cancer by boosting the costs of cancer-preventing fruits and vegetables so that fewer are bought and eaten.

8. *Evidence of toxic hazards is ambiguous, inconclusive, or incomplete; we therefore need "more research" to clarify ambiguities, improve estimates of risk, elucidate mechanisms, and so forth.* This has become one of the more effective industry arguments in recent years. The tobacco industry, for example, has long used this tactic to convince consumers that it is premature to conclude that cigarettes cause cancer. The Tobacco Institute's paper on "The Cigarette Controversy" thus calls for "more research" (or "far more research" and the like) half a dozen times in as many pages, the purported goal being to resolve a smoking and health "debate" or "contro-

versy."[163] Internal tobacco industry documents brag about the success of this "brilliantly conceived and executed" strategy to create "doubt about the health charge without actually denying it."[164]

Given the inherent uncertainties of the scientific process, it is always possible to delve ever deeper into the mechanisms of pharmacokinetic action, to elucidate risks with ever greater precision, to demand ever larger animal or human epidemiological studies. The call for more research often translates into a call for "less action"— and, specifically, less regulation. Reagan administration scientists used this strategy effectively with regard to the problem of acid rain—though its cynicism eventually became so transparent that even William K. Reilly, President Bush's Environmental Protection Agency administrator, stated in the fall of 1989 that such a strategy had given research a bad name.

Calls for regulatory delays on the grounds of insufficient or incomplete science are a regular part of the PR package of the tobacco, petrochemical, and other industries.* The net effect is to shift the focus of attention from the need to eliminate a probable hazard to the need to resolve a certain ambiguity. (The same effect is achieved by reference to "potential" hazards, "possible" risks, and so on.) A high premium is placed on not being wrong: the scientist's desire for accuracy is substituted for the consumer's desire for product safety. Animal studies are a classic target of industry skepticism in this regard. Tobacco spokesmen in the 1950s and 1960s argued that tobacco-induced tumors in the lungs of rats were profoundly different from tumors in the lungs of humans. Shell Oil in the early 1970s was able to argue during congressional hearings on dieldrin that, though the pesticide had indeed caused cancer in mice, it had not been shown to cause cancer in rats. (Shell, of course, preferred to believe that humans are more like rats than like mice.) The net and desired effect of such arguments is to insinuate doubt: trade association scien-

---

*Research as delay was explicitly recognized as one of the "strategies for established firms and industries" by Bruce M. Owen and Ronald Braeutigam in *The Regulation Game: Strategic Use of the Administrative Process* (Cambridge, Mass., 1978), one of the most brazen expositions of how industries should manipulate science to avoid unfavorable regulatory outcomes. The authors suggest that:

> The ability to control the flow of information to the regulatory agency is a crucial element in affecting decisions. Agencies can be guided in the desired direction by making available carefully selected facts. Alternatively, the withholding of information can be used to compel a lawsuit for "production" when delay is advantageous. Delay can also be achieved by overresponse: flooding the agency with more information than it can absorb. Sometimes, when a specific item of information is requested and it is difficult or impossible to delay in providing it, the best tactic is to bury it in a mountain of irrelevant material. This is a familiar tactic of attorneys in antitrust suits. It is also sometimes useful to provide the information but to deny its reliability and to commence a study to acquire more reliable data. Another option is to provide "accurate" information unofficially to selected personnel of the agency who are known to be sympathetic. If another party has supplied damaging information, it is important to supply contrary information in as technical a form as possible so that a hearing is necessary to settle the issues of "fact" (p. 4).

Compare also the authors' assertion that:

> Regulatory policy is increasingly made with the participation of experts, especially academics. A regulated firm or industry should be prepared whenever possible to coopt these experts. This is most effectively done by identifying the leading experts in each relevant field and hiring them as consultants or advisors, or giving them research grants and the like. This activity requires a modicum of finesse; it must not be too blatant, for the experts themselves must not recognize that they have lost their objectivity and freedom of action (p. 7).

tists cite and twist Lewis Thomas's gloss on Heisenberg—that "science is founded on uncertainty"—and Aristotle's view that "you can't prove a negative."[165]

Lest I be misunderstood, I want to make clear that my goal is not to argue that trade association science is necessarily bogus. Thresholds may well exist, and animal studies are no doubt often flawed. Costs can be profitably weighed against benefits, and chemophobia is no doubt a real problem for some people. Doubts and uncertainties are often real, and some environmentalists do bully.[166] And life certainly would be impossible without chemicals.

My point is rather to suggest that science plays an underappreciated ideological role in the industrial context. Trade associations appeal to science to give themselves a semblance of neutrality, balance, and levelheadedness in questions of environmental risk assessment. The appeal to science creates an impression that the company is making progress, keeping up with the times. It also buys time. The call for more research often works to delay implementation of regulatory standards—paralysis by analysis, as a 1980 OSHA document put it. Whatever level of proof can be demonstrated linking cancer with asbestos or Alar or tobacco, the demand can always be put forward that more studies are needed to eliminate uncertainties—perhaps tests with larger sample sizes run over longer periods in situations more closely approximating the conditions of the workplace or in animal species more closely approximating the human body. By sheer force of funding, trade associations can create the illusion (or the reality) that since experts disagree, there must be room for disagreement.

Trade association science must be understood in this context. In an era of forceful environmental regulation, manufacturers of carcinogens have come to recognize the relevance of risk assessment and risk perception to their economic well-being. Trade associations make it their business, among other things, to exploit, disseminate, and, if need be, produce divergent expert opinion in matters of environmental carcinogenesis. This has important consequences for public perceptions of science. The production and management of uncertainty has facilitated what might be called the MacNeil-Lehrerization of science: the media display of equal and opposing views on apparently settled questions, generating the sense that endless research is needed to resolve guilt or innocence in questions of carcinogenesis. Anthony Ramirez of the *New York Times* captured the essence of the process when he characterized news coverage of the cellular phone–brain cancer controversy as "evidence Ping-Pong," where "one scientist serves up a potential peril and another expert lobs back a calming reassurance."[167]

Bias in trade association science, though, is sometimes subtle. Bias in this context does not necessarily mean the science is phoney: support even of "good science" may assist an industry in deflecting attention from the hazards of a product—as when the Tobacco Institute promotes genetic research. The net effect is to jam the scientific airwaves with true but trivial work, distracting from what is going on more fundamentally. Bias in such cases lies typically not in the falsification or misrepresentation of research (though both of these do occur, as we have seen) but rather in the diversion of attention from one problem to another—from causes onto mechanisms, for example, or from questions of health onto questions of free speech. Bias also emerges, however, from the fetishization of certain of the self-avowed ideals of conventional science. Science requires empirical verification, but the rabid distrust of a well-founded hypothetical can be used to dismiss a plausible point—as when animal evidence of carcinogenicity is dismissed by industry-financed skeptics. Science usually requires some degree of precision, but a misplaced call for precision can also be wielded as a political tool—as when calls are made to delay regulation until endlessly more elaborate studies confirm that a substance is hazardous. Scientists are supposed to seek to reduce uncertainty, but the demand that regulators hold back until all uncertainty is eliminated can be used as an excuse to do nothing.

The manufacture and distribution of uncertainty are insufficiently recognized social functions of science. This is the lesson of the tobacco company concession that "doubt is our product." We don't yet know enough about the use of science as an instrument of public relations, perhaps because this is a relatively new historical phenomenon—at least in the form I have described it. Industrial support for a theory does not make it wrong, any more than government support makes it right. But since it is not just the righteous who have recognized that scientific facts are (among other things) social constructs, the question of origins and intentions is a vital one. Bright lights may be used to cast confusing shadows, even when they cannot dazzle us to the point of blindness.

Chapter 6

Natural Carcinogens and the Myth of Toxic Hazards

*Warning: this next report may be hazardous to your beliefs.
For two decades now, we of the media have brought story
after story where experts warn of links between all kinds of
pollutants and cancer. But tonight a distinguished research
scientist makes a case that many of the warnings we hear are
unnecessary, that all the concern about this toxin, that pesticide, is "Much Ado About Nothing." Wouldn't it be nice if he
was right?*
— Report on Bruce Ames, ABC-TV's *20/20*, 1988

THE 1980s was a decade of conservative administration, industrial triumph, and radical retreat. Environmentalists became nervous as evidence mounted that lifestyle—especially smoking—is a more important contributor to cancer than industrial pollutants. Radon emerged as a major health hazard, as did domestic indoor air pollution, salt in foods, dietary fat, and the sedentary lifestyle. Tobacco was increasingly blamed for the epidemic of cancer, insofar as an epidemic is admitted. Animal studies came under attack as an imperfect predictor of human cancer hazards, and the idea that there are thresholds of exposure below which chemical exposures are safe gained a new measure of scientific respectability.

As an icon of this transformation, it would be difficult to find a more representative figure than Bruce N. Ames, chairman of Berkeley's biochemistry department, who shocked the world in 1983 with his thesis that natural carcinogens are likely to pose a far greater hazard than industrial pollutants. Ames's paper ignited a firestorm of controversy that is still burning today; environmentalists have challenged him on a number of important points, but the question of whether and to what extent he is right remains hotly contested. The heat of the debate has been further fueled by the fact that since his ideas have been made known, there has been unprecedented growth in public environmental concern—ranging from the much-publicized fears of global warming and ozone depletion to worries over the loss of biodiversity and the myriad hazards of nuclear, medical, and other forms of waste. Ames's thesis that industrial pollutants pose a trivial risk to human health struck a powerful chord with people tired of hearing about the hazards of industrial life. However the debate is resolved, Ames may well be the most powerful anti-environmentalist of the century.

THE ARGUMENT:
NATURE IS NOT BENIGN

Ames's argument, published in *Science,* the official organ of the American Association for the Advancement of Science, is simple: plants synthesize toxic chemicals largely as a defense against potential bacterial, fungal, and insect enemies, and many of these chemicals turn out to be carcinogens. ("All of plant evolution is chemical warfare.") Safrole, for example, found in many edible plants (and the major component of oil of sassafras), has been shown to cause cancer in rats, as has piperine, one of the major components (about 10 percent by weight) of black pepper. Rancid fat contains many carcinogens, as do most burned meats and browned foods. Brown mustard, certain edible mushrooms (such as the commercially popular *Agaricus bisporus*), and many foods contaminated by mold contain cancer-causing compounds. The substances are in some cases quite potent: the aflatoxin found in many moldy foods (corn, nuts, peanut butter, bread, cheese, and certain fruits, for example) is one of the most powerful of all known carcinogens. Ames also reviewed evidence linking alcohol consumption with cancers of the mouth, esophagus, pharynx, larynx, and liver.[1]

Equally provocative was his claim that many of the most celebrated "health foods" contain substantial carcinogens. Some herbal teas and honeys contain pyrrolizidine alkaloids, many of which are carcinogenic, mutagenic, and teratogenic (causing birth defects). Many injured plants—bruised celery, for example—produce toxins as a response to insects and other predators; even uninjured foods, though, may have such components. Oils derived from okra and cottonseed contain toxic cyclopropenoid fatty acids, known to cause cancer in trout, atherosclerosis in rabbits, mitogenesis in rats, and a variety of other toxic effects in farm animals. Alfalfa sprouts contain about 1.5 percent by dry weight of canavanine, a highly toxic substance that appears to cause a severe lupuslike syndrome in laboratory monkeys. The fava bean contains about 2 percent by weight of the toxins vicine and convicine; Ames noted that the Greek philosopher Pythagoras forbade his followers to eat the bean, "presumably because he was one of the millions of Mediterranean people with a deficiency of glucose-6-phosphate dehydrogenase"—a genetic disorder that confers increased resistance to malaria but also makes one vulnerable to a hemolytic anemia caused by the hydrolysis of vicine. Gossypol is a major toxin in cottonseed, accounting for about 1 percent of its dry weight; unrefined cottonseed oil, consumed in countries such as Egypt, may pose a reproductive as well as a carcinogenic hazard. (The substance has been shown to be so effective in interfering with reproduction that it has been tested as a male contraceptive in over 10,000 people in China.) Beets, celery, lettuce, spinach, radishes, and rhubarb all contain high levels of nitrates—about 200 milligrams per 100 grams of food—which are known to form cancer-causing nitrosamines in the acidic environment of the stomach.[2] Carcinogens have been found in coffee, in caramelized sugars, and in the toasted crust of breads.

Ames also provided a novel explanation for the mechanism of cancer causation. Cancer, he suggested, may be like aging insofar as both processes involve damage to DNA by oxygen radicals and lipid peroxidation (fat combustion, in other words).

The odds of coming down with cancer increase with about the fourth or fifth power of age; cancer may simply be a natural consequence of growing old.

The presence of carcinogens in common foods was only one part of Ames's argument. The Berkeley biochemist also pointed out that the human body has a number of defense mechanisms against carcinogens, including DNA repair and continuous shedding of the surface layers of our skin, stomach, cornea, intestines, and colon. Dietary antioxidants are a major source of cancer-fighting agents: Vitamin E, for example, traps oxygen radicals in lipid membranes and therefore may have an anticarcinogenic potential; beta-carotene may also protect body fat and lipid membranes against oxidation. Dietary selenium has been shown to inhibit the onset of cancers of the skin, liver, colon, and breast in experimental animals, and low selenium may be a contributing cause of cancer susceptibility. Glutathione may guard against cancer, as may ascorbic acid (Vitamin C), uric acid, and a number of other substances found naturally in edible plants. Ames suggested that the geographic variability of cancer widely attributed to environmental factors may actually be due, at least in part, to variations in the consumption of anticarcinogens and protective factors in the diet.[3]

The argument that nature could be carcinogenic was not, of course, entirely new. Sunshine had long been known to cause skin cancer, and many other notorious carcinogens were in some sense natural—uranium dust, chimney soot, and asbestos dust, for example. Wilhelm Hueper in his 1952 testimony before the Delaney Committee had stressed that there were natural carcinogens in foods, though he was quick to add that this didn't mean we were justified in adding new ones.[4] The peanut butter threat was also nothing new, having been first discovered in 1960, when 100,000 ducks and turkeys died on British farms after eating peanut meal contaminated with the aflatoxin-producing mold *Aspergillus flavus*. Dietary aflatoxin was later associated with high liver cancer rates in Mozambique, Thailand, Kenya, and several other nations.[5] In the southern Chinese province of Guangxi, nearly half of all male cancer deaths were found to have been caused by aflatoxin/hepatitis B–associated liver cancer. The National Research Council published its first report on natural toxicants in foods in 1966, identifying several different fungi, molds, and plants—bracken fern, for example—as carcinogens.[6]

By the late 1970s, natural carcinogens were commonly being cited to diminish the role of industrial carcinogens.[7] John Cairns, director of an Imperial Cancer Research Fund laboratory in London, argued that "it seems more likely that the main determinant of cancer is diet rather than industry, and that we should be looking for mutagens (or perhaps promoters) that are formed in the body from the normal ingredients in our diet." [8] Elizabeth Whelan featured "naturally occurring cancer-causing agents" as a central focus of her newly formed American Council on Science and Health,[9] and Ames himself was already beginning to suggest that "much of the cancer occurring today" was likely due to "natural carcinogens in our diet."[10] Monsanto played the natural carcinogen card in its 1977 Chemical Facts of Life campaign: under the heading "Rhubarb Can Kill," the company pointed out that "lettuce contains nickel, a metal which causes cancer," and that four ounces of caffeine "are enough to kill a person."[11]

In his 1983 paper, Ames went beyond previous critics by suggesting that natural carcinogens in foods are, generally speaking, far more dangerous than industrial hazards. Especially disconcerting (or refreshing, as some regarded the matter) was his claim that human consumption of industrial pesticides is minuscule compared with our consumption of natural pesticides—pesticides produced naturally by plants to defend against insect or fungal pests–which may account for as much as 10 percent of a plant's dry weight. The human dietary intake of nature's pesticides is therefore on the order of several grams per day—"probably at least 10,000 times higher than the dietary intake of man-made pesticides."[12] Nature is "not benign," though the hazards of natural pesticides are often overlooked as a result of our single-minded focus on industrially derived carcinogens. Ames noted that, when faced with apparent poisonings, clinicians often fail to consider the possibility of nonindustrial contamination. In one notable case in rural California, a baby boy, a litter of puppies, and goat kids all developed "crooked" bone birth defects after drinking milk from goats that had foraged on lupine. The defects were at first mistakenly blamed on exposure to the pesticide 2,4-D, when in fact the lupine was to blame. (Milk derived from animals foraging on lupine can become contaminated with teratogenic substances such as anagyrine.) Ames cautioned that the natural pesticides in our foods can be expected to increase, as breeders move to create greater natural resistance in crops—an ironic consequence of our distrust of artificial pesticides. In one possible harbinger of things to come, breeders were forced to withdraw a strain of potatoes bred for natural insect resistance after it was found that the natural protective toxins enhanced by the process could be dangerous to humans.[13] A new insect-resistant celery is similarly much higher in natural carcinogens than the garden variety.[14]

Ames also challenged the idea put forward by Epstein and others that cancer rates are rising as a result of increased exposures to toxic pollutants. (Chemical plants, as he puts it, pose less of a threat than green plants.) With the exception of melanoma and lung cancer, Ames argued, cancer rates are not on the rise. Growing skin cancer rates are caused by increased exposure to the sun (primarily as a consequence of sunbathing fashions), and growing lung cancer rates are caused almost exclusively by smoking. Industry is therefore not to blame for the growth of overall cancer rates. Air pollution, for example, is a relatively minor hazard in comparison with smoking. Los Angeles smog contains far lower levels of soluble organic matter than those to which smokers expose themselves: one would have to inhale smoggy Los Angeles air for a year to get the same level of particulate contaminants that two-pack-a-day smokers get in one day. Epidemiological studies have furthermore shown, to Ames's apparent satisfaction, that there is no evidence of a significant cancer risk from urban air pollution.[15]

Ames is not easy to dismiss, given his credentials in the field of environmental carcinogenesis. In the mid-1970s he developed a powerful new technique—the so-called Ames test—to estimate carcinogenic potentials using bacterial bioassays rather than animal tests. The problem with animal studies (most commonly, exposing rats or mice to suspected carcinogens) is that, while generally reliable, they are expensive and take months or even years to produce results. Bacteria, by contrast, grow fast and cost relatively little. Ames's achievement was to develop a test that appeared to correlate bacterial mutagenesis with mammalian carcinogenesis. The idea that cancers might arise through mutation of the genetic material dates back to the early years of the century; Ames himself had been one of the first to suggest that cancer could be caused through the mutagenic action of a single molecule.[16] Experimental methods, however, had never been able to demonstrate a good correlation between mutagens and carcinogens. Ames was one of the first to appreciate why: in a path-breaking 1973 paper, he argued that human carcinogens often do not show up as bacterial mutagens because human and bacterial physiology are so different. To solve the problem, Ames proposed growing bacteria in a medium to which human liver extract had been added, in addition to the suspected carcinogen in question. The liver extract would metabolize the carcinogens in a manner similar to that in the human body, and the resulting metabolites would mutate the bacterial DNA. Ames was able to show that, when prepared in this manner, known human carcinogens did in fact show up as bacterial mutagens.[17]

Ames's now famous test revolutionized carcinogenic bioassays. The new process was not only cheaper but promised to reveal potentially dangerous chemicals much faster. By the end of the 1970s, the Ames test had become a standard rapid assay for carcinogens. Ames also gained national attention when, in 1977, he published results of a study showing that the flame retardant known as Tris, commonly used in children's pajamas, was a carcinogen. Tris was eventually banned, partly as a result of Ames's studies (data from animal tests were also a strong factor).[18] About this time, the Ames test was used to show that tar-based hair dyes contained carcinogens; this, too, led to an FDA ban. Ames later played a prominent role in strengthening the State of California's regulation of ethylene dibromide: Ames had found that workers were being exposed to high levels of the fumigant (levels shown to cause cancer in half the rats exposed); his 1981 testimony before a state commission investigating the matter resulted in a lowering of the permissible occupational exposure by a factor of more than a hundred.[19] (Ames, interestingly, has never been averse to the notion that occupational exposures pose a hazard. This is consistent with his view that large doses are disproportionately hazardous compared with low-level exposures.) From these and other investigations, especially in the area of mutagenesis, Ames gained a reputation as one of the nation's foremost cancer researchers. Between 1973 and 1984, Ames was the twenty-third most frequently cited scientist in any field.

He continued his argument in a 1987 article for a special issue of *Science* devoted to risk assessment, in which he unveiled a new and controversial way to rank the health risks of specific carcinogens. The article, co-authored with Renae Magaw and Lois S. Gold of the Lawrence Berkeley Laboratory, introduced an index of carcinogenic exposure, which they named HERP, for Human Exposure /Rodent Potency.[20] The HERP index presents a ranking of the relative hazards of specific substances, taking into account both the carcinogenic potential of the substance (determined by animal tests) and the extent of human exposure. The logic behind the ranking is that it would be useful to be able to compare the effects of low-dose exposures to potent carcinogens against high-dose exposures to much weaker carcinogens. Substances commonly regarded as hazardous (such as DDT and PCB's) may thus score low on the index if they are sufficiently rare in terms of the numbers of people exposed and/or the extent of those exposures. Less feared substances may score high on the index, if they are consumed in sufficient

Bruce Ames's HERP (Human Exposure/Rodent Potency) Ranking Showing the Relative Severity of Possible Carcinogenic Hazards

Relative Cancer Hazard HERP (%)	Daily Human Exposure	Carcinogen, Dose per 70 kg Person
.0002	PCBs: daily dietary intake	PCBs, .2 µg (U.S. average)
.0002	AF-2: daily dietary intake before banning	AF-2, 4.8 µg
.0003	DDT/DDE: daily dietary intake	DDE, 2.2 µg (U.S. average)
.0004	EDB: daily dietary intake from grains and grain products	Ethylene Dibromide, .42 µg (U.S. average)
.0004	Well water, 1 liter contaminated, Woburn	Trichloroethylene, 267 µg
.001	Tap water, 1 liter	Chloroform, 83 µg (U.S. average)
.003	Sake (250 ml)	Urethane, 43 µg
.004	Well water, 1 liter contamined (worst well in Silicon Valley)	Trichloroethylene, 2800 µg
.006	Bacon, cooked (100 g)	Diethylnitrosamine, .1 µg
.008	Swimming pool, 1 hour (for child)	Chloroform, 250 µg (average pool)
.03	Comfrey herb tea, 1 cup	Symphytine, 38 µg
.03	Peanut butter (32 g, one sandwich)	Aflatoxin, 64 ng
.06	Dried squid, broiled in gas oven (54 g)	Dimethylnitrosamine, 7.9 µg
.06	Diet cola (12 ounces)	Saccharin, 95 mg
.07	Brown mustard (5 g)	Allyl isothiocyanate, 4.6 mg
.1	Basil (1 g dried leaf)	Estragole, 3.8 mg
.1	Mushroom (*Agaricus bisporus*), one raw (15 g)	Diverse hydrazines
.2	Natural root beer (12 oz, now banned)	Safrole, 6.6 mg
[.3]	Phenacetin pill (average dose)	Phenacetin, 300 mg
.6	Conventional home air (14 hours/day)	Formaldehyde, 598 µg
2.1	Mobile home air (14 hours/day)	Formaldehyde, 2.2 mg
2.8	Beer (12 oz)	Ethyl alcohol, 18 ml
4.7	Wine (1 glass = 250 ml)	Ethyl alcohol, 30 ml
[5.6]	Metronidazole (therapeutic dose)	Metronidazole, 2000 mg
5.8	Formaldehyde: Workers' average daily intake	Formaldehyde, 6.1 mg
6.2	Comfrey-pepsin tablets (nine daily)	Comfrey root, 2700 mg
[14]	Isoniazid pill (prophylactic dose)	Isoniazid, 60 mg
16	Phenobarbital, one sleeping pill	Phenobarbital, 60 mg
17	Clofibrate (average daily dose)	Clofibrate, 2000 mg
140	EDB: Workers' daily intake (high exposure)	Ethylene dibromide, 150 mg

Note: The calculations assume a daily dose for a lifetime; a high percentage indicates a high potential cancer risk, as indicated by animal tests. HERP values for drugs taken only for a brief period of time are bracketed.

Source: Abbreviated from Ames et al.'s "Ranking Possible Carcinogenic Hazards," *Science* 236 (1987): 273.

quantities by sufficient numbers of people (see table).

The purpose of the ranking, again, was to discover which among the various carcinogens to which we are exposed are most dangerous and which are least dangerous. It would be useful to know, for example, whether traces of the banned fumigant ethylene dibromide (EDB) in food (used to prevent moldy contamination of grain) pose more or less of a threat than the alternatives—food irradiation or moldy food, for example. Given the widespread misconception that "everything causes cancer," it is important, they argued, to be able to compare the severity of different types of hazards. At stake was the question of priorities for research and regulation. The authors suggested that the public has been led to believe that synthetic chemicals are a major health hazard when, in fact, natural substances pose a far greater risk. A single daily glass of wine, for example, is—according to their calculations—more than 10,000 times as dangerous as the average American's exposure to EDB, and more than 20,000 times as dangerous as typical exposures to the highly feared PCBs. People who worry about the pesticide but not the wine are uninformed about the true hazards (or safety) of what they consume.

As Ames repeatedly emphasized, his primary goal was not to introduce new fears but to alleviate old ones. "Our world is full of carcinogens. . . . Fortunately, almost all of these are present in tiny doses which pose no real danger." His aim, in other words, was not to make people worry about "an occasional raw mushroom or beer, but to put the possible hazard of man-made carcinogens in proper perspective." The HERP index would allow us to focus on the true dangers facing us: it was important not to divert society's attention from "the few really serious hazards, such as tobacco or saturated fat (for heart disease), by the pursuit of hundreds of minor or nonexistent hazards."[21] News reports made it clear that Ames wanted to exonerate false threats, not stir up new ones. The *Los Angeles Times* reproduced his 1986 testimony before a California Senate Committee as an op-ed piece headlined "Cancer Scares over Trivia: Natural Carcinogens in Food Outweigh Traces in Our Water"; the *Wall Street Journal* printed an equally dismissive essay by Ames.[22] A widely watched ABC-TV *20/20* program on Ames spoke of "Much Ado About Nothing," and a long interview in the health magazine *Hippocrates* reassured readers "Not to Worry." The words *trivial* or *exaggerated risks* were often featured in these and other reports, including a *60 Minutes* segment.

Ames himself made it clear that his work was motivated by a larger philosophical conviction that nature is not benign, and that technology should not be blamed for all that ails the human species. In 1988, in a tour through Berkeley's Natural Grocery (the lead-in to his *Hippocrates* interview), Ames argued that such stores are basically selling "fear of modern technology" to consumers misled into believing that "organic is better" when the opposite is more often the case (organic produce may have mold, which is frequently carcinogenic). Even Rachel Carson's fable in *Silent Spring* celebrating the beauty of "laurel, viburnum, and alder, great ferns and wildflowers" along the traveler's path misleads: "Here's this nice sylvan view of ferns, and one of the most common ones, the bracken fern, is absolutely full of carcinogens." The chemicals in such ferns, which can pass into cow's milk or be reabsorbed into soil, might be even deadlier than DDT, the biocide par excellence in Carson's book. It was strange, Ames argued, that people have come to believe that evil can come only from human hands: "It's a very funny way of looking at things. Only in the modern Western world, where everyone is so wealthy and healthy, can you even think of the idea that nature is benign. Whereas in all of previous history, nature was something to be fought."[23]

Ames had applied these themes in 1987, in response to a call by Ellen Silbergeld for a moratorium on incinerator construction until we can be assured that dioxin poses no harm even to people who smoke, sunbathe, or otherwise engage in risky behaviors. (Dioxin is one of the many compounds produced by burning cardboard and other wastes.) Ames countered that such a zero-hazard approach was impractical and "an invitation to paralysis." It ignored "the benefits of technology, and the hazardous side effects of the alternatives when some technology is eliminated." Ames suggested that the risks of dioxin from incinerators might be low when compared to the other risky activities people are already engaged in: HERP index values, for example, indicate that drinking one beer everyday would pose about the same birth-defect hazard as "eating a daily kilogram of dirt contaminated with 1 part per billion of dioxin."[24] The greatest thing to fear was apparently fear itself.

THE CRITICS

The national press picked up on Ames's thesis shortly after the publication of his 1987 article in *Science*. Jane Brody reported in the *New York Times* that the ranking system he and his colleagues developed showed that "some cancer dangers are overrated and others ignored"; she graphically told how the new ranking suggested that water from thirty-five polluted wells shut down in California's Santa Clara Valley for trichloroethylene contamination was in fact "no worse than ordinary tap water and far less dangerous than an equal volume of cola, beer or wine, all of which contain carcinogens." Brody reported Ames's calculation that the alcohol in a single bottle of beer was nearly 3,000 times more dangerous than a quart of tap water contaminated by chlorination-derived chloroform, and that the formaldehyde commonly breathed in homes (from building materials) was 60 times more carcinogenic than drinking a quart of tap water per day. Average residues of DDT and PCBs in food and EDB in water were less dangerous than a daily cup of coffee and only one-tenth as dangerous as eating a daily peanut butter sandwich. The HERP rankings indicated that pollutants in drinking water and pesticide residues in foods were likely to pose only a "minimal carcinogenic hazard relative to the background of natural carcinogens."[25] Surely this was good news for those who had come to regard life itself as a kind of treachery: as Brody put it, "With danger seeming to lurk everywhere, many people are tempted to give up trying to protect themselves at all." The new ranking seemed to indicate that most industrial hazards are really quite mild.[26]

Television also picked up on the story. The MacNeil/Lehrer NewsHour did a segment on Ames, as did several other stations. The most provocative was the March 18, 1988, report produced by ABC-TV's *20/20* entitled "Much Ado About Nothing." The program opened by suggesting that viewers were about to see something "hazardous to your beliefs"; its general thrust was that fear and hysteria had replaced common sense on the cancer issue, and that environmentalists and sensation-seeking reporters were largely to blame. Ames himself, identified only as a "cancer researcher," went on camera to suggest that whereas a "whole movement of people" was committed to the idea that man-made chemicals were causing a lot of cancer, the "science" was in fact "all going the other way." Ames repeated his claims that vegetables by weight were about 5 percent toxic chemicals and that "celery, alfalfa sprouts and mushrooms are just chock full of carcinogens." Asked whether he was suggesting that we avoid fresh produce, Ames responded that his point was rather that industrial hazards have been exaggerated, given the amount of man-made pesticide residues consumed by people: "We get more carcinogens in a cup of coffee than we do in all the pesticide residues you eat in a day." As host Hugh Downs's lead-in suggested, "all the concern about this toxin, that pesticide" was much ado about nothing.[27]

The appeal of Ames's account went far beyond that of your average science story. Here was welcome news for people tired of being reminded that industrial life is cause for constant vigilance. "Is living in modern society really so hazardous?" Jane Brody of the *Times* asked. Apparently not. The policy implications were also clear: as Brody reported, the new HERP figures suggested that "banning wine would have a far greater beneficial effect, at least in terms of preventing cancer, than did banning EDB." The "hidden lesson" of Ames's analysis was that industrial chemicals present both benefits and risks; in the case of EDB, for example, one had to balance the hazard of the substance against its value for preventing carcinogenic molds in food.

Environmentalists and consumer advocates were understandably less convinced by the new data, and were particularly worried that Ames's argument would be used to thwart or relax environmental legislation. Samuel Epstein and Joel Swartz, in a long letter to *Science,* denounced the argument for its "substantial errors, omission of relevant data, and reliance on tenuous hypotheses."[28] *Consumer Reports* charged that Ames's thesis was essentially a scientific rubber stamp for industry efforts to absolve itself of guilt for toxic pollution. *Public Citizen,* the Washington-based Nader magazine, complained that Ames had offered "a soothing, seductive fantasy" for "a scandal-weary public eager to forget about toxic dumps, pesticides, and food additives."[29]

The presentation of Ames's views on national television generated a predictably harsh response from alternative agriculture groups. One week after the *20/20* broadcast, Terry Gips, executive director of the Minneapolis-based International Alliance for Sustainable Agriculture, wrote to Samuel Epstein to thank him for his "unflagging struggle against the forces that are leading us (and have already led us) to disaster"; Gips agreed to do what he could to alert other groups to the need for a response.[30] Shortly thereafter Gips sent a memo to 100 pesticide activist and sustainable agriculture groups asking for the formation of a national campaign to respond to *20/20s* "imbalanced and dangerously misleading analysis" of the Ames argument.[31]

Gips and John Clark, a Berkeley-trained biochemist, convened a meeting of the newly formed Bruce Ames Action Committee to discuss possible responses to the program. (The meeting took place at a conference organized by the National Coalition Against the Misuse of Pesticides in Washington, D.C.) The committee prepared a document repudiating Ames's assertions that cancer rates are not increasing, that natural carcinogens pose a greater hazard than synthetic carcinogens, and that organic is no better than conventional produce. The committee also rejected Ames's assertion that cancer clusters occur merely "by chance." Ames had compared correlations between cancer and pesticide usage to the coincident declines of stork migrations and the European birthrate. The committee countered by pointing to strong evidence linking pesticides to cancer, including: a 1986 study showing that Kansas farmers who used the herbicide 2,4-D more than twenty times per year (and mixed their own) were eight times more likely to contract non-Hodgkin's lymphoma; a 1987 study linking regular use of indoor and outdoor pesticides to a sixfold increase in childhood leukemia; a California study linking agrichemical exposures at the Kesterson Wildlife Refuge to excess cancers; a study showing that children raised in homes treated with chlordane have a higher incidence of brain tumors; and several others. The committee argued that it was premature to dismiss pesticides as health hazards, given that most of the 50,000 pesticide products registered for U.S. use had never been tested, and only about 20 percent of the 600 most commonly used pesticide ingredients had been evaluated by the EPA. The group cautioned that Ames had had a "chilling" effect on universities, "pushing them to the right."[32]

Environmentalists were also disturbed by the fact that, in the mid-1970s, Ames had been a leading critic of environmental pollution. In his 1977 testimony before a California Department of Industrial Relations inquiry into the safety of dibromochloropropane, for example, Ames expressed his worry "about the modern chemical world and whether food additives and environmental chemicals might start to increase mutations in the human gene pool, and result in many more defective children, and we wouldn't know about it for a long time."[33] Ames had been particularly concerned about the brominated chemicals widely used in agriculture and manufacturing—the flame retardant Tris fell into this group, as did the fumigants methyl bromide and EDB. In his 1977 testimony, Ames had warned that EDB was a powerful carcinogen, that, indeed, one had to suspect the entire class of chlorinated or brominated compounds: "I think a very high percentage of them are going to turn out to be mutagens and carcinogens."[34] Ames pointed out that pesticide residues had been found in body fat and mothers' milk; he also stated that, though it was difficult to assess the risks of very low doses, "it's unlikely there is going to be a safe dose." He speculated that the cancer rate might suddenly start "shooting up" in the next five or ten years as a consequence of increased exposures to industrial pollutants. Most of the elements of environmentalist rhetoric are here: that "most of the cancer we have now is environmentally caused" ("maybe even 70 or 80 percent"), that high-dose testing is a valid way to compensate for the statistical limitations of animal studies, that there is unlikely to be a "safe dose" for carcinogens, and so forth. Ames asserted that in the absence of evidence to the contrary, it was best to be conservative and assume that carcinogens could cause cancer even at very low levels of exposure.[35]

By the 1980s, however, Ames had reversed many of these positions and was arguing that industrial hazards were being exaggerated and that the real culprits were natural carcinogens, faulty lifestyles, tobacco, and high-fat diets. In 1987, having already questioned the value of his own rapid bioassay (the Ames test), he further challenged the value of animal tests for estimating human carcinogens—especially at very high doses.[36] By the end of the 1980s, his work would be called "probably the greatest blow ever to the environmental movement."[37] Edith Efron wrote that Ames was the first to breach the "ideological wall of silence" surrounding the idea that nature, not industry, was primarily to blame for human cancers; he was the first great "defector" from the apocalyptic cause.[38] Ames, incidentally, returned the compliment in a review of Efron's *The Apocalyptics,* which he characterized as "the *Silent Spring* of the counter-revolution."[39]

Environmentalists troubled by Ames's reversal have provided a number of different explanations for the change. According to a member of the Citizen's Clearinghouse for Hazardous Wastes,

> Ames, who always had a reputation for having a big ego, finds his testing method slipping in prestige. Around the same time, he is befriended by political ideologues who introduce him to a new way of looking at the world. He develops a "new" method for ranking carcinogenic risks. As his research and public pronouncements take a new slant, focusing on "natural" carcinogens, industry starts to give him the rewards and homage he was missing from his peers and the public. The Reagan Administration adds fuel to this change by abruptly shifting federal policies and research funding *away* from industry-generated hazards and towards lifestyle and blame-the-victim theories of cancer causation.[40]

Members of the committee formed to counter the *20/20* broadcast suggested privately a number of other possible reasons for Ames's reversal:

> (1) Ames had had a bad run-in with some environmentalists; (2) He is afraid of nature, a "lifeaphobic"; (3) He is simply shifting the focus of his limited, reductionist, concepts from synthetic carcinogens to naturally occurring ones without a holistic context; (4) He has been lured by the industry through various awards and distinctions. . . . ; (5) Despite his lack of credentials in epidemiology, he has promoted his theories and, once they were advanced, has been so attacked that he has dug in his heels even further to defend them.[41]

Forbes magazine had a simpler explanation, characterizing Ames as "not so much a turncoat as an unflinchingly honest man."[42] One wonders whether the overheated, sometimes crazy atmosphere of Berkeley may have also played a role.

I myself asked Ames about his change of heart, and he willingly conceded that, in the late 1970s, he underwent a political conversion from a kind of liberal environmentalism to a kind of free-market libertarianism. A Polish scientist working in his lab, Tadeusz Klopotowski, introduced him to ideas of antibureaucratic, free-market libertarianism, and a geneticist friend, Bill Havender, convinced him to read the libertarian works of Friedrich A. Hayek. (Klopotowski later became a leader in the Solidarity movement and a Polish senator; Ames helped to support his underground newspaper prior to the fall of communism.) He now describes himself as a market liberal "in the Madison/Jefferson tradition."[43]

Many critics have held this reversal against him. (It would be easy to make too much of his change: I asked him whether he had read Rachel Carson's *Silent Spring* when it first came out in 1962, and was surprised to hear him say that he had—and that even then it had made him "want to puke.") Epstein and Swartz, for example, in their 1984 letter to *Science,* argued that Ames's backtracking had undermined his credibility.[44] But others countered that such an argument could easily backfire. Marvin Schneiderman advised Epstein not to make too much of Ames's change of mind, given that such a reversal could be used as a "credentialing" device saying, in effect, "I must be right now, because the data has led me to switch my position. See what an open-minded person I must be?" After all, science is supposed to be self-correcting; what shame should there be in admitting one was wrong?[45]

Ames has indeed managed to present this self-correcting aspect of science as a key to his broader social philosophy. In his 1989 "Science and the Spontaneous Order," originally presented as a graduation address at the University of California, San Francisco, he draws from Hayek and others to argue that the power of science derives from its being spontaneously formed rather than centrally planned. He quotes Adam Ferguson that civilization is "the result of human action, but not the execution of any human design." Ames supports Robert Nozick's preference for the minimal state over centralized and utopian bureaucracies; he also endorses Hayek's opinion that, in science as in society, competition gradually produces a natural selection of superior over inferior forms.[46]

Ames therefore opposes governmental intrusions in the name of environmental regulation. He stresses repeatedly that the costs of regulation generally exceed the benefits. As he put it in an April 1992 lecture at Penn State University: "A lot of our recession is a green recession" caused by excessive environmentalist zeal in promulgating regulations. His advice in the realm of cancer control is purely personal: stop smoking and "eat your veggies." (Vegetables contain natural carcinogens, but these are more than counterbalanced by the anticarcinogens they contain.) He is disappointed that federal environmental and food protection agencies have been slow to test his ideas: natural substances are apparently not tested unless they are produced industrially, and only then because the people responsible for making them can be sued in a court of law.

Ames also makes the more complex argument that environmental regulations—of pesticides, for example—can actually *harm* health by increasing the price of fruits and vegetables. If environmental regulations were relaxed, he suggests, fruits and vegetables could be grown more cheaply. These lower costs could then be passed on to consumers, who would then eat more of these healthful foods, lowering their cancer rates. Environmental regulations exacerbate the cancer problem by sapping the nation's wealth. In a 1991 interview with *Reason* magazine, a libertarian publication devoted to "free minds and free markets," Ames recalls a dinner with Milton Friedman, where the Chicago School economist asserted that it would be a good thing to abolish "the FDA, the EPA, and government funding of science." Ames apparently countered that, while he could certainly consider the first two suggestions, it would be harder to do without government funding of science—"as long as it's competitive."[47]

One can, of course, dispute whether the savings to be gained by deregulating agricultural industries would be passed on to consumers; one might also question whether, even if they were, this would result in the consumption of more cancer-fighting vegetables. Wealth may or may not be the cure for cancer: as Ames himself points out, the optimal diet—from a cancer prevention point of view—would probably be something like what Chinese peasants ate in the nineteenth century.[48] Ames still believes today that DDT should never have been banned. The pesticide for him is still a big net plus in the history of world health: by combating malaria and other insect-borne diseases, DDT "saved tens of millions of lives, more than any substance in history with the possible exception of antibiotics."[49]

Quite apart from political objections, critics have sought to challenge the Ames thesis on technical grounds. Epstein and Swartz charged that the HERP ranking underestimated the potency of industrial carcinogens by presuming a linear extrapolation from high-dose tests (see chapter 7) and ignored sensitive groups such as pregnant women.[50] William Lijinsky of Bionetics Research, Inc., testified before the Senate Committee on Labor and Human Resources that one could hardly argue that natural carcinogens in food were "much more important" than industrially produced pesticides, given that vegetarians have a considerably lower cancer risk than others.[51] Devra Lee Davis of the National Research Council wrote to *Science* to note that, when prepared properly, many of the natural toxicants in foods are rendered harmless; she cited a 1966 study showing that cooking reduces or destroys the cyanogenic glycosides in lima beans, thiaminase in fish, goitrogens in vegetables, and avidin in eggs.[52] Thomas Culliney of the USDA Forest Service pointed out that many of the foods singled out by Ames as hazardous (including fruits such as apples, oranges, and bananas and vegetables such as broccoli, brussels sprouts, and cabbage) are outstanding sources of vitamins A and C—both of which may play a role in reducing human cancers. Culliney conceded that there were natural toxicants in many foods and that several of these were quite potent (the mycotoxins in moldy cereals and nuts, for example), but there was little or no nutritional evidence that the fruits and vegetables discussed by Ames had ever posed a health risk to American consumers.[53]

Edward Groth of the Consumers Union further charged that, even if it were true that there were natural carcinogens in foods, this was hardly grounds for complacency about industrially derived contaminants. If natural toxicants promote the growth of cancer by initiating cellular proliferation, then the fact that there are natural carcinogens in foods does not mean that we should forget about the synthetics: "indeed, we should worry *more* about the presence of carcinogenic initiators in foods." If natural foods contain natural cancer promoters, all the more reason to avoid synthetics that appear to initiate cancer.[54] David Pimentel of Cornell pointed out that the nitrates naturally present in spinach and lettuce can be increased to dangerous proportions by commercial fertilizer applications; he also noted that herbicides and other pesticides can stimulate a plant to increase its production of natural toxins by a very large margin—a point that Ames never mentions.[55]

Critics also pointed to difficulties in Ames's discussion of the peanut butter threat. Leonard Stoloff noted that there were doubts about whether the contaminant in question, aflatoxin, was in fact a carcinogen. Mice develop cancers when injected with aflatoxins into the gut, but not when given orally. Aflatoxin exposures have been correlated with high liver cancer rates in Africa, but the very same aflatoxins appear to cause very little liver cancer in the United States. The reason, as Stoloff pointed out, is that aflatoxins apparently cause cancer only when the liver has been damaged by other agents, such as the hepatitis B virus. Hepatitis is far more common in Africa than in the United States, which may account for the higher rates of (perhaps aflatoxin-induced) liver cancers.[56]

Several critics suggested that over millions of years of evolution, humans may have developed physiological mechanisms to defend against natural carcinogens.[57] This might be true, for example, for substances that have been common in the human diet for thousands of generations. Devra Lee Davis, in her letter to *Science,* presented archaeological evidence that fire-cooked wild game meats have been consumed by humans for at least 700,000 years. Natural selection would probably not weed out agents that induce chronic diseases in people past the age of reproduction (unless there were kin-selection effects), but since most carcinogens are also teratogens, long-term exposures to food-borne toxicants might have weeded out genotypes that produce sperm, eggs, embryos, and fetuses vulnerable to the toxic components of foods. Davis also noted that diet is of limited value in explaining shifting cancer patterns. Breast cancer may be linked to patterns of fat consumption, but there are other patterns with no known nutritional basis. Davis speculated that the overall rise in cancer incidence may have to do with the exponential growth of chemical production from 1945 through the late 1970s: humans have been exposed to dramatically increased levels of synthetic organic carcinogens, and exposures to these new compounds may enhance the toxicity of older kinds of exposures.[58]

Peter Weiner suggested similarly that human DNA-repair mechanisms may be better able to handle the "familiar" damage caused by natural agents than the injuries caused by the novel petrochemical agents introduced over the last century or so.[59] Marvin Schneiderman added that humans may have evolved resistance mechanisms that work only during the ages of reproductive capacity. Given that men often retain the ability to reproduce past the age of sixty, this might explain why the average age of onset of cancer is higher in men than in women. Women, even though they live longer, get cancer (on average) at significantly earlier ages than men.[60]

Ames and Gold responded to these and other criticisms in great detail. In response to Davis's evolutionary argument, they noted that humans have developed general—rather than specific—defenses to the natural toxins in plants. These include the constant shedding of the surface layer of cells of the digestive tract, the glutathione transferases for detoxifying alkylating agents, the excretion of hydrophobic toxins from liver or intestinal cells, a variety of defenses against oxygen radicals, and DNA excision repair. Ames and Gold argued that these general defenses work against both natural and synthetic compounds, since the carcinogenic mechanisms are the same in both cases. They also took issue with Davis's implicit claim that the human diet has remained constant over many thousands of years. In his 1987 review, Ames had argued that the modern diet is "vastly different" from that of primitive humankind. The human diet is still changing: "For example, as part of the back-to-nature movement we are eating canavanine in alfalfa sprouts, carcinogenic hydrazines in raw mushrooms, and carcinogens in herb teas." Cooking food destroys some carcinogens, but it also produces others—the nitrosamines produced in gas ovens, for example.

The key issue, Ames and Gold suggested, was not whether the production of synthetic chemicals has increased in recent years but whether the pesticide residues that enter food or water supplies are significant. They repeated their claim that humans ingest "about 10,000 times more of nature's pesticides than man-made pesticides," and that the biological effect in either case is exaggerated by animal studies that depend on extrapolations from maximally tolerated doses. They called for the compilation of a teratological equivalent of the HERP index, to see how the risk of birth defects from commonly ingested substances (such as alcohol) compare with the risks of residual synthetic teratogens. They disputed Davis's claim that multiple myeloma and brain cancer rates have risen sharply among the elderly, citing Richard Peto's suggestion that the increases may be diagnostic artifacts. Finally, they questioned the wisdom of the U.S. spending $80 billion a year on pollution abatement and control, given the uncertainty of whether environmental pollutants "at parts-per-billion levels have public health significance." They reasserted that technology has contributed in important ways "to our steadily increasing life-span," and that critics of a particular technology must always consider the hazards of the alternatives.[61]

Ames often makes the point that it is better to worry about large risks to many than about small risks to a few. This is consistent with the utilitarian notion that it is better to reduce a large harm done to many than a small harm to a few. (He doesn't pay much attention to small risks faced by many.) And of course it is true that if governments really wanted to reduce cancer, and to do so efficiently, they would strengthen efforts to reduce smoking. In terms of social policy, the easiest and most direct way to reduce American cancers would be to triple the price of cigarettes, ban tobacco advertising, and eliminate smoking in the workplace. Most cancer hazards pale by contrast with the tobacco menace, and little has been done to address the problem.[62]

One drawback of the utilitarian approach, however, is that risks are often unequally shared. A single hierarchy of total average risks masks the fact that toxic burdens are suffered unequally—and not just by rich and poor but also by young and old, white and black, city dweller and rural resident, and so forth. It is not enough, in other words, to examine the magnitude of risk; one also has to ask who is suffering, why, and how. The problem is the same as that encountered in efforts to tax pollution or to allow the purchasing of pollution rights. Such plans can be effective in reducing overall pollution levels but they may not be very fair, allowing industries to concentrate their polluting in one area while keeping another area clean. The greatest good for the greatest number is not always nice for those at the bottom of the heap.

Several of Ames's critics have stressed the importance of distinguishing between voluntary and involuntary risks—a distinction lost in the HERP rankings. Stephen Lester of the Citizen's Clearinghouse for Toxic Wastes has argued, for example, that even if it turns out to be true that the hazards posed by natural carcinogens are greater than those posed by synthetic carcinogens, it would be wrong, for purely political or ethical reasons, to devote too much attention to this fact. Lester argues that people should have a *right to choose the risks* to which they are exposed, and that this choice is foreclosed when they are forced to drink polluted water or breathe polluted air—however large or small the actual threat. Natural and industrial carcinogens are different in this regard: people can choose whether or not to smoke cigarettes or eat peanut butter, "but they cannot choose not to breathe air."[63]

Edward Groth, associate technical director for the Consumers Union, put the matter as follows in his long reply to Ames's criticism of the *Consumer Reports* article on his work:

> My perspective is that the public responds not only to the *size* of various risks, but to how *acceptable* they are, which is determined by many attributes in addition to size of a hazard. A very large risk, such as cigarette smoking or driving a car, is acceptable to most consumers because these are self-chosen activities, and the benefits accrue immediately and directly to the person at risk. On the other hand, a small risk, such as UDMH [Alar] in apple juice or radiation from nuclear power plants, is unacceptable to many people, because they have no choice in the matter (which is seen as ethically intolerable) and because the distribution of risks and benefits is perceived as unfair.

Groth noted that the importance of such a distinction was borne out by the recent history of public policy on cigarette smoking:

> As long as it was perceived that only smokers were at risk, public opinion favored giving freedom of choice greater weight than protecting health. But once smokers were seen as threatening the health of others via "second-hand smoke," smoking in public places, on planes and trains, and in offices has been aggressively restricted, even though the risk to passive smokers is a small fraction of that borne by smokers themselves. . . . In a democracy, the public has a right to pick and choose among risks, to find some morally acceptable and others morally abhorrent, and to ask government to manage risks based on their acceptability, not only on their size.[64]

It is wrong, in other words, to treat risks of equal probability as having equal political significance. People who willfully engage in smoking or risky sports (skydiving or motorcycle racing, for example), can still claim the right to be able to drink clean water—even if their self-imposed risks vastly exceed those that are foisted upon them. The right to drink clean water should not be contingent upon one's giving up smoking. Daniel Wartenberg echoes Groth when he points out that it is wrong to rank all possible health hazards along a single scale and then to focus only on those that head the list: such a procedure ignores the fact that the source of the risk and whether it is voluntarily assumed can have much to do with how people feel about it.[65]

Finally, one should not forget that petrochemical pesticides—unlike "nature's pesticides"—are hazardous not just in their consumption but also in their production and distribution. Natural pesticides may pose a danger when they are eaten, but they cannot pose any other type of danger. Natural pesticides are never spilled, dumped, or left lying around for children to play with; they do not soak into groundwaters or accumulate in the food chain; they are not stored in body fat for months or years. The same cannot be said for petrochemical pesticides. Petrochemical pesticides must be manufactured, transported, stored, purchased, and then sprayed. At every step of the way, there are potential hazards. The World Health Organization in 1990 estimated that there were 3 million cases of acute severe pesticide poisonings each year throughout the world, responsible for about 220,000 annual deaths. Ninety-nine percent of these deaths are in the developing world.[66]

In the United States, the American Association of Poison Control Centers records some 67,000 annual pesticide poisonings, 27 of which lead to death. Pesticides and herbicides have had profoundly destructive effects on many forms of wildlife: David Pimentel and colleagues recently calculated that the environmental and public health costs of pesticide use in the U.S. exceed $8 billion per year.[67] One should also recall that it was a pesticide ingredient—methyl isocyanate—that killed some 3,000 people in the world's worst industrial accident in Bhopal, India, in December 1984; it was also pesticides that leaked into the Rhine from a burning Sandoz chemical plant in November 1986, devastating an entire ecosystem and several cities' water supplies. These and other disasters must be figured into the social cost of using pesticides. It is misleading, in short, to consider only the risks of consumption but not those of production, distribution, and application, along with the ever-present possibility of accidents.

THE POLICY IMPLICATIONS

Ames has always been careful to point out that he is not an industry hack. He has never consulted for industry, nor has he ever accepted corporate fees for lectures or publications—he is untainted. ("If I give a talk at Du Pont, I have them send the honorarium to charity.")[68] Given his background as a pioneer in the field of environmental carcinogenesis, the Berkeley biochemist must surely appear as a godsend to anyone seeking to exonerate industry from charges of toxic malice. In their wildest dreams, Monsanto or Dow could never have imagined such a man. Here was an independent and well-respected scientist—in his early incarnation, in fact, a respected environmentalist—with a message that appeared to pull the scientific rug from under the environmentalists' feet.

As one might imagine, industry advocates immediately seized upon the story. The Asbestos Information Association featured press reports on Ames in its monthly newsletter; Edith Efron welcomed him as "the first great apocalyptic defector" in her critique of *The Apocalyptics*.[69] When the national press picked up on the story, trade associations stepped up their coverage. On April 1, 1988, one week after ABC-TV's *20/20* report, the National Agricultural Chemicals Association (NACA) issued a news release asserting that Americans enjoy "the safest, highest quality, and most abundant food supply in the world due largely to the judicious use of agricultural chemicals." Ames had shown that the pesticide residues consumed by people were "absolutely trivial"; environmentalist scare stories linking pesticides to cancer had resulted in the misallocation of "billions of dollars to reduce trivial risks." It was "re-freshing to see a respected network news program" handle such a controversial issue in a responsible manner. The release was accompanied by a transcript of the *20/20* report and an invitation to discuss agrichemical residue issues with NACA's director of scientific affairs. The release also noted that a forty-five-minute videotape featur-ing Ames lecturing on "Carcinogens, Anticarcinogens and Risk Assessment" was available from the Council for Chemical Research in Washington, D.C.[70]

Ames has been popularized by other industry groups. In the summer of 1989, the American Council on Science and Health (Whelan's organization) released a film entitled "Big Fears, Little Risks: A Report on Chemicals in the Environment," fea-turing Bruce Ames refuting charges that pesticide residues pose a human health haz-ard. The film, narrated by Walter Cronkite (for which Cronkite was paid $25,000), was first aired on PBS, and since then has been widely distributed by the California Grape Commission and several pesticide manufacturers (the Mobay Chemical Com-pany, for example).[71] The Styrene Information and Research Center has reprinted a 1990 *Science* essay by Ames and Gold suggesting that humans are "well buffered" from the possible carcinogenic effects of styrene, the molecular building block for polystyrene cups and plates.[72] Jeffrey B. Kaplan, a molecular biologist with Ameri-can Cyanamid, has argued that pesticide bans can do little to reduce cancer because "humans ingest about 10,000 times more natural than synthetic pesticides."[73] Ames has also received sympathetic coverage from journals like *Pest Control* and *National Review;* the *American Spectator* ran an opinion piece by Ames originally requested by (and then refused by) *Sierra* magazine to celebrate the hundredth anniversary of the founding of the club.[74]

Ames has also begun to figure in debates on the course of environmental policy. In 1986, he led the scientific fight against California's Safe Drinking Water and Toxic Enforcement Act (Proposition 65), a bill requiring that "No person in the course of doing business shall knowingly and intentionally expose any individual to a chemical known to the state to cause cancer or reproductive toxicity without first giving clear and reasonable warning to such individual." The proposition was over-whelmingly approved by voters in November of that year, but not before Ames had testified that it was irresponsible and a waste of taxpayers' money. Ames was joined in this effort by the Grocery Manufacturers of America, which pointed out that most foods contain natural arsenic and other compounds known to cause cancer. A *Sci-ence* magazine editorial by Philip Abelson endorsed both Ames and the grocery manufacturers by predicting that salt, wine, beer, and most roasted and broiled foods would all have to carry labels under the new law, as would 15,000 other grocery items and most foods served in restaurants.[75]

When Proposition 65 received the overwhelming approval of California voters, Governor George Deukmejian sought to narrow its interpretation by restricting it to only 29 chemicals. (A 1987 ruling by a Sacramento judge, however, added another 201 to the list.) Deukmejian appointed Ames to the scientific advisory panel respon-sible for determining which chemicals should be added to the list, prompting Carl Pope, the Sierra Club's political director and a co-author of the proposition, to de-nounce the appointment as an act of "sabotage" on the part of the governor.[76] In

1990, Ames spoke out against California's Proposition 128, which would have banned many pesticides. He also opposed the Alar ban: in the May 19, 1989, issue of *Science,* he and Lois Gold defended the ripening agent on the grounds that it could actually prevent cancer:

> Without Alar, the danger of fruit fall from leafminers is greater, and more pesticides are required to control them. Also, when apples fall prematurely, pests on the apples remain in the orchard to attack the crop the next summer, and more pesticides must be used. Since Alar produces firmer apples and results in fewer falling to the ground, treated fruit may be less susceptible to molds. Therefore, it is possible that the amounts and variety of mold toxins present in apple juice . . . will be higher in juice made from untreated apples.[77]

Ames has supporters in the FDA[78] and in other branches of government: in 1990, for example, the Bush administration's Office of Management and Budget published a major critique of risk assessment and risk management, especially the practice of relying on "conservative" or "worst-case" assessments to guarantee a margin of safety. The report suggested that the use of high-dose animal experiments to estimate biohazards tends to exaggerate human health effects by several orders of magnitude; Ames et al.'s "Ranking Possible Carcinogenic Hazards" was the most frequently cited source. The text also incorporated Ames's views on the perils of aflatoxin in peanut butter and other "natural hazards."[79]

The force of Ames's argument stemmed partly from the fact that, in the wake of widespread attention to industrial carcinogens, the contribution of lifestyle factors such as smoking or sexual and eating behavior had been somewhat neglected. At least part of the neglect was ideological: as a later jargon might have it, it was simply not politically correct to believe that barbecue and wild mushrooms could be as dangerous as Exxon and Shell.

Bruce Ames has become the nation's foremost scientific defender of the claim that industrial pollution is a minor threat to human health. On December 14, 1990, when Philip H. Abelson, one of the most powerful figures in American science, wrote a *Science* editorial arguing that the EPA was embarked on programs "that will cost hundreds of billions of dollars but will have little impact on human health," Ames was the chief authority he cited to support his challenge. Ames had questioned the "cornerstone" of the EPA effort, its reliance on animal studies involving the administration of huge levels of chemicals to rats; Ames had produced "substantial evidence" that such studies were often misleading. Abelson argued that the EPA was setting guidelines on the basis of the science of the 1970s; what was needed was regulation based on the science of the 1980s and 1990s.[80]

Abelson was referring not to the 1983 or 1987 papers comparing natural and synthetic carcinogens but rather to a series of more recent articles in which Ames had shown reasons to doubt the value of traditional high-dose animal studies for predicting human carcinogenic risks. Animal studies had long been challenged by industry advocates, and those challenges could generally be dismissed as self-serving strategies to stave off regulation. Ames and his colleagues now gave new life to these claims by showing that high dose exposures could stimulate cancerous growths merely by initiating cellular replication.

Hanging in the balance was not simply a question of the biochemical mechanisms of carcinogenesis but one of the central theoretical questions of cancer theory: namely, whether there are thresholds of exposures below which carcinogens are safe. If chemical irritants could produce cellular growth, and if cellular growth could magnify mutations, then it was likely that a broad range of common chemical irritants could cause cancer when administered at sufficiently large doses. Animal studies relying on such large doses were probably flawed, insofar as they exaggerated carcinogenic potentials. Most important, there were probably thresholds of exposure below which most carcinogens were actually safe. The postulate of thresholds was by no means new, but Ames had managed to give it a respectability it had never before had—and to which we now turn.

Chapter 7

The Political Morphology of Dose-Response Curves

The connection between low-level insult and bodily harm is probably as difficult to prove as the connection between witches and failed crops.
—Alvin M. Weinberg, 1985

If you poison your boss a little bit each day it's called murder; if your boss poisons you a little each day it's called a Threshold Limit Value.
—James P. Keogh, M.D.

OF ALL THE CANCER WARS, perhaps none has been more hard-fought than the one over whether there are thresholds below which exposure to a given carcinogen—chemical or radionuclear—is safe. For decades, fierce battles have been waged over the shape of the dose-response curve. The *Science* reporter Thomas H. Maugh has described the disagreement over the body's ability to detoxify carcinogens as one that "has raged for years with all the intensity of a jihad."[1] An equally contentious dispute has divided radiation experts. Though there are plausible grounds for belief on both sides, there is often little empirical evidence to adjudicate the matter, so it is difficult in principle to resolve. None of this has prevented the protagonists from holding very strong views; even Richard Peto has been surprised by the longevity of the principle of thresholds, unsubstantiated as it is by any empirical evidence.[2]

The significance of the issue of thresholds derives primarily from its policy implications. If exposures to toxic chemicals below a certain dose are safe, then there is no public health reason to scrub the environment clean beyond a certain point. If, however, even very low level exposures are proportionately dangerous, then it is at least an open question how much of a particular hazard we should allow ourselves or others to endure. Governmental agencies worry about exposures that raise one's lifetime chances of contracting cancer by 1 in 100,000 or even 1 in 1,000,000; it is often difficult to know, however, which among the many carcinogens to which we are exposed are likely to produce a risk of this magnitude.

The question is highly charged because low-level exposures are much more common than high-level exposures—indeed, the majority of cancer deaths may well be traceable to low-level exposures to carcinogens. The EPA estimates that radon gas in homes, for example, kills between 7,000 and 30,000 Americans every year, the majority of whom were exposed at levels below the EPA's own limit for "acceptable" exposures (4 picoCuries/liter).[3] The same could well be true for many other cancer-causing agents. If low-level exposures are responsible for the bulk of cancer, this has obvious implications for public health and environmental policy. It also has potentially serious economic implications, given that it is often difficult or even impossible to reduce a low-level exposure to a zero-level exposure—short of an outright ban on the offending substance. Even then, of course, the substance in question may linger in the environment for years after the end of its official use. Finally, there are substantial political implications, because it would obviously be in many people's interests to maintain that, below a certain dose, a known carcinogen is completely safe. If, however, as many environmentalists believe, low doses can be proportionately or even more than proportionately dangerous, there are good reasons to worry about the production, distribution, and consumption of even small amounts of toxic substances.

Policy, though, is not the only thing at stake. As we shall see, the "jihad" over thresholds often boils down to different conceptions of the body and different conceptions of regulatory prudence. It is therefore important to realize that allegiances in such matters spring not just from the "facts" but from deeply held political and ethical commitments that, by focusing attention and organizing ambiguity, can shape the dose-response curves preferred by cancer-risk assessors. Little is known about the health effects of very low doses of most potential carcinogens, but the high stakes involved have made it difficult to avoid acting on the basis of ignorance.

EARLY CONCEPTS OF CHEMICAL THRESHOLDS

The idea that something good for you in one dose may be bad for you in another is old—much older, in fact, than the comparatively recent idea that a single molecule or photon might cause cancer. The medical use of the term *threshold* dates only from the 1930s,[4] though forerunners of the modern notion (tolerance dose, maximum allowable concentration, safety level, and so forth) date from the end of the nineteenth century.[5] In a more figurative form, the idea that something may be good at one dose and bad at another has been conventional pharmacopedic wisdom for half a millennium: Paracelsus is the author of the maxim that "the dose makes the poison"—though the frequent use of this slogan by industry apologists forgets that Paracelsus was talking about the medicinal use of things that, in small doses, might be good for you.

The dose-response curve, the mainstay of modern toxicology and risk assessment, is a twentieth-century invention—a consequence of efforts by industrial and public health officials to quantify the harms done to industrial laborers by substances like lead or radium. The whole idea of a stable relation between dose and response, one should note, is much less clear in the context of infectious disease, since germs are biological organisms that grow and reproduce in the body. The course of an infectious disease can be independent of how severely one is exposed; it is first with industrial diseases that one gets the notion that the probability or severity of a disease can be understood as a well-behaved function of the frequency, intensity, and duration of exposure. Industrial hygienists in the 1920s and '30s sought to quantify the injurious effects of chemical and physical agents in a way that was both explanatory and predictive; the longer-term goal was to establish levels of safe or at least acceptable exposure. Pressures to monitor hazards increased, especially as workers' compensation laws gave legal force to workers' claims that they had been injured on the job.

Early concepts of thresholds were developed as part of these efforts to determine how much of a given chemical could be ingested or inhaled without "undue" injury. German toxicologists were the first to conduct systematic research to determine tolerance levels for specific chemical substances. K. B. Lehmann in the 1880s began a series of experimental studies that would eventually win him the title "father of toxicology."[6] Lehmann's problem was to determine how much of a particular irritating gas—sulfur dioxide, ammonia, or a given halogen, for instance—could be tolerated by workers in Germany's nascent chemical industries. The modus operandi of the German professor is shocking by today's standards:

> Lehmann ordered his laboratory servant to stay enclosed for one hour in the housekeeper's laundry room, where he was to pour out a calculated amount of the volatile test fluid, to move a sheet of newspaper vigorously until complete vaporization and distribution were achieved, and to handle an analytical sampling device for determination of the concentration in the room air. The experimenter checked the condition of the test person by observing him occasionally from outside through a window. If the servant left the premises with significant signs of survival, the concentration was rated "just tolerable for short-term exposure."[7]

Lehmann's work formed the basis for early-twentieth-century efforts to develop workplace exposure standards, efforts that accelerated with the frenzied production of explosives, dyes, reagents, solvents, fuels, and other synthetic substances during World War I. German toxicologists were again at the forefront of this work: by 1912, Rudolf Kobert was publishing tables of "The Smallest Amounts of Noxious Industrial Gases Which Are Toxic and the Amounts Which May Perhaps Be Endured"— data that were later used by the American Bureau of Mines to formulate guidelines for American miners.[8] The first efforts to construct mathematical models of dose-response relations appeared about this time: a 1921 article in a German medical journal showed that animals exposed to hydrocyanic acid below a certain concentration showed no ill effects; this contrasted with other lung irritants, such as phosgene—a poisonous gas used in the manufacture of glass, dyes, and resins—which evidenced no apparent threshold. Animals breathing phosgene died at a rate proportional to both the concentration *and* the duration of exposure.[9] Phosgene, in other words, presented a cumulative hazard with "no safe dose," while hydrocyanic acid appeared to be "safe," so long as a certain minimal exposure was not exceeded. The physiological reason for the latter type of response was not understood until the 1930s, when it was found that hydrocyanic is detoxified in the body as the cyanide ion is bound to sulphur with the production of a thiocyanate. The reaction is limited by the availability of sulphur, hence the threshold response.[10]

Interest in such questions was not purely academic. In the United States, health scandals and labor activism after World War I focused attention onto industrial hygiene. In the spring of 1920, for example, two Du Pont rubber workers died after inhaling benzene, used to make coated rubber fabric; a physician hired by the company's safety division found that workers were wearing cotton nasal plugs to stop the nosebleeds caused by the fumes. Another worker died two years later. The use of benzene was curtailed, though no effort was made at that time to establish a tolerance level for the substance.[11] Lead was another concern, especially after 1924, when 80 percent of the workers in a Bayway, New Jersey, Standard Oil laboratory either died or were driven insane by exposure to tetraethyl lead—the antiknock compound then beginning to be used in gasoline. Animal models were used as early as 1921 to develop a voluntary lead-exposure standard, and a 1924 study did the same for mercury.[12] The U.S. Public Health Service and several state boards of health began to issue recommendations for "maximum allowable concentrations" of particularly hazardous substances, but these did not yet have the force of law behind them. Efforts to establish standards encountered resistance: Lothar E. Weber of the Boston India Rubber Laboratory claimed as early as 1921 that a few guinea pig experiments were not enough to convince him that benzene was dangerous.[13] Employers at this time were more likely to weed out workers with lung damage than to monitor and remediate safety hazards.[14]

Early efforts to develop exposure standards were always open to the charge of ambiguity. Was exposure to a chemical below the recommended level a guarantee that it would not make you sick? Management's and labor's interpretations of such questions were not always in harmony. In 1926, when the U.S. fuel industry established voluntary rules on the maximum exposure of workers to tetraethyl lead, the presumption of management was that exposures below this level would pose no risk to workers. Robert A. Kehoe of the University of Cincinnati's Kettering Laboratory of Applied Physiology (and a leading lead industry consultant) argued that the metal was naturally present in the body, and that excesses ingested above this level would simply be excreted. Kehoe maintained that the human body was able to tolerate continuous exposure, so long as it was ingested at or below a rate at which an "equilibrium" could be established between ingestion and excretion.[15] Alice Hamilton, a leading champion of occupational health reforms, disputed the possibility of safe levels of lead as early as 1925, arguing that "where there is lead some case of lead poisoning sooner or later develops, even under the strictest supervision."[16]

Ambiguities persisted in efforts to define exposure limits. Problems of inadequate testing, nonexistent monitoring, and insufficient attention to latent maladies plagued the influential reports issued by the American Conference of Governmental Industrial Hygienists (ACGIH), a voluntary organization formed in 1938 to develop industrywide standards for chemical hazards. The ACGIH developed the trademarked concept of threshold limit values (TLVs), defined as airborne concentrations of particular substances to which "all workers may be repeatedly exposed, day after day, without adverse effect." TLVs differed from the previous era's maximum allowable concentrations[17] in being defined in terms of "maximum *average* atmospheric concentration . . . for an eight-hour day." The earlier notion of maximum *allowable* concentrations had been understood as a ceiling that was never to be exceeded; the new concept implied an upper bound above which values were occasionally allowed to fluctuate. There was also a recognition under the new system that some workers, due to individual variation, might experience discomfort or become sick as a result of exposures below the threshold value. At the tenth annual meeting of the ACGIH in 1948, the chairman of the Committee on Threshold Limits asserted that "it is a figment of the imagination to think that we can set down a precise limit below which there is complete safety and immediately above which there may be a high percentage of cases of poisoning among those exposed."[18]

The ACGIH published TLVs beginning in 1950 and continues this practice today. In 1969, anticipating the establishment of OSHA, most of the threshold limit values on the ACGIH's 1968 list became enforceable under federal law. Critics both before and since this time have pointed out that the ACGIH's Committee on Threshold Limits has always relied heavily on unpublished industry data unchecked by independent authorities: a 1966 committee of the Industrial Medical Association expressed its concern that TLVs were being circulated with little scientific data to support them, and a 1984 German review of that country's standards, based heavily on U.S. TLVs, found that not even one in ten was based on "sufficient animal tests and/or field experience." The basic problem has been that, as a voluntary organization, the ACGIH has had no way to force individual companies to release toxicological data on new chemicals being introduced into the workplace. The chairman of the ACGIH's own Committee on Threshold Limits recognized this in 1969, when he characterized the chemical industry's role in the standard-setting process as "pathetic."[19] A 1990 review by two scholars from the University of California's School of Public Health in Berkeley found that "adverse effects" were common at or below many of the recommended TLVs.[20]

Exposure limits were not as controversial in the first half of the twentieth century as they would become in the second. Exposures might or might not fall within a particular definition of "maximum allowable concentration," but the point was generally moot since toxic chemicals were unregulated during those years and there were few laws governing their use.[21] Cancer was not mentioned in the American Chemical Society's 1927 publication on short-term limits ("maximum concentrations allowable for prolonged exposure") for twenty-five "noxious gases." Neither was it mentioned in 1940, when the State of Massachusetts published maximum "suggested" concentrations for forty-one toxic substances used in industry.[22] Carcinogens were not barred from foods until the 1950s, and no thought was given to cancer in the promulgation of the Federal Insecticide, Fungicide, and Rodenticide Act of 1947. Cancer was only a marginal concern even at Du Pont's Haskell Laboratory, where Hueper worked, the first major industry-financed toxicology laboratory in the country. (Dow and Union Carbide soon set up comparable facilities.) Prior to 1962, the only substance recognized as a carcinogen in ACGIH reports was nickel carbonyl. The ACGIH report for 1962 added three to its list: benzidine, betanaphthylamine, and nitrosodimethylamine. By 1969, when the ACGIH's TLVs became federal law, the list of known carcinogens had grown to about a dozen, with asbestos finally included.[23] Since few of the hundreds of other substances classified by the council had ever been tested for carcinogenesis, however, even the OSHA adoption of council-approved TLVs did not do enough to protect workers.

THE QUESTION OF REPAIR

From one point of view, one could probably argue that it is the no threshold hypothesis that has to be defended as something worth taking seriously. Common sense tells that a large bump on the head might do a great deal of damage, but a small bump might do none whatsoever. One in ten bottles might break when dropped from a height of one foot, but one doesn't expect one in a hundred bottles to break when dropped from a tenth of a foot, or one in a million bottles to break when dropped from 1/100,000th of a foot. Bottles simply don't break when dropped from below a certain height, and the same appears to be true for certain kinds of human bodily insults. Quantity transforms quality. Substances that are deadly poisons in large doses may be vital nutrients in smaller doses: this is true of selenium, for example, and it is also true of several vitamins. Vitamin A is believed to protect against cancer in moderate dosages, but there is evidence that at high levels it may promote cancer.[24] The same appears to be true of many other nutrients—and even of ubiquitous, necessary substances such as estrogen or salt. Ingestion of a great deal of salt can kill a person, though it is harmless or even beneficial in smaller doses. Salt appears to contribute to stomach cancer in places like Japan, where high-salt soy sauce, dried fish, pickles, and salty soybean-based foods are a regular part of the diet. And new theories suggest that a lifetime of hormonal fluctuations can result in the development of breast, ovarian, or endometrial cancer. Once again, the dose makes the poison.

There are certain kinds of poisons, though, that do appear to pose a danger even at very low doses. Radiation is the most famous example. Radiation studies were in fact the first major source of evidence against thresholds. Hermann J. Muller in 1927 had shown that X rays could induce mutations in the genes of fruit flies, and that even a very tiny dose could be potent in this manner.[25] The Russian geneticist N. Timoféeff-Ressovsky in the 1930s found that a single photon could cause the death of a cell.[26] Builders of bombs in the United States in the 1940s learned that minute particles of ingested or inhaled plutonium could cause cancer, leading the National Committee on Radiation Protection to adopt the no-threshold theory in 1948—the policy version of which held that exposures should be "as low as possible." In the 1950s Alice Stewart showed that the frequency of leukemia among children was closely correlated with how many X rays their mothers had received while pregnant; Stewart later showed that even a single X ray increased the likelihood of developing the disease.[27] Recognition of the potency of very low doses led to the replacement of the concept of "tolerance dose" with the concept of "maximum permissible dose" for external radiation and "maximum permissible concentration" for internal emitters.[28]

Not everyone, though, accepted the linearity of dose-response with regard to radiation. Physicists involved in the development of the new field of radiation biology tended to be brave in the face of radioactive hazards. Health data for very low levels were difficult to come by, and in the absence of evidence one way or another, many radiation biologists were led to take an optimistic attitude with regard to potential health effects. This tended to fit well with the policy goals of their primary employer, the U.S. Atomic Energy Commission. In private, atomic officials conceded a potential danger, even of very low doses: in 1953, for example, the AEC's director of biology and medicine, John Burgher, wrote a memo conceding that he knew of "no threshold to significant injury in this field." The AEC was eager to promote the safety of the atom, though, and therefore went out of its way to convince the public that, in proper hands, atomic energy was safe. A 1953 AEC pamphlet insisted that low-level exposure to radiation "can be continued indefinitely without any detectable bodily change."[29] AEC scientists did eventually show that mice were apparently able to repair genetic damage, so long as it occurred below a certain rate. William L. Russell of Oak Ridge National Laboratory in Tennessee, for example,showed in 1958 that radiation delivered at low-dose rates produced fewer mutations than when administered all at once.[30] This so-called protection by dose fractionation was later used to argue that radiation emissions from nuclear power plants posed less of a health threat than commonly thought.[31]

Prior even to the establishment of the AEC, radiologists had developed radiation "tolerance levels" in the course of wrestling with the problem of how to administer X rays safely. The presumption in some such efforts was that injuries below a certain level of severity could be healed by rest or vacation. Until the 1950s, accordingly, the National Committee on Radiation Protection required that X-ray workers be given at least four weeks of vacation a year.[32] As in the case of chemical carcinogens, however, there was often confusion over whether the standards were supposed to be regarded as "safe" or simply "tolerable." In 1957, for example, when the U.S. Public Health Service proposed a "working level" (equivalent to 100 pCi/l) of radiation for the nation's uranium miners, a margin of safety was said to have been included "sufficiently great to prevent recognizable damage to the lung tissue in the normal healthy worker." The very same document, however, stressed that the goal was a standard "which appears to be safe, yet not unnecessarily restrictive to industrial operations."[33] As in the case of chemical hazards, the standard represented a compromise between safety and financial cost to industry.

Following the recognition in the 1960s that most carcinogens are probably mutagens, radiation carcinogenesis became a model for carcinogenesis more generally. Conventional scientific opinion began to hold that even very low doses of chemical exposures could cause cancer, and that the threshold hypothesis was improbable. The key assumption was that since carcinogens were mutagens, a single "hit" (or single molecule) could cause cancer (see chapter 10). Epidemiological studies had found a close correlation between numbers of cigarettes smoked and one's odds of contracting cancer, and antismoking activists by the mid-1960s were able to use this to argue that there was no such thing as a safe cigarette or safe level of cigarette consumption. In 1967 Richard Doll suggested that "no dose of a carcinogenic agent, no matter how small, can be assumed to be safe."[34] Nathan Mantel of the National Cancer Institute's Biometry Branch asserted that a carcinogenic hazard "never disappears with diminishing dose but rather becomes indefinitely small."[35] U.S. environmental agencies by and large agreed: in the 1970s, the EPA, OSHA, Consumer Product Safety Commission, and FDA all adopted no-threshold models on the grounds that, given the uncertainties inherent in animal studies, it was prudent to assume low-dose linearity.[36] In 1982 Samuel Epstein declared that "no safe levels or thresholds are recognized for chemicals inducing carcinogenic, teratogenic, or mutagenic effects."[37]

Those who stress the linearity of dose-response tend to emphasize the potency of carcinogens. Threshold advocates respond by pointing to the fact and frequency of repair. What is surprising, after all, as Richard Peto and others have pointed out, is not that some of us get cancer but that most of us do not. Every day, legions of toxic molecules assault the billions of nucleotides that make up the DNA molecule in each of our 100 trillion cells. Mutation rates (base-pair substitutions) at any given locus are normally on the order of 10^{-6} to 10^{-10} per cellular generation. This is a significant amount of potential damage, considering the size of the genome and the fact that every cell in the body has a full complement of DNA. Some nucleotides are more easily altered than others: roughly 10 percent of all our cytosine, for example, is deaminated to uracil every day. This requires repair. Rapidly dividing cells—such as those that make up the epithelium—are especially vulnerable. The opportunities for carcinogenesis are enormous. One can imagine a kind of microscopic Darwinian struggle for existence among the various separate cells of our body: cells with altered genes allowing them to reproduce and migrate to other locations are favored in such a competition, and it is not hard to envision the disastrous neoplastic consequences.

One interesting question, then, as Peto points out, is, "By what conceivable mechanisms can we keep Darwinian evolution among the cells of our bodies in check for our three score years and ten?" Fortunately for us, as for the life of the species, most of our cells do not survive longer than a few days. Every day, 100 billion or so of our trillion-odd blood cells will perish, to be replaced by another hundred billion fresh cells. Instructions for self-destruction may be planted in cellular genomes to ensure that cancer-prone cells do not get off on the wrong carcinogenic foot. Peto suggests that this constant cellular death and renewal may be one way the body deals with cellular damage that could otherwise lead to cancer: most cells simply do not live long enough to mutate, replicate, and begin the deadly process of oncogenesis.[38]

The fact of repair is important for those who defend the existence of thresholds. The body, according to this argument, has evolved to be able to respond to environmental assaults. Even if a single hit may in theory cause cancer, this is very unlikely, given our natural ability to repair genetic damage. Bruce Ames points out that DNA repair is an ongoing activity, and that many of our organs undergo continuous renewal through surface-layer shedding. Cellular shedding in the stomach, colon, and epithelial tissues keeps most altered cells from flourishing. Cancer-prone cells are singled out for elimination, and potentially dangerous toxins are neutralized into harmless metabolites. Were it not for these and other protective mechanisms (dietary anti-oxidants, for instance), the body would soon be a mass of tumors. The body is equipped to deal with the myriad of insults incessantly attacking it—continuous oxygenation (Ames suggests we live in a sea of toxic oxygen), omnipresent mutagens, exposure to free radicals, and so forth. Evolution has equipped us with mechanisms to tolerate, sequester, or disarm the most serious threats up to a point—the threshold—beyond which our defenses are overwhelmed.

Environmentalists generally concede the fact of repair, but they also point out that repair is never perfect, and that the immune system—responsible for mopping up cells that have undergone the early stages leading to cancer—may itself be vulnerable to toxic injury. (Ingested radionuclides may damage the bone marrow that produces the T-cells vital for immunologic function, for example.) Certain repair mechanisms may be generalized (cell shedding, for example), but not all cells shed. Mature brain cells do not replicate at all. Generalized repair mechanisms also presumably cannot work very well against synergistic effects produced by widely divergent types of injuries—radiation and job stress, for example, or hormonal flux and damage to the immune system. The body has the capacity to heal itself, but why add insult to injury? And who repairs the repairers when they become disabled?

DOSE AND RESPONSE

Given the ambiguity of the evidence and the high stakes of the outcome, it should come as no surprise that the question of thresholds has become a fertile ground for imaginative speculation and political posturing. A 1983 survey showed that 80 percent of industry scientists believed in carcinogenic thresholds, compared with 60 percent of academic scientists and only 37 percent of scientists employed by regulatory agencies.[39] Had environmental activists been polled, we surely would have seen an even lower percentage.

There are several different ways the body may respond to low-dose carcinogenic exposures. Four of the more commonly expressed models—greatly simplified—are the following:

1) Low doses may be hazardous in direct proportion to the dose (linear, no-threshold hypothesis).
2) Low doses may not be hazardous at all (threshold, or "hockey stick," hypothesis).
3) Low doses may be good for you (hormesis thesis).
4) Low-level doses may be disproportionately hazardous (supralinear hypothesis).

These various responses can be represented by four different, and rather idealized, dose-response curves, to which I have (somewhat lightheartedly) given political labels to indicate the divisiveness on this issue (see figure 7.1). In an ideal world, or rather in a world of infinite resources and a willingness to experiment willy-nilly on humans, the question of which model is the correct one for a given carcinogen and a given kind of cancer could presumably be determined empirically. One would

FIGURE 7.1

The Political Morphology of Dose-Response Curves

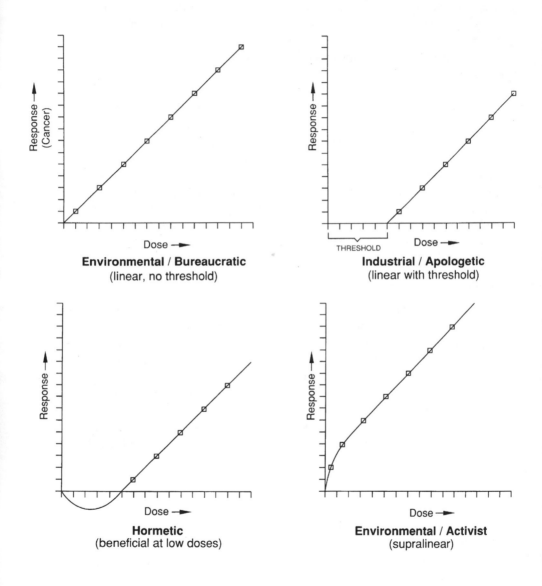

Environmental / Bureaucratic
(linear, no threshold)

Industrial / Apologetic
(linear with threshold)

Hormetic
(beneficial at low doses)

Environmental / Activist
(supralinear)

Uncertainty concerning how to extrapolate from high-dose animal studies leads to a free rein of the imagination in what a dose-response curve should look like at low-dose exposures. These are some of the possibilities.

identify all possible synergistic effects, all differences in susceptibilities (by age, state of health, genomic idiosyncrasies, and so on), and all the ways that humans differ from the organisms used in animal studies. Since we do not live in such a world, people with conflicting agendas and apathies tend to lean one way rather than another. Let's look a little more closely at what each model entails.

The *linear, no-threshold* theory is the dominant model used by federal agencies such as the EPA, the FDA, and the Consumer Product Safety Commission to evaluate potential cancer hazards. The presumption is that the dose-response curve extends linearly to the origin (at least for low-level exposures), that there are no thresholds, and that a single hit is sufficient to induce cancer. Radiation-induced mutation is the model here, though there are some who will argue that low-level radiation can actually prevent cancer (the hormesis thesis) and others who will argue that low-level radiation can exert a more than proportionate harm (the supralinear hypothesis).

Arthur Upton, director of the National Cancer Institute during the Carter administration, in a 1988 review listed several of the main reasons to believe that even trace exposures—down to a few molecules or even one—can contribute to the changes that lead to cancer. These include: (1) the fact that most malignancies can be traced back to a single cell of origin, implying that damage to a single cell is all it takes to launch a cancer; (2) the fact that most carcinogens are also mutagens; (3) the association between cancerous growths and specific mutations or chromosomal aberrations; and (4) experimental evidence indicating a no-threshold dose-response relation.[40] The most important defense of linearity, however, is political. Linearity is conceived as the "prudent" or "conservative" presumption, given that the main alternative, the threshold hypothesis, is potentially dangerous. Marc Lappé, a professor of health policy and ethics at the University of Illinois, has argued that the idea of thresholds is one of ten environmental health myths that "endanger us all."[41]

The *industrial/apologetic* model presumes a threshold "dose" below which there is no cancer "response." This is a popular view among industry trade associations, pro-industry groups such as the American Council on Science and Health, neoconservatives such as Efron, and cancer conservatives more generally. The key idea is that, up to a certain point, the body has an ability to detoxify foreign substances. Beyond that point, a toxin overwhelms the body's capacity to metabolize or eliminate the toxin in question and the excess is expressed as a tumor or death. The situation has been compared to filling a bucket with water, the bucket representing the body and the water an ingested toxin (see figure 7.2).[42] Holes have been punched in the bucket, allowing the inflowing water to drain off. So long as the water does not enter faster than it can escape through the holes, there is no overflow problem. But if water flows in faster than it can escape, the bucket overflows. This is comparable to the body's defenses being overwhelmed, leading to a toxic response or tumor. A Dow Chemical spokesman who testified before hearings on dibromochloropropane in 1977 invoked a different analogy to drive home the same point: 100-mile-per-hour winds might be devastating, but 10-mile-per-hour winds might do no damage whatsoever.[43]

FIGURE 7.2

A Schematic Representation of One Source of Thresholds

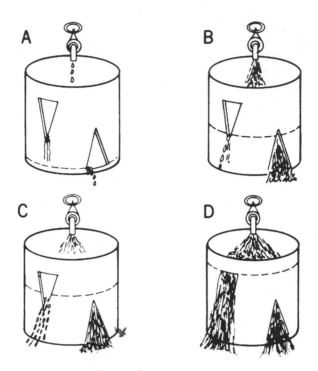

As increasing amounts of fluid flow into the bucket (as greater quantities of a chemical are ingested), elimination (detoxification) via the lower slit becomes overwhelmed. This results in disproportionate increases in the amount of fluid in the bucket (the amount of carcinogen in the body) and elimination via the upper slit (induction of a tumor).

Source: From P. J. Gehring, Dow Chemical Company. Reprinted with permission from Thomas H. Maugh II, "Chemical Carcinogens: How Dangerous Are Low Doses?" *Science* 202 (1978): 41. Copyright 1978 by the AAAS.

The *hormesis* thesis suggests that low-level radiation (or potential chemical carcinogens) can stimulate bodily defenses, resulting in a lowered cancer risk. Proponents of this theory cite several human and animal studies apparently confirming the effect and a variety of plausible mechanisms. Some suggest that low-level radiation may stimulate DNA repair, others that low-dose exposures may fire up the immune system or produce chemical scavengers that could help rid the body of harmful radicals produced by radiation. The net result would be a lower incidence of cancer.[44] The idea has been put forward for chemical toxins as well as for radioactive hazards: a 1994 publication by the American Industrial Health Council listed thirty studies showing beneficial effects ("seemingly paradoxical" U-shaped dose-response curves) for lead, nicotine, PCBs, toluene, alcohol, jet fuel, methyl mercury, and several other poisons. The review cites "some provocative data" suggesting that certain kinds of dioxin may actually provide "a protective effect for low levels vis-à-vis breast cancer."[45] The idea is reminiscent of the logic of homeopathy or vaccination—a small dose of a poison actually healing or preventing the very harm that a large dose is known to cause. There are commonsense examples: if you are going to fry yourself in the sun, for example, it is surely better to protect yourself first by getting a tan.[46]

The idea that low-level radiation may be good for you is frequently encountered on the conservative flank of the nuclear industry.[47] The French Academy of Sciences considered the question of hormesis in a 1989 report, and Dixy Lee Ray, chairman of the Atomic Energy Commission under Nixon, endorsed it in *Trashing the Planet*.[48] Hormesis was the primary focus of a 1985 conference co-sponsored by the Palo Alto–based Electric Power Research Institute (EPRI) and the Northern California chapters of the American Nuclear Society and the Health Physics Society.[49] Speakers included several eminent radiation specialists, notably Jacob I. Fabrikant of the Donner Laboratory at Berkeley and Hiroo Kato of the Hiroshima-based Radiation Effects Research Foundation, though not all of those present were unabashed fans of hormesis. Hiroo Kato noted that atomic-bomb survivors who received between 1 and 50 rads of exposure from the blast could hardly be said to have benefited from the experience. *Science* magazine editors later allowed the organizer, Leonard A. Sagan of EPRI, to express his views in its Policy Forum, but even they felt compelled to publish a skeptical rebuttal. Sheldon Wolff, director of the University of California's Laboratory of Radiobiology and Environmental Health, compared the persistent but irreproducible claims for hormesis to T. D. Lysenko's pseudoscientific genetics in Stalin's Soviet Union, where numerous reports—later unsubstantiated—claimed that irradiated plants matured faster, grew bigger, and so forth.[50]

The *supralinear* model tends to be favored by environmental activists. Rachel Carson provides an excellent early exposition of this view in *Silent Spring,* explaining how repeated small doses of a carcinogen can be more dangerous than a single large dose: "The latter may kill the cells outright, whereas the small doses allow some to survive, though in a damaged condition. These survivors may then develop into cancer cells. This is why there is no 'safe' dose of a carcinogen."[51] High doses, in other words, may kill a cell, but low doses can mutate without killing. Extrapolation from high-dose studies might therefore underestimate the hazard of a particular substance. A strong argument has been made that this is true not just for internal alpha radiation but also for vinyl chloride and several other carcinogens.

A supralinear model for low-level radiation has also been put forward by John Gofman, a veteran of the Manhattan Project and a long-standing critic of nuclear power.[52] Gofman, who co-discovered Uranium-233 and proved its fissionability while a graduate student at Berkeley, argues that for internal emitters such as radon or bone seekers such as plutonium, the linear model "may actually underestimate the risk of getting cancer and leukemia." For alpha emitters such as these, he explains, the cancer risk per unit of exposure is higher in the low-dose range than in the high-dose range. Gofman opposes the idea that a dose that causes harm when received all at once may be harmless when spread out over a longer period of time; he rejects, in other words, the principle of protection by dose fractionation, an idea that has emboldened polluters for whom the solution to pollution is dilution.[53] Uranium miner epidemiology is displayed to support supralinearity, and Hiroshima data collected by the Atomic Bomb Casualty Commission—for breast cancer and leukemia, but also for cancers overall—are cited to suggest that people exposed to lower levels of radiation had a higher cancer rate per unit of exposure.[54] Gofman uses these data to argue that the assumption of linearity (which he and others use to estimate cancer hazards) is more likely to underestimate than to overestimate the risks of low-level radiation.

One last point deserves attention in this context. The question of thresholds has become intertwined with the question of what fraction of all chemicals are carcinogens. Surprisingly, perhaps, environmentalists have tended to give low figures for the fraction ("not all chemicals cause cancer"), while anti-environmentalists have tended to give high estimates ("most anything can cause cancer if consumed in sufficient quantities"). Of course, it matters a lot whether one is talking about a percentage of all pesticides or a percentage of all chemicals. In either case, the entire notion of a certain fraction of all chemicals being carcinogenic is problematic, as is any numerical estimate of how many chemicals we are exposed to. Chemicals ingested are rarely inert, and a given food or food additive may have hundreds or even thousands of different breakdown products. Chemicals break down in the body—that is the whole point of digestion. It is therefore hard to say whether there are 70,000 chemicals to which we are exposed or 7 million (both figures have been put forward); citing numbers in such cases makes little pharmacokinetic sense.

In *The Apocalyptics,* Edith Efron points out that environmentalists tend to underestimate the fraction of chemicals that are known carcinogens in order to "fight off attacks on high-dose testing." If the proportion of chemicals that are carcinogens is vanishingly small, this bolsters our confidence that high-dose studies are reliable and that it is not the case that "if you give a mouse *anything* in sufficient doses, it will get cancer." This is why environmentalists have been so eager to argue that not everything causes cancer. If, however, a substance is a carcinogen depending on how much of it you ingest, then high-dose animal studies may not be very good indicators of human cancer risks. Environmentalists, according to Efron, have been doing "sentry duty" to defend the use of high-dose animal studies to determine carcinogenic potentials.[55]

The significance of animal studies is often what is disputed in quarrels over the existence of thresholds. Controversies swirl around the relevance of rodent bioassays for human populations: a Ruckelshaus policy adviser conceded in the mid-1980s that "even a single rat study can be interpreted two ways, depending on your ideology."[56]

THE PROBLEM WITH ANIMAL STUDIES

In the 1970s, the linearity of dose-response curves—and the corollary absence of thresholds—was often expressed in terms of the so-called one-molecule theory: the idea that exposure to as little as one molecule of a carcinogen (or one particle of ionizing radiation) could cause cancer. The principle was established early on in radiation studies and by the 1950s had a popular following in chemical toxicology as well. The 1958 Delaney Amendment to the Food and Drug Act was implicitly based on the one-molecule theory, barring any measurable trace—no matter how small—of synthetic carcinogens added to foods. Subsequent regulatory acts operated from a similar principle. Bruce Ames endorsed the one-molecule theory in an influential paper of 1971, and the idea was subsequently embraced by environmentalists eager to argue that there was no safe level of chemical or radioactive carcinogens.

Critics of the one-molecule theory have recently noted that while some carcinogens may act in such a manner, many do not. Bruce Ames, an early champion, is now one of the theory's strongest critics. Ames currently argues, contrary to his earlier views, that not all carcinogens work by damaging DNA.[57] Cancer may arise through the stimulus of *mitogenesis* (cellular proliferation) and not just by *mutagenesis* (mutation). The distinction is important because, whereas mutagens may incite cancer through a single hit, nonmutagenic carcinogens may work by quite a different route, overwhelming the body's natural repair mechanisms, forcing or allowing the development of a tumor. A much broader class of chemicals may be carcinogens than are mutagens. And if it is the high-dose exposures themselves that are responsible for inducing cancer, by inducing mitogenesis, animal studies may not be very good indicators of human cancer hazards.

Animal studies remain one of the most commonly used measures of human cancer risks. The justification for their use is both an analytic and a practical one. Analytically, the presumption is that chemicals that cause cancer in mice or rats can also cause cancer in humans. Vinyl chloride, for example, was shown to cause angiosarcomas in rats and mice long before it was found to cause the same kind of cancer in humans exposed to comparable doses.[58] Animal studies are also often the only practical way to determine whether a particular substance causes cancer, given that one cannot—perhaps I should say should not—deliberately expose humans to suspected carcinogens in order to gauge their danger. As mentioned earlier, animal tests are usually far cheaper than long-term epidemiological studies; the results are also usually available much sooner.

This is not to say, though, that animal studies are infallible. Arsenic causes cancer in humans, but not in most laboratory animals. Azo dyes are carcinogenic in rodents, but there is no evidence of a cancer hazard in humans. The compound 2-naphthylamine produces bladder tumors in humans, monkeys, dogs, and hamsters, but apparently has no effect on rats or rabbits. Coal-tar cancer is relatively easy to induce in rabbits, but is nearly impossible to induce in guinea pigs. Tobacco smoke has never conclusively been shown to cause lung cancer in experimental animals, though the link is virtually indisputable in humans. Different species often metabolize toxins in dissimilar ways—and that is why the same substance can produce different effects in different animals. Wilhelm Hueper pointed out in the 1950s that a given chemical could have very different effects in different species, having to do with how easily the substance is dissolved in the body, how quickly it is removed from the site of primary contact, and its route of secretion and place of retention. The only universal carcinogen, in his view, was ionizing radiation.[59]

Even radiation, though, has been shown to affect different species in different ways. In 1950, for example, Shields Warren of the Atomic Energy Commission learned from a member of his scientific staff that rat and mice studies were showing a far greater plutonium hazard than previously thought, leading to the proposal of new tolerances that, if applied consistently, could "shut down Los Alamos" (J. Newell Stannard's words, in his official history of this period.) The commission responded by launching a major study of the health effects of ingested radionuclides on beagles, an animal purportedly chosen on the grounds that it would give a more accurate indication of the human response to radioactive damage. More than a thousand beagles were eventually irradiated at the University of Utah's Laboratory of Radiobiology and four other labs—a project that began in 1950 and continues with Department of Energy support even today. Beagles were already known to have a low natural rate of bone and other radiogenic cancers, leading later critics—such as Elizabeth Hanson of the University of Pennsylvania—to suggest that this particular animal may have been deliberately chosen in order to diminish the apparent threat of fallout from atmospheric testing.[60] (The AEC later moved to pigs and eventually to humans, hoping, scandalously, that by experimenting on soldiers, prisoners, and the "mentally enfeebled," one could discover things that could not be found with nonhuman models.)[61]

Animal tests are not, in other words, fail-safe predictors of the safety of a given chemical substance to humans. The Food Protection Committee of the National Academy of Sciences recognized this in the 1950s, asserting that laboratory animal studies "cannot provide irrefutable proof of the safety or carcinogenicity of a substance for the human species." Neoconservative publicists such as Efron and Whelan have made a great deal of the fact that a positive animal test is not always proof of a human health hazard, but the second half of the principle should not be forgotten: a negative result is not necessarily proof of a substance's safety. Sigismund Peller pointed out in his 1952 *Cancer in Man* that if Yamagiwa and Ichikawa had rubbed coal-tar extract onto the skins of rats instead of rabbits and mice, the human carcinogenic potency of tar "would have been 'expertly' denied for another seventeen to twenty years, and the whole development of chemistry of cancerogens would have been retarded."[62] Efforts to induce cancer in animals using tar had been made since the 1880s: these early efforts to isolate the offending agent in chimney sweeps' and tar workers' cancer failed, largely because animal models were chosen—mainly dogs and rats—that turn out to be resistant to the carcinogenic action of tar.[63]

The traditional justification for high-dose animal studies is that they are necessary to avoid the expense involved in conducting lower-dose studies that might require the administration of chemicals to thousands or even hundreds of thousands of animals. From high-dose studies, one can calculate the effects of low-dose exposures by assuming some stable relation between dose and response. At the EPA, for example, mice or rats are typically exposed to what are known as maximum tolerated doses (MTDs)—defined as the highest doses than can be given without causing immediate and obvious harm to the animal (through weight loss, toxic illness, or some such.). If, say, 10 percent of all experimental animals contract cancer in a high-dose study, then it might be assumed that 1 percent of all experimental animals will contract cancer from exposure to one-tenth of the original dose. The assumption is that response (cancer in this case) declines linearly with dose, and that high-dose studies can therefore be used to estimate low-dose hazards.

Ames, however, argues that the high-dose studies are inherently biased, insofar as it is the high doses per se that are inducing cancers. The argument runs as follows. Toxicological studies by the EPA or FDA typically involve exposing fifty or a hundred animals to very high levels of chemicals—by injection into the gut, inclusion with food, or painting onto an animal's shaved back (as was done with nicotine in early studies of cigarettes). The dose provided is as close as possible to the maximum tolerated dose. MTDs are not supposed to be toxic but, according to Ames, the doses may nevertheless be high enough to initiate cellular proliferation, a kind of emergency response undertaken by the body to repair damage caused by the offending irritant. Cellular proliferation can greatly increase the odds of contracting cancer, because every time a cell divides the DNA in the nucleus of that cell must be copied. Copies made from an often-copied text are more likely to be corrupt than are copies closer to the original; the same is true of genetic scripts. Cellular proliferation thus provides a greater opportunity for mutagenesis, a key initial step in carcinogenesis. High-dose animal studies may overestimate a potential cancer hazard, insofar as the irritation caused by high-dose exposures can result in cellular proliferation.

Ames was not the first to question whether high-dose toxicity might be distorting efforts to ascertain the human cancer potency of specific substances,[64] H. F. Kraybill of the National Cancer Institute pointed out in 1977 that overdosing might result in a chemical being metabolized, stored, or secreted in a manner different from what one would expect in a lower-dose context. Since even an otherwise "useful" substance—Vitamin A or D, for example—could be toxic at very high doses, might it not also be true that a substance benign at one dose could be carcinogenic at another?[65] Kraybill pointed out that extrapolations from studies where rats were consuming the human daily equivalent of 552 bottles of soft drink (in a cyclamate test), 100 million cups of coffee (in a trichloroethylene experiment), and 5 million pounds of liver (in a diethylstilbestrol test) were probably better regarded as science fiction than as science. Richard Peto in his 1980 critique of Epstein argued similarly that "any chemical which causes proliferation or necrosis in any organ that is subject to spontaneous cancers is likely to modify the onset rate of tumours in that organ"; it was therefore "not surprising" that many high-dose animal tests were yielding positive results for carcinogenic potency.[66] Ames built upon these earlier works, drawing out the policy implications in a starker form than anyone had previously.

Industry scientists have welcomed the new theory. James D. Wilson of Monsanto wrote to *Science* in 1991 in support of Ames's critique of the bioassays on which much of federal cancer policy is based. Wilson proposed that substances can be "situational carcinogens," causing cancer in some circumstances but not in others. He also complained that neither the EPA's Guidelines for Carcinogen Risk Assessment nor the FDA's Delaney Clause had recognized this situational aspect, and that it was high time for regulatory policy to catch up with science.[67] The theory has also gained support from the BELLE Advisory Committee, a group of conservative scientists formed in 1990 to study the biological effects of low-level exposures (hence the acronym), with representatives from the tobacco industry, the DOE, and the U.S. Army, and financial support from the EPA.[68]

Critics of Ames's approach have countered that the evidence for a correlation between toxicity and carcinogenicity is weak. A 1988 study by David Hoel of the National Toxicology Program found, after examining tissues from the rodents used in its carcinogen testing program (the largest U.S. program of this sort), that there was little or no correlation between toxicity and tumor formation: cancers were found in organs with little apparent toxic damage, and tissues with extensive damage were often tumor-free.[69] This study, conducted for the NIEHS, has been cited by critics of Ames in support of the argument that toxic or near-toxic exposures do not by themselves induce tumors.

Critics have also argued that there are cases even where the linear, one-hit model of carcinogenesis may not be conservative in the sense of erring on the side of caution. John Bailar III, former editor of the *Journal of the National Cancer Institute,* showed that there were circumstances where the one-hit model might actually underestimate risks. Analyzing data from more than a thousand separate bioassays, Bailar showed that extrapolations from high-dose studies underestimate risks almost as often as they overestimate risks.[70]

Bailar suggested several reasons why this might be true. High-dose exposures may kill or sterilize cells, diminishing the opportunity for a carcinogenic response. High-dose exposures may also work as cancer suppressants, as implied by the fact that certain chemotherapies cause cancer at the same time that they obstruct it (8 percent of patients successfully treated with dihydroxybusulfan for ovarian cancer, for example, develop leukemia within a few years). Pharmacokinetic pathways must also be considered in this context, since chemicals are sometimes converted to carcinogens by metabolic processes in the body—a kind of inverse detoxification. Enzymes that perform this conversion may become saturated, preventing further production of the carcinogen. High-dose experimental procedures might also inadvertently lower cancer risks by diminishing appetites. Rats fed on a near-starvation diet are known to have lower cancer risks and this, too, could push a dose-response curve in the direction of supralinearity. Finally, high-dose studies may underestimate risks at low-level exposures by failing to take into account variations in susceptibility. If most of those who suffer from a particular kind of cancer (say, sunlight-induced skin cancer) are sensitive by virtue of a particular genetic constitution, then a linear extrapolation from high-dose exposures would obscure the fact that low levels of exposure might be producing relatively high rates of cancer among that sensitive subpopulation. A properly drawn dose-response relation in this case would show a rapidly rising (supralinear) curve, followed by a linear and eventually sublinear region. On these and other grounds, Bailar concluded that the one-hit model, widely regarded as conservative, "may substantially understate true risks at low exposures." His recommendation: carcinogenic risk assessments should "incorporate some measure of additional uncertainty."[71]

It is impossible—at present—to know with any degree of certainty how many people are killed in any given year by X rays, saccharin, or air pollution. Efforts to estimate the magnitude of a given cancer hazard rely on a number of different techniques—including animal tests on different species, prospective and retrospective epidemiology, bacterial bioassays, reconstruction of historical experience, and so forth. Different techniques are useful for different purposes. For chemical products newly arriving on a market, or where epidemiological evidence is unavailable or inappropriate, animal tests are often the only practical way to predict whether a given food dye or pesticide or growth hormone is carcinogenic.

Far too often, however, animal tests are never done in the first place, or else performed with inadequate controls or on inappropriate models.[72] It is a not infrequent tragedy that chemicals are found to cause cancer in humans when properly administered animal studies could have alerted us to the hazard. In the absence of better alternatives, the attack on animal studies serves simply to weaken regulation and to increase our chances of suffering yet another round of environmentally induced cancer. The call to wait and see, to complement every animal study with human epidemiology prior to taking action, can have an Alice in Wonderland quality about it. The danger is the kind of justice the queen issued to Alice: "Sentence now, verdict later."[73]

BODY VICTIMOLOGY, BODY MACHISMO, AND REGULATORY PRUDENCE

Fights over whether or not to tolerate low-level exposures to carcinogens usually erupt not just because evidence of low-dose effects is lacking but also because there are often hefty financial stakes in the outcome. Opinions expressed often boil down to what, in the face of ignorance, one is willing to do in the name of prudence. Environmentalists generally argue that, in the absence of evidence one way or another, it is prudent to assume that even small doses of carcinogens may cause cancer. Industry advocates and their academic allies remind us that there are financial costs associated with removing every last trace of a chemical from the environment, and that a risk-free society is an impossibility. How much is one going to pay for what may be negligible or even nonexistent benefits? From this side of the debate, environmental prudence looks more like ideological zealotry. In neither case, though, is the appeal to evidence convincing. Action in the face of this kind of ambiguity must root itself in an ethical vision, and this is one of the things that has kept the cancer wars so heated.

Different visions of regulatory prudence, though, are only part of the picture. There are also different conceptions of the human body, especially its ability to withstand toxic insults. Supporters of the threshold concept tend to regard the body as tough, resilient. The macho body is able to detoxify potential carcinogens or repair genetic damage at rates sufficient to neutralize low-dose chemical or radioactive insults. Environmentalists, by contrast, tend to regard the body as passive in the face of environmental insults. The assumption is that even minute traces of a carcinogen can cause cancer—hence the preference for a no-threshold dose-response relation. The carcinogen is the active agent, the body the passive victim.

The environmental Left's "body victimology," then, focuses more on the vulnerability of the body (and the potency of carcinogens) and less on the possibility of repair, while the political Right's "body machismo" has a confidence in the body's ability to detoxify carcinogens and to repair genetic damage. Conflicting evolutionary narratives are at work here, with environmentalists emphasizing the novelty of synthetics hazards (and hence our likely unpreparedness to deal with them), and anti-environmentalists stressing their similarity to the natural toxins to which our species has always been exposed (and to which we presumably have become adapted). The extreme version of body machismo is the hormesis thesis, which maintains that low-dose exposures may actually prevent cancer. Hormesis is the homeopathic, technocratic analogue of Nietzsche's "that which doesn't kill me makes me stronger."

The images deployed to defend thresholds or their absence are revealing. Threshold advocates tend to stress simple mechanical analogies: wind on a house, water filling a bucket. Environmentalists may use mechanical metaphors, but their emphasis is more often on the complexity and vulnerability of the human corpus. The body is a balanced whole that is easily upset. Hermann J. Muller, an early defender of the no-threshold hypothesis for radiation, compared the odds of a mutation being good for you to the likelihood that smashing a fine Swiss watch with a hammer would improve its ability to tell time—the chance is there, but it is, indeed, rather small! Those who defend thresholds, by contrast, tend to see the body as capacious and resilient, constantly repairing injuries and replacing defective parts. The threshold model presumes a reductive, cybernetic conception of the body—self-correcting, error-checking. The focus is on genetic damage and genetic repair, but other sites of potential harm—epigenetic and synergistic impairments, for example, or damage to the immune or error-checking system—are slighted.

The contrast between these two visions of the body (as of the environment) cuts across traditional political distinctions. The Old Left and technocratic Right, for example, are generally united in their confidence that nature and the body can be engineered, and that environmental hazards have been exaggerated. Both support pesticides and nuclear power; both are wary of antinuclear protesters. Both are confident that the rule of enlightened experts can guarantee social progress without messy democratic interference from the public. Both maintain that the planet, like the body, is tough, plastic. This was the message of Ray's book accusing environmentalists of "trashing the planet" by impugning its resilience in the face of environmental insults. The New Left and religious (or fascist) Right, by contrast, tend to be more wary of environmental hazards. Rachel Carson and the Nazis were united in their fears of environmental hazards.[74]

What we find are different conceptions of what is fragile and what is strong— what can and should be changed and what is better left alone. The environmentalist's "body victimology" assumes a passive bodily response to carcinogenic insults. This is consistent with a larger view of death as an insult, and every cancer (like every death) a potential tort. All cancers have causes, and all causes have culpable agents. Someone, or some social institution, is to blame. The individual is weak but society is strong; remove the causes of death and we could live forever. Anti-environmentalists, by contrast, tend to see the individual (body) as strong, and society as brittle. Nature, if not indestructible, is manipulable, malleable. Nature (or the body) can take care of itself. These divergent philosophies embody divergent fears—contrary attitudes toward what is endangered and what is secure, what is complex and what is simple. The politics involved in such prejudices get written into the curves used to display our purported knowledge of cancer hazards.

In pointing to the political morphology of dose-response curves, I don't mean to imply that there is no way to know the truth behind such matters or—still worse— that the truth is entirely a matter of perspective. Low levels of exposure have very real consequences, however difficult or even impossible these may be to fathom. Surely a hazard can exist without our ever being able to know it: there is an ostrich-like naïveté in the positivist presumption that what you don't know can't hurt you. The politics of low-dose hazard evaluation shows that much depends upon whom, what, and where one grants a benefit of doubt, but also that much depends on how one understands the phrase "to err on the side of caution." What is at stake is not just the construction of curves or the power of the body to respond to insults but what it means for citizens and collectivities to act prudently in the face of ignorance.

Chapter 8

Nuclear Nemesis

Danger is foremost in our minds when we think of atomic radiation.

—Heinz Haber, *The Walt Disney Story of Our Friend the Atom,* 1956

NOTHING EXCITES FEAR LIKE RADIATION. That fear is grounded in the fact that radiation is ubiquitous, invisible, and potentially dangerous.* Radiation comes from reactor waste and uranium tailings, from X-ray machines and fallout from nuclear testing. Uranium is added to paper to make it whiter, to porcelain dentures to make them sparkle. Gemstones (topaz, for instance) are irradiated to enhance their color; foods are irradiated to kill fungal and bacterial pests. Radioactive isotopes are found in the phosphates used for fertilizers and in the tobacco used for cigarettes (smoking may well provide people with a larger dose of radioactivity than any other single source).[1] Exposures in many cases have been lowered through technical advances—the introduction of high-speed X-ray film, for example, which requires a lower dose—but the use of some radioactive materials has actually been increasing. Americium is commonly used in smoke detectors (a circuit produced by the ionizing radiation is interrupted in the presence of smoke), and though these are supposed to be returned to the manufacturers when they break, most probably end up in landfills or incinerators. Demand for uranium has fallen with the stagnation of the U.S nuclear industry, but the number of reactors on line throughout the world continues to increase. China, for example, has begun to export reactor technologies to several Third World nations.

Radiation occupies a special place in the history of cancer causation—both for those who suffer from it and for those who study it. Radiation causes cancer, and the questions of why and to what extent have been objects of controversy since radiation cancers were first observed in the early years of this century. Much of the controversy has to do with numbers. Controversies swirl over the magnitude of the hazard

---

*In this chapter I am dealing only with ionizing radiation—radiation, that is, with energy levels high enough to knock electrons from the orbits of atoms (so to speak) and induce chemical reactions. Recently, speculations have arisen that nonionizing radiation of the sort produced by household appliances, high-voltage wires, and cellular phones may also cause cancer; I ignore these here partly because of the recency of the debates and partly because the outlines of the debates are already accessible in Brodeur's *Currents of Death.*

posed by X rays, over the cancer toll for the Japanese exposed to atomic bomb radiation at Hiroshima and Nagasaki, and over the cancers to be expected from exposure to nuclear plants, atomic fallout, and household radon. A shroud of secrecy has left critics in the dark, victims unaware of exposures, and scientists unable to pursue open-ended inquiries.

This chapter begins by tracing the early history of radioactive hazards, then looks at how building the bomb raised the scale of radiation exposures and how the postwar nuclear boom allowed cancers to be planted in the lungs of thousands of Mormon and Navajo uranium miners. Finally, I discuss the shocking story of Czechoslovakian and Soviet uranium-mine concentration camps.

EDISON'S ASSISTANT

Radioactivity was discovered at the end of the nineteenth century in the course of efforts to explain the curious displays of light produced inside electrical discharge or vacuum tubes, forerunners of the modern television screen and X-ray tube. In the 1870s, the English physician William Crookes had found that, when an electrical current is passed between two electrodes inside an evacuated tube, a greenish light is produced as a result of the movement of what he called *radiant matter* (later renamed *electrons*) from the cathode (negatively charged wire) to the anode (positively charged). In 1895, Wilhelm Roentgen of Würzburg discovered that a new kind of ray could be produced from the vacuum tube when sufficiently high voltages were applied to its electrodes. Unlike the electron rays found by Crookes, these new "X rays" (as Roentgen called them) could penetrate cardboard and even certain kinds of metal. Placing his hand between the tube and the screen, Roentgen produced the very first images of the bones of a live human. The diagnostic value of the new rays was immediately obvious, and within a year of their discovery every major city in the United States had at least one medical X-ray center.

Only a few months after the discovery of X rays, Henri Becquerel of Paris found that similar rays were naturally emitted from the uranium ore known as pitchblende. Becquerel had been trying to determine whether X rays might have something to do with the fluorescence observed in uranium salts after exposure to sunlight or electron radiation. In a famous example of serendipitous discovery, Becquerel found that uranium salts could fog photographic plates even without having first been exposed to sunlight (in his earlier experiments, he had placed the samples on photographic plates only after exposing them to the sun). Becquerel rightly concluded that uranium was the source of a new and invisible kind of radiation that, like X rays, could penetrate matter and fog photographic plates.[2] Marie Curie subsequently found that uranium was not the only element possessing "radioactivity" (she and her husband, Pierre, coined the term): around the turn of the century, the Polish-born chemist isolated from pitchblende two new elements—polonium and radium, the latter having more than a million times the radiant discharge of uranium. The newly discovered radium was so radioactive that pure samples were always warmer than room temperature and glowed in the dark.

Faced with an entirely new property of matter, scientists in the early years of the century sought to explain how a pure element could release such enormous quantities of energy. It had long been recognized that the electrons emerging from a discharge tube could be bent by forcing them through a magnetic field. When the rays produced by radium were focused into a beam and shot through a magnetic field, however, three very different sorts of rays were observed. Rays that were bent to the right were called *alpha rays*. Rays that were bent to the left were called *beta rays*. Beta radiation was quickly identified as electrons moving close to the speed of light; alpha radiation was eventually recognized as a stream of slower, positively charged helium nuclei, each about four times heavier than the hydrogen atom. A third sort of radiation appeared not to bend at all under the influence of the magnet. These *gamma rays* were later found to be similar in virtually every respect to X rays, the only difference being in the manner of generation (gamma rays come from nuclear processes, X rays from extranuclear electron shell events). The discovery of alpha, beta, and gamma radiation from elements such as uranium, polonium, and radium eventually made it clear that atoms were not—as the name implied—indivisible: some were constantly decaying into so-called daughter products: uranium into radium, radium into radon, radon into polonium, and polonium into radioactively stable lead.

Scientists and physicians eager to explore these powerful new rays soon realized that they could have adverse health effects. The potential for commercial application, however, blinded X-ray advocates to possible dangers. For many years after their discovery, the rays were more commonly hailed as a wonderful cure-all. X rays were used to treat ringworm and acne and prescribed as a cure for criminality, impotence, and intemperance. Women had their ovaries X-rayed as a treatment for depression, and X rays became a popular (and disastrous) treatment for the removal of facial hair.[3] They were first used to treat cancer in 1896, and by 1902 the editor of the *American X-Ray Journal* could boast that there were "100 named diseases" yielding favorably to X-ray treatment.[4] Few people worried about the hazards of the rays—the more common fear was that men would peer through women's undergarments and into their bodies (one theater went so far as to ban "X-ray opera glasses").

X rays were popular with the public. In May 1896, when Thomas Edison exhibited his newly invented fluoroscope at the Electrical Light Association Exhibition in New York's Grand Central Palace, thousands of people queued up to view their hands, legs, and heads through the device.[5] A half-century later, thousands of people would have their feet X-rayed in shoe stores advertising the ray as an exact way to measure their feet. The rays were used to examine eggs for freshness, grain for the presence of weevils, castings for blowholes, cables for corrosion, and welding joints and concrete blocks for cracks. Human fetuses were X-rayed to detect potential difficulties in delivery, and X rays were used to straighten out bowlegs in children by slowing bone growth.[6] People exposed included not just dentists, nurses, and physicians but manufacturers of X-ray equipment, metallurgic engineers, plumbers and builders, ammunition experts, textile chemists, crystallographers, physicists, rubber chemists, automobile tire testers, and experts using the rays to authenticate objects such as pearls, diamonds, and paintings.[7]

One of the first recorded X-ray injuries took place in 1896 at Thomas Edison's Menlo Park, New Jersey, lab during work on an X-ray lightbulb. A glassblower by the name of Clarence Dally became ill from the work: his hands began to ulcerate and his hair fell out. Edison was horrified and soon dropped the project, noting in his diary that the product was in any event not likely to be "a very popular kind of light." By the end of 1896, at least twenty-three cases of severe X-ray injury had been recorded in scientific journals, most of them suffered by the operators or manufacturers of equipment. Late that same year a General Electric physicist named Elihu Thompson deliberately burned the little finger of his left hand with the rays to see how much harm they could cause; his widely reported account of the flesh slowly peeling from his finger led radiologists to control the strength and focus of the beam better and to design protective gear.[8] In 1899, the first malpractice award for an X-ray injury was settled in favor of the plaintiff.

Early reports of ill effects describe what we today would call *acute radiation sickness*—typically, burns and inflammation, often accompanied by hair loss and skin ulcerations. Evidence that X rays might also cause longer-term chronic illness did not emerge until they had been in use for several years. The first German report of a radiation-induced cancer was of a thirty-three-year-old worker at a roentgen tube–manufacturing plant. The man had used his right hand to test X-ray devices for four years. In 1901 he developed ulcers on his right hand, later diagnosed as cancer.[9] His arm was amputated at the shoulder in 1902, and he died from a recurrence of the disease in 1906. Subsequent reports of other German X-ray cancers appeared in 1903 and 1904.[10] In the United States Edison's assistant, Dally, became the first American known to have died from exposure to X rays. He had developed cancers in his arms and hands and had both arms amputated shortly before his death in 1904 (he also received X-ray treatments for the disease in the hope "of undoing what the ray itself had done").[11]

For many years, skeptics refused to admit that X rays could cause cancer. The International Commission on Radiological Protection's 1934 occupational standard (.1 roentgen per day) entirely ignored the risk of delayed cancer—the commission's concerns were limited to short-term burns and ulcerations. Long-term statistical studies, though, eventually nailed down the link. From 1935 to 1954, British physicians provided radiation treatments to some 14,000 people for ankylosing spondylitis, a crippling arthritis of the lower spine. Radiation therapy was discontinued in 1955 when it was realized that a disproportionate number of those irradiated were coming down with leukemia. Subsequent studies showed that this treatment caused approximately 400 excess cancer deaths among the treated population.[12] Other studies revealed similar patterns. In 1946, a *New England Journal of Medicine* study showed that, from 1935 and 1944, radiologists died of leukemia about eight times more often than other physicians.[13] In 1958, Alice Stewart of Oxford showed that mothers who had had pelvic X rays while pregnant were nearly twice as likely to give birth to children who would develop leukemia. Another widely publicized case concerned 20,000 children in Israel and New York who, between 1949 and 1960, had had their scalps irradiated as part of a treatment for ringworm (grotesque as it seems today, the purpose of the radiation was simply to cause the hair to fall out to facilitate conventional treatment). Studies published in the 1970s found that the irradiated children showed six times as many cancers as nonirradiated children, and even higher rates of brain cancers and leukemias.[14] Excess cancers were also found among the thousands of people who, as children in the 1920s and '30s, received X-ray treatments for a pseudodisease known as "status lymphaticus"—an enlarged thymus condition at one time believed to impair the breathing of infants. No one has precise figures, but the total number of children and young adults irradiated for these and other innocuous conditions (tonsils, adenoids, acne) may be as high as several hundred thousand or even a million.[15]

A different sort of danger was revealed in the use of radium in popular patent medicines. Most of these "tonics" were harmless, being devoid of radium, but long-term consumption of the genuine article could prove fatal. An expensive tonic sold under the brand name "Radiothor" was marketed in the 1920s; a well-known Pittsburgh industrialist, Eben Byers, drank a bottle a day from 1926 until 1931 when he contracted cancer of the jaw. His death in 1932 took much of the wind out of the radium-cure sails, though treatments of this sort continued on a smaller scale into the post–World War II era. Radon inhalation therapy remains a legal and occasional practice both in the United States and abroad. In certain Montana mines, for example, you can still breathe radon gas as an arthritis "therapy"; radon therapy also remains very popular in certain resort areas of the former Soviet Union. The Pyatigorsk Superior Radon Center in Georgia still prepares radon drinks, baths, enemas, and vaginal and nasal irrigations for more than 1,000 patients per day.[16] The "thermal galleries" of the radioactive spas in Badgastein, Austria, have radon concentrations averaging nearly a thousand times higher than the guideline established by the EPA in the 1980s for U.S. indoor air.

THE RADIUM-DIAL PAINTERS

Radiation-induced cancers increased rapidly in the 1930s and '40s, as the effects of both X-ray diagnostics and radium cures took their toll. Marie Curie and her daughter, Irene, both eventually died of leukemia, most likely caused by their work with radioactive materials. The most notorious early radiation scandal, however, was the tragedy of the radium-dial painters. Radium had become a military necessity and popular novelty after it was discovered that, by mixing a tiny amount of the element with a larger volume of a phosphorescent salt (zinc sulphide, for example—30,000 parts of which were mixed with one part radium to make Undark, one of the more popular products), one could produce a cheap and brightly luminescent paint to light up clock dials and watches, doll's eyes and fish bait, gunsights and the lids and handles of chamber pots. The largest and most successful commercial venture of this sort was the U.S. Radium Corporation, founded in 1915 in West Orange, New Jersey, only two blocks from Thomas Edison's famous laboratory. This company manufactured luminous instrument dials for airplanes and other military gear; by World War I, one in six American soldiers wore a radium-lighted watch. Painting the numerals on these devices required a great deal of skill and a very fine brush: the 250 workers at U.S. Radium—almost all women and girls—put the tips of their brushes in their mouths to bring them to a point. Each time they licked a brush, they ingested a small quantity of radium. So did many of the 2,000 other American workers in the industry.[17]

Health problems began to appear among the radium-dial painters in the early 1920s. Between 1922 and 1924, nine of these young women died after painful bouts with mysterious illnesses diagnosed as stomach ulcers, syphilis, trench mouth, or "jaw rot." The U.S. Radium Corporation assured health inspectors that radium was not toxic, and that the women suffering from chronic ulcerations around their teeth had simply practiced "poor dental hygiene." Cecil Drinker of Harvard was brought in to evaluate the problem, but when he found that radium was indeed the culprit, the company barred him from publishing his findings and misrepresented his research to government investigators. The company eventually got what it wanted in the form of Professor Frederick Flinn of Columbia University, an industrial hygienist who wrote a report for the *Journal of the American Medical Association* exonerating U.S. Radium and attributing the workers' deaths to a minor epidemic of communicable diseases. Flinn's authority served to delay recognition of the epidemic; he went on to serve for many years as the company's primary scientific spokesman.

Evidence was meanwhile accumulating that radiation was the cause of the epidemic. In May 1925, autopsies of two former workers at the plant showed high levels of radioactivity in their bones and internal organs; examinations of living dial painters found that their contaminated breath could light up a fluorescent screen. Radium was obviously accumulating in the body. It also became apparent that inhaling radium dust was dangerous when, in the spring of 1925, the company's chemist died of acute anemia without ever having swallowed any paint. At least fifteen women had been identified as victims of radium poisoning by 1928, though many more were no doubt wrongly diagnosed with syphilis, trench mouth, and anemia. Autopsies showed that even the burned bones of the victims contained enough radioactivity to expose photographic plates in only two or three days.[18]

A local secretary of the New Jersey Consumers' League, Katherine Wiley, eventually broke the story. After learning of the epidemic from a public health official in the spring of 1924 and failing to interest labor or health officials in the matter, she turned to Frederick Hoffman of the Prudential Life Insurance Corporation. Hoffman conducted a statistical analysis of the deaths and concluded that this was "an entirely new occupational affliction," caused probably by radium poisoning. His report, published in 1925, was the first to indicate that radioactivity was indeed at the root of the epidemic.[19] (Skeptics continued to deny a link: Marie Curie apparently accused Harrison Martland of "charlatanism" for asserting that radium was responsible for the dial-painters' bone cancers.)[20] In 1927 Wiley, then working with Alice Hamilton (famous for her crusade against the occupational hazards of lead) and Cecil Drinker, persuaded five former radium workers to file suit against the U.S. Radium Corporation for $250,000 each. The suit failed when lawyers were able to show that New Jersey employers were liable only for claims reported within two years of the onset of an injury. This was a perennial problem for workers seeking compensation for chronic illnesses such as cancer, where symptoms often do not appear until ten or twenty years after exposure. Many protested the ruling: the liberal publicist Walter Lippmann called it "one of the most damnable travesties of justice that has ever come to our attention." The plaintiffs were eventually able to appeal the decision and settled out of court for a lump sum of $10,000 each plus a small pension and medical expenses. Within six months of the settlement, the founder of the company himself had died of radium poisoning.[21]

The dial-painters' tragedy is usually regarded as a phenomenon of the 1930s, but it is important to realize that many of those exposed did not die for years, sometimes decades. At Ottawa, Illinois, another major site of luminescent dial manufacture, more than 30 head and bone cancers have shown up among 700 women who painted dials at Luminous Processes, Inc., a plant that was shut down in 1978. The practice of brush licking was stopped in the 1930s, but workers who began even after this time have proved to be at risk for cancer. A study by the Argonne National Laboratory's Center for Human Radiobiology of 463 women who worked at the Ottawa plant and did not put brushes in their mouths found more than twice as many breast cancer deaths as expected. Women interviewed for a 1983 *Wall Street Journal* article testified that safety precautions at the plant were almost nonexistent: radium, they had been told, was safe to handle, so they were somewhat cavalier in dealing with the paint. Pearl Schott, who worked at the Ottawa plant from 1946 to 1977, reported that women sometimes painted their fingernails "for kicks"; she and others wiped their paint-covered hands on their clothes and sometimes got up during the night to find their hair glowing green in the mirror. Mrs. Schott returned to work in 1964 after a mastectomy; still unaware of the link, she developed more tumors on her feet and was forced to walk with a cane. Another former worker who developed breast cancer in 1978 protested: "It never entered anyone's mind that we were doing something dangerous; why didn't they say something?" Warren Holm, vice president of Luminous Processes for forty-three years until his retirement in 1979, told the *Wall Street Journal:* "Breast cancer is thought to be a hormonal problem, not a radioactivity hazard."[22]

Former workers who have died are brought to a local funeral parlor, where they are laid out on a bed of lead and inspected by scientists from Argonne National Laboratory as part of a Department of Energy study of how low-level radiation affects the body.[23]

A different kind of legacy continues even today at West Orange, New Jersey, home of the original radium tragedy. In December 1983, the New Jersey Department of Environmental Protection announced that about 100 homes had been found with dangerously high levels of radon, a consequence of having been built over radioactive waste dumped sixty years earlier by the U.S. Radium Corporation. Two years later, New Jersey environmental officials announced they would spend $8 million to remove the soil from under twelve of those houses. Fifteen thousand drums of contaminated soil were stockpiled in Kearny and Montclair, but citizen protests and governmental indecisiveness led to a long-drawn-out scandal over how to dispose of the soil permanently. Meanwhile, the number of houses identified as contaminated rose to 200, 300, and then 400, with estimated cleanup costs rising to $250 million. At the end of 1993, a final resting place for the soil had not yet been found.[24]

THE SHADOW OF THE BOMB

The most important impetus for the study of radioactive hazards in the United States was the Manhattan Project of 1942 to 1945, the goal of which was to build, test, and deliver an atom bomb for use in World War II. The success of the project and the subsequent industrial-scale manufacture and testing of atom bombs permanently changed the extent to which people were exposed to ionizing radiation. Hundreds of radiologists and several hundred radium workers had suffered pathogenic doses of radiation prior to the war; during the making of the bomb, many times that many scientists, technicians, and miners were exposed. The dropping of two atomic bombs on Japan at the end of World War II subjected hundreds of thousands more to potentially cancer-causing radiation: victims included not just the Japanese survivors of the blasts but thousands of American soldiers brought in afterward to clean up. The postwar weapons buildup further increased the numbers exposed by several orders of magnitude; the fifty years since the bombs were dropped can be viewed as a kind of laboratory for the study of the health effects of radiation on atomic workers, atomic veterans, and a diverse class of "downwinders."

Manhattan Project officials were aware of many of the potential hazards of nuclear work. Arthur H. Compton, the Nobel Prize–winning head of the project's metallurgical laboratory at the University of Chicago, established a health division in the summer of 1942; the division distributed pocket ionization chambers and film badges to monitor gamma-ray exposure and devised means of protecting workers against the "special hazard" of radiation (*radiation* was a classified word, as was *plutonium*). The term *health physics,* incidentally, dates from this time—originally referring to the physics section of the health division, though its meaning was later generalized to include the entire science of radiation protection. One of Compton's goals in establishing the division was to reassure the project's staff that their safety was in good hands; this was also a major goal of the military directors of the project, who were especially concerned with heading off injury-based lawsuits. The man chosen to head the health division, Robert S. Stone, had been chairman of the radiology department at the University of California Medical School at San Francisco; his new assignment, as he saw it, was essentially "one vast experiment." Never before had so many people been exposed to so much radiation.[25]

It was not an easy task to monitor radiation hazards at that time. Many of the isotopes produced by the nuclear chain reaction had never been studied, and it was often difficult or even impossible to tell how much of a particular isotope had been ingested or inhaled.* Most of the isotopes of concern in the Manhattan Project were not as easily monitored as radium, with its powerful gamma emissions. Internally deposited radium could be monitored by measuring the gamma radiation emerging

---

*A given element may have different isotopes, according to how many neutrons there are in its nucleus. Carbon 12, for example, has 6 neutrons and 6 protons; radioactive carbon 14, by contrast, has 8 neutrons and 6 protons. The *half-life* of a radioactive isotope is the time it takes for half the mass of that particular isotope to decay: highly radioactive isotopes will have a short half-life, even fractions of a second; less radioactive isotopes (uranium 238, for example) can have half-lives in the billions of years. The biological hazard of a particular isotope depends on the rate of decay, the type and energy of emission, where it is deposited in the body, and how long it is retained.

from a contaminated body (along with exhaled decay products such as radon), but that was not the case for uranium and thorium, two of the most commonly handled materials. A urine test developed in 1945 for uranium and thorium gave a rough idea of internal contamination, but it was often unclear how a particular radionuclide would travel through the body, whether some people were more vulnerable than others, and so forth. Monitoring difficulties were heightened by the fact that, contrary to original expectations, several of the new materials—plutonium, for example— turned out to be far more hazardous than radium. Plutonium is retained longer in the body and tends to deposit selectively in the gonads and the blood-forming marrow of the bone, where its radioactive emissions can do severe damage. Its half-life of 24,000 years (for plutonium 239) also made it difficult to dispose of. Chemists purifying the element for the Manhattan Project refused to work without extra insurance, though critics such as Cyril Stanley Smith, head of metallurgical chemistry for the project, denounced the Project's insurance policy as inhumane and unfair, covering as it did "only illnesses or disabilities that appeared within 90 days of an accident or 30 days of leaving the project, even though radiation damage might take thirty years to surface." A policy of "high amputation" was established for workers unfortunate enough to be contaminated through a cut or scratch. Veterans who became ill from exposure to atomic bomb radiation often had difficulty obtaining compensation from the Veterans Administration—the most common excuse being that the illness in question had appeared only after the GI had left the service.[26]

Military urgencies ultimately subordinated health concerns. Fallout from the first nuclear test at Alamogordo, New Mexico, for example, was tracked for only a day after the blast. And little effort was made to study biological repercussions: some mice were hung by their tails as part of a halfhearted effort, but they all died of thirst in the desert before the bomb even went off. Fallout was also not originally considered a serious problem. Film badges designed to monitor individual exposures were issued to post offices near the test site, but the primary goal, again, may have been to protect the army against the possibility of lawsuits rather than seeing to it that citizens in the surrounding areas were adequately warned and protected. J. Robert Oppenheimer, the project's scientific director, was worried that people claiming to be injured by fallout might sue the government; the health group's reports (on the effects of fallout on farm animals, for example) were therefore held in strictest secrecy. Even after the bombs were dropped and Japan had surrendered, physicians were sent to Hiroshima and Nagasaki, among other things, to prove that "there was no radioactivity from the bomb." Journalists visiting the cities under military escort were seduced by army propaganda: the *New York Times,* for example, published headlines announcing "No Radioactivity in Hiroshima Ruin." Wilfred Burchett, the first independent journalist to enter Hiroshima without an army escort, gave quite a different account, warning that thirty days after the bomb fell, people were dying, "mysteriously and horribly," from "an unknown something which I can only describe as the atomic plague."[27]

The hazards of radiation were well concealed in the months following the Japanese defeat. Catherine Caufield in her history of this period notes that, among the 200 letters published by U.S. newspapers immediately following the dropping of the bombs, not a single one mentions the possibility of a radiation hazard. Oppenheimer and General Leslie Groves, representing the scientific and military leadership of the Manhattan Project, had successfully camouflaged the radiation hazards lingering in the shadow of the bomb. Many of those who did eventually worry about health effects were more concerned about the prospect of long-term genetic damage to the species than about the more familiar danger of cancer. Health officials tracing the effects of the bomb found little evidence of long-term genetic damage, and both popular and scientific accounts of the period sometimes used this to belittle the effects of radioactive exposures.

Studies of the health effects of atomic bomb radiation were begun in 1950 by the Atomic Bomb Casualty Commission and continue today, under the authority of its Hiroshima-based successor organization, the Radiation Effects Research Foundation (RERF).[28] The methods and results have been somewhat controversial, but in 1985 the study reported that among a group of 76,000 Japanese survivors, about 5,940 had died of cancer. Control groups indicated that in the absence of radiation from the bomb, about 5,600 cancer deaths would have occurred over this time from other causes. Roughly 300 people were therefore calculated to have died from cancer caused by radiation from the bombs (90 of these were lung cancers). Otherwise put, atomic bomb radiation increased the likelihood that survivors of the bomb blasts would die of cancer by about 5 percent.[29]

Critics still worry, though, that even these carefully researched figures may underestimate the hazards of low-level radiation. Alice Stewart points out that RERF studies have failed to take into account what she calls the "healthy-survivor effect"—the fact that many of those who survived the initial blast were likely to be exceptionally fit with a higher-than-average ability to fend off cancer. Studies ignoring this selection effect would tend to underestimate the cancer risk of radioactive exposures. RERF epidemiologists looked for effects such as these and failed to find any. Stewart, though, points out that the healthy-survivor effect may have been masked by the fact that atom bomb radiation may have caused immune system damage leading to higher rates of noncancer deaths (from infection or heart attack, for example) among survivors. If this is true, then cancer rates would be pushed down, washing out the healthy-survivor effect.[30]

Stewart charges that something similar is going on at nuclear factories like the Hanford Nuclear Reservation in southeast Washington State, where weapons-grade plutonium was produced from 1943 to 1992. She found excess cancers among certain classes of workers at the plant as early as 1977,[31] but subsequent studies appeared to indicate, strangely, that Hanford nuclear workers as a whole actually had *lower* rates of cancer than would have been expected even if they had never been exposed.[32] Industry scientists interpreted this to mean that cancer was not an occupational hazard at such facilities, but Stewart and her colleagues countered by suggesting that the lower cancer rate was simply a consequence of the fact that sickly persons were excluded from work at such plants (the so-called healthy-worker effect). Similar accusations of bias were leveled against a 1985 National Academy of Sciences study on the health of U.S. servicemen exposed to radiation from atomic bomb tests. In 1992, the General Accounting Office and the Office of Technology

Assessment ordered the academy to redo the study when it was found that thousands of exposed servicemen had been excluded and thousands who were nowhere near the tests had been included. Charges were also raised that the healthy-worker effect ("healthy-soldier effect," as one report put it) had not been taken sufficiently into account. In July 1992, the academy agreed to do a $3.6 million follow-up study, hoping to avoid the previous report's flaws.[33]

Disputes persist concerning how to interpret the Hanford data. Until recently, a major problem was the refusal of the Department of Energy to release health records of nuclear workers. On July 17, 1990, the DOE finally released the health records of 44,000 Hanford nuclear reservation employees, and records of more than a half a million other radiation workers are promised for the future.[34] It remains to be seen how epidemiological analysis of these records will affect the climate of the debate.[35] In the meantime, uncertainties continue to plague efforts to determine how much radiation was released during the atomic explosions over Japan. In the early 1980s, RERF studies were thrown into disarray when scientists at the Lawrence Livermore Laboratory showed in computer simulations that the radiation received by survivors at Hiroshima was *lower* than previously thought. Conventional wisdom had held that much of the bomb's radiation had been in the form of neutrons, but the new study indicated that neutron release was minimal and that most of the bomb's output was in the form of gamma rays. This meant that gamma radiation—and, by implication, X rays—was a more potent cause of cancer than previously imagined. RERF scientists had to revise their estimates of the radiation hazard upward, though not without substantial debate and hair pulling. Los Alamos physicists seeking to test the new theory built an exact replica of the Hiroshima bomb—using old materials found in storage—and found that the simulations were probably accurate.[36] When the new results were finally accepted, in the mid-1980s, calls were heard to strengthen occupational radiation standards.

It was not long, though, before even these revisions were called into question. In 1992, a biophysicist at Livermore found that by measuring isotopic variations in granite and concrete materials subjected to the Hiroshima blast, he could tell how much neutron radiation had been released by the bomb. These new measurements seemed to indicate that the simulation-based revision was flawed, and that the number of neutrons released by the bomb had been underestimated by as much as a factor of ten. This meant that Hiroshima survivors had actually suffered *higher* levels of radiation than in the previous revision; it began to look again like the cancer hazards of bomb blast radiation had been exaggerated.[37] The fallout from this latest revision has not yet settled, but it is already clear that the $100 million spent thus far on trying to figure out how many cancers were produced at Hiroshima and Nagasaki has not been enough to establish a consensus.

Compounding the difficulties of the atomic bomb data are ambiguities in how to interpret evidence of cancer clusters surrounding British nuclear power plants. The most recent round came in February 1990, when Martin J. Gardner, a respected British epidemiologist, reported unexpectedly high levels of leukemia among the children living in the village of Seascale, in northwestern England, home of the Sellafield nuclear reprocessing plant. The report, published in the *British Medical Jour-*

nal, aroused a storm of controversy, primarily because the low-level exposures from the plant had not been considered enough to produce detectable levels of cancer. A great deal of effort has gone into finding alternative explanations for the findings. Leo Kinlen of the University of Edinburgh has proposed the "new town hypothesis," arguing that the clusters might be due to the fact that, during the building of such installations, construction workers bring in cancer-causing viruses. (One would presumably therefore expect, according to this argument, *all* "new towns" to have higher leukemia rates.) Richard Doll of Oxford has suggested, alternatively, that the leukemias may be just a statistical fluke. Diseases are never evenly distributed throughout a population: clusters invariably emerge simply as a result of random fluctuations, and some of these will turn out, by chance, to be in nuclear areas. In support of his hypothesis, Doll notes that there are leukemia clusters even in some of the "phantom sites" considered for nuclear facilities—but never built.[38] The idea has received predictably favorable coverage from nuclear energy trade associations like the Council for Energy Awareness.[39]

Gardner questions both of these interpretations and suggests instead that the correlation he has found between a father's exposure and childhood leukemia is more likely due to radiation-induced mutations in sperm cells passed on as a congenital predisposition to leukemia. Gardner's explanation is controversial, primarily because the radiation exposures of the men working at Seascale were well within the British occupational limit (50 milliSieverts a year; the highest doses were in the 10–20 milliSievert range). Moreover, the study appears to contradict the RERF studies of atom bomb survivors, which have never found any evidence that the *children* of the survivors suffered excess leukemias (even though their parents were exposed, on average—according to early 1990 calculations—to about 450 milliSieverts of radiation). Conventional wisdom has most often held that the sudden burst of radiation generated by an atomic bomb should be more damaging than the long-term low-level exposures typical of the occupational setting, but Gardner suspects that the opposite may in fact be the case. Interviewed by a *Science* magazine reporter, Gardner speculated that DNA may be better able to repair itself after a single instantaneous exposure (as in Hiroshima) than during the slow and steady damage typical of occupational exposures.[40]

Gardner's hypothesis gained a measure of support in February 1992, when two teams of scholars from Britain and the United States independently discovered an entirely new form of radiation damage that showed up only in the descendant cells, several generations removed, of rodent tissues irradiated with X rays or alpha particles. This so-called delayed mutation effect was significant because it meant that radiation damage might not show up until the cells of exposed tissues have undergone several generations of cellular division after exposure. The phenomenon might explain why cancers may develop long after radioactive exposures; it might also lead to calls for more stringent regulation of exposures.[41]

*   *   *

URANIUM MINES

Uranium mining is unlike most other kinds of mining in that, in the course of blasting and digging for ore, radioactive radon 222 gas is released. Radon 222 is a natural decay product of uranium, with a half-life of about 3 1/2 days. Radon gas by itself poses no real danger: as a noble gas it is chemically inert and is simply exhaled. But its radioactive daughter products can settle in the lungs and injure those tissues. The primary hazard comes from polonium 218 and 214, alpha-emitting radionuclides that lodge in the lining of the lung. Uranium miners are also bombarded by gamma radiation, but the primary danger, again, stems from the ingestion and inhalation of alpha emitters. Smoking in combination with uranium mining can prove especially dangerous: Geno Saccamano, a resident of Grand Junction, Colorado, and an internationally known pathologist, showed in the 1960s that uranium miners who smoke cigarettes increase their odds of getting lung cancer as much as thirty times. More recent studies have shown that nonsmokers are also at risk: Robert J. Roscoe of the National Institute for Occupational Safety and Health has shown that nonsmoking uranium miners followed from 1950 to 1984 were thirteen times more likely to die from lung cancer than a comparable group of nonsmoking U.S. veterans.[42]

The health hazards of what we today recognize as radon gas were appreciated prior even to the discovery of X rays, primarily through studies of the silver and uranium miners in the *Erzgebirge* (ore mountains) of Germany and northern Czechoslovakia, near the towns of Schneeberg in Saxony and Joachimsthal in Bohemia. These mines had been worked since the fifteenth century, first as a source of silver (the word *dollar* is derived from *thaler,* a shortened form of *Joachimsthaler,* a coin made from the very pure silver of this region) and later as a source of nickel, cobalt, bismuth, and arsenic. In the second half of the nineteenth century, mining was again revived in this area as the search went on for uranium, used in the manufacture of high-quality pigments.

Miners from this region had long been known to suffer from a painful and "consumptive" lung disease known locally as *Bergkrankheit* ("mountain sickness"),* correctly identified by F. H. Härting and W. Hesse in 1879 as lung cancer. (Härting and Hesse were in fact the first to document the fact that internal cancers could be caused by environmental carcinogens: previous discussions—such as Percivall Pott's—presented evidence only of environmentally induced *skin* cancers.) Using a cleverly designed device with wax paper to capture the dust particles, along with a machine to measure breathing volume, they showed that miners were inhaling up to six grams of dust in a seven-hour shift. Autopsies showed that after a lifetime of dust inhalation, miners' lungs were usually black with pigmentation. Most frightening of all, Härting and Hesse found that a miner could expect to live only about twenty years after entering the mines, and that about 75 percent of all miners in the region were

---

*Georgius Agricola practiced medicine in Joachimsthal from 1527 to 1533; his famous *De Re Metallica* (1556) described the health effects of silver mining in this region, where miners inhaled a corrosive dust that "eats away the lungs, and implants consumption in the body; hence in the mines of the Carpathian mountains women are found who have married seven husbands, all of whom this terrible consumption has carried off to a premature death" (trans. H. H. Hoover and L. H. Hoover [London, 1912], p. 214).

dying from lung cancer—an extraordinary rate, but one that would be even higher, they suggested, were miners not also dying prematurely from other causes, such as accidents.[43]

Further investigations along these lines were not performed until 1913, when an Austrian physician, Alfred Arnstein, showed that from 1875 to 1912 about three-quarters of all the miners in Schneeberg had died of lung diseases, and that lung cancer alone accounted for nearly half of all deaths in this period. Arnstein was the first to show that the miners were dying from cancers of the so-called squamous-cell type.[44] A rash of corroborating studies soon followed, and lung cancer was finally recognized as a compensable occupational disease for miners in 1926 in Schneeberg (Germany) and in 1932 in Joachimsthal (Czechoslovakia).[45] The pro-worker political climate accompanying the triumph of Social Democracy in the postwar years was crucial in this regard. The disease by no means disappeared, however: a 1935 study showed that the majority of Schneeberg miners were still dying of lung cancer; the author of this report marked out the locations of 266 cancer deaths on a map and found that cancer rates were three times higher in radium-rich areas than in radium-poor areas.[46] A Czechoslovakian report in 1932 noted that the miners themselves believed that "the discovery of a rich uranium ore vein is always followed some years later by a strong mortality among them."[47]

For most of the nineteenth century, the uranium miners' *Bergkrankheit* had most commonly been attributed to foul air, consumption, or "miner's phthisis" (probably silicosis or silico-tuberculosis). Although the ailment was correctly diagnosed as cancer in the 1870s, it was not until the 1920s that the specific cause was identified as the inhalation of radioactive gases. Even after the disease was recognized as cancer, scientific belief commonly blamed the inhalation of arsenic (Härting and Hesse's view), fungal spores on wooden mine supports, or genetic predispositions fostered by inbreeding in these remote mountain regions.[48] "Radium emanation" was measured in Joachimsthal in 1905 and in Schneeberg in 1909, though a link was not yet drawn to cancer. Apparently the first person to suggest that radiation was at the root of the miners' disease was an employee (whose name we do not know) in the local office of the mines. A medical student by the name of Margarethe Uhlig ran across the suggestion in the Schneeberg mining archives and, in a 1921 article published in Germany's leading journal of pathology, argued that the idea should be taken seriously.[49] Prompted by Uhlig's suggestion, further measurements of radioactivity in the Schneeberg mines were published in 1924 in one of Germany's leading physics journals. The authors, to their surprise, found that radon in the Schneeberg mines varied from as little as 410 to as much as 18,200 picocuries per liter.[50] Studies eventually confirmed that the radioactivity in any given mine could vary by several orders of magnitude but was generally highest, as one would expect, near the drill holes where new ores were ripped from the rock face. According to one calculation, workers were inhaling an average of 12.4 mg of radium per year, an extremely toxic dose. Wilhelm Hueper pointed out in 1942 that mine workers had long been aware that certain shafts were more hazardous than others: the shaft found to have the highest radioactivity, for example, was popularly known as the *Todesschacht,* or "death mine."[51] A 1924 survey had found between 2,000 and 9,000 picocuries per liter in this mine; this was convincing evidence, in their view, that radon inhalation was the primary cause of the miners' cancer.[52]

In the United States, concerns about the health effects of exposures to radon emerged largely after World War II in response to the rapid growth of military demand for uranium. U.S. bomb production, driven at a frenetic pace by cold war fears and military-industrial enthusiasm, required the extraction of huge quantities of ore, all of which—by law—had to be sold to the Atomic Energy Commission (AEC). Ores were eventually found throughout the world, but the most important U.S. source for many years was the arid Four Corners region best known as the Colorado Plateau, stretching across Utah, Colorado, New Mexico, and Arizona. Skyrocketing demand set off a uranium rush, though smaller investors were soon driven out by larger companies and, by the end of the 1950s, five or six large corporations—notably Kerr-McGee, Union Carbide, Anaconda, United Nuclear, and Homestake Mining—dominated the industry. After the Arab oil embargo of 1973, oil companies joined in the search for uranium, including Exxon, Gulf, Mobil, Conoco, Getty Oil, Atlantic Richfield, and Standard Oil of Ohio. In 1966 there were 600 uranium mines in the United States (all on the Colorado Plateau) employing 2,500 miners, 90 percent of whom worked underground.[53] Most of them were Mormons; many of the non-Mormons were Navajos.

Given the European experience, the emergence of a radon hazard in U.S. uranium mines should not have been difficult to predict. Gaseous emissions from radium had been recognized as a potential health hazard since early in the century, and by the end of the 1930s efforts had already been made to establish a "tolerance concentration" for inhaled radon.[54] In 1942 Wilhelm Hueper presented an elaborate historical review of the discovery of the link between radon and lung cancer in European mines, along with a convincing refutation of efforts to attribute the disease to other factors, like inbreeding.[55] A 1944 review by Egon Lorenz for the *Journal of the National Cancer Institute* concluded that "the radioactivity of the ore and the radon content of the air in the mines are generally considered to be the primary cause" of the European *Bergkrankheit.*[56] An expert on occupational cancer predicted that "one will see cases of cancer and of leukemia in our newest group of industrialists, workers in the field of fissionable materials."[57]

In the late 1940s, several scientists tried to alert the federal government to the uranium-mining hazard, but without success. Bernard Wolf and Merril Eisenbud, the first and second director of the AEC's medical division, traveled to Colorado in April of 1948 to inspect the mines and make recommendations for protection of the workers, but were rebuffed when they tried to alert their superiors to the problem: "Much to our surprise, we were told by Washington that the health problems of the mines were not the responsibility of the AEC, and that they should be left to the jurisdiction of the local authorities." The AEC had been assigned by Congress the responsibility for radiation safety in the nuclear program but, according to a bizarre interpretation of the 1946 Atomic Energy Act, the commission was bound only to regulate radiation exposures *after* the ore had been mined. Responsibility for the health and safety of uranium miners had been left up to individual states, a situation Merril Eisenbud rightly recognized as "absurd," given their lack of equipment and expertise to deal with the expected health problems.[58]

This was not a good time to cross the AEC on matters of radiation safety: we already saw in chapter 1 how Hueper had begun an investigation of the health of uranium miners in 1948, only to see his research blocked when the AEC heard about his plans to publicize the work. There are other examples: the radiobiologist William F. Bale, for example, was also ignored by the AEC when he presented evidence in 1949 that U.S. miners were being exposed to levels high enough to cause lung cancer. Even H. J. Muller, the Nobel laureate celebrated as both the founder of radiation genetics and (later) the first president of the American Society of Human Genetics, was barred from speaking at the International Conference on the Peaceful Uses of Atomic Energy at Geneva in 1955. Muller had been involved in publicizing the threat posed by radiation to the genetic integrity of the species and, for that reason, was feared by the commission, which twisted his ideas and sought to present a more benign view of the health effects of radiation.[59]

The U.S. Public Health Service (PHS) began to study the environmental hazards of the mines and the health of underground miners in 1949 and 1950, but this agency, too, was hamstrung by apathy, bureaucratic conservatism, and government censorship. In the early 1950s, the service launched an educational program to inform mining companies and state agencies of the danger and how it might be lessened, but—incredibly—the miners themselves were never informed of the hazard.[60] The service recommended improved ventilation and the wearing of masks, though neither was mandatory. In the summer of 1952, a PHS team measured radon levels in 157 mines in Colorado, Utah, Arizona, and New Mexico, and found median concentrations ranging from 8 pCi/l (at Eastern Slope) to 9,000 pCi/l (at Bull Canyon). The highest levels recorded were at Bull Canyon and Marysvale, where peak readings of 59,000 pCi/l and 48,700 pCi/l were obtained.[61] In 1957, the service proposed a standard of one "working level" (WL)—equivalent to 100 picocuries of radon daughters per liter of air—as the maximum safe level of radioactivity in a working mine; by its own admission, 64 percent of all workers were exposed to levels at least ten times this high.[62] As late as 1961, nearly a third of all U.S. uranium miners were still breathing more than 1,000 pCi/l at work.[63]

Victor Archer and three of his colleagues at the Public Health Service published the first major study of U.S. uranium miner epidemiology in 1962. The miners had been followed for over a decade, and the first excess cancers were beginning to appear. Accidents and heart disease were still the primary hazards, but 5 of 907 miners with more than three years' experience underground had already died from lung cancer, when only one such death was expected.[64] These preliminary results were confirmed a year later, when an updated analysis of the PHS data showed that uranium mine workers were more than five times as likely than the normal population to die from respiratory cancer, and that men who had been exposed to the highest levels of radiation showed lung cancer rates up to forty times higher than expected. These and subsequent studies confirmed what had been known to European physicians for the better part of a century: improperly ventilated uranium mining causes cancer.[65]

It is sometimes said that firm proof of a radiation hazard in U.S. mines emerged only in the 1960s, and it is true that published reports of actual cancer deaths did not appear until this time. But, as we have seen, good evidence that such deaths would occur was available much earlier and was ignored or even suppressed by the Public Health Service, acting on orders from the Atomic Energy Commission. The Public Health Service Report of 1957 called for "the immediate application of corrective measures,"[66] and half a dozen scientists had warned—since the late 1940s—of an impending public health disaster.[67] Duncan A. Holaday, the PHS's health physicist responsible for monitoring the radiation in U.S. uranium mines, became frustrated by the failure of the AEC to clean up its act: in 1953, he complained that "the only way we can ever get anything done is to collect [dead] bodies and lay them on somebody's doorstep." Archer used similar language to express his own frustration. In the mid-1980s, at the trial of former Navajo miners and their families suing the federal government for compensation, Archer testified that he and his colleagues had caved in to AEC and PHS pressures not to publicize the hazard: "We did not want to rock the boat . . . we had to take the position that we were neutral scientists trying to find out what the facts were, that we were not going to make any public announcements until the results of our scientific study were completed."[68] Official pressures to "monitor" the disaster without informing those at risk or forcing the companies to reduce the hazard led PHS scientists to characterize their study as a "death watch" or "dead body approach." A federal judge involved in the Navajo case charged that U.S. atomic authorities had failed to warn the miners in order to guarantee "a constant, uninterrupted and reliable flow" of uranium ore "for national security purposes."[69]

Local physicians were certainly not unaware that something was amiss. Lung cancer had always been rare among the Navajos, as it still is, relatively speaking, today. By the mid-1950s physicians on the plateau began to see increasing numbers of Navajos with the disease: "It got so I didn't need to wait for the tests. They would come in—the wife and the six kids were always there, too—and they would say they had been spitting blood. I would ask and they would answer that they had been in the mines. Then I didn't need to wait to make the diagnosis, really, I already knew." But as late as 1979, Kerr-McGee spokesmen were still denying that Navajos who had worked in the mines had contracted cancer from their work.[70]

The Atomic Energy Commission and Public Health Service were not the only agencies unwilling or unable to tackle the problem. The Bureau of Mines had long regulated health and safety standards in the nation's coal mines, but it was not until 1966 that it was authorized to regulate health and safety in uranium mines. The Labor Department was equally negligent, despite the powers granted its secretary under the Walsh-Healy Act of the 1930s to stop government purchases from hazardous mines. During the years of federal neglect, responsibility for monitoring and regulating the radiation hazard in mines was left up to individual states. Utah state officials in 1959 tried to have one company, Vanadium, install ventilation equipment, but without success.[71] Colorado was the only state to recognize radon-induced lung cancer as a compensable disease: an actuarial firm hired by the state in the mid-1960s calculated that by 1985, 1,150 cases of lung cancer would have arisen from working in the mines. These figures assumed, rather optimistically, that exposures would cease after 1966.

Federal action finally came on May 5, 1967, when President Johnson's Secretary of Labor, W. Willard Wirtz, announced that the U.S. government would purchase no more uranium from mines in which there was more than .3 WL (working levels) of radiation. Wirtz's decision immediately ran afoul of the government's Joint Committee on Atomic Energy, a Senate body charged with both supervision and promotion of the nuclear industry—a relationship labor union critics characterized as "the fox guarding the henhouse." The committee, siding with the AEC, made it clear it wanted to retain the Public Health Service's standard of 1 WL. At hearings before the Joint Committee on Atomic Energy, Paul C. Tomkins, executive director of the Federal Radiation Council, testified that the Labor Secretary's recommendation "could do nothing but close up the [nuclear] industry." A Kerr-McGee physicist testified that miners could sustain up to 3 WL of radiation without harm. Wirtz was ultimately forced to accept the AEC/PHS standard; critics of the sequence of events observed that miners, the weakest of all the participants in the drama, would have to continue "their deadly love affair with the daughters of radon."[72]

Controversies continue to ebb and flow over the precise shape of the dose-response curve, but no one doubts that miners are still dying. By 1978, 205 of the 3,400 white miners followed by the Public Health Service since the early 1960s had died of lung cancer; by 1990, the deaths totaled nearly twice that figure. Given that 30,000 to 40,000 men have worked underground in U.S. uranium mines since the 1940s, it is probable that 4,000 or 5,000 Americans have died (or will die) from lung cancer caused by that line of work. Hundreds of others have had a lung removed or been otherwise debilitated.[73]

Why were systematic efforts to monitor the hazard not set up until the 1960s? Why, even after this time, were miners allowed to suffer ten to thirty times the legal levels? Why were the miners never warned? Many of these mines were on Navajo lands and many miners were Native Americans; some have suggested that racism may have played a role in the failure to establish effective standards early on. The fact that the 3,400 miners originally studied by the PHS were all white lends credence to such a theory, as does the fact that Navajo miners were not studied until the 1960s.[74] Fear of losing a reliable labor source was probably an additional factor. In 1985, in a case brought by non-Native American miners against the federal government, Judge Aldon Anderson of the U.S. District Court in Utah ruled that the AEC failed to warn uranium miners of a health hazard out of a misguided sense of national security: "The AEC feared that informed miners would flee the mines and thereby threaten the nation's uranium supply."[75]

Duncan A. Holaday, the Public Health Service scientist who warned federal officials about the danger early on, has testified that "by the late 1940s, there was no question in anybody's mind that radiation in the mines was a real problem. . . . But nobody in the Atomic Energy Commission wanted to pay attention." Holaday calculated in 1954 that mine operators could install ventilation equipment at a cost of only five or six cents per pound of processed uranium—a cost that easily could have been passed on to the AEC. (The AEC bought more than three million pounds of the metal that year at about \$12 a pound.) Holaday's advice was ignored by the AEC and mine operators, however, on the grounds that ventilating the mines was "unnecessary and too expensive."[76]

The circle of blame for the U.S. uranium mining tragedy is large. Hueper at the NCI blamed his superiors at the PHS; PHS scientists such as Holaday and Archer were able to blame the AEC. The AEC's division of biology and medicine blamed the division of raw materials (to which it was subordinate), while lawyers for the division of raw materials claimed that responsibility for the miners' health lay with the individual states. Even those who did realize that something was wrong felt impotent: Duncan Holaday once told a mining official that the PHS was going into the fight against the lung cancer epidemic armed only with a feather duster. Wayne Owens (D-Utah), one of the congressional sponsors of the 1991 Radiation Exposure Compensation Act (and whose own brother-in-law died from lung cancer caused by uranium mining) has put the matter rather simply: "These people were used as guinea pigs . . . in the life-and-death struggle with 'godless communism.'"[77]

URANIUM-MINE CONCENTRATION CAMPS

Historians often treat it as a disease of the distant past, but uranium-induced lung cancer is primarily a phenomenon of the late twentieth century. In the United States, this has to do with the fact that uranium mining is almost entirely a post–World War II industry, but the same is true even for Europe, where the industry is much older. The *Schneeberger Krankheit* was discovered in 1879 but—scandalously—more people have probably died from the disease since the 1950s than in the previous 200 or 300 years of mining. The reason, of course, is that prior to the development of nuclear weapons and atomic power, uranium mining was a very small-scale operation. The metal was not even known as a separate element until 1879; before then, miners simply discarded the ore as refuse (hence the name *pitchblende*—that which you just pitch, or throw away). After 1879, uranium was used as a fire-resistant coloring for glass and ceramics, though even then the number of active miners was never more than several hundred. At Schneeberg there were 663 miners at the height of uranium production in 1880; the numbers fell steadily thereafter to about 200 prior to the outbreak of World War I and only 54 in 1923.[78]

In Eastern Europe, as in America, the postwar call to nuclear arms forced a rapid expansion of the uranium mines of the *Erzgebirge,* employing tens of thousands and ultimately hundreds of thousands of miners. (The weapons potential of the mines was apparently unappreciated by American intelligence officials, who allowed these regions to fall into Soviet hands after the collapse of Nazi armies.) In Czechoslovakia, a secret agreement of November 23, 1945, granted the Soviet Union exclusive rights to all uranium mined in the country. The Schneeberg mines of Saxony, now under Soviet occupation, were revived; uranium ore figured as part of the war reparations paid by the East Germans to the Soviet Union.

Outside North America, the largest single producer of uranium in the world was the German Democratic Republic, where, from 1945 to the end of the 1980s, half a million workers produced some 200,000 tons of enriched uranium for Soviet atomic bombs and reactors. The authority responsible for coordinating the East German atomic effort was code-named Wismut (literally "bismuth") in order to disguise the fact that uranium was the primary object of interest.[79] Wismut had been organized by the Soviets shortly after the war, and by the mid-1950s the organization employed 150,000 workers—plus 5,000 Soviet supervisors—in underground mines, mills, and enrichment plants built to produce the "yellow cake" needed for atomic bombs. With its own Communist party organization, hospitals, secret police, and propaganda apparatus, Wismut was, as a 1993 *Science* news feature put it, "an all-powerful state within a state," shrouded in secrecy.[80]

Much to the apparent surprise and pleasure of postunification German health physicists, detailed medical records were kept on the health of the East German nuclear workers. As one Munich physicist recently put it, the archives may well constitute "the world's biggest treasure chest of data on radiation and human health."[81] Western epidemiologists hope that the archives will help to clear up long-standing debates about the hazards of low-level radiation, though it is not yet clear who is going to be granted access to the files. Nor is it even clear whether or to what extent they may have been sanitized of politically embarrassing information. A staff member of the East German Cancer Registry has admitted that the uranium workers of Schneeberg were excluded from the registry, as were high-ranking members of government.[82]

However useful such records turn out to be for the construction of dose-response curves, the episode clearly constitutes one of the most grotesque mistreatments ever of workers, both women and men. The files indicate that by 1989, when East German environmental activists* revealed the existence of the records to the public, more than 20,000 miners had already died from lung disease—15,000 from silicosis and 6,000 from lung cancer. Many more will die as the delayed effects of exposures take their toll. More than a thousand square kilometers have been polluted: one lake near Oberrothenbach, west of Chemnitz (formerly Karl-Marx-Stadt), contains an estimated 22,500 tons of arsenic, plus comparable amounts of sulfuric acid, cadmium, and lead, mixed in with various and sundry radioactive materials. Five hundred million tons of chemical and radioactive waste are scattered over the mountains and valleys south of Leipzig, and it is difficult to imagine what the long-term health effects will be. Cleanup costs have been estimated at between $3 billion and $8 billion.[83]

Patricia Kahn, a science reporter in Heidelberg, brought the Wismut story to the attention of an English-speaking audience in 1993,[84] but did not mention that a comparable program existed in Czechoslovakia, on the southern slopes of the *Erzgebirge*. About the same time that U.S. health authorities began to take the problem seriously, Czech officials began to study the health effects of that nation's uranium

---

*A key figure in this story was a zoological preparer by the name of Michael Beleites from the town of Gera in the heart of the mining country, who in 1989 exposed the scale of Wismut's misdeeds in a church document cranked out on a smuggled, hand-operated press. Joachim Krause, a chemist and environmental activist for the Protestant church in Saxony, gathered health files on 450,000 Wismut workers, now in the possession of German atomic authorities. See Kahn, "Grisly Archive," p. 448.

mining. Health records and vital statistics were gathered on tens of thousands of uranium miners, coordinated by the Health Institute of the Uranium Industry (HIUI), established in 1954 in the mining town of Příbram. As in the U.S., Czechoslovakian efforts to publicize the health problems of miners ran up against cold war fears that releasing such information would compromise national security. In 1960, Dr. Vladimír Řeřicha of the HIUI was asked to prepare a comprehensive overview of lung cancer among Czech miners. Between 1960 and 1965, he and his staff compiled epidemiological evidence that miners at Joachimsthal and Horní Slavkov were dying from lung cancer at about five times the rate of coal miners, results that were similar to U.S. findings at this time. Řeřicha prepared a paper detailing his findings for the institute in 1966, and simultaneously sought to publish his results in a more conventional scientific journal.[85]

Řeřicha's efforts to publicize the carnage in Czech uranium mines were blocked by orders of the State Security Police. The report and its contents were classified and publication was barred in foreign or Czech periodicals—a ban that was not lifted until the "velvet revolution" of 1989. The logic behind the ban was extraordinary: as Řeřicha today recalls it, Czech security authorities were apparently afraid that, from uranium health statistics, one could calculate uranium production levels and the quality of uranium being mined. The cynicism of such a ban was made apparent in the 1970s when Řeřicha was again denied the right to publish, even though the administrative chief responsible for all Czech uranium mining (Karel Boček) had defected to West Germany in 1970. With details of the nature and scope of Czech uranium mining no longer secret from the West, why were the health hazards of uranium mining still kept secret? One can only conclude that Czech authorities feared that revelation of the sacrifice of its miners for Soviet atomic power would cause political embarrassment for Czechoslovakian leaders.[86]

Czech exploitation of uranium is widely known; what is not widely known (even to Řeřicha, judging from what he revealed to me in a 1992 interview) is that the Czech government organized seventeen camps to provide labor for the three major uranium mining regions, in which tens of thousands of political prisoners were forced to work in the 1950s and '60s.[87] The number of political prisoners working in these mines was precisely recorded in secret police archives:

Year	Number of Political Prisoners
1946	64 (all Germans, presumably Nazis)
1947	1,739
1948	3,446
1950	5,500 (all Czechs by this time)
1951	9,029
1952	11,280
1953	11,816
1954	9,655
1955	7,474
1960	2,600

As those who have survived today recall the situation, the number of "politicals" in the mines was approximately equal to the number of civilians. At Joachimsthal and Horní Slavkov, there were about 65,000 civilian miners, and probably a comparable number at Příbram, the largest of the three major uranium centers. (Řeřicha estimates that, all together, 500,000 people may have worked at one time or another for the uranium industry in Czechoslovakia.) All uranium mine political prisoners were released as part of a more general amnesty granted in 1963; after this time, the only prisoners working in the mines were nonpolitical criminals. Officials at the Příbram-based Health Institute of the Uranium Industry estimate that there were probably about 50,000 political prisoners in Czech uranium mines from the beginning of the camps in 1946 until the amnesty in 1963.* It is not yet known whether there were similar camps in East Germany, but there is good evidence that forced labor of this sort was used in the Soviet atomic project.[88]

Both American and East European atomic authorities were quick to disregard the value of human life in their eagerness to win the arms race. In the United States, hundreds of miners were allowed to die in order to keep up the feverish pitch of weapons production. In the Soviet Union and in other parts of Eastern Europe, prisoners were literally worked to death in mines, apparently as part of a deliberate plan to kill them. In both cases, secrecy allowed those administering the crimes to operate undetected; secrecy also served to bar those who should have known from finding out. Czechoslovakian scientists in the 1970s published several reports on the hazards of uranium mining, but never mentioned the forced labor.[89] A 1993 update notes that a total of 95,000 "volunteer" uranium miners were registered by Czech authorities, but no figures are given on how many "nonvolunteer" miners were involved.[90]

Americans like to think that things like this cannot happen in this country, and it is true that, unlike in Eastern Europe, no one in this country was ever forced to work in a uranium mine. People were allowed to work, though, without knowledge of the harms they were likely to suffer, and compensation has been difficult or even impossible to obtain. In 1991, after decades of unsuccessful efforts to secure federal compensation for injured miners, Congress finally passed the Radiation Exposure Compensation Act, setting up a $30 million trust fund to compensate uranium miners, along with other victims of radiation exposure—notably, downwinders of southern Utah, participants in nuclear tests, and certain workers at atomic plants. The text of the act included a formal apology: "The Congress apologizes on behalf of the Nation to the individuals . . . and their families for the hardships they have endured."[91] Congress in 1991 and 1992 voted to increase the total fund to $200 million. Miners who suffered from lung cancer are now, theoretically, entitled to $100,000 in compensation from the fund. Families may be compensated if the victim is no longer living.

*Many of these prisoners continued working in the mines even after their release from the camps. Civilian miners were paid extremely well—roughly ten times the average salary of physicians. As of the summer of 1992, Czech uranium miners were eligible for health and hardship compensation. Political prisoners became eligible for compensation only after the revolution of 1989. (Interview with Dr. Jan Jerabek, May 27, 1992.)

Bureaucratic red tape has thus far made it difficult, however, to obtain compensation. The forms one must fill out are daunting, running to more than 100 pages. Written proof must be provided that one was exposed and got cancer from that exposure. Only lung cancers are compensable, and smokers must prove that their exposure was "excessive." Most of those who worked in the so-called dog holes (very small mines with little or no ventilation) have no way of proving it, and medical records may be even harder to come by. The Indian Health Service, to whom many Navajo miners turned for treatment, generally destroyed records after twenty-five years, and elderly widows of victims may not even have written evidence of the marriage. As a result of these and other problems, only half a dozen claims had been successfully paid out by mid-1992.

Money has finally begun to trickle into the communities where lung cancers were bred, but the process is still slow. Stewart L. Udall, the former interior secretary who has served as a lawyer for many victims, recently characterized the administration of the process as "a bureaucratic legal maze designed to prevent compensation to Navajo miners. There's no pity for what happened to these people. No understanding. You have a compassionate program administered in an utterly uncompassionate manner." Jim John, a Navajo man who lost his father to lung cancer when his father was only thirty-four, put the matter bluntly: "It's a bad thing that happened here. It's a bad thing for anybody to do. It destroyed lives. It took the beauty away from this mountain."[92]

Chapter 9

Radon's Deadly Daughters

Man is silly; he suspects his neighbor, never the sky.
—Marcel Griaule, 1942

RADON IS A COLORLESS, odorless, radioactive gas formed by the natural decay of uranium in rocks and soils. It is a noble gas—meaning that, like helium or neon, it is chemically inert and therefore cannot easily form a chemical bond. What it can and does do, though, is seep into homes through foundation posts and basement cracks, where it decays into a number of radioactive "daughter products" that, when inhaled into the lungs, can cause cancer. The Environmental Protection Agency estimates that radon gas is responsible for the lung cancer deaths of 7,000 to 30,000 Americans every year[1]—in other words, about 5 to 20 percent of all U.S. lung cancer deaths. Radon is, by many accounts, second only to cigarette smoking as the leading cause of cancer death in the United States. Among nonsmokers it is probably the single most important cause of lung cancer, surpassing even the 9,000 annual lung cancer fatalities attributable to asbestos. The primary source of the danger is from soils under foundations, though radon can also be released from water, especially while showering and washing dishes. (Like most gases, radon is more soluble in cold water than in hot; heating the water releases the gas).[2]

The case of radon is interesting for a variety of reasons. First and foremost, there is the question of the circumstances surrounding its discovery. Radon is commonly imagined to have come to the attention of scientists in the mid-1980s, when the home of Stanley and Diane Watras of Boyertown, Pennsylvania, was found to have levels of the gas nearly a thousand times higher than the EPA's "action guideline." In fact, household radon has a much older history. High levels of household radon were measured in Germany in the 1930s, and the phenomenon was periodically rediscovered in the 1950s, 1960s, and 1970s, both in the United States and abroad. Until the late 1970s, though, the phenomenon appears to have been regarded more as a curiosity than a threat. Radon had been found in houses built on uranium mine waste and on lands reclaimed from phosphate mines, but these were isolated occurrences, confined to specific areas of the country. A broader hazard was recognized in the late 1970s, primarily in response to fears that the weathertight insulation of homes following the energy crisis was creating unprecedented levels of indoor air pollution. Radon was elevated to the status of a national peril in the 1980s, when the Watras house and other surveys showed that the problem was much more widespread than previously thought.

Radon is also interesting because there remains much disagreement over the magnitude of the hazard. Many experts question the EPA figures; some go so far as to deny the threat altogether, based on studies that seem to indicate that low-level exposures may actually be good for you. One primary reason for doubt is that the majority of deaths are presumed to result from low-level exposures—lower even than the EPA's action guideline (4 picocuries per liter).* In 1992, the EPA estimated that roughly 6 million American homes have radon levels this high or higher, a risk equivalent to smoking about half a pack of cigarettes a day.[3] Evidence for risk estimates of this magnitude comes from epidemiological studies of the Colorado Plateau uranium miners exposed to high levels of radon for many years; the assumption is that the hazards of household exposures can be extrapolated linearly from these high-dose studies. As we have already seen, however, the question of the health effects of low-level carcinogens is hotly debated. Direct epidemiological evidence has been difficult to come by, and the secretive intrigue surrounding nuclear weapons work and commercial nuclear power has made the true effects of the gas even more difficult to fathom.

Part of the problem has to do with what might be called the political provenance of the gas. Radon is often described as a crime without blame. The fact that radon comes primarily from the ground, and not from smokestacks or waste dumps, has made it difficult to rally political support to investigate and fight the hazard. Environmentalists like to concentrate on problems with litigious possibilities—but whom do you sue when the culpable party is the earth itself? The picture is complicated by the fact that some of those who advertise the radon hazard have been rightly accused of seeking to divert attention from other dangers, especially chemical toxics and nuclear pollution. In the early 1980s, the radon hazard was occasionally invoked in defense of nuclear power: Weren't the exposures from nuclear plants minuscule compared with the natural radiation breathed by the neighbors of such plants on a daily basis? Might not cheap nuclear power even *save* consumers from excess exposure to radiation, given that we would no longer have to bottle ourselves up in our homes, sealed in with our radon and other indoor air pollutants? Nuclear advocates brandished increased exposure to radon gas as the ineluctable consequence of energy conservation. Faced with such arguments, environmentalists (with some notable exceptions)[4] were often less than eager to publicize the hazard.

The effort to sort out fact from fantasy in the estimation of the radon risk is further complicated by the fact that there are several separate and competing industries with stakes in the magnification or diminution of radioactive hazards. For the nuclear industry, represented by power utilities and trade associations such as the Council for Energy Awareness, radon appeared for a time to be a kind of radioactive blessing in disguise, though it was ultimately not in this industry's interest to high-

---

*A picocurie is one-trillionth of a *curie* (honoring Marie Curie, discoverer of radium), defined as the radiation emitted by 1 gram of radium, or 37 billion disintegrations per second. A home with 1 picocurie per liter means that about two radon atoms are decaying every second in every quart of air.

light yet another radioactive hazard. Also, there is a not insignificant radon remediation industry, whose business it is to manufacture and process the test kits used to monitor indoor radon. (Such companies also sometimes manufacture and install the equipment used to remedy the hazard.) Radon remediators sometimes find it in their interest to inflate the hazard: they are the ones advising you to be safe and measure radon in your basement, where levels are invariably highest; they are the ones who suggest an "action guideline" twice as strict as the EPA's. Finally, there are the real estate and home builders' associations, which have repeatedly protested efforts to require radon testing prior to construction of new homes or sale of old ones. Home builders have worried about the additional costs, and realtors have worried about the impact of radon on sales.

In light of these complex and conflicting interests, coupled with the fact that radon has no natural political nemesis, it may come as no surprise that the American public has tended to remain indifferent toward the problem. Lesser hazards—like dioxin or lead—with simpler political contours attract equal or even greater public attention. People have been remarkably slow to test their homes, and environmental authorities have been granted very little power to deal with the hazard. Responsibility for radon has been left to individual states or, more commonly, to individual homeowners, who more often than not disregard the problem. This despite the fact that estimates of 10,000 or more people killed annually by the gas nationwide are based on epidemiological inferences that are virtually unsurpassed in terms of both quality and quantity.

Why did it take so long for the danger to be appreciated? As we shall see, household radon became an object of sustained scientific attention because of a curious confluence of political circumstances, including the insulation of homes following the 1973 oil embargo, an upsurge of interest in indoor air pollution, growing fears of radiation in the wake of the Three Mile Island accident, and the public relations goals of the nuclear industry. Radon then became a win-win situation for legislators, who found themselves able to crusade against a terrible scourge without fear of endangering their business and industry support.[5]

THE TAILINGS PROBLEM

The element we today call *radon* was discovered in 1900 by Friedrich Ernst Dorn, a German physicist who gave the name *radium emanation* to the radioactive gas he found spontaneously emitted from radium.[6] Radon was chemically isolated in 1908 and given the name *niton*, rechristened *radon* in 1923.[7] As in the case of X rays, questions about the health effects of radium emanation were raised almost as soon as it was discovered. In 1903, for example, a British physician cautioned against the indiscriminate use of the gas as a treatment for tuberculosis:

> It has been suggested . . . that consumptive patients should inhale the emanation
> so as to bring the powerful agent into as close contact as possible with the diseased lung. A few milligrams of radium bromide might be placed in water in a

suitable inhaler, and the emanation drawn into the lungs with a deep breath; what effect if any this treatment might produce is not yet known, but a strong word of warning to patients not to attempt such a remedy without previous medical sanction may not be out of place. Owing to the property the emanation has of imparting radio-activity, the effect of the inhalation would probably continue for a considerable time, for the walls of the air passages would for the time become radio-active themselves.[8]

Radiation in these early years, however, was still regarded more often as a cure for cancer than a cause of it. It is hardly surprising, then, that little notice was taken when, in 1931, a German physicist reported low levels of natural radioactivity (about 10 pCi/l) in the air of a Berlin basement.[9] Nor is it surprising that, when natural indoor radon was periodically rediscovered—in Sweden and the United States in the 1950s and in Austria in the 1970s[10]—there was apparently never a concern that the phenomenon might pose a health risk. Natural indoor radon was regarded either as a curiosity, observed in the course of calibrating equipment,[11] or even as a potential health resource with commercial possibilities.

Radon first came to be known as a cancer hazard in the occupational context of uranium mining. Only later were suspicions of a broader hazard carried over into the domestic sphere. The first important studies demonstrating a household radon hazard were in the 1960s, following revelations that houses built over uranium tailings were unsafe. *Tailings* are the enormous piles of fine, gray, radioactive sand discarded as a waste product during the milling process, where uranium ore is crushed and processed to produce the "yellow cake" used (after enrichment and reduction) to make the fuel for nuclear reactors or the explosive charge for atomic bombs. Thirty such mills operated at one time or another across nine western states (Arizona, Colorado, New Mexico, Oregon, South Dakota, Texas, Utah, Washington, and Wyoming), dumping more than a hundred million tons of radioactive waste in several dozen towns.

The most dramatic case was that of Grand Junction, Colorado, a town of 58,000 people where, from 1952 to 1966, tailings from the Climax Uranium Company were used to fill the foundations of homes, schools, churches, and other buildings. By 1966, when the hazard was discovered, the company had allowed more than 300,000 tons of radioactive waste to be carted from its property, and about a third of the city's residents were living in homes built on tailings. Housing was not the only source of exposures, though: people used the tailings in flower boxes, golf course sand traps, and children's sandboxes; people went for hikes on the hundred-foot-high "dunes." A story in a 1970 issue of *McCall's* called the town (with some hyperbole) "America's most radioactive city" and its residents "the largest single group of people exposed to radiation since Hiroshima."[12] The doses in many cases were substantial: according to calculations by the Colorado State Health Department, residents in 500 of the worst-hit homes were being exposed to the equivalent of 553 chest X rays per year.

Responding to these findings, the Colorado State Health Department sent letters to the city manager and chamber of commerce recommending that real estate sales be suspended until it could be determined which properties were affected. (The Atomic Energy Commission had washed its hands of the whole matter in 1966, ruling—without outside consultation or published evidence—that the tailings presented "no hazard to the environment, either short term or long term." The AEC also discouraged the Public Health Service from monitoring radiation in schools.)[13] The Board of County Commissioners about this time required removal of tailings as a condition for obtaining a building permit, and the Health Department sent letters to 5,000 homeowners warning them that their houses may have dangerous levels of radiation. State health authorities also began testing residences in other towns.[14] The federal government then jumped into the act, holding congressional hearings on the use of uranium mill tailings at Grand Junction. Congress authorized funds to fix the problem, and federal remedial efforts finally began in 1973. In 1978, the U.S. Congress passed the Uranium Mill Tailings Radiation Control Act, requiring the EPA to establish radon standards for houses built on uranium waste. The 1983 report produced by the agency established 4 pCi/l as a target level and 6 pCi/l as a tolerance maximum. The Department of Defense, the Nuclear Regulatory Commission, the Health Physics Society, the American College of Nuclear Physicians, and the American Mining Society all entered briefs opposing the standard, though several environmentalists argued that it was actually too lax.[15]

Ironically, as we shall see, the EPA had already by this time recognized that there were many homes with "natural" radon levels higher than this, but no effort was made to establish policy for these other homes.[16]

Uranium tailings in fact turn out to be a relatively minor source of overall radon exposures compared to naturally distributed uranium. Homes contaminated by tailings number in the thousands; homes contaminated by naturally occurring radon number in the millions. Household radon was first discovered in physical proximity to uranium mines because it was there that people first suspected they would find a hazard. But there is more to the story. Uranium mines had owners and operators with a history of negligence, so it was easy to identify them as culprits. This would not be true for the naturally occurring radon from the soil and rock that underlie our homes and workplaces.

THE HOUSEHOLD HAZARD

Historians of science generally recognize that the timing of a particular discovery often has to do with the social or political forces moving to create interest in that field of inquiry. Discoveries made without a framework within which to lodge them usually remain mere curiosities. Radon had been detected in ordinary homes as early as 1931, but why was radon recognized as a household hazard in 1985 and not in, say, 1955 or, for that matter, 1935?

The discovery of a widespread natural household radon hazard emerged from an accidental confluence of political forces. The most prominent of these was the effort to seal up homes in the wake of the Arab oil embargo of 1973, which sent fossil-fuel prices soaring, created long lines at the gas pump, and shot up the price of heating oil. The Energy Reorganization Act of 1974 required the Department of Energy to undertake efforts to promote conservation, but also to study the health effects of energy conservation and other energy technologies (solar, along with nuclear fission and fusion, for example). The U.S. Congress passed the National Energy Conservation Policy Act in 1978, requiring electric utilities to help their customers insulate their homes. The speed limit was reduced to 55 miles per hour on the nation's highways, and the Energy Department required all states to submit residential conservation plans.

The DOE began investigating the presence of radon in homes in the mid-1970s, as part of its program on energy conservation. The DOE had long been interested in the health effects of energy technologies, dating back to earlier AEC work on the health effects of atomic power and nuclear war. Air quality had been of concern, with regard to both the effects of atomic fallout and outdoor air quality more generally. As early as 1967, the AEC had surveyed radon in homes in Tennessee and Florida, ostensibly as part of its effort to demonstrate the safety of homes built on uranium tailings.[17] With the onset of the oil crisis, a number of DOE scientists became concerned about the impact that energy conservation might have on indoor air quality. In 1975, the Lawrence Berkeley Laboratory (LBL) set up an indoor air quality program to investigate whether conservation efforts might be exacerbating indoor air pollutants—especially nitrous oxides and carbon monoxide, but also soot and tar from cigarette smoke and fireplaces, and formaldehyde from insulating foam, particle board, and carpeting. Other agencies began to worry about indoor air pollution: the Consumer Product Safety Commission began developing warnings for consumers, and OSHA began studies of the effects of energy-efficient offices on workers. The EPA began an indoor air quality study in 1976, but as late as 1979, an Office of Technology Assessment report could note that all U.S. government agencies combined were spending only about $1 *million* per year on indoor air quality.[18] That same year, by contrast, the U.S. federal government and private corporations spent more than $25 *billion* to abate outdoor air pollution.[19]

LBL scientists began testing homes for radon in 1978, after learning of Swedish work suggesting a household hazard from natural building materials. Bengt Hultqvist of the Department of Radiation Physics in Stockholm in 1956 had published evidence of a surprising number of homes with very high radon (up to 50 picocuries per liter),[20] though initial suspicions were that the problem was confined to houses built from a specific type of construction material—a lightweight, aerated concrete using alum shale as an aggregate. The National Institute of Radiation Protection in Stockholm sponsored a series of studies on the problem, arousing public concerns that resulted in a ban on the manufacture of alum shale concrete in 1974.[21] Subsequent Swedish studies, most notably by Gun Astri Swedjemark of the National Institute of Radiation Protection, stressed the aggravation of the problem by energy conservation measures. Swedjemark estimated that the increased radiation in insulated, energy-efficient homes was causing from 100 to 600 extra lung cancers per year in Sweden, accounting for an extra 5 to 30 percent in the total population's risk for lung cancer.[22] LBL scientists used this to argue that conservation might be causing cancer in the United States, too.

The Lawrence Berkeley Lab was only one of several U.S. groups puzzling over the question of household radon at this time. The DOE's Environmental Measurements Lab (EML) in Manhattan had been founded in 1948 as the Health and Safety Laboratory (HASL) with a mandate to monitor local atmospheric fallout from nuclear testing and airborne radioactivity in atomic mills, enrichment facilities, and weapons assembly plants. The lab also began a project to monitor global fallout in 1951, after radioactive debris from a Nevada atomic test turned up thousands of miles away in a Kodak film processing plant in Rochester, New York. Technicians at the plant had found that unexposed film was being fogged by an unknown agent. The problem was eventually traced to fallout from the Nevada test site, which was contaminating the snow that provided the water for the laboratory, ruining the film. Alterted to the danger, the AEC's Health and Safety Laboratory set up a worldwide network of detectors to monitor fallout, and stepped up its analysis of how radioactive isotopes move through the food chain and are expelled from or retained in the human body.

As part of these efforts, AEC scientists measured radioactive isotopes in human foods and in human tissues—in bones and breast milk, for example. Radon and other radioactive gases were measured in the stratosphere, over the ocean, and underground. Little attention was given to household radon, because the laboratory's primary mission was to monitor occupational exposures and atomic fallout. Natural radioactivity was occasionally looked at for purposes of comparison or calibration, but never as a potential health risk in its own right. Thus in 1957, when two scientists from the lab—A. J. Breslin and Harold Glauberman—measured indoor radon levels of about .2 picocuries per liter in several Manhattan buildings and in one subway station and tunnel,[23] no one seems to have found the results very interesting. Things were different twenty years later when, Breslin and another colleague from the EML found radon levels averaging .83 pCi/l in twenty-one New Jersey and New York homes.[24] After two decades of environmental activism and extensive news media coverage of the tailings problem, this "new" result had new meaning.

Anthony Nero and his colleagues at the Lawrence Berkeley Lab's Energy Efficient Buildings Program were among the first to realize the health impact of these figures. In August 1978, the group published a paper using the Manhattan lab estimates to suggest that radon in U.S. homes could produce "an added annual risk of lung cancer in the vicinity of 100 cases per million." That turns out to be about 20,000 extra U.S. lung cancers per year for the entire United States, assuming a population of 200 million. Nero suggested that "the current radiation levels in conventional homes and buildings from radon daughters could account for a significant portion of the lung cancer rate in non-smokers."[25] The Berkeley group recommended that radon health effects should be taken into account when adopting energy conservation standards.

Coincidentally, evidence of high indoor radon began to emerge from yet another source. In December 1978, J. Rundo and two colleagues from the DOE's Argonne National Laboratory near Chicago submitted a brief note to *Health Physics* alerting readers to their discovery of houses in the Chicago area with up to 26 picocuries per liter of radon. Rundo had been trying to correlate radioactive exhalation rates with the natural radioactivity to which people were exposed in their homes; it was in the course of these studies that the problem (again) was accidentally discovered. Among twenty-two houses studied, six showed radon of 10 picocuries per liter or more; one showed 26. This was particularly surprising, given that even the Swedish homes built of alum shale concrete were nowhere near this high. Rundo and his colleagues imagined at first that the radon might be coming from the smoke detector (such devices normally contain radioactive elements) or the ceramic tile in the house; it turned out to be coming from the earth below an unpaved crawl space under the kitchen and living room. The Argonne group concluded that "the average exposure of man to radon daughters in houses in this country may be substantially higher than has hitherto been thought" and urged further research to determine the scope of the hazard.[26]

Nero's and Rundo's observations were published in rather obscure journals of the health physics profession, attracting little attention. What did attract attention was the failure of the main reactor at Three Mile Island on the Susquehanna River on March 28, 1979—initiating the nation's most infamous, if not most deadly, nuclear accident. (*The China Syndrome,* a popular movie dramatizing the possibility of a reactor core meltdown, was released about the same time as the accident.) Though it proved not to have been as serious an accident as many originally believed, the fears it generated provided a powerful stimulus for further work on radioactive hazards, including radon. Responding to the incident, Pennsylvania environmental officials began to look for radioactive contaminants in nearby homes and found, to their great surprise, that many had very high levels of ambient radioactivity (20–100 picocuries per liter) quite independent of the accident.

DOE officials were also continuing to find houses with very high levels of radon. About two weeks after the Three Mile Island mishap, DOE scientists working on the health effects of indoor air pollution discovered that several of the energy-efficient homes they had been monitoring on the East Coast showed high levels of indoor radon. Anthony Nero received a call from a colleague reporting that a house in Maryland had been found with 20 picocuries per liter. This was a specially designed energy-efficient house, the entire outside of which had been sprayed with a plastic sealant, and it appeared to confirm the fears of LBL scientists that energy conservation might well produce a radon hazard. Nero recalls his astonishment at the time, given that this was "the radiation equivalent of having a Three Mile Island accident . . . occur in the neighborhood once a week."[27] Radiation in the Maryland home equaled the occupational limit for nuclear workers recognized by both the National Council on Radiation Protection (NCRP) and the Mine Safety and Health Administration (MSHA).[28]

The EPA picked up on the issue about this time. It had become involved in the mid-1970s, in response to Florida officials' concerns about radioactivity in homes built over land reclaimed from phosphate mines. Florida was the nation's leading producer of phosphate fertilizers: over 80 percent of the nation's output—and one-third of the world's—came from the strip mines in and around Polk County in the central part of the state. Phosphate by itself is not normally radioactive, but phosphate in the ground is commonly associated with uranium. (More uranium is indeed discarded in the course of phosphate mining than is obtained through uranium mining itself: early EPA studies identified phosphate tailings as a potential uranium resource.)[29] In May 1976 President Gerald Ford ordered the EPA to investigate the problem, after a preliminary EPA study indicated that houses in the area had enough radon to double one's risk of lung cancer after only three years of residence.[30] In the course of those studies it was found that many Florida homes *not* built over phosphate tailings had very high radon. The same was found to be true in Montana. Responding to these findings, the Department of Housing and Urban Development in 1979 required that houses in Montana and Florida be tested for radon before homeowners could obtain federally guaranteed mortgages.[31]

Nineteen-eighty was a banner year for radon. David Rosenbaum of the EPA's Office of Radiation Programs asserted that radon was "by far the highest radiation danger that the American public faces. It's certainly up there with the top dangers the EPA deals with. It's easily more serious than Love Canal."[32] The General Accounting Office released its *Indoor Air Pollution,* indicating that 10,000–20,000 Americans were being killed every year by conservation-induced radon (a misconception, as we shall see in a moment).[33] President Carter's Council on Environmental Quality reported that indoor radon could be causing as much as 10 percent of all U.S. lung cancer deaths, and the *New York Times* ran its first feature story on the hazard.[34]

A very different kind of publicity came from the nuclear industry, which quickly recognized the public relations value of the threat. The nuclear industry sought to portray the radon hazard as an energy-conservation issue: nuclear energy could now be seen not just as economically superior to fossil fuels but as a more healthy alternative as well. Nuclear energy would save us not just from foreign oil but from the poisons trapped inside our weathertight homes. Some of the more creative efforts were those by Henry Hurwitz, a staff scientist at the corporate research and development department of General Electric, the company that had designed and built one of the nation's most widely used reactors, the boiling-water reactor. In a 1981 brochure published by General Electric, and in a 1983 article in *Risk Analysis,* Hurwitz argued that the risk posed by conservation-induced lung cancer far exceeded any risk the industry might pose. Lifetime exposures to indoor radon were probably in the 5–15 WLM (Working Level Month) range, but efforts to seal homes to save energy could produce exposures "more than an order of magnitude higher." Hurwitz cited EPA estimates that 10,000–20,000 lung cancers per year could be caused by air-tightening homes and offices to conserve energy, and drew attention to the "surprising" fact that indoor radiological exposures could be "comparable to the average exposure that would be received by the imputed victims of a hypothetical uncontained nuclear meltdown."[35]

Bernard L. Cohen, a Pittsburgh physicist and former Group Leader for Cyclotron Research at Oak Ridge National Laboratory, was another who argued that nuclear energy might actually save us from radioactive hazards.[36] In 1980, Cohen estimated that energy conservation might already be causing 1,000 extra lung cancer deaths per year and that long-run plans for conservation could cause as many as 10,000 extra fatalities per year.[37] Six years later, Cohen reaffirmed his belief that, "from the standpoint of radiation exposure, 'energy conservation' is by far the most dangerous energy strategy." The Pittsburgh physicist pointed out that by offering conservation incentives, the U.S. government might well be implicating itself in some rather nasty legal liabilities: "If a nonsmoker follows government recommendations and takes tax credits to insulate his home, and if he later gets lung cancer (as 7,000 nonsmokers do each year), there is a substantial probability—perhaps 40 %—that the government is responsible."[38]

By 1980, however, it had become clear, at least to DOE scientists such as Nero, that energy efficiency was only one—and not even the major—factor indicating a hazard. Tightly sealed homes had been found with very little radon, and much looser structures had been found with very high levels. The most important factor in a given house was not how tightly it was sealed but, rather, what kind of soil it was sitting on. Improved ventilation might well be the cure, but the absence of ventilation was not the major cause. Improper ventilation might even exacerbate the problem: the "chimney" or "stack effect" produced by a fireplace, for example, could lower air pressure in living spaces, drawing radon-filled air up from the basement. Even an open window could have similar consequences, if winds moving past the window produced a Bernoulli effect (reduced air pressure), drawing air up from the basement. Energy-efficient housing was not, in short, the cause of the radon problem. For the struggling nuclear industry, however, the continued misrepresentation of the problem served as a useful public relations tool. As Nero put it, "the industry ideal would be that we all live in a drafty tent heated by nuclear energy."[39]

For a time, the question of the contribution of energy conservation to the radiation burden led to interagency tensions between the DOE and the EPA. In the late 1970s, the EPA sought to postpone the DOE's residential conservation program on the grounds that it was likely to exacerbate indoor air pollution. As David Berg of the EPA's energy processes division put it: "We may be asphyxiating in our own homes." DOE scientists pointed out that the EPA had confused the overall effects of radon with the (much lower) marginal effects of conservation. Ruth Clusen, assistant secretary for conservation at the DOE, suggested in 1980 that there might be only 2,400 additional cancer deaths as a result of energy conservation, a far cry from the EPA's original estimate of 10,000 to 20,000.[40] EPA officials demurred, regarding this as a self-serving effort to shore up the energy department's conservation program. The General Accounting Office's 1980 report on indoor air pollution noted the disagreement and suggested that the Clean Air Act be amended to shift authority for indoor air quality in the "nonworkplace" from the DOE to the EPA.[41]

For a time, these squabbles were brushed aside by the election of Ronald Reagan, who declared the energy crisis over and put radon along with all other environmental issues on the back burner. It would be several years before radon would reemerge into public view, this time prompted by the fortuitous discovery that the radon hazard could be much more serious than anyone had ever imagined.

THE WATRAS FAMILY

The turning point in terms of media attention to the radon threat was the winter of 1984–85, when a dramatic event caught the public eye. In December, officials at the Limerick nuclear plant near Boyertown, Pennsylvania—about 35 miles northwest of Philadelphia—were astonished to find that a nuclear worker, Stanley Watras, was setting off radiation alarms while passing through detectors on his way *to* work. Radiation monitors had recently been installed at the entrance to the plant, and Watras "constantly tripped every alarm on all zones." Plant officials first assumed that he had somehow been contaminated from the plant. Upon investigation, however, it was found that he had been exposed to extraordinarily high levels of radiation from naturally derived radon in his home. Measurements in his living room found as much as 3,200 picocuries per liter of radiation—hundreds of times higher than the EPA standard (only one other home has ever been found with such high levels).[42] By one calculation, it was as if Watras and his family were smoking 136 packs of cigarettes per day, or receiving 100,000 times the radiation received by people immediately surrounding Three Mile Island during the 1979 accident.[43] Had the house been a uranium mine, it would have exceeded federal radioactivity standards by a factor of about 50. By EPA measures, Watras and his family were breathing the radioactive equivalent of about a million chest X rays per year.

The national press picked up the story, and *radon* soon became a household word. The *New York Times* carried a front-page report ("Major Peril is Declared by U.S.") on May 19, 1985, and this released the floodgates. The *Atlanta Constitution* announced that "Radon in Homes Could Kill 30,000 Yearly," and the *Chicago Tribune* warned about "Radon Gas Tied to Cancer Deaths." *Newsweek* magazine headlined "Radon Gas: A Deadly Threat—A Natural Hazard Is Seeping into 8 Million Homes."[44] As reports of high-radon residences filtered into the press, scientific and commercial bodies responded. In 1987 the National Council on Radiation Protection raised its estimate of the radon component of natural background radiation from 28 to more than 200 millirems.[45] Radon remediators, sensing a boom, banded together to form professional societies such as the American Association of Radon Scientists and Technologists and the American Radon Association. Anthony Nero predicted in 1986 that radon would "soon be the focus of the largest environmental search that this country has ever undertaken." A writer for the *Journal of Real Estate Development* predicted that radon might well become the hot environmental issue of the 1990s, just as asbestos had been during the 1980s.[46]

The publicity given the Watras family's house prompted several state governments to deal with the problem.[47] In 1985, Pennsylvania's Bureau of Radiation Protection established the nation's first state Radon Monitoring Program, with a budget of $6.5 million to test and mitigate homes. A door-to-door search of 2,800 homes in the township where the Watrases lived showed that nearly 50 percent of the homes had basement radon concentrations exceeding 4 picocuries per liter. Suspicions began to arise that the ultimate source of the contamination was the Reading Prong, a narrow, uranium-rich rock formation extending from eastern Pennsylvania through northern New Jersey and part of New York in a great arc. The state's Department of Environmental Resources (DER) offered free alpha track detectors to anyone living on the Prong, and about 60 percent of the 22,000 residents who took up the offer found radon in excess of federal guidelines.[48] The DER was authorized to establish radon standards for construction materials and to provide low-interest loans for remediation. A hot line was established to respond to queries about the hazard, and the department published a brochure for homeowners outlining the nature of the problem and how best to fix it. The EPA's own *Citizen's Guide to Radon,* published in 1986, was modeled on the Pennsylvania guide.

For a time, it appeared that eastern Pennsylvania and western New Jersey would turn out to have the highest levels of radon in the nation—following the assumption that the Reading Prong geological formation was the root cause of the problem. The Prong had been identified as a high radon region in the mid-1970s, when the National Uranium Resources Evaluation Program conducted a nationwide aerial survey in search of nuclear fuel sources.[49] Later it was discovered, though, that homes could have very high readings even if they were far from the Prong, and that many houses well inside the formation have very low readings. Granite bedrock now appears not to be a very good predictor of high radon. Many high-radon homes overlie sedimentary rock such as limestone, where the clay layers separating the strata have concentrated water-insoluble uranium. Uranium—and therefore radium and radon—turns out to be much more widely dispersed than once thought: the metal is apparently more common in the earth's crust than even mercury or silver. Recent studies have shown that eastern states are not even national leaders in the radon department: as of 1993, Iowa, Colorado, and North Dakota shared this distinction.[50]

Snapshot accounts of the history of radon policy often present the discovery of the Watras house as the turning point in the public's appreciation of the hazard. It is also important to recognize, though, that evidence of a hazard had been building—and neglected—for some time. Even in Pennsylvania, the Watras house was not the first high-radon house to attract media attention. Four years earlier, a physician by the name of Joel Nobel had measured 55 picocuries per liter in his new home in suburban Philadelphia. Nobel, who also ran a nonprofit biomedical engineering laboratory, had been testing for a variety of pollutants when he found the radon. He subsequently had 100 of his employees test their homes, and found that 15 percent had readings above 4 pCi/l. Nobel contacted Harvey Sachs, an indoor-pollution expert at Princeton University, to remediate his home. Sachs in turn contacted the Pennsylvania DER to try to interest it in a study of the problem, but the department was not receptive. Sachs eventually persuaded the *Philadelphia Inquirer* to do a story on the topic. The *Washington Post* picked it up, and there was even a brief moment of national recognition on December 10, 1984, when the *CBS Evening News* ran a piece on the "radon gas threat," featuring Harvey Sachs, Joel Nobel, and a rather dismissive EPA spokesman. Sachs later accused Pennsylvania environmental authorities of "covering up" the radon threat in that state.[51]

If Pennsylvania prior to Watras was slow to respond, the federal government was even slower. President Ronald Reagan's philosophy of government was penurious, especially when it came to environmental issues. Nothing of any consequence was done for radon in his first term of office, and his second term was only slightly better. The Radon Gas and Indoor Air Quality Research Act of October 1986 assigned research responsibility to the EPA, but offered little in the way of practical incentives to mitigate the hazard. The Indoor Radon Abatement Act of 1988 required the EPA to update its *Citizen's Guide* and to conduct further research, but this amendment to the Toxic Substances Control Act was also short on practical incentives to remediate. It also attracted ridicule for its rather fanciful long-term goal that indoor air "should be as free of radon as the ambient air outside of buildings"—a goal that critics calculated could cost on the order of $1 trillion.[52] (Outdoor radon in the U.S. averages about one-twentieth of the EPA's action guideline.) The U.S. General Services Administration required that rental properties under its control be tested for radon and that landlords undertake and document any necessary remedial action,[53] but apart from these and a few other measures—including a much-criticized EPA media campaign organized in cooperation with the Ad Council[54]—little was done by federal health or environmental authorities.

Political representatives from high-radon districts didn't take this quietly. In October 1985, Senator Frank Lautenberg of New Jersey charged the EPA and the Office of Management and Budget (OMB) with having tried to "cover up" an internal EPA report projecting thousands of preventable deaths from household radon. The document in question, dated July 23, 1985, had recommended a five-year, $10 million program to reduce the public health risk by 30 to 50 percent, "avoiding several thousand lung cancer deaths per year." The OMB refused to release the entire report, however, and Lautenberg suggested that EPA officials had come under "enormous pressure" from the OMB not to propose a radon-control program.[55] Sheldon Meyers, acting director of the EPA's Office of Radiation Programs, denied the charges, but accusations of negligence continued nonetheless. At hearings on the Radon Pollution Control Act of 1987, Congressman Thomas A. Luken of Ohio excoriated EPA officials for their "passive" approach to this "killer"—involving little more than the distribution of brochures, answering phone calls, and the like.[56] The Bush administration did only slightly better: the Indoor Radon Abatement Act had authorized $13 million to be spent on state grants and technical assistance for each of the fiscal years 1989, 1990, and 1991 (to maintain a radon information hot line and so forth), but spending in each case fell below these already meager levels.

PUBLIC APATHY AND REGULATORY IMPOTENCE

Why were federal agencies slow to act in this sphere? Some have suggested that the physical properties of the gas—odorless, colorless—may have had something to do with the failure of the public and its representatives to get excited about the issue. People apparently have trouble taking seriously the idea that "clean" air in their homes can kill them.* A 1986 survey of 1,000 New Jersey residents found that most people were more concerned about the effects of radon on their real estate values

---

*Anne R. Jackowitz, in a 1988 article for the *Boston College Law Review,* suggested several different reasons for the public's apparent apathy, including the fact that radon is odorless and invisible and the tendency to view the home as a sanctuary from the troubles of the outside world. Jackowitz also suggested that the recency of the gas's discovery may play a role, as may the fact that the federal government has taken few steps to regulate the problem. The low cost of remediation may ironically make it seem even less of a hazard. Homeowners may also fear that a radon test could reduce their real estate values. See her "Radon's Radioactive Ramifications: How Federal and State Governments Should Address the Problem," *Boston College Law Review: Environmental Affairs* 16 (1988): 329–81, esp. p. 356. Not mentioned by Jackowitz is the fact that people may distrust expert estimates of the magnitude of the hazard.

than on their health. A 1989 survey of Denver residents found that only about 2 percent were aware that radon could cause lung cancer.[57] As mentioned earlier, radon
has also not yet fit easily into American political traditions. In the United States, environmental issues have usually attracted public attention when there is a culpable
agent responsible for the generation of the hazard. Radon is unlike asbestos or dioxin
or most other toxic hazards in this regard—though this has also begun to change.
Property owners were for a time afraid to test for fear of finding a problem, though
now the greater fear, apparently, is of being sued for failing to have tested.[58] Radon
may cease to become a crime without blame if landlords or employers can be held
responsible for failing to identify and fix a problem.

Another part of the failure to deal with the problem, though, must be traced to the
fragmented regulatory history of radiation. Radiation has always been viewed as a
distinct type of hazard (early bomb builders called it "the special hazard"), and radiation does inflict injuries in a different manner than chemical hazards. More important than the distinct nature of the hazard, however, is its peculiar regulatory
taxonomy. Chemical hazards have generally fallen under the rubric of the Public
Health Service, the Labor Department, or the Environmental Protection Agency. Radioactive hazards, by contrast, were until recently policed by the Atomic Energy
Commission and its successor organizations (DOE and NRC), established after
World War II with broad and exclusive powers to regulate anything having to do
with radiation.

This curious arrangement meant that public health officials often found it difficult
to monitor radiation hazards. It also meant that radiation hazards were often omitted
from general surveys of carcinogenic hazards. In the 1940s, for example, when
Jonathan Hartwell began cataloging carcinogens for the National Cancer Institute,
he excluded radioactive substances on the grounds that they were totally separate
types of hazards.[59] Samuel Epstein excluded analysis of radiation from his *Politics
of Cancer* on the grounds that this would require "a book in itself."[60] Radiation was
not discussed in the Califano Estimates Paper of 1978, nor was it a topic of delibera-
tion in the Carter administration's Council on Environmental Quality. The historical
connection between nuclear war and radioactive hazards led to dramatically different professional competencies for chemical and nuclear medicine: chemical toxicology tended to be the sphere of public health–oriented epidemiologists; radioactive
hazards were largely the province of radiologists and health physicists.

The origins of this arrangement go back to the long-standing regulatory tradition
according to which new agencies are not supposed to encroach on one another's regulatory turf. When atomic bomb factories were set up in the 1940s and '50s, the
Atomic Energy Commission was granted governance over nuclear worker health
and safety, but its responsibilities did not extend to the mining of uranium—which
was left to the individual states. When the EPA and OSHA were formed in 1971,
both agencies were deliberately barred from including radiation as objects of their
concerns. The AEC (later the DOE) was responsible for radiation safety in the workplace, an arrangement that persisted until 1991, when responsibility for all "inside
defense" studies of workers exposed to radiation was transferred to the Public
Health Service (NIOSH). Even today, OSHA is still unable to govern radiation,

which is one reason why radon, arguably one of the most deadly of all occupational exposures, is entirely unregulated in the workplace (except in mines).

When household radon was finally discovered in the late 1970s, it was not at all obvious which—if any—among these various regulatory agencies should assume responsibility. The Department of Energy was responsible for protecting workers from radiation hazards in nuclear power plants and weapons facilities, but no one had ever imagined that this authority might be extended to the home—except in the event that contamination had been spread from the workplace to the home. The EPA might have been the natural body to address the hazard, but it, too, ran into problems. The agency, after all, was never intended to address *all* environmental problems that might pose a threat to human health. Radiation was specifically excluded, as were hazards of an obviously occupational nature. Carcinogens in foods were regulated by the FDA or the Bureau of Alcohol, Tobacco, and Firearms. Certain carcinogens of an apparently "personal" nature were also excluded: a 1976 law, for example, expressly barred the EPA from regulating tobacco smoke. Indoor domestic radon is something like cigarette smoke, insofar as both are substances over which the EPA has never been able to exercise any control. The FDA can ban a food additive that might kill one in a million, but the EPA is powerless to bar the sale of a house filled a radioactive gas that has a one-in-twenty chance of killing its occupants.

The larger political climate is relevant here. Radon was not widely publicized, one should recall, until Ronald Reagan's presidential tour of duty. The link is not a simple one: Reagan officials did little to combat the problem, and radiation programs suffered along with environmental spending as a whole. In his first months in office, Reagan fired seventy-one employees from the EPA's Office of Radiation Programs, effectively gutting the agency's nascent radon effort. Reagan Republicans were suspicious of anything that looked like government intrusion in private lives, and barred any kind of aggressive action in this sphere.[61]

On an ideological level, however, radon was in fact the kind of environmental hazard that fit fairly well with conservative political sensibilities. Reagan environmental officials were eager to downplay the hazards of industrial pollution; administration officials were therefore not disappointed to hear that indoor domestic pollution could be as dangerous as outdoor or occupational exposures.[62] Radon was politically "safe": a 1989 notice in *Lancet,* Britain's leading medical journal, suggested that a politically conservative EPA had seen the development of radon programs as a way to appear responsive to environmental problems without confronting industrial polluters.[63] Edward Groth of *Consumer Reports* was somewhat more colorful, calling radon a "Republican's environmental version of a wet dream."[64]

THE PROBLEM OF SYNERGY

Disagreement persists about the magnitude of the radon hazard. As with many other carcinogens, the debate centers on the hazards of low-level exposures. This is especially significant because, as already noted, low-level exposures may well account for the majority of deaths. Though higher levels are obviously more dangerous, homes with lower-level exposures are far more numerous—in fact, according to the EPA, it is in homes with radon levels satisfying federal guidelines that the majority of radon-induced deaths are expected to occur. Estimates vary concerning how many homes are affected: the NCRP in 1984 estimated that about 3 percent of all homes exceeded the EPA's target; the EPA itself says that the true figure is closer to 7 or 8 percent (down from earlier estimates of about 12 percent). The *Journal of Air Pollution Control Association,* published to serve the remediation and waste management industry, puts the figure much higher, at about 23 percent—an estimate Nero derides as "junk science."[65]

Part of the ambiguity can be traced to inherent difficulties in modeling the risk. The EPA today estimates that 7,000 to 30,000 Americans are killed every year by radon-induced lung cancer; this and other estimates are based on extrapolations from uranium-mining epidemiology. Skeptics have long pointed out, though, that the uranium-miner data are suspect by virtue of the fact that many of these miners were smokers—up to 70 percent, in some studies. There is also the complicating fact that radon is only one of several things that might cause cancer in people whose work takes them underground. Mines have always been dirty places, filled with dust and smoke from drilling and blasting. It is not yet clear what role dust plays in this story: radon daughters apparently attach themselves to dust particles, but does this increase or decrease the danger? Dust may trap the daughters and make them easier to inhale; alternatively, it may plate them out as the dust settles. The same may even be true of tobacco smoke. How fair, therefore, is it to compare radon in the home with radon in mines? How important is it that miners exert themselves more (and therefore breathe more heavily) at work than at home? Some critics maintain that the mine and the home are such different places that it is difficult to extrapolate from one to the other.[66]

The possibility of a smoking-radon synergy is a particularly vexing one. Synergistic risk is when two or more factors (smoking and radon, for example) combine to produce a risk that is greater than the sum of each considered separately. The EPA, presumably to avoid complicating the issue, at one time presented radon risk equivalents without assuming any synergistic effects. The EPA's 1986 *Citizen's Guide* thus stated that the inhalation of 4 picocuries per liter was equivalent to smoking about half a pack of cigarettes per day. But was this for smokers or nonsmokers? Critics of the EPA's approach noted that its action limit might be equivalent to half a pack per day if you already smoke, but nonsmokers might increase their risk by the equivalent of only, say, two or three cigarettes. Anthony Nero, for example, concedes that 10,000 Americans may be killed by radon every year (he was one of the first to produce such estimates), but he also believes that it is wrong to ignore evidence of a smoking-radon synergy. Most of these 10,000-odd deaths may well be among smokers, for whom radon is a contributing or promoting cause.[67] Given these synergistic effects, one simply cannot equate radon levels with cigarettes smoked in any direct manner. Edward Martell of the National Center for Atmospheric Research in Boulder, Colorado, is convinced that the EPA overestimated the hazard for nonsmokers by 600 percent and underestimated it for smokers by 50 percent.[68] The question is significant because if synergistic effects are large enough, the "radon problem" becomes largely a "smoker's problem."[69]

Epidemiological studies could theoretically go a long way toward resolving the question of indoor radon. The first sketchy evidence of a direct link between lung cancer and residential radon came in 1984, in a study sponsored by the Swedish Radon Commission. Drs. Christer Edling, Hans Kling, and Olav Axelson reported an interesting case of an island off the coast of Sweden, where part of the community lived on uranium-rich rock and was exposed to about three times the indoor radon as the other part of the community. The Swedish investigators found that the more highly exposed population suffered about twice as much lung cancer as the less exposed group, though the sample sizes were too small to be convincing.[70] A couple of subsequent studies have confirmed the link, though again not very strongly.[71]

The problem as it now stands is that the few studies demonstrating a nonmining hazard have been contradicted by a somewhat larger number of reports showing no ill effects. Early studies by the Pennsylvania Department of Health showed no observable increase in cancer in the township in which the Watras family lived. Two separate Chinese studies failed to find a correlation between lung cancer and radon levels. In the first, published in *Science* in 1980, the lung cancer incidence of 73,000 inhabitants of regions with high-background radiation was compared to that of 77,000 inhabitants of low-background radiation regions. No significant difference in lung cancer mortality was found between the two groups.[72] A subsequent study of women in the industrial city of Shenyang controlled for smoking, but still no evidence was found of a correlation.[73] In January 1994, however, a Swedish study led by the Karolinska Institute epidemiologist Göran Pershagen seemed to indicate a positive correlation: examining 1,360 lung cancers and 2,847 controls, Pershagen found that people exposed to 4–11 picocuries per liter increased their lung cancer risk by about 30 percent by comparison with people exposed to less than 4. People exposed to more than 11 increased their risk by about 80 percent.[74] A *Science* report pointed out that the Swedish results would soon be contradicted by a Canadian study showing no correlation between radon levels and lung cancer in the homes of 738 lung cancer victims and an equal number of controls.[75]

Most provocatively, Bernard Cohen has argued that low levels of radon may actually be good for you. His impressive data set of tens of thousands of household measurements adjusted for several dozen socioeconomic variables seems to indicate that people who live in counties with moderately high radon tend to live longer than people who live in low-radon counties.[76] But critics point to certain confounding variables in his data: people who live in high-rise apartments, for example, tend to have low radon, but they also tend to live in built-up urban areas, which may have higher baseline cancer rates from other causes. There is also the notorious problem that epidemiological studies based on countywide radon statistics are often unreliable because house-to-house variation is great. Critics have also faulted him for failing to take into account variations in smoking. None of this has prevented the *Science* editor Philip Abelson from citing Cohen's unpublished work in support of the claim that the EPA has exaggerated the hazard.[77] The *New York Times* reported on Cohen's work in 1988 under the headlines, "Scientist Says Low Radon Levels May Be Harmless."[78] Cohen himself has been sufficiently convinced by his findings to take them seriously in his own home: in the spring of 1992, the Pittsburgh physicist turned off the radon fan in his basement, allowing radon levels to rise to 50 picocuries per liter. He told me in an interview that he felt he simply had to practice what he was preaching.[79]

RADON REDUX

Controversy persists in this area not for lack of scientific talent or money devoted to the problem: radiation health studies have absorbed billions of public tax dollars. The problem is not a poverty of dollars but an abundance of conflicting interests. Professional and intellectual allegiances are arguably more convoluted here than in other public health issues, since environmentalists are not simply pitted against industry interests. Nor is the situation like that of smoking, where there is a professional consensus against an entrenched and powerful industry. The interests are more dispersed, and the "victims" have rarely been very visible, vocal, or well organized.

One root of the problem, as already noted, has to do with divergent professional competencies. At one extreme, health physicists schooled in AEC traditions have tended to argue that the hazard has been exaggerated. (In a recent survey, fewer than half the 6,000 members of the Health Physics Society had tested their own homes for radon.)[80] This particular slant has to be understood in light of the fact that the discipline, originally called "lung dynamics," grew out of the fight with fascism and the cold war sense of pronuclear urgency. Health physicists were charged with exploring the hazards of radiation, but they were also usually employed by institutions whose mission was, among other things, to mollify the public's fears of radiation. For decades, health physicists were forced to live with a kind of moral schizophrenia, insofar as the primary demand for their talents came from the Atomic Energy Commission, charged with both promoting nuclear power and safeguarding nuclear health. The AEC was broken up in 1974 into what eventually became the Department of Energy and the Nuclear Regulatory Commission: this was supposed to separate the regulatory and promotional activities of the agency, but much of the expertise and ideology of the AEC survived the breakup. Health physicists have tended to monopolize the field of radiation studies, making it difficult for dissenting views to gain a hearing. Traces of this legacy can be seen in the reluctance of many DOE scientists, by contrast with EPA officials, to move aggressively in the area of radon regulation. DOE officials are generally conservative in the scientific sense; EPA officials tend to be conservative in the public health sense (see conclusion). Congressional staffers in 1994 expressed their exasperation with the "pissing match" between the two federal agencies over the severity of the problem and how best to fix it.[81]

Environmentalists have tended to downplay the radon hazard for very different reasons. The principal fear here is that radon is being used as a red herring to deflect attention from the hazards of nuclear power. Samuel Epstein has settled on this view, as have a number of other defenders of a progressive public health point of view (David Ozonoff, for example). The *East West* journal recently gave prominent coverage to a report by Ernest Letourneaux, head of Canada's Radiation Protection Program, that the radon threat has been grossly exaggerated, the real danger being roughly comparable only to that of air travel (trivial, in other words). A broad spectrum of environmentalist, left-wing, and otherwise alternative journals have published similar reports, placing them in a curious alliance with conservative publications like the *National Review* and the *Washington Times*. Ernest Sternglass, author of the apocalyptic thesis that fallout has killed hundreds of thousands of babies, has stated that the federal government is using radon "as a way to divert the public's attention from the problems associated with nuclear power emissions." John W. Gofman, another influential antinuclearist, admits the reality of the threat but maintains that the EPA's focus on radon is "a highly effective way to deflect concern about the man-made sources of radiation."[82]

Yet another aspect of the problem, though, is that the nuclear industry is not the only corporate body with a stake in the matter. Radon detection and remediation have become big business, and hundreds of new companies have sprung up to meet the demand. Radon remediation got a big boost after the discovery of the Watras house in 1984, and then again in 1988 when the EPA and the surgeon general recommended that all houses be monitored. Radon piggybacked on a surge of interest in indoor air pollution, and by 1988, according to EPA estimates, about 500,000 radon measurements had been made nationwide. *Venture* magazine that same year suggested that building diagnostics had become an industry with "colossal" potential; one entrepreneur recalls that in the first few days after the 1988 EPA report, he had to turn down 2 million dollars' worth of business.[83]

There are now hundreds of commercial firms in the United States involved in distributing and processing the test kits used to monitor household radon; hundreds of other firms provide counseling and install fans to lessen the hazard. These are often boosters of the hazard, a not surprising consequence of the fact that their economic livelihood depends on people recognizing a danger and doing something about it. Trade association journals such as the *Indoor Air Review* support predictably high-end estimates of the hazards posed by the gas. The industry also supports an action level about twice as stiff as that of the EPA.[84] The differences are significant, given that state government officials are being increasingly called upon to determine which of these values should be used to determine whether a given real estate property is fit to be sold with or without remediation.

Finally, there is one last group with interests at stake in the hazard. Home builders and realtors have both expressed concerns about governmental policy in this area, based on financial interests they have in not overcomplicating the process of building and selling homes. Barry Rosengarten, representing the National Association of Home Builders (NAHB), testified before a Senate subcommittee in May 1988 that, because of uncertainties in short-term radon tests, such tests "should not be included as a condition of sale in real estate transactions."[85] Rosengarten, in testimony before a similar committee, denounced publicity techniques "designed to frighten the general public about the overall health risks posed by radon." He was worried, as were many DOE scientists, that the EPA's radon program, mounted with the support of the Ad Council, was frightening homeowners into taking actions they might not actually need to pursue. Real estate and home-building interests are obviously not equipped to mount the kinds of technical challenges prepared by Energy Department scientists or nuclear trade associations, but their views have been regularly consulted at congressional hearings concerning matters of radon policy impact.

One curious aspect of the radon debate is that it is generally carried on as if indoor air meant exclusively air in the home. Obviously people also breathe air at work, but there is little discussion of indoor radon as an occupational problem that affects jobs other than mining or commercial cave tour-guiding[86] and the like. This was brought to my attention during the writing of this book, when I measured the radon in my office on the ground floor of Dickinson Hall at Princeton University. Using a professional-quality radon detector, I found that the office was radioactive to the tune of more than 20 picocuries per liter. I subsequently measured the offices of several other colleagues and found levels as high as 60. The problem has since been remediated, but the incident reminds us that by ignoring radon and other indoor air pollutants, we run the risk of underestimating the proportion of cancers caused by occupational exposures.

One might have imagined that radon, issuing innocently as it does from the ground, would be difficult to politicize. But, as we have seen, the question of health effects is politically charged, for several different reasons. Recall that radon was first recognized in industrial settings because it was difficult for people to imagine an environmental crime without a culpable agent—as hard, one could say, as it was to imagine a toxic spill without a negligent corporation.[87] The tailings scandal and the phosphate fiasco located radon in familiar political space—but household radon was a kind of environmental oxymoron. The two dominant forms of environmentalism, enshrining alternatively (and sometimes agonistically) the wilderness and the workplace, left no room for an indoor toxic hazard. Radon was difficult to face for those who believed that nature is innocent and that hazards are industrial artifacts.

Radon became a domestic hazard when the uranium-mine cancer crisis spilled over from the military-industrial arena into the domestic sphere. A wider danger was discovered in consequence of the weathertight insulation of homes, following fears of an energy crisis. Escalating fears of radiation compounded the interest, especially in the wake of Three Mile Island. Radon was recognized as a hazard in consequence of both the dependence on fossil fuels and the promotion of nuclear power.

One cannot distinguish sharply, though, between the discovery of the hazard and its reception. The same forces pushing the one were forming the other. Attitudes toward radon have been difficult to dissociate from debates over the hazards of radiation more generally—radon controversies often serve as surrogates for controversies over nuclear power. Radon became a marketing ploy for the nuclear industry when it seemed that conservation could be blamed for the household hazard. Environmentalists were reluctant to concede a menace, recognizing the nefarious uses to which such a concession might be put. It is wrong, in short, to think of the hazard as entirely "natural" in every sense of that term. The point is not just that estimates of the magnitude of the threat continue to be shaped by partisan affinities. Whatever those magnitudes turn out to be, radon is likely to remain a political construct, insofar as its continued presence or absence depends as much on political action as on the fissures and isotopes that underlie our homes and places of work.

Chapter 10

Genetic Hopes

If livin' were a thing
That money could buy,
Then the rich would live
And the poor would die.
—American folk song

MOLECULAR GENETICS is changing the way many people think about life. Genetics can tell us the sex of our kids before birth and their prospects for certain kinds of disease. It has brought about new forms of identification (in rape trials and paternity suits, for example), new forms of medical diagnostics (amniocentesis and chorionic villi sampling), new forms of therapy, and even new forms of life (transgenic animals). Genetically engineered organisms are being patented and marketed (the oncomouse, for example, which invariably develops breast cancer), and DNA data banks are being assembled by armies and security forces throughout the world to identify corpses and suspected felons. Genetics now tells who can play in the Olympic Games (in cases of ambiguous sexual identity) and whether the whale meat served in Japanese restaurants has been obtained legally.[1] The new techniques have brought new challenges to the courts, new concepts of patenting and privacy, and a host of new ethical protests. Ours has been heralded—perhaps with some hyperbole—as the Age of Genes.[2]

Genetics is also bringing about dramatic changes in our understanding of cancer. Evidence is growing that people can differ significantly in their susceptibility to particular cancers, confirming a long history of folk and scientific wisdom that cancer runs in families. New genetic diagnostics are already allowing physicians to gauge the likelihood of particular individuals developing cancer; the time may not be far off when genetic tests may reveal a broad spectrum of differential susceptibilities to the disease. Some 200 separate inherited disorders are known in which cancer is a regular or occasional feature.

Most of these confer only a slightly increased risk, but in several the risk approaches 100 percent. There are cancers from which only men suffer (prostate and penis) and cancers from which only women suffer (vaginal and uterine). People with blood type A appear to be about 20 percent more likely to contract stomach cancer than people with type O or B, and there are other kinds of susceptibilities that seem to have a genetic basis. Melanoma of the foot, for example, is more common in African ethnic groups with pigmented spots on the soles of their feet.[3] Virtually every kind of cancer has a known or suspected familial aggregation.[4] As new disorders are identified and underlying mechanisms are identified, there will probably be no cancer for which there is not a predisposing or protecting gene of one sort or another. Statisticians have traditionally ascribed to chance the fact that someone in a vulnerable group does or does not develop cancer, but that is changing. Predisposing genes are being looked to, more and more, to answer the question posed by *Science* reporter Jean Marx: "Why doesn't everybody get cancer?"[5]

In the preceding chapters, I have discussed how the slow pace of conquering cancer can be understood as the outcome of conflicting interests, structural apathies, ideological gaps, and socially constructed ignorance and impotence. Here, I want to look at the origins and outcome of the idea that cancer is *a disease of the genes*. The topic is not an uncontroversial one: as with many other human talents and disabilities, cancer has been pulled within the orbit of the nature-versus-nurture dispute. There are hopeful signs that recent breakthroughs in the identification of cancer genes will yield new therapies, but there are also signs that much of the determinist hype that has long accompanied genetics is still with us.

Genetics has an ideological history that extends through the present; tracing that history will make it easier to understand the potential of genetics both to clarify and to cloud our understanding of cancer.

EARLY RACIAL AND FAMILIAL THEORIES

People have long been aware that cancer sometimes runs in families. In the nineteenth century, scientists in England, France, and Germany published histories of many of these so-called cancer families. Paul Broca, the famous French surgeon and anthropologist, identified no fewer than sixteen cancer deaths within a single family (his own wife's) over a period of three generations from 1788 to 1856. Broca believed that a cancer in one generation could be inherited as a different form of cancer in another: carcinoma might appear in the mother, and then enchondroma (a tumor having a structure resembling cartilage) in the daughter.[6] A British medical survey in the 1880s found that, among 184 cases of breast cancer, cancer had also occurred in other members of the family in 68 cases. The cancer line could generally be traced through either the mother or the father, but not both. The author of this study drew support from other studies—Sir James Paget's, for example—suggesting that one in three or four cancers occurred in families with a previous history of malignancy,[7] but one cannot say there ever was a consensus concerning the magnitude of the hereditary factor. W. Harrison Cripps of St. Bartholomew's Hospital reported that the parents of cancer patients were no more likely to have cancer than the general population, and Herbert Snow, a surgeon at Middlesex Hospital, used Cripps's statistics to argue that the belief in cancer heredity was "derived merely from popular tradition."[8]

The classical problem with heritability studies, of course, is that it is often difficult to say whether familial patterns are due to shared nature or shared nurture. This is as true of cancer as it is of IQ, schizophrenia, criminality, cravings for sugar,[9] or couch potatoism. People used not to move very far from where they were raised, and when children took up the same occupations as their parents or suffered from the same diseases, it was often difficult to distinguish the effects of common environments from those of shared genetic heritage. The son of a farmer who himself became a farmer might develop a certain kind of cancer, but it was difficult to say whether this was due to family biology or some environmental coincidence.[10]

In the nineteenth century, familial disease clusters were sometimes traced to heredity, sometimes to environment, sometimes to chance. Tuberculosis, for example, tended to run in families, and protracted debates were fought over whether the disease was inherited or acquired. Heritable predispositions were postulated even for diseases that, to the modern ear, sound like open-and-shut cases for an environmental interpretation. Thus Henry Earle, a grandson of Percivall Pott, in 1823 conceded that while chimney soot was doubtless the cause of the chimney sweeps' cancer, the disease also tended to run in families, suggesting a "constitution predisposition" that "renders the individual susceptible of the action of the soot."[11] Numerous other authors postulated a heritability for various kinds of cancers,[12] and by the end of the century reports of identical twins with similar kinds of cancers had begun to appear in scientific journals.[13]

Hereditary studies were given a boost in 1900, in the wake of excitement over the rediscovery of Gregor Mendel's particulate theory of hereditary transmission. Xeroderma pigmentosum, a childhood disease that results in multiple cancers of the skin, was recognized as heritable, both because it tended to run in families (three siblings might be affected, for instance) and because the tender age at which the disease appeared seemed to argue against an environmental etiology. As one commentator put it in 1906, most of the victims "have never been subjected to the accidents of adult life; to the exposures of the day laborer; to the frictions incident to toil; to the artificial habits of the consumer of tea, coffee, alcohol, and tobacco; to the habitual stress and strain of most men and women of mature years."[14] Otto Jüngling in Germany showed that multiple polyposis of the colon—a rare familial cancer—was inherited as a dominant trait,[15] and by the end of the 1920s evidence was beginning to accumulate that von Recklinghausen's disease (now known as neurofibromatosis type 1) and multiple cartilaginous exostoses (tumors growing on the surface of bone) were also heritable.[16] Aldred S. Warthin in 1913 published his first of several studies tracing the inheritance of colon cancer in several Michigan families, suggesting that perhaps 15 percent of all malignancies were in families with a history of cancer.[17] In 1916 Theodor Leber of Heidelberg reviewed more than a dozen published reports of families with a history of retinoblastoma—a rare childhood cancer of the eye—one recording ten out of sixteen children in a single family afflicted by the disorder.[18]

The first experimental manipulation of cancer heritability took place in 1907, when Ernest E. Tyzzer of Harvard University showed that by selective breeding of cancerous mice, family lines could be produced that show an abnormally high incidence of certain tumors.[19] Inspired by Tyzzer, Maud Slye of the Sprague Memorial Institute in Chicago bred and dissected more than 67,000 mice to show that cancer is not contagious, and that both the "tendency to be exempt from cancer and the tendency to be susceptible to it" are heritable.[20] A 1942 German review noted that different mice within the same strain, and even different parts of the body on the very same mouse, could be differently susceptible to carcinogenic agents.[21]

Early heredity studies were also resisted by those who felt that chronic irritation or some kind of environmental or infectious agent was the true cause of cancer. Tar extracts, X rays, wax, viruses, and a veritable zoo of other agents had been shown to induce cancer by the 1920s, and this dampened enthusiasm for the heredity hypothesis. Geneticists also had to deal with the objection, occasionally expressed, that heredity seemed to imply a distasteful fatalism: If the demon seed of cancer followed an unalterable course from parent to offspring, what were physicians to tell the children of cancer victims? A 1915 editorial in *Scientific American* denounced the sense of shame that was likely to follow from the idea that "there must be some taint handed down from one generation to another which causes cancer to flourish in certain families."[22]

Slye encountered similar resistance in her campaign to demonstrate heritable cancer predispositions. Her methods were eventually questioned, but the Chicago geneticist was also rebuked by physicians who regarded the whole notion of heritability as morally pernicious. One physician wrote to her: "To destroy all hope in patients with cancer or supposed cancer, as well as to get their children morbid on the subject as to the likelihood of their children having cancer, is to engender an untold amount of mental torture in the supposed victims and you cannot be too roundly condemned for such doings."[23] Slye herself was convinced that genetic and environmental theories were not incompatible. Coal tar, for example, could induce cancer, but individuals might be differentially susceptible to its cancer-causing agency. She pointed out that no one denied the inheritance of our ability to grow hair, even though everyone could see that coal tar spread onto tissues could destroy that ability. In a similar fashion, certain individuals might inherit a resistance to cancer, but that did not mean that such resistance could not be overcome by a sufficiently powerful carcinogen.[24]

Part of the early interest in cancer genetics was fostered by the eugenics movement, one guiding assumption of which was that cancer and many other diseases were the consequence of heritable predispositions.[25] Cancer research in the 1920s was often practiced in the name of racial health: George Papanicolaou's famous and eponymous Pap smear test, for example, was originally announced at the Third Race Betterment Conference of 1928, alongside papers on "The Menace of the Half Man" and "The Menace of the Melting Pot Myth."[26] Carl Weller's otherwise admirable 1941 review of retinoblastoma genetics concluded by recommending that parents of afflicted offspring should stop having children, and that children successfully treated for the disease should be sterilized. Weller, a professor of pathology at the University of Michigan, advised the sterilization even of "sporadic" (nonfamilial) cases in order to avoid the possibility of founding a new "retinoblastomatous family." The need was especially great, he suggested, since more and more children with the disease were being operated upon and cured, allowing them to survive and reproduce their kind.[27] German racial hygienists during the Nazi period voiced similar concerns.[28]

Eugenicists, generally speaking, were much more worried about poverty, criminality, and "feeble-mindedness" than about cancer, but assertions about the different races of the world being unequally resistant to cancer are not hard to find. The "negresses of Senegambia" were said to suffer only rarely from cancer of the uterus,[29] and a 1937 statistical review indicated that cancer struck "coloreds" less often than whites.[30] Jews were said to be particularly susceptible to tumors of the "neuromyo-arterial glomus,"[31] but were also held to be constitutionally immune to cancer of the penis. At the First International Conference on Cancer in London in 1928, Alfredo Niceforo of Naples and Eugene Pittard of Geneva claimed that *Homo alpinus* and *Homo nordicus* (the Alpine and Nordic types) had the highest rates of cancer among the "native" anthropological types in Europe, due to their higher cancer "receptivity."[32] Racial idiosyncrasies were said to explain the rarity of breast cancer among the Japanese and the prevalence of gastrointestinal cancer among Jews. A 1942 book on "the biology of the Negro" pointed to evidence that uterine fibroids and cancer of the jaw were particularly common among American blacks, while rhabdomyomas of the heart and intracranial meningiomas were particularly rare in that population.[33]

In the interwar period, ethnic or geographic differences in cancer rates were commonly discussed in terms of racial or constitutional predispositions. Differential cancer rates were said to derive as much from "blood" as from, say, climate or diet or workplace chemicals. Such assumptions were especially common in Germany in the 1930s, where Nazi social policies emphasized the primacy of race over class, nature over nurture. Hans R. Schinz and Franz Buschke in their comprehensive *Krebs und Vererbung* (*Cancer and Inheritance*) endorsed Pittard's thesis that cancer was more common among Nordic than among Mediterranean types, and Johannes Schottky in his 1937 *Rasse und Krankheit* (*Race and Disease*) included several essays surveying ethnic variations in cancer rates.[34] Prominent Nazi geneticists such as Otmar Freiherr von Verschuer, Mengele's mentor, spent a lot of time comparing identical and nonidentical twins to determine the relative significance of heredity and environment in carcinogenesis.[35]

In the United States, suggestions were even put forward that genetic or racial susceptibilities could be used to screen out vulnerable workers. In 1934, Du Pont's medical director noted that applicants for employment in the company's dye works, where bladder cancer was rampant, were refused if they admitted a family history of cancer.[36] In 1946, *Business Week* announced that the nation's silicosis problem was persisting because a small minority of workers were hypersensitive to dust. The magazine recommended that only dark-haired workers should be employed in the dusty trades: "Brunettes, who generally have more hair on the body, naturally have more hair in the nostrils, which tends to keep silica dust from reaching the lungs."[37] Even the otherwise astute Wilhelm Hueper wrote a long article exploring the racial specificity of cancer and other diseases.[38] As late as 1956, the German émigré published a paper claiming that "Negroes have been found to be more resistant to the carcinogenic action of coal tar, petroleum oils and ultraviolet radiation than are white persons"; Hueper therefore advised that "colored workers might best be selected for employment in operations in which contact with these carcinogenic agents is unavoidable."[39]

Antiracist critics throughout this period were dutifully exposing flaws in such studies. In 1931 Maurice Sorsby, a surgeon at London's Jewish Hospital, tried to show that cancer mortality among both Jews and non-Jews followed a cultural rather than a racial distribution. Cancer of the penis, for instance, was rare among Jews not by virtue of any racial predisposition but, rather, because Jewish men were circumcised. Male circumcision also explained the lower incidence of uterine cancer among Jewish women.[40] Suggestions also began to appear that the peculiarities of American Negro cancer rates might have a nongenetic explanation. An important breakthrough came in 1944, when it was found that U.S. blacks suffered far lower liver cancer rates than West Africans of comparable age.[41] Here was proof that a disease widely regarded as racially distinct was actually caused by environmental factors. (It was later shown that infection by the hepatitis B virus in conjunction with consumption of aflatoxin-contaminated food was the primary cause of African liver cancer.) Similar results were obtained in the 1950s and 1960s, when it was shown that the Japanese who migrated to America began to fall victim to the same kinds of cancer—and nearly to the same degree—as their American neighbors.[42] Whatever was causing the large-scale geographic differences in cancer rates had little to do with inborn "racial" traits.

By the 1960s, the most exciting work in cancer genetics had little to do with race, and often nothing to do with heritability. In 1960, a deletion in what we now identify as chromosome 22 (the "Philadelphia chromosome") was identified as the most common cause of chronic myeloid leukemia.[43] This was the first genetic alteration identified with a specific form of cancer; the condition was genetic, but apparently not heritable. In 1967, a defect known as chromosome 22 monosomy (loss of one of the two chromosome 22s normally present in a cell) was found to be associated with meningioma, a cancer of the membranes lining the brain and spinal cord. This was the first solid tumor traced to a specific chromosomal anomaly, an illness later tied to the rarer form of neurofibromatosis.[44] In 1968, xeroderma pigmentosum was recognized as the product of a heritable defect in the enzymes that repair DNA damaged by ultraviolet light.[45] The emphasis of cancer genetics by this time—especially in the wake of the horrors of Nazi racial hygiene—had clearly shifted toward defects potentially shared by all human populations in common.

MULTISTAGE MODELS AND THE
MUTATION THEORY

Cancers may be "genetic" in two very different senses. In one sense, all cancers are genetic. All cancer involves the runaway replication of cellular tissue; carcinogenesis involves the switching on or off of genes that normally would not act in this fashion. Prompted by an environmental insult—ionizing radiation or a chemical mutagen, for example—the malignant cell begins to reproduce itself in an antisocial or hypersocial manner, invading other tissues (metastasis), expanding enormously in size (tumorigenesis), sapping the body's strength by blocking vital metabolic processes.

Only a very small fraction of cancers, though, are genetic in the second sense, being heritable, passed from generation to generation via the DNA contained in the fertilized egg that gives rise to an embryo. These are "germ-line" cancers, passed from parent to offspring via the germ cells (eggs or sperm), sometimes regardless of the external environment within which the individual develops. Cancers with very high penetrance (expression regardless of environmental exposure) are fairly rare, and include childhood cancers such as retinoblastoma, familial forms of colon and breast cancer, and the cancers associated with certain rare genetic syndromes.

The difference between heritable and nonheritable cancers is essentially in their point of origin: in the former, the defect responsible for the onset of malignancy (or one of several defects predisposing to malignancy) is inherited; in the latter, the defect is acquired through an alteration in the DNA of the "somatic" cells of the body—the nonheritable cells that make up the overwhelming bulk of our tissues (everything but the germ cells, in fact). Since somatic mutations are responsible for the overwhelming majority of cancers, cancers are much more commonly acquired than inherited. As Walter F. Bodmer of the Imperial Cancer Research Fund in London once put it, cancer is "essentially a genetic disease at the cellular level, but not a disease with a major inherited component."[46]

The distinction is not quite as sharp, though, as just presented. Germ-line defects may be passed down that make an individual particularly susceptible to environmental carcinogens: the term *ecogenetics* is sometimes used to describe the study of heritable variations in susceptibility to environmental agents.[47] There are cancers that have both heritable and nonheritable forms, and the same genetic locus may be involved in each. Clear-cut germ-line cancers with strong penetrance are fairly rare, but there are probably many cancers that develop from people inheriting more or less competent means of metabolizing or expelling carcinogens, organizing an immune response, repairing genetic damage, and so forth. Cancers may be made more likely through exposure to environments in which genetic instability is increased.[48] Certain areas of the genome are more susceptible than others to mutation (the so-called chromosomal fragile sites, perhaps); people may vary, genetically, in how they turn noncarcinogens into carcinogens or in how those carcinogens are delivered to cancer-onset sites on the genome.[49] The nature-versus-nurture merry-go-round does not capture the complexity by which cancers develop: most cancers probably exist in a kind of gray zone—facilitated by genetic predispositions, stimulated by environmental carcinogens.

The first to propose a somatic mutation theory of cancer was Theodor H. Boveri, a German zoologist best known for his demonstration of the separate and continuous existence of chromosomes, the "colored bodies" of the cellular nucleus now known to contain the bulk of the body's genetic information. In 1914, Boveri postulated that the abnormalities observed in chromosomes might somehow be responsible for the development of cancer. Cancer, in his view, was at root a cellular disease: tumors typically arise from a single mutated cell, which then gives rise to other cells. Normal cells become malignant by a process of chromosomal transformation: organisms with large numbers of chromosomes are particularly vulnerable to breakage and hence tumors; organisms with few chromosomes are generally less vulnerable or even invulnerable. Boveri was the first to notice that chromosomal aberrations accumulate with the aging of a tumor, and that the probability of cancer developing in a given kind of tissue is proportional to how rapidly cells reproduce in that tissue.[50] He advised the study of "nuclear pathology" as the key to understanding cancer,[51] though as late as 1973 it was not yet clear whether chromosomal aberrations were the cause or the effect of malignant transformations.[52] The rapid progress of cancer genetics in the late 1970s and early 1980s solidified support for the theory, and a 1983 call to "rediscover" Boveri ranked his chromosomal cancer theory nearly on a par with Mendel's contribution to genetics.[53]

As we now understand the situation, tumors are indeed more likely than healthy cells to have chromosomes with multiple copies of a particular gene, or to be "hyperploid"—that is, having more than the usual number of chromosomes. Metastatic tumors are more likely to have these and other abnormalities than cells at the original site of tumorigenesis; tumor cells tend to be genetically unstable, having different properties from the original line. This is one reason immunotherapies such as monoclonal antibody treatment have had only limited effectiveness. Tumors are genetically and therefore immunologically heterogeneous, almost as if separate organisms were living in one and the very same body, invisible to the body's immune defenses.

Prior to the development of an adequate theory of mutation, however, it was difficult to say much more than Boveri did about the role of chromosomes. In 1927, the somatic mutation theory of cancer got a boost from Hermann J. Muller's demonstration that ionizing radiation could cause mutations. Muller, a student of Thomas Hunt Morgan's "Drosophila genetics" school at Columbia University, bombarded male fruit flies with X rays and found that lethal mutations had been induced in the sperm cells of those flies. His work,[54] for which he was awarded a Nobel Prize in 1946, opened an entirely new field of experimental genetic manipulation. But it also suggested that carcinogens might be mutagens. (Recall that X rays were already known to cause cancer.) Further evidence for the somatic mutation theory came from the fact that tumors continued to grow even after the tumor-inducing factor (a chemical carcinogen, for example) was removed. This seemed to indicate that whatever was causing a tumor to grow was "inherited" from one tumor cell generation to the next. Evidence also began to emerge that chemical carcinogens such as methylcholanthrene and mustard gas could cause mutations in plants and in fruit flies, giving further reason to believe that cancer was the outcome of a defect in the local genetic information controlling cellular growth.[55]

The idea that every cell in the body contains a full complement of genetic material (the so-called totipotency of somatic nuclear material) would not be empirically demonstrated until the 1950s,[56] but by the 1920s some scholars were already suggesting that cancer cells contain mutant marching orders. In 1928, a German surgeon, Karl H. Bauer, wrote a whole book postulating that cancer arises in mutant somatic tissues.[57] Bauer's book is the most comprehensive early defense of what is now generally accepted as true, namely, that cancer begins through a mutation in the human genome. Nongenetic mechanisms were still being tossed around in the 1940s and '50s (the protein-deletion hypothesis, for example, according to which carcinogens acted by binding to proteins—also the idea that tumors grew from misplaced embryonic remnants left over from early fetal development), but by 1960 or so it was generally agreed that most mutagens were carcinogens and vice versa.[58]

An important breakthrough in understanding cancer genetics came in the 1940s with the idea that tumors most commonly arise through a multistage process of two or more mutations. Prior to this time, and indeed for a long time afterward, most experimental attention was directed toward so-called complete carcinogens: chemicals or physical agents (X rays, for example) that contain within themselves sufficient power to "crash the gates of ordinary insusceptibility" and induce cancer.[59] The first solid evidence for a two-step process came from experimental work showing that the sequence in which carcinogens were applied to tissues was often crucial to how easily a cancer could be induced. Isaac Berenblum of Oxford showed in 1941 that benzpyrene and croton oil (a plant-derived purgative) were far more powerful carcinogens when applied in that order rather than the reverse. William F. Friedewald and Peyton Rous observed three years later that skin tumors induced by polycyclic hydrocarbons first regress, but can later be made to reappear by applying croton oil.[60] Berenblum and Shubik expanded on this notion in 1947 with experimental work suggesting that there were two distinct aspects to carcinogenesis: initiation and promotion. Some substances act as initiators, launching the carcinogenic process; others promote or accelerate the growth of a cancer once it has begun.[61] In 1949, Berenblum and Shubik argued that carcinogenesis was "at least a two-stage process."[62]

Peter Armitage and Richard Doll devised a multistage model of carcinogenesis in the 1950s, primarily in order to explain the curious fact that cancer is mainly a disease of the elderly.[63] Cancer incidence grows much more rapidly with age than one would expect from a single mutation model of cancer causation. If cancer were a product of single-event "hits" by carcinogens such as X radiation or asbestos, one would expect cancer incidence to grow with age, but only linearly. Instead, what one sees is that cancer incidence grows with about the fourth, fifth, or sixth power of age.

The odds of dying from cancer of the large intestine, for example, increase about a thousandfold between the ages of twenty and eighty. What this suggests is that several mutations may be required to transform a healthy cell into a malignant one, and that the probability of having acquired any given mutation is proportional to age. According to this model, every cell of the body contains several defenses against cancer, all of which must be knocked out for cancerous growth to begin.[64]

Further evidence for the multihit model emerged in 1971, when Alfred G. Knudson of the M. D. Anderson Hospital in Houston hypothesized that retinoblastoma was caused by two mutational events—the difference between hereditary and nonhereditary forms of the disease being only in the timing of the mutations. In the inherited form, he suggested, one mutation is inherited via the germinal cells and a second occurs in a somatic cell. In the nonhereditary form, both mutations occur in a somatic cell. The disease could originate in either manner, but people who inherited the first, predisposing mutation were much more likely to receive the second "hit" that would lead them on the path to cancer. Children inheriting the first mutation tended to develop tumors earlier than those who contract the nonhereditary form of the disease; children inheriting the predisposing mutation were also much more likely to develop tumors in both eyes. This made sense, given that anyone inheriting the first mutation would harbor the defect in every cell of the body. It was therefore much more likely that such children would experience the second requisite hit in more than one location. By contrast, children acquiring both mutations somatically only rarely developed more than one tumor. This, too, made sense, given that the probability of acquiring two independent hits at the exact same genomic site in more than one cell was "vanishingly small."[65]

One of the most intriguing things about Knudson's two-hit model of cancer genetics was that the same segment of DNA was presumed to be involved in both inherited and acquired versions of the disease. For several years after publication, though, Knudson's theory didn't get a lot of attention. Retinoblastoma affected only about 200 children per year in the United States, and it was not yet obvious how it might be relevant to cancer more generally. In 1976, however, Knudson noticed that the disease was occasionally associated with mental retardation, and that when the two conditions appeared together, one also found visible damage to chromosome 13. Knudson concluded from this that the genes responsible for both the retinoblastoma and the retardation were almost certainly located on that chromosome.[66] Knudson's second hit was eventually shown to be the result of cellular events (including crossovers) causing the loss or defection action of the retinoblastoma gene. The year was 1983, and the discovery was particularly noteworthy because it demonstrated a new kind of gene action—a gene whose *absence* was required for the development of cancer.[67] Knudson originally called the gene an "anti-oncogene";[68] others used expressions such as "anticancer gene" or "recessive oncogene," but "tumor-suppressor gene" was the name that eventually stuck.[69] In subsequent years, tumor-suppressor genes would be discovered whose loss or malfunction is responsible for Wilms' tumor (a rare infant kidney cancer), neurofibromatosis, Li-Fraumeni syndrome, and several of the more common cancers.

Multistage models of cancer are important because they provide an explanation of how some people may be genetically predisposed—but not necessarily predestined—to develop cancer. Knudson recognized this as early as 1971, when he pointed out that people born with one or more of the requisite mutational events will have a head start in the development of cancer, since a carcinogen need only activate the second or third or however many other hits are required to initiate oncogenesis. Knudson's model also presupposed that the same genetic sites involved in the expression of hereditary cancers like retinoblastoma might also be involved in the much more common nonhereditary cancers. This would eventually prove quite useful, because it meant that the search for cancer genes could concentrate on sections of the human genome suspected of coding for hereditary cancers. By the mid-1980s, a consensus appeared to have emerged that most cancers arise through two or more hits, each of which is necessary for a tumor to begin to grow.[70]

The multistage theory also allows an explanation of some of the more subtle aspects of the age-dependence of cancer. We've already seen how the likelihood of one's contracting cancer increases rapidly with age. This is as one would expect, since the longer we live the more likely we are to have been exposed to cancer-causing chemicals and other irritants. The pattern is not as simple as that, though. For one thing, babies are much more likely to die of cancer than are ten-year-olds. Another puzzling fact is that once one reaches the age of sixty-five or seventy, the incidence rate basically levels off. Everything else being equal, ninety-five-year-olds are not much more likely to die of stomach or lung cancer than are sixty-five-year-olds.

William G. Thilly and his colleagues at MIT's Center for Environmental Health Sciences have argued that the reason for this curious phenomenon is that mutations that occur early in fetal development play an important role in whether and when one is likely to get cancer. Three factors are crucial: the number of cancer genes one has inherited, the timing of any cancer mutations during fetal growth, and the timing of mutations after birth. Take the case of liver cancer. By the time a child is born, its liver has grown from just a few cells to some 45 billion. If any given gene—cancer genes included—has about a 1 in 10 million chance of mutating per generation, then the average liver has about 450 mutations in every gene by the time it is born. Not everyone is average in this sense, however. Some people suffer mutagenic insults earlier in fetal development than others, and these early mutant cells reproduce the genetic error with each cellular replication. As a result, people with these so-called jackpot mutations have many more cells with mutant cancer genes than other people. If a single jackpot mutation occurs when the fetal liver has reached 2 million cells, this will increase by twenty times the number of mutated cells one would normally have at birth. Thilly suspects that people who die at a relatively young age from liver cancer— say, twenty-five—may have acquired a jackpot mutation, whereas people who die at age sixty-five probably have not. A similar argument can be used to explain why cancer deaths level off after about age sixty-five and why most people, even if they live to one hundred, do not die of cancer. Most people at birth have inherited relatively few cancer mutations, and the odds of them accumulating the number needed to induce cancer is fairly small—unless, of course, one is exposed to high levels of carcinogens at some point in one's life. Very young children who die of cancer may simply have had the misfortune of inheriting—at conception—all but one of the necessary mutations to induce cancer. This would explain why cancer death rates are higher for newborns than for children aged ten or even twenty; for some cancers (kidney and liver, for example), adult rates don't exceed newborn rates until age thirty-five or so.[71]

Multistage models are also significant because they offer additional insights into the diversity of ways one might go about preventing cancer. If cancers arise through several transformations, each of which is necessary for a tumor to develop, this means that intervention at any one of these stages might prevent that cancer from developing. Marvin Schneiderman has suggested that cancer causation may be regarded as a chain, any link of which might be broken to halt the progress of the disease. If the multistage model is correct, then rapid reductions in cancer incidence should follow from reducing exposure to late-stage promoters. (He cites the dramatic drop in endometrial cancers following the cutback in sales of postmenopausal estrogens as an example.) Blockage of initiators, by contrast, should have a longer time lag before cancer rates fall.[72] Radiation and asbestos appear to effect early-stage transitions; cigarette smoking and nickel appear to effect mainly late-stage transitions. Preventing exposure to late-stage carcinogens, according to this argument, would be the most effective way to see an immediate reduction in cancer incidence; there is also the frightening implication, though, that by the time a human hazard may be discovered, those exposed may have already accumulated sufficient hits to guarantee a cancerous outcome.[73]

Finally, there is what might be called the ideological significance of the multistage model. The multihit model allowed cancer geneticists to distance themselves from an earlier era's fixation with germ-line heritable cancers. The new model focused attention not just on the familial inheritance of a few rare cancer syndromes but rather on the mechanisms common to both germ-line and somatic cell genetic expression. One should recall that the multistage model rose to prominence at a time—the 1970s—when the environmental view of cancer was dominant: the multihit model allowed one to maintain that all cancers are genetic, while simultaneously conceding, without contradiction, that 80 to 90 percent of all cancers are initiated by environmental agents. Geneticists captured the field of environmental carcinogenesis by shifting their focus from germ-line inheritance to somatic cell inheritance. Heredity itself was redefined to include the replication of the somatic cell. Freed from the hereditarian, biological determinist baggage that had prompted such skepticism among environmentalists, the new approach opened up the possibility of regarding all cancer as genetic.

The ideological break with the past was not, however, a perfect one.

GENETIC SUSCEPTIBILITY:
RARE SYNDROMES

Human genetics is historically a science of human defects, or at least of human inequalities. This is as true for cancer genetics as for most other fields of medicine: it
is easier to study that which is broken than that which is working normally. Genetic
replication is usually stunningly perfect, considering the vastness of the clerical duties required to prevent chaos in the files of the human genetic text. The human body
contains some 100 trillion cells, most of which are issuing like-formed progeny
every several days. Each of these cells contains a two-meter-long DNA molecule, divided among some twenty-three pairs of chromosomes containing 3 billion nucleotides—so the opportunities for mistakes are great. The overwhelming majority
of these mistakes will be corrected, and many of those that are not will be of little
consequence. Most cells that acquire major defects will simply die. Occasionally,
though, a cell containing mutant genetic information will survive and reproduce its
in-built errors, and some of these will give rise to cancers.

Much of what we know about human genetics concerns the working of various
syndromes—rare disease complexes caused by the improper replication or processing of genetic information, traits that travel together to excess. Several of these syndromes, often characterized by gross chromosomal breakages, deletions,
rearrangements, or other genetic mixups, confer an increased susceptibility to cancer. The best-known syndromes tend to be dominant (requiring only a single mutated
gene to show up), non-sex-linked traits with fairly high penetrance; cancer for the affected individuals tends to strike earlier than in the rest of the population.

Ataxia telangiectasia (AT or Louis-Bar syndrome), for example, is a neurological
disorder characterized by disturbances of muscular coordination; sufferers are unable to repair DNA damaged by ultraviolet radiation and have unusually high rates
of cancer. *Bloom syndrome,* found mainly among Ashkenazi Jews, predisposes to
leukemia and stomach cancer, while *Gardner syndrome* predisposes toward cancer
of the colon. People with *multiple endocrine neoplasia syndromes* (MENS type 1
and 2) often have tumors in hormone-secreting organs, while sufferers from *Von
Hippel-Lindau syndrome* tend to develop cancers of the eye, brain, kidney, and other
organs. Down syndrome—once known as "mongoloid idiocy" and today known as
trisomy 21—has also been linked to certain kinds of cancer.[74]

Neurofibromatosis (NF) is one of the more common syndromes of this sort, affecting about 100,000 people in the United States. For many years, NF was thought to
be the "Elephant Man's disease," though it is now believed that the nineteenth-century Englishman, Joseph Merrick, suffered instead from a disease known as Proteus
syndrome. There are two different forms of the disease with two different genes responsible, both autosomal dominant. In 1987, Ray White and Mark Skolnick of the
University of Utah mapped the gene for the more common form, NF1—von Recklinghausen's disease—to chromosome 17; that same year James Gusella of Massachusetts General Hospital in Boston identified a gene for the second type (NF2),

about ten times rarer than NF1, on chromosome 22.[75] The NF1 gene was isolated and cloned in the summer of 1990;[76] the NF2 gene was cloned in 1993.[77] NF1 turns out to be one of the largest of all known genes, spanning roughly one million nucleotides; it is also unusual in that several other genes are "nested" inside this megagene. It appears to have been conserved in organisms as evolutionarily distant as yeast and humans, suggesting that it plays some vital role in normal cellular function.[78]

As tragic as such diseases are for the people afflicted, it is important to recognize that these and other heritable cancers account for only a tiny fraction of all cancers. Retinoblastoma affects about one in every 20,000 children. Von Hippel-Lindau syndrome affects about one in 50,000 or 60,000 babies. One in 40,000 babies is born with the gene coding for neurofibromatosis type 2, and even the much more common NF1 gene appears only once in every 4,000 births. (In about half of these cases neither parent has the disease, meaning that a new germ-line mutation has been acquired—most likely during spermatogenesis.) The gene or genes predisposing for ataxia telangiectasia affect about 1 in 100,000 (some say 2 or 3 per 100,000), and Gardner's syndrome strikes roughly 1 in 16,000. Down syndrome is not that rare, but the predisposition to cancer is fairly slight: the proportion who develop leukemia, for example, is on the order of 1 percent.[79] One million Americans have hemochromatosis, and 1 in 14 Americans is a carrier of the recessive gene linked to the disease (mapped to chromosome 6), but the increased risk of liver cancer is rather small. All together, the syndromes mentioned account for less than 1 percent of all cancers. Several of the best known are quite rare indeed: as of 1991, for example, fewer than 100 families with Li-Fraumeni syndrome had been identified worldwide.

GENETIC SUSCEPTIBILITY: MORE COMMON CANCERS

One of the most exciting things about cancer genetics in the 1980s was the discovery that some of the same mechanisms involved in several of the most esoteric inherited cancer syndromes also operate in much more common cancers.[80] Knudson's two-hit model provided the most important theoretical insight, allowing cancer geneticists to rise above the nature-versus-nurture battle, in which nature had almost always come up short. The tumor suppressor genes (anti-oncogenes) found in the course of elaborating Knudson's model did not even have the viral taint of oncogenes, which had emerged rather suspiciously from the viral research environmentalists had never really trusted. The newfound focus on somatic cellular inheritance made it difficult to argue with the claim that "all cancers are genetic." This, in turn, allowed a reversal of classical metaphors, a recognition that environmental agents plant the seed of cancer but that genetic predispositions have a big say in whether that seed is nourished or allowed to wither.

Theoretical insights, of course, are useless without the wherewithal to put them into practice: most of the rapid advances of cancer genetics in the last decade or so were made possible by the development of powerful new techniques in molecular biology. Twenty years ago, cancer geneticists fixated on the extent to which cancer runs in families—a research programmatic largely unchanged from the nineteenth-century fascination with "cancer families." Since the late 1970s, however, cancer genetics has changed its primary focus from germ-line inheritance to somatic nuclear expression. The key to this transformation has been the movement from a breeding, or pedigree-based, genetics to a biotechnology-based genetics based on the tools of cloning, transfecting, synthesizing, mapping, and sequencing that allow you to identify where on a chromosome a particular gene lies and the nature of its protein products. Restriction endonucleases, developed around 1980, enabled geneticists to discover new genetic markers—especially restriction-fragment-length polymorphisms (RFLPS)—allowing the rapid mapping of the human genome.[81] Polymerase chain reaction, introduced in the mid-1980s, allowed geneticists to amplify trace amounts of DNA, making it possible to analyze minute samples of blood or tissue preserved in pathologic specimens. Biotechnology has opened up entirely new areas of inquiry: molecular epidemiology, molecular cytology, molecular paleontology, and pharmacogenetics, just to name a few (the last dates back at least to the 1950s, but has been greatly strengthened by the new techniques). Genetics has become a branch of biotechnology. Some of the most dramatic advances have been made in knowledge about colon cancer. Scientists had known since Warthin and Jüngling's work earlier in the century that multiple polyposis—a rare colon disease—was highly heritable: people born with the disease develop a carpet of polyps in their colons at a very early age and by thirty or forty, if untreated, usually die from colon cancer. (People with early signs of the disease can have their colons removed, extending their lives dramatically.) Scientists could also see, using light microscopy combined with appropriate staining and radiographic tracers, that the chromosomes in the tumors of patients with heritable colon cancer were often broken, translocated, or otherwise defective, showing multiple copies, deletions, and so forth. It was difficult, though, to say which of these defects was responsible for the disease. There was also little reason to believe that people developing colon cancer without the carpet of polyps (polyposis) might have inherited a susceptibility to the disease, or that the rare familial form might have anything to do with the more common, nonhereditary variants of colorectal cancer.

By 1990, this picture had very much changed. In 1988, Bert Vogelstein of the Johns Hopkins Oncology Center in Baltimore showed that the stages through which colorectal tumors develop (polyp, tumor, metastasis) are associated with a well-defined progression of genetic mutations and chromosomal defects—including mutation of the *ras* proto-oncogene on chromosome 11 and deletions on chromosomes 5, 17, and 18.[82] Vogelstein's paper was important because he showed that tumor development followed the same kind of stepwise development postulated by Knudson and other multistage modelers. The paper was also significant because it showed that some of the same genetic events involved in somatic cell tumorigenesis were also involved in heritable cancers. Walter Bodmer of London had already mapped the gene for familial polyposis to the long arm of chromosome 5 (as had Ray White of Utah and Yusuke Nakamura of the Cancer Institute in Tokyo),[83] and Vogelstein suspected that this same gene was probably involved in the defects he was observing in actual tumors. Vogelstein had been hoping all along that an analysis of tumor cell genetics might shed light on heritable cancers.[84] The new gene seemed to operate in the early stages of tumorigenesis; there was also reason to believe it was a tumor suppressor, since deletions of the region containing the gene had been found in a majority of colon cancer tumors.

In August 1991, Vogelstein, Nakamura, and White published simultaneous reports in *Science* and *Cell* announcing the isolation of the gene responsible for familial polyposis.[85] The new gene—dubbed APC, for adenomatous polyposis coli—was different from previously discovered oncogenes. Oncogenes are control genes gone wrong, genetic deregulators that lead to uncontrolled cellular proliferation.[86] Oncogenes have been found on the genomes of viruses that infect chickens, monkeys, and most other vertebrates; they occur naturally in certain animals and may be picked up by these viruses and transmitted in the course of infection. Oncogenes were in fact discovered in the context of the search for a tumor virus: Robert Weinberg and his collaborators at MIT in the late 1970s found that the DNA extracted from tumor viruses was capable of transforming normal cells into cancer cells; later studies identified the region of the viral DNA containing the genes that, when mutated, can cause cancer. A hundred separate oncogenes have been identified, several of which coexist in species as far apart as humans and yeast. The conservation of such sequences over a billion years of evolutionary history suggests, again, that in their nonmutant state, oncogenes are important in normal cellular function, probably involving regulation of cellular growth.[87]

APC, by contrast, is a tumor suppressor: you need it working properly to *prevent* the transformation of normal tissue into cancer. In the automotive analogy of Harold Varmus and Robert Weinberg, if oncogenes are like a cancer accelerator, tumor suppressor genes are like cancer brakes.[88] If you lose a tumor suppressor, there is nothing to stop the runaway growth of cancer. Evidence for the existence of tumor suppression genes had emerged in the 1950s and 1960s from several different lines of research. In Germany, Hans Breider of Würzburg postulated in 1952 that melanoma in certain species of fish resulted from the loss of what he called "inhibitory" genes (*Hemmungsfaktoren*) that suppress melanin-generating genes.[89] In the United States, a kind of tumor suppression was observed in the course of experiments on hybrid cells formed from the fusion of normal and cancerous cells. Contrary to what one might expect, the cells containing nuclei from both cancerous and noncancerous cells tended to behave like normal cells. The same was true of the descendants of the fused cells. This seemed to indicate that something in the nuclear material of the normal cell was capable of suppressing cancerous growth, and since that capability was passed from one cellular generation to the next, the suspicion immediately arose that the suppression was genetic.[90]

APC was not the only tumor suppressor gene involved in Vogelstein's tumors, however. Vogelstein had earlier observed that chromosome 17 was often missing an arm in the tumor cells of colon cancer patients. A tumor suppressor gene by the name of p53 was known to lie on that chromosome, and Vogelstein hypothesized that p53 might be involved in colon tumorigenesis. Arnold Levine at Princeton had co-discovered the p53 gene (named for the molecular weight of its protein product, 53,000 daltons) in the late 1970s, in the context of an effort to understand how a notorious viral contaminant of polio vaccines (a simian virus known as SV40) was able to transforming normal mouse cells into cancer cells.[91] Levine cloned the p53 gene in 1983, and within two years it had been mapped onto chromosome 17.[92] P53 was originally thought to be an oncogene, but that turned out to be, as Robert Weinberg has described it, "a mirage." (The p53 clones used in the early experiments were found to have been mutants.) Levine in 1989 showed that p53 acted as a "suppressor of transformation"—a tumor suppressor gene, in other words.[93]

P53 prior to this time had not yet been recognized as a human tumor suppressor. The p53 gene had been shown to be altered or even absent in rat tumors (truncation of the p53 protein had also been observed),[94] and evidence was beginning to accumulate that p53 expression increased in response to DNA damage, but the relevance of such findings for human carcinogenesis had not yet been appreciated. (Levine's experiments were on rat embryos.) The turning point in the recognition that p53 might play an important role in human carcinogenesis came in April 1989, when Vogelstein showed that the gene was consistently lost in human colorectal tumors. The whole arm of chromosome 17 was sometimes missing, but even a single base-pair change could do the trick. The gene could be lost in any cell, but it seemed to be more often absent in cells that were already well on the road to cancer. This again was consistent with the multihit theory, according to which cancers progress through a well-defined series of stages. In the case of colon cancer, cells in the polyps might already have several of the requisite hits; end-stage metastatic cells might have all. P53 appeared to be a late-stage change in tumorigenesis: in precancerous polyps, for example, the gene was rarely abnormal, but in late-stage tumors it was almost always broken.[95] APC, by contrast, seemed to play a role in the earlier stages of tumorigenesis—the transformation of normal epithelium into polyps, for example.[96]

What was not yet clear, however, was whether there was a germ-line version of the p53 gene. Could p53 defects be inherited? The answer came in November 1990, when Frederick Li and Joseph Fraumeni of the NCI, along with several colleagues from the Massachusetts General Hospital Cancer Center (notably, Stephen H. Friend) and elsewhere, found that several of the cancer families they had been following had inherited a mutation in the p53 tumor suppressor gene.[97] From a clinical point of view, locating the gene meant that people could now be tested to see whether they carry the mutation. This itself, of course, was not of great significance from a policy point of view since, as already noted, Li-Fraumeni syndrome is one of the rarest known cancer syndromes. Of much greater consequence was the fact that the cancers to which these families were susceptible were not rare childhood cancers like retinoblastoma and Wilms' tumor but several of the most common adult malignancies—brain and breast cancer, for example—in addition to rarer types (soft-tissue sarcomas, leukemias, and cancers of the bone and adrenal cortex).

The implication of p53 in this broader spectrum of cancers led to a search for p53 defects in other kinds of cancer, with astonishing results.[98] P53 inactivation was found in heritable breast tumors and in the lung cells of uranium miners. P53 mutations have been found in the tumor cells of skin cancer patients, confirming the mutagenic effects of ultraviolet light.[99] P53 mutations eventually turned up in more than half of all cancers (including the archival tissues of former vice president Hubert Humphrey's bladder tumor),[100] making oncogenes look insignificant by comparison. Even Robert Weinberg, a principal architect of the oncogene paradigm of the early 1980s, was conceding by 1991 that tumor suppressors had replaced oncogenes as the most promising topic of cancer gene research.[101] Vogelstein's work with p53 had put "an obscure gene right in the cockpit of cancer formation."[102] A 1993 study by the Philadelphia-based Institute for Scientific Information showed that the number of scientific papers with titles mentioning p53 had doubled every year since 1989; the magazine *ScienceWatch* dubbed Vogelstein "the hottest scientist in the known universe," with more than ten thousand citations during the period 1981 to 1990 and several thousand more in the early '90s.[103]

P53 is apparently the most widely mutated tumor suppressor gene, though it is not yet clear how it works. We know that its breakage or absence can lead to cancer, but no one yet knows how or why. Experimental manipulation of mice bred not to have the gene has led to some interesting suggestions: in 1992, Lawrence A. Donehower of the Baylor College of Medicine bred such "p53 knockout mice" (using the gene targeting technology that had previously been used to give mice cystic fibrosis) and found, surprisingly, that they develop normally and are fertile.[104] This seemed odd, given that the gene was thought to be needed to hold back the tumorigenic floodgates. How could mice survive without the gene? A hint came in 1993, with experimental work suggesting that the gene might be involved in "emergency response" or "damage control." Experiments at MIT and the University of Massachusetts, Amherst, showed that p53 knockout mice are much more resistant to the lethal effects of ionizing radiation than "wild type" mice with their p53 genes intact. What this seems to indicate is that the absence of p53 may lead to "inappropriate cell survival" after irradiation. P53 in normal cells may act as a kind of suicide squad leader—initiating a process of programmed cellular death following damage by irradiation or some other insult. This would have an obvious anticancer value, allowing injured cells to die that otherwise might go on to produce cancer. Cells with their p53 genes intact would not live long enough to succumb to cancer, but cells with damaged or absent p53 would fail to be eliminated from the cellular breeding stock. With nothing to stop them, the immortalized cells would multiply out of control, eventually killing the organism.[105]

Perhaps the p53 gene activates transcription of death-inducing genes; it may also interact more directly with genetic repair machinery, "perhaps increasing its fidelity but delaying repair," enabling damaged cells to die an early death.[106] P53 may instruct cells to halt DNA synthesis when the cellular genome becomes damaged;[107]

p53 protein products may inhibit cell growth by blocking some vital phase of the cell cycle, giving the cell's genetic-repair machinery time to work.[108] P53 defects are often associated with genetic instability elsewhere on the genome, which may further increase the likelihood of cancer. However it works, there is little doubt that it plays a major role in the development of many kinds of cancer. A 1993 article in *Nature* concluded that p53 was "the most widely mutated gene in human tumorigenesis," involved in up to 50 percent of all cancers.[109] By the end of the year, p53 was being hailed as the "guardian of the genome"; *Science* magazine editors voted it the "molecule of the year."[110]

Excitement about the p53 gene derives not just from the fact that it is so common; there is also the interesting fact that it mutates in a predictable fashion, according to the particular carcinogen to which it has been exposed. Exposure to ultraviolet light produces transition mutations at dipyrimidine sites, while dietary aflatoxin produces a characteristic codon 249 mutation and consequent serine substitution in liver cancer cells. Cigarette smoke appears to be associated with G:C to T:A transversions in tumors of the lung, and future research will surely uncover many other patterns. Analysis of how p53 in a particular tumor has mutated may eventually allow one to tell whether a given cancer resulted from an environmental carcinogen or an "endogenous" mutation,[111] and one can imagine the legal consequences. It may also tell you which among a group of, say, women treated for breast cancer are most likely to suffer a relapse, allowing chemotherapeutic recommendations to be adjusted accordingly. There is the prospect of earlier detection than has heretofore been possible (Mary-Claire King of Berkeley envisions a "molecular mammogram");[112] there is also the promise of genetic therapy—the ultimate *raison d'être* of cancer genetics.

The fact that different kinds of carcinogens lead to characteristic and predictable alterations makes the p53 gene an ideal candidate for use in the new field of molecular epidemiology. In traditional epidemiology, the goal is to correlate patterns of disease incidence or mortality with potential causes of those patterns. Molecular cancer epidemiology, by contrast, looks at the tissues of persons exposed to carcinogenic agents—and especially the genes and gene products of those tissues—to determine who is likely to contract cancer.[113] In March 1992, for example, researchers at the National Cancer Institute reported evidence linking radon exposure to damage of the p53 gene in a study of New Mexico uranium mine workers.[114] The focus of such studies, again, is not on the visible health of the person but rather on the molecular changes in that person's cells. The presumption is that early molecular changes associated with the development of cancer can be detected prior to its symptomatic expression and that this may be used for diagnostic purposes.[115] Since genes may bear the traces of the specific form of insult to which they have been subjected, analysis of such patterns may give rise to a kind of molecular archaeology, revealing the past history of insults and potential future harms.

Familial polyposis, retinoblastoma, and most other cancer syndromes are rare diseases, but in December 1993 two separate research teams led by Richard Fishel of the University of Vermont and Richard Kolodner of Harvard, and by Kenneth Kin-

Heritable Cancer Predisposition Genes

Disease	Gene	Chromosome	Mapped[a]	Cloned
Retinoblastoma	Rb	13	1976	1986
Wilms' tumor	WT1	11	1984	1990
Li-Fraumeni syndrome[b]	p53	17	1985	1983
Neurofibromatosis type 1	NF1	17	1987	1990
Neurofibromatosis type 2	NF2	22	1987	1993
Familial polyposis of the colon	APC	5	1987	1991
Von Hippel-Lindau syndrome[c]	VHL	3	1988	1993
Multiple endocrine neoplasia type 2	RET	10	1987	1993
Multiple endocrine neoplasia type 1	?	11	1988	(not yet)
Hereditary nonpolyposis colon cancer	MSH2	2	1993	1993
	MLH1	3	1994	1994
Early onset familial breast cancer	BRCA1	17	1990	1994
	BRCA2	13	1994	1994

[a]Gene mapping may take several years: the dates listed here indicate the first time a chromosomal location is established, rather than the first time a chromosomal location is suspected (due to observation of breakages, for example).

[b]P53 was mapped and cloned prior to its recognition as the germ-line tumor suppressor gene involved in Li-Fraumeni syndrome; that connection was not made until 1990.

[c]Von Hippel-Lindau syndrome predisposes to kidney, pancreatic, and several other cancers.

zler and Bert Vogelstein of Johns Hopkins and Albert de la Chapelle of Helsinki, announced the discovery of another colon cancer gene that appears to account for a much larger fraction of colon cancers—up to 10–15 percent, according to some estimates. Colorectal cancer is the second most common cause of cancer death in the United States, claiming an estimated 56,000 people every year. In 1994, 149,000 new cases will be diagnosed, according to American Cancer Society projections. The new gene, coding for what is known as hereditary nonpolyposis colorectal cancer (HNPCC—also known as Lynch syndrome II), is expected to account for 15,000 to 22,000 of these new cases, making it the most important cancer-predisposition gene thus far discovered (see table).

The HNPCC gene, dubbed *MSH2*, is interesting from a theoretical point of view because it seems to work by an entirely new mechanism. Oncogenes regulate cellular division, and tumor-suppressor genes block runaway cellular growth, but the new gene seems to be involved in genetic repair, acting as a kind of genetic spellchecker, correcting mismatched nucleotide base pairs on the DNA molecule. When damaged, this mutator or housekeeping gene allows mismatched base pairs to accumulate on the genome—guanine bound to thymine instead of to cytosine, for example—giving rise to the genetic instability that can lead to cancer.[116] Discovery of the new gene was made possible by the development of a new set of genetic probes known as microsatellite DNA—short sequences of DNA repeated throughout the genome—which geneticists used to search for markers shared by the suspected carriers of the gene but not by anyone else.[117] Evidence of microsatellite instability in colon cancer DNA had been demonstrated by the late spring of 1993, by which time the gene for hereditary nonpolyposis colon cancer had been traced to chromosome 2.[118] The gene responsible for failing to correct that instability—the spellchecker gene, *MSH2*—had been cloned and sequenced by the end of the year.[119]

Discovery of the new gene is also significant because it opens up the possibility of diagnostic screening. One in 200 Americans is suspected of harboring the gene, which confers a 70 to 90 percent risk of colon cancer and a number of other malignancies (ovarian and uterine cancer, for example) over the lifetime of the carrier. A diagnostic test to identify carriers, which should cost on the order of $1,000, should be available commercially within a few years. Colon cancer is treatable if detected early, and people found to carry the gene could be encouraged to undergo regular colonoscopic examinations beginning in their twenties or thirties, allowing precancerous polyps to be detected and removed. Carriers will also presumably be advised to adopt a high-fiber, low-fat diet to reduce their vulnerability, though it is not yet clear whether it will ever be fair to claim, as Natalie Angier does in a report on the discovery, that "people deemed free of the mutation could be relieved of their worries."[120] Since the new gene accounts for only 10 to 15 percent of all colon cancers, most of those who contract the disease are probably acquiring it by exposure to dietary, occupational, or some other kind of environmental carcinogens. Other predisposing genes may be at work, but there is no reason to believe that a negative judgment of genetic predisposition is a guarantor of colon cancer invulnerability. Prudence in matters of diet and exposure to carcinogenic hazards is still going to be sound advice, regardless of whether or not one is identified as a carrier of a predisposing gene.

Breast cancer has long been one of the diseases suspected of involving a substantial hereditary component. Anecdotal evidence of a familial aspect dates from the ancient world, and by the nineteenth century medical journals were full of examples of extended "breast cancer families." Mid-nineteenth-century hospital surveys reported that breast cancer was much more likely to occur among relatives of people with breast cancer, and more recent studies corroborate these early suspicions.[121] Few people have ever suggested, however, that these familial cancers account for a sizable proportion of all breast cancers. Neil Risch of Yale's School of Medicine in 1991 estimated that roughly 5 to 10 percent of all cases of breast cancer are to be accounted for by inherited factors, and that the distribution of the disease is consistent with the existence of an autosomal dominant gene affecting about one-third of 1 percent of all women.[122]

The first concrete evidence of a breast cancer gene came in 1990, when Mary-Claire King of the University of California, Berkeley, used conventional linkage studies to identify a rough location for "the" gene (there turns out to be more than one). Working with some 100 families in which breast cancer seemed to be unusually common (afflicting two or more sisters, for example), King and her students compared the DNA of the families with unaffected controls to see whether genetic markers could be found that "traveled with" the cancer families. King began her work in the 1970s with only thirty protein markers scattered across all twenty-three chromosomes, and if the tools of genetics had remained frozen, her quest would probably have failed. The development of new DNA markers in the early 1980s allowed her to refine her search, and the introduction of polymerase chain reaction (PCR) in 1985 allowed her to amplify—by several orders of magnitude—the DNA present in very small tissue samples.[123] PCR remains today one of the premier tools of molecular genetics—allowing rapists to be identified from very slight traces of semen, for example, and DNA from insects preserved in amber to be magnified and manipulated.

King eventually found a marker that seemed to indicate a linkage in a 50 million base-pair region of the long arm of chromosome 17. The consistency wasn't great (only seven of the twenty-three families she was studying showed linkage), but a good sign was that younger women with the disease—whom one would expect to be much more likely to have inherited it—showed a very high linkage.[124] She announced her discovery in October 1990 at the annual meeting of the American Society of Human Genetics in Cincinnati, and though some were at first skeptical (gene finds for complex traits have often been advanced and then rescinded), scholars elsewhere quickly found that the marker in question was also linked to ovarian cancer. This was good reason to believe that a cancer gene was near at hand. Scientists from Europe and several other major American laboratories joined in the hunt, and by the end of 1993 the search had been narrowed to a 300,000 base-pair region.[125] The *BRCA1* gene, as it is now called, was finally cloned and sequenced in September 1994 by a team of forty-five scientists led by Mark Skolnick of Utah. Another gene, dubbed *BRCA2*, accounting for a comparable proportion of heritable breast cancers, was identified about this same time.[126] Other breast cancer genes will no doubt be located, though expectations are that these two will account for the large majority of heritable tumors.

Present estimates are that about 5 percent of women with breast cancer will be found to harbor a gene strongly predisposing them to the disease. One in 200 women is presumed to be a carrier, making this the second most common kind of cancer-predisposition gene known.[127] (Recall that 1 in 200 men *and* women are suspected of carrying the *MSH2* colon cancer gene.) Both *BRCA1* and *BRCA2* are autosomal-dominant, meaning that one needs only one copy (from either one's mother or father) to contract the disease. Women who carry *BRCA1* also have a higher risk of ovarian cancer, but there seems to be no such additional risk for *BRCA2* carriers. Both genes seem to be fairly highly penetrant, which means that most women who are carriers will develop breast cancer. Estimates indicate that about half of all women with one of these genes will develop breast cancer before age fifty, and about 85 percent will eventually get the disease over the course of a normal lifetime, in the absence of prophylactic mastectomy. As in the case of retinoblastoma, women born with a breast cancer gene may need fewer "hits" to get the disease, which would explain why carriers tend to contract cancer earlier than women without the genes. Roughly 40 percent of all women who get breast cancer in their twenties are getting it by virtue of an inherited predisposition; by contrast, among women who contract the disease after the age of eighty, only about 1 percent will be getting it by virtue of a predisposition.[128] Sibling studies have confirmed the power of the genes: sisters of women with the genes who are not themselves carriers have the same risk of developing breast cancer as anyone else in the population—about 10 percent by age sixty. It is not yet clear how the new genes work.

The prospect of widely available diagnostic tests for cancer genes raises a number of disconcerting political and ethical questions.[129] It is not yet clear what the cost of such tests will be, or who will pay. It is also not clear how genetic counselors will deal with people who want to be tested: Will parents be able to test their teenage sons or daughters? Or their infant children? Or their unborn fetuses? No one knows what the demand for such tests will be, but if it is high, one can presumably expect testing for *BRCA1, BRCA2, MSH2,* p53, and other genes eventually to become part of routine clinical practice.[130] Efforts are already under way to commercialize such tests: in December 1993, for example, a diagnostics firm based in Gaithersburg, Maryland, announced plans to provide "genetic risk profiles to the healthcare community," capitalizing on the discovery of the breast and colon cancer markers. The service is to be offered by OncorMed, Inc., drawing upon the resources of the Cancer Family Genetic Database compiled by the Hereditary Cancer Institute of Omaha, Nebraska, containing medical records of more 200,000 people from 2,300 cancer families throughout the world. The National Advisory Council for Human Genome Research responded with a position paper maintaining that it was "premature to offer DNA testing or screening for cancer predisposition outside a carefully monitored research environment."[131] Commercial development may be difficult to slow, however, given the possibility of sizable profits from such tests—and from whatever therapies may eventually emerge. In the fall of 1994, within weeks of the cloning of *BRCA1,* a bitter dispute arose over who owns the property rights to the gene.[132]

No one knows what the psychological impact of such tests may be: there is a novel element, insofar as these are diagnostic technologies being developed in the absence of any appropriate therapy (short of watchful waiting or prophylactic coloectomy, ovariectomy, and mastectomy). Will women found not to have a breast cancer gene feel themselves not to be at risk? Will people not predisposed to colon cancer feel not obliged to exercise or cut down on fat and increase dietary fiber?

Colon cancer and breast cancer are not the only common cancers for which there may be heritable predispositions. Efforts are under way to demonstrate differential susceptibilities to lung cancer, focusing either on the inheritance of "the addictive personality"[133] or on differently inherited capacities to metabolize the carcinogenic components of (for example) tobacco smoke.[134] Predispositions will no doubt be found for many other kinds of cancer.

There are those who argue, however, that too much emphasis is being given to nucleotide sequences as the be-all and end-all of cancer causation. Donald S. Coffey of the Johns Hopkins Oncology Center argues that insufficient attention has been given to the spatial organization of DNA (especially folding patterns)—alterations of which may be as important as sequence variations in determining who gets cancer and in what part of the body. One out of ten men develop prostate cancer, for example, but why does cancer never appear in the bulbourethral gland, lying just below the prostate? The cells of the two glands contain DNA molecules with identical nucleotide sequences—why is the one organ vulnerable, the other resistant? Why do dogs and humans get prostate cancer, but never horses, cats, or bulls? Why do cows never get breast cancer? Coffey points out that nucleotide sequences alone cannot explain such differences—one must look to the nuclear matrix, the "dynamic structural subcomponent of the nucleus that directs the 3-dimensional organization of DNA into loop domains and provides sites for the specific control of nucleic acid intranuclear and particulate transport," to understand how the very same DNA sequence may do different things in different cells. This is a nontrivial problem, given the sheer length and complexity of the DNA molecule: Coffey points out that if the nucleus were magnified to a 3-foot sphere, the DNA molecule would extend for one hundred miles within that sphere. The possibilities for tangles, misreadings, and the like are great, and what gets read or not read is going to depend on how the whole is packaged. Genes are important in the regulation of other genes, but many events in development—and carcinogenesis is a kind of development—are *epigenetic:* controlled by situational events that include access clues, packing structures, membrane properties, protein activity, chemical salinities, immune response, and a thousand other factors not readily deducible from nucleotide sequences alone. Cancer is a genetic disease, but it is also an epigenetic disease.[135] The conception of genetic organization in terms of linear and sequential bits of code allows certain important kinds of manipulation, but it also grossly oversimplifies the beauty and complexity of organic processes. Genes are not the "self-moved movers" publicists and scientists sometimes like to have us think they are.[136]

The more serious problem, though, may be unwarranted generalizations about the frequency and implications of predisposing genes. Mark Skolnick, one of the country's foremost cancer geneticists, speculates that the genes predisposing toward colon cancer are inherited "by as many as one-third of all Americans," but a 1993 review in *Science* states that germ-line genetic inheritance accounts for perhaps 4 to 13 percent of all colorectal cancers.[137] Breast cancer heritability is even more contentious. Mary-Claire King gives a figure of 5 to 10 percent, but Henry T. Lynch in 1981 estimated that 15 to 20 percent of all breast cancer would turn out to be familial, and the 1991 *Encyclopedia of Genetic Disorders* states that "between 15 percent and 30 percent of all breast cancers are thought to be genetically determined."[138] Other studies suggest a much smaller role for heredity. A 1993 study analyzing 118,000 women ages thirty to fifty-five found that a woman's risk of developing breast cancer increases by 80 percent if her mother has had the disease, and by 130 percent if a sister has had it. Since a woman is as close—genetically speaking—to her mother as her sister, the higher correlation with one's sister suggests that a common familial environment may be responsible for part of the sibling concordance. The study thus concludes that "only 2.5 % of breast cancer cases are attributable to a positive family history."[139] A 1992 study of 989 pedigrees collected by the Icelandic Cancer Registry, the oldest and most complete breast cancer registry in the world (dating back to 1910), reached a similar conclusion.[140] Lung cancer genetics also continues to arouse controversy: a 1994 analysis of 16,000 male twin pairs by the National Cancer Institute and National Academy of Sciences found—contrary to the expectations of most molecular epidemiologists—that there was "little if any effect of inherited predisposition on development of lung cancer."[141]

Such discrepancies can arise from reporting errors, differences in modeling methods, or just plain hype. The science journalism in this area is often overwrought, fostering the mistaken impression that cancers are often hereditary, passed from parent to offspring regardless of the environment to which one is exposed. That there are some such cases of highly penetrant cancer, no one can deny. But again, the frequency of such cases remains in dispute.

Also in dispute is what should be done. Skolnick draws clear policy implications from the prospect of predispositions: if suitable markers can be found, "we could use a simple blood test to screen the entire U.S. population. Those with the gene or genes would know they are carrying within them a potentially dangerous genetic defect. They would be warned to get regular checkups, and avoid the kinds of high-fat foods thought to trigger the cancer." Skolnick also states that "one-third of Americans stand some risk of developing colon cancer, while the other two-thirds probably aren't at risk at all."[142]

Quite apart from the logistical difficulties of screening "the entire U.S. population," there are a number of questions one can raise about such statements. Once again, the figures commonly given for the frequency of cancer predisposing genes are somewhat speculative. The statistical models used to generate such figures (most recently for breast and lung cancers) are designed to measure the extent to which cancer runs in families,[143] but, as previously mentioned, it is notoriously difficult to control for the fact that families often share common exposures to carcinogens (through the "heritability" of occupation or household environment, for example). There is furthermore little evidence for the claim that two-thirds of Americans "probably aren't at risk at all" to a disease such as colon cancer. Such a claim presumes a small number of predisposing genes when the actual number might be quite large, producing a continuum of differential susceptibility rather than a simple yes or no, susceptible or not. Such a claim also ignores the role of chance. Doll and Peto in their 1981 *Causes of Cancer* make a convincing case that most people who are exposed to carcinogens and yet do not develop cancer are probably just lucky.[144]

There is no evidence that a sizable fraction of the American population is invulnerable to cancer—though we are likely to hear more expressions of what might be called the ideology of invulnerability in years to come.[145] We will also probably continue to see genetics being used to explain away cancer clusters, as was recently done when high rates of breast cancer—15 percent above the national average—were found on Long Island. The Centers for Disease Control and Prevention reported in December 1992 that the excess was due primarily to the high percentage of Jewish residents, known to have higher rates of breast cancer than other Americans. Critics attacked this exercise in genetic exculpation and demanded further studies, which are ongoing at this time.[146] Outrage over the government's complacency led to the formation of One in Nine, a breast cancer activist group that has called for a shift in research focus from treatment and early detection to exploration of causes and possible means of prevention.[147]

The practical implications of having a widely available test for breast cancer susceptibility remain to be seen. Fears of genetic disease are likely to increase in any event. Herbert Snow as early as 1885 recognized that the presumption of a heritable predisposition could hang over a person's head like a sword of Damocles; Snow believed that the mental depression resulting from such a belief might itself lead to cancer, as yet another illustration of "the well-known tendency of a prophecy to work out its own fulfilment."[148] In our own times, fear of cancer has led many women to change how they view their bodies. Many women diagnosed with a tumor choose to have the entire breast removed, even though study after study has shown that women with small- to moderate-size tumors fare just as well with a lumpectomy—removal of just the tumor, leaving the breast intact. Why are so many women willing to undergo more radical procedures needlessly?[149] Gina Kolata in a recent *New York Times* report suggests that physicians may have psychological difficulty "letting go of the methods and philosophy of treatment that they learned in medical school," and that some may even be subtly influenced by the extra money they earn by performing a mastectomy, often presented as the gold standard of breast cancer treatment. Fear of radiation, a common postoperative therapy, may be involved, but Kolata suggests that women are often "irrationally afraid of keeping their breasts." Fear of cancer has led many women to ask to have their breasts removed, thinking this will increase their chances of survival.[150] Ruth Hubbard, professor emeritus of biology at Harvard University, believes that the fear is partly physician-fostered: the traditional medical focus on female breasts as "time bombs waiting to go off" has led many women to devalue or even fear this part of their bodies. Juliet Wittman, author of a moving account of her own struggle with cancer, writes that "my breasts were dangerous. They could harbor my death. At times I wished I'd had the courage to have them cut off" (she chose lumpectomy instead).[151] No such fear is attached to the lungs, even though lung cancer is the most common cause of cancer death among both men and women. Men have not yet begun to fear their prostates, though early signs of such fears are on the horizon.

The prospect of identifying genetically anchored cancer predispositions promises to increase these fears. Jane Brody of the *New York Times* tells the story of a forty-three-year old New York publishing executive who had both breasts removed after a precancerous growth was discovered in one breast. Her physician suggested waiting and watching, but "After a few months, this watch-and-wait technique was making me increasingly nervous. I felt like I was sitting on a powder keg and I just couldn't live with that." She had both breasts removed and replaced with implants, and apparently has never regretted her decision.[152] Prophylactic breast removal is not yet common, but it may become so if predisposition diagnostics become available and if surgeons promote the operation. Women in the breast cancer families being tracked by laboratories across the country have, on occasion, demanded to be told whether they are at risk: in one famous case at the cancer clinic run by Barbara Weber and Francis Collins at the University of Michigan, a young woman from such a family showed up at Weber's office and announced that, since her mother and sister had already died from breast cancer and another sister had just been diagnosed, she could no longer stand the fear and had scheduled a prophylactic bilateral mastectomy for the next week. Weber and Collins knew the woman was not a carrier and decided to tell her, contrary to their experimental protocol barring disclosure. The woman canceled the operation and celebrated with her family, though *Science* magazine reported that the woman was later "devastated by survivor guilt" similar to that felt by survivors of Nazi concentration camps. Collins himself, the newly named director of the National Center for Human Genome Research, reported that even he was sometimes "terrified" by the newfound powers he and his colleagues were unleashing.[153]

NATURE VERSUS NURTURE AND THE IDEOLOGY OF INVULNERABILITY

Scientific attention always comes at a certain cost: the decision to investigate one area is simultaneously a decision to ignore another, and this is as true in cancer research as anywhere else. One problem with the recent fascination with predisposition studies is that they too easily distract from the fact that cancer is a disease whose incidence varies according to diet, occupation, socioeconomic status, and personal habits such as smoking. Genetics may eventually tell us who is more likely to get certain kinds of cancer—everything else being equal, which it rarely is—but it will never tell us anything interesting about large-scale shifts in cancer patterns, patterns that have to do with changing exposures to carcinogens. Lung cancer rates, for example, have risen dramatically over the course of the century; genetic propensities can have little to do with such increases, or with the slight decline observed among males in recent years. Genetic propensities will not tell us why stomach cancer was more common in the first half of this century than in the second.[154] Barring diagnostic artifacts, such dramatic changes lead one to suspect environmental changes as the primary culprit. Carcinogens work on predispositions, but this should not overshadow the fact that the majority of cancers are environmentally induced and therefore preventable.[155]

A similar argument can be made for the differences in cancer rates found among racial groupings. American blacks, for example, suffer significantly higher cancer rates than American whites. But, as a National Cancer Institute study revealed in the spring of 1991, poverty and racial discrimination—not biological race—are the primary causes of that difference.[156] And surely there are many other patterns for which genetics will be irrelevant. Genetics is not going to explain why asbestos miners suffer higher mesothelioma rates than people who work in air-conditioned offices, or why people who live in homes with radon seepage are more likely to contract cancer. Even if individuals vary in susceptibility to such agents, it is probably wishful thinking to imagine that physicians may one day be able to assure people, from their genetic profiles alone, that they are or are not at risk for common diseases such as cancer. Nancy Wexler, president of the Hereditary Disease Foundation and chair of the ethics group of the Human Genome Project, has recently suggested that:

> As geneticists learn more about diabetes or hypertension or cancer, at some point they will cross an important line. Instead of saying, as they do now, "Lung cancer

runs in your family and you should be careful," physicians will be able to ask their patients, "Would you like to take a blood test to see if you are going to get lung cancer?"[157]

But if the majority of lung cancers are environmentally induced, then physicians are unlikely ever to be able to predict cancers of this sort at an early age—except perhaps for the small percentage for whom a clear, inborn, germ-line genetic defect with fairly high penetrance can be discovered. It is misleading to suggest that physicians will ever have this power.[158]

Recall again that there are two types of cancers, or rather two different ways cancers may originate: somatic cancers arise from genetic transformations in some particular bodily tissue (caused by exposure to mutagens, for example); germ-line, or heritable, cancers are passed from generation to generation through the genetic information in the germ cells. The genetic defect is distributed differently in the two cases. In somatically induced cancers, the genetic malfunction lies only in the injured cells of the tumor. In heritable cancers, the genetic defect will be found in every tissue of the body. The distinction is not always clear-cut: some genes confer an increased susceptibility to cancer, the ultimate trigger for which is some environmental mutagen or cellular proliferating agent. A given cancer (retinoblastoma, for example) may have both familial and sporadic (somatic) forms—the distinction being only in the timing and location of the mutations required to launch the malignancy. The appropriate therapies for heritable and somatic cancers may be indistinguishable. Still, the root cause of cancer in the two extreme cases is quite different. Highly penetrant germ-line cancers will be expressed regardless of the particular environment to which one is exposed; somatic cancers are triggered by some postnatal environmental insult.[159]

The important question of policy interest, though, is: What causes a gene to mutate in the first place? Much has been made of the fact that carcinogenesis involves genetic changes: J. Michael Bishop states that oncogenes may represent "the final common pathway" by which carcinogens act, and Robert Weinberg of the Whitehead Institute states that "the roots of cancer lie in our genes."[160] Improved therapies may well emerge from investigating precisely what biochemical functions are curtailed by the loss of tumor-suppressor genes, but there is little evidence for Renato Dulbecco's optimistic hope that knowledge of genetic mechanisms may soon lead to "a completion of our knowledge about cancer, closing one of the most challenging chapters in biological research."[161] If, as most often appears to be the case, cancer begins through some kind of environmental insult—exposure to ionizing radiation or to one of the many chemical carcinogens in our air, food, or water—then the fact that oncogenes must be activated and suppressor genes turned off does not alter the fact that, from a long-run societal point of view, prevention is probably going to remain the best way to approach the problem of cancer.[162]

One potential danger, then, of an imprudent focus on genetic mechanisms is that attention to disease causation may be shifted from carcinogens in the environment to biological defects in the individual. The language by which the heritable is in contrast with the environmental reveals this shift: colon cancers, for example, are commonly classed as either "familial" or "sporadic." The latter term, according to my Word-Perfect 5.0 dictionary, calls to mind the "erratic, infrequent, intermittent, irregular, and occasional." But if only 10 to 15 percent of all colon cancers are inherited, how do the remaining 85 to 90 percent that are not inherited come to be regarded as "sporadic"? Jerry Bishop and Michael Waldholz, in their book celebrating the sequencing of the human genome ("the most astonishing scientific adventure of our time"), contrast the well-behaved cancers produced by genes with cancers arising "as a result of some random series of events, possibly the consumption of a high-fat diet or some other exogenous factors"[163]—but why, again, is diet considered random? The presumption is apparently that heredity is orderly, while environmental causation is chaotic, perhaps even indecipherable.[164] There is a genetic hubris involved here, as when Bert Vogelstein in 1988 stated that "only a decade ago cancer was thought of as a black box in which nothing could be seen."[165] *Avant moi le néant!* The presumption is that environmentalism has been tried and failed. Mary-Claire King confesses that her interest in genetics springs from her conviction that cancer is an unavoidable accompaniment of an affluent society: "If there were something we could do to prevent breast cancer, I would not be doing genetics." Genetics offers hope for new forms of therapy, but it also seems to imply a resignation with regard to the possibility of prevention.[166]

There is also the practical question of how knowledge of genetic susceptibilities will be used. Geneticists often claim that identification of persons especially at risk will allow us (who are we?) to determine "who will benefit maximally from treatments designed to manipulate the environmental causes of those conditions."[167] All of us will be urged to stop smoking, but people already on the genetic road to cancer will apparently get an extra dose of advice. John Minna, director of the National Cancer Institute's Naval Medical Oncology Branch, in 1989 asserted that "if we could tell people that they have a 100 percent chance of getting lung cancer if they smoke, we could focus our anti-smoking, anti-radon, and anti-asbestos efforts on them."[168] One can wonder how efficacious such fine-tuned advice will be; one can also wonder whether advice will be the only consequence of the push to identify susceptibilities. *Consumer Reports* in its July 1990 cover story on genetic screening warned that "The danger is that industry may try to screen out the most vulnerable rather than clean up an environment that places all workers at increased risk."[169] The dangers of discrimination in jobs and insurance have been persistent themes in the literature produced by the NIH Working Group coordinating research on the ethical, legal, and social implications of genome mapping and by the Boston-based Council for Responsible Genetics.[170]

Apart from the prospects of job discrimination and of persons diagnosed with cancer genes being denied or priced out of life insurance and health insurance (genetic defects may be considered "pre-existing conditions"), there is the more subtle phenomenon of genetic exculpation, exemplified in the efforts of tobacco lobbyists to argue that smoking causes cancer only in those persons for whom there is already a genetic predisposition.[171] The recent discovery of a gene triggering the onset of lung cancer (by converting hydrocarbons into carcinogens when exposed to cigarette smoke) prompted the discoverer to assert: "If we could identify those people in whom this gene is easily activated, then we could counsel them, not only not to smoke, but to avoid exposure to certain environmental pollutants."[172] The danger, again, is that attention will be drawn to defects ("genetic lesions") in the individual rather than to shortcomings of the industrial product or environment. The risk is what might be called an ideological one: if the (mis)conception grows that nature is more influential than nurture in the onset of certain diseases, lawmakers may find themselves less willing or able to enact pollution-prevention or consumer-protection legislation. Massive financial support for the effort to map and sequence the human genome is likely to enhance the perceived role of genes; human geneticists and counselors may take on increasing obligations of medical micromanagement, ushering in an era of medically managed genophobia and a risk of coercive public health.

It is not always easy, of course, to separate out nature and nurture in such matters, given that genetic predispositions can operate in a very roundabout or indirect fashion. Dark skin, for example, is not directly associated with cancer, but people with dark skin—people of African or South Asian or Australian aboriginal descent, for example—generally suffer less from skin cancers because their darker skin protects them from ultraviolet radiation. People without a Y chromosome (women, in other words) are far more likely to contract breast cancer and cannot, of course, get cancer of the prostate. The causal chain may be circuitous: people with dark skin in the United States are more likely to live in areas exposed to high levels of carcinogenic toxics, and people with Y chromosomes are more likely to work outdoors and therefore fall victim to skin cancer (cancer atlases show that the disease is more geographically localized for men than for women, apparently because men perform more outdoor labor and are therefore more likely to get skin cancer). Asbestosis is obviously not a genetic disease, but whether it progresses to lung cancer might certainly be shaped, at least in part, by genetic factors influencing how well one is able to expel the fibers or repair the damage caused by those fibers. The same is likely to be true for people responding to dietary carcinogens (fat, for example): it is wrong, as a recent review put it, to regard people as "genetically identical rodents."[173]

Finally, there is the problem that diagnostic capabilities are likely to far outstrip therapeutic possibilities. The long-term hope is that mapping and sequencing cancer genes will allow the development of therapies to counter the action of wayward genes, but genetic therapy is a much more daunting prospect than mapping and sequencing genes and identifying protein products. Tumor cells tend to be genetically heterogeneous; that is one reason they succeed so well in the competitive "struggle for existence" that includes overcoming the harsh threats of therapeutic agents.[174] There is also the thorny question of how to deliver a corrected gene or protein product to its proper site. How many years will it be before successful cancer therapies are developed? How much of the promise will pan out, and how much is just the latest version of the cancer community's perennial just-around-the-cornerism? The promise of a cure—or even a modestly successful treatment–may well come true,[175] but until then our condition will remain what one might call a state of enlightened impotence—with ever more precise knowledge of our fate but little knowledge of how to control it.

From *Herblock on All Fronts* (New American Library, 1980). Reprinted with permission from Herblock Cartoons.

Conclusion

How Can We Win the War?

*While much is known about the science of cancer, its preven-
tion depends largely, if not exclusively, on political action.*
                                                    —Samuel Epstein, 1978

THE PEOPLES OF WEALTHY industrialized nations suffer from diseases that
are unlike those suffered by poorer peoples. Prior to the twentieth century, as is still
the case in many of the poorer parts of the world today, the most common cause of
death was infectious disease. Most people died young, though survival past childhood usually meant one had a reasonable chance of living to a ripe old age. Today, at
least in wealthier parts of the world, people suffer more from chronic maladies like
heart disease, cancer, mental illness, and stroke. Infant mortality is a rarity, and most
people die old, from diseases that spring from a lifelong accumulation of bodily insults. There are several important exceptions to this generalization: AIDS, for example, is a global infectious plague that strikes both young and old, and accidents are
still distressingly common. Homicide tends to strike the relatively young (especially
males), but not the very young or very old. As serious as such misfortunes are, they
account for only a small percentage of deaths in industrial nations. Cancer and heart
disease are the two most common causes of death among Europeans, Americans,
and the urban elites of Third World nations, accounting for about 60 percent of all
deaths. Ours is an age of chronic disease. Cancer is the emblem of this situation, a
disease of old age, of last resort, of First World poverty and First World affluence.

Cancer, though, is also unlike many other diseases in that it has proved largely impervious to the scientific and political assaults launched against it. Heart disease mortality in the United States declined by 55 percent from 1950 to 1987; age-adjusted
cancer death rates over this same period climbed about 6 percent. Improvements with
regard to heart disease are generally attributed to changes in lifestyle and progress in
medical treatment,[1] but where is the comparable success story for cancer?

Part of the problem must be traced to the failure to realize early on that cancer
differs in fundamental ways from the terrible infectious scourges of the nineteenth
century. Late-nineteenth-century expectations were that a "cancer germ" might be
found against which a cancer therapy could be devised. Antibiotics would be deployed to fight those germs, just as powerful new vaccines eventually allowed millions to live free of the fear of polio, smallpox, and other former killers. Comparable
heroic acts were expected of enlightened cancer specialists. In the 1950s and 1960s,
hopes were still high that cancer viruses might be identified against which a cancer

vaccine could be produced. Huge sums of money were poured into these hopes, with little payoff.

As we now know, cancer is unlike infectious disease in its mode of attacking the body, its latency prior to expression, and its circumvention of the body's normal lines of immunological defense. Few infectious diseases show a latency period of ten, twenty, or even thirty or more years between the onset of disease and evidence of symptoms (AIDS being the most notorious exception). The two kinds of disease also exploit different aspects of the body. Infectious germs attack the weakened body, but cancers seem to grow as well in healthy as in infirm bodies. In the absence of treatment, most cancers will eventually kill their otherwise vigorous hosts. This inexorable, juggernaut quality is one of the most fearsome aspects of the disease.

One of the great achievements of medical historians is to have shown that the infectious scourges of the previous century were conquered mainly by improvements in nutrition and sanitation, shorter working hours, and a general improvement in living standards. The more rapid transport of foods, the systematic separation of drinking water from sewage, and improved work conditions were far more important to halting the spread of disease than novel forms of medical intervention.[2] Some of these changes have become the stuff of legend. John Snow removed the handle from the public water pump on London's Broad Street in 1854, easing the cholera epidemic that had plagued the city. Physicians attending childbirth began to wash their hands, ending an era of childbed fever. Great Britain's colonial armies were saved by provisions of fresh meat and vegetables. Sailors were given lime juice, ending the threat of scurvy (hence the nickname "limeys"). Increased leisure time relieved work stress and allowed recovery from otherwise crippling illnesses. Medicine, as it is normally understood, appears to have played a relatively minor role.

Medical historians have inferred from such examples that the best hope for the world's chronic ailments is prevention rather than cure. But there are many obstacles to its effective practice.

ARE WE WINNING THE WAR?

Official cancer bodies in the United States have long tried to appear upbeat in their presentation of how we are faring in the war against cancer. The American Cancer Society glows regularly with news from the front, assuring potential contributors that the "revolutionary discoveries" of molecular biology have left us "on the threshold of a kind of 'golden age' of medicine."[3] The society often reminds the war-weary public that cancer can be cured, though the paths out of the darkness are always just around the corner. Optimism usually follows breakthroughs in basic biology: with the discovery of oncogenes, for example, Dr. Lewis Thomas, chancellor of Memorial Sloan-Kettering Cancer Center in New York, said that the end of cancer could come "before this century is over."[4] Governmental bodies are equally sanguine. In 1986, the National Cancer Institute celebrated the fact that the U.S. cancer death rate had fallen by about 7 percent between 1974 and 1984 among people under the age of fifty-five. Vincent T. DeVita, director of the institute at that time, called the decline "one of the most encouraging cancer statistics we see this year," one of many "signs of progress."[5] That same year, the NCI announced a goal of reducing U.S. cancer mortality by 50 percent by the year 2000; DeVita expressed his hope that "by the year 2000 we may not have cancer. . . . The speed of advance has been enormous."[6]

Cancer institutions tend to highlight the good news and downplay the bad. Critics counter that the good news may be "seriously skewed" by virtue of selective reporting. This was the judgment of John C. Bailar and Elaine M. Smith of the Harvard School of Public Health and the University of Iowa Medical Center in their 1986 review of progress against cancer for the *New England Journal of Medicine.* Bailar and Smith pointed out that the NCI had correctly noted the decline in death rates for people under fifty-five but ignored the fact that rates were rising for people fifty-five and older: "by making deliberate choices among these measures, one can convey any impression from overwhelming success against cancer to disaster." They were equally skeptical of the NCI's goal of reducing cancer mortality by 50 percent by the year 2000. Plotting cancer death rates over time, they showed that the NCI goal presumed a dramatic break with previous trends—an unlikely proposition. The authors concluded rather starkly that "some 35 years of intense effort focused largely on improving treatment must be judged a qualified failure."[7] A 1989 review of West German statistics concluded similarly that "little benefit" could be attributed to improvements in cancer treatment, and that medical interventions were "unlikely to have an impact on overall cancer mortality statistics" in the near future. Prevention, the review concluded, was likely to be more successful in reducing mortality.[8]

Cancer rates have continued to rise, and there is little reason to believe the trend will soon reverse itself. From 1958 to 1988, incidence rates for several U.S. cancers showed dramatic increases: lung cancer is up 425 percent for males, 121 percent for females; kidney cancer is up 35 percent in males and 22 percent in females. Men have experienced dramatic increases in cancers of the colon (up 21 percent), esophagus (up 20 percent), prostate (up 12 percent), and skin (up 126 percent). Incidence rates have outpaced death rates: according to the National Cancer Institute, the age-adjusted death rate for all forms of cancer rose by 3 percent between 1950 and 1988; over this same period, the age-adjusted incidence rate rose an alarming 44 percent.[9]

The situation is even worse for African Americans: from 1956 to 1986, African-American cancer death rates rose 66 percent for males and 10 percent for females. Cancer of the mouth was up 70 percent for men and 30 percent for women; lung cancer was up 259 percent for males and 440 percent for females. These are all age-adjusted rates, and are therefore unaffected by the fact that people are now living slightly longer.[10] Incidence rates show an even larger rise. Breast cancer incidence, for example, increased 29 percent among U.S. whites and 41 percent among U.S. blacks over the period 1979 to 1986. A 1991 review in the *American Journal of Public Health* concluded that increasing use of mammographic screening accounted for only 40 percent of the reported increase among whites and only 25 percent among blacks.[11]

Smoking is no doubt responsible for much of the overall upsurge, but there are also increases that do not seem to be linked to smoking. Incidence rates for multiple myeloma, non-Hodgkin's lymphoma,[12] and cancers of the brain, breast, and testicles have all risen in the past twenty years, though none of these has ever been tied to smoking. Smoking doesn't seem to be responsible for the recent upsurge in esophageal adenocarcinoma, whose incidence tripled between 1976 and 1990.[13] Diagnostic improvements and better access to hospitals must account for part of the increase, but studies of what is known as ascertainment and enumeration consistency indicate that not all the increases are diagnostic artifacts.[14] John Bailar pointed out in 1994 that, among the fourteen types of cancer showing increasing mortality rates over the period 1973 to 1990, the second largest increase was in "site unknown," suggesting that improvements in determining the site of origin are probably not inflating the death rates for specific kinds of cancer.[15]

If the case is so clear that cancer is on the increase, why do disagreements persist about whether we are winning the war?

Part of the problem stems from conflicting standards for progress against the disease. Medical optimists like to see progress measured in term of five-year survival rates—defined as the percentage of patients who manage to live for five years after diagnosis. By this measure, some slight progress has indeed been made. The American Cancer Society reminds us that in the 1930s, not even one in five persons treated for cancer would still be alive five years after treatment. By the 1960s this had risen to one in three, and by 1993 to two out of five. Adjusted for normal life expectancy (people no longer die as often from other causes such as heart disease or accidents), the "relative" survival rate is even higher.[16]

Critics remind us, though, that survival rates tend to give an overly rosy view of the war on cancer. The higher cure rates are not much to celebrate if more and more people are getting the disease. Survival rates can also be deceptively high if more people are going to doctors for easily treated cancers. Skin cancer, for example, is the most easily treated form of cancer, and its increasing prevalence, thanks to tanning fashions and the abandonment of hats in the 1960s and 1970s (Kennedy caused cancer!), is one reason overall survival rates have "improved" over previous decades. The National Cancer Institute estimates that since President Nixon launched the war on cancer in 1971, survival rates have improved by about 4 percent, but Bailar and others claim that even this number overestimates the actual improvement, since the same kind of bias that inflates incidence rates also produces a deceptive growth in survival rates. An increasingly thorough registration of easily treatable cancers will augment the proportion of overall cases that survive for five years after diagnosis.[17]

Survival rates are only one of several ways to measure progress against a disease. Survival rates will not tell you if chemotherapy or radiation is less unpleasant than it once was, or whether surgery leaves the typical cancer patient less disfigured than it once did. Survival rates do not tell you how attentive physicians are to the creature comforts of cancer patients, or whether patients are less likely to die from a cancer caused by medical treatment (such as X rays or powerful chemotherapeutic poisons). They also don't tell you whether a person born into the world today is more or less likely to die of cancer than that child's parents or grandparents. Death rates may actually be rising at the same time that survival rates are improving—which pretty well summarizes the twentieth-century history of cancer. Incidence rates have grown so fast that they have outpaced improvements in survival rates. Of course, growing incidence rates would not be so troubling if cure rates were growing even more rapidly. Most of the nineteenth-century infectious diseases that killed children are still around, but most are easily cured or preventable by vaccination. This is not the case, however, with cancer. Lung cancer incidence rates are very close to lung cancer death rates: 90 percent of those who contract the disease die from it. Mesothelioma incidence and death rates are even closer: very few people (about 2 percent) survive for five years after diagnosis. Childhood cancers are a thankful exception: incidence rates are up more than 20 percent since 1950, but death rates have dropped by about 60 percent—mainly due to improved methods of treatment.

Death rates are important, but they don't necessarily tell us much about how we're doing in the struggle to eliminate the *causes* of cancer. For this, incidence rates—if accurate—are more revealing. Incidence rates give us a relatively pure measure of cause, regardless of how many people live or die after treatment. Incidence rates tell us how many people are getting cancer in the first place. (Comparing incidence and death rates also tells us something about the quality of treatment people receive: African Americans, for example, have a lower incidence of breast cancer than women of European descent, but are more likely to die of it.[18]) Again, if cancer were easily cured, then rising incidence rates would not be so disturbing; but for a disease where cure rates are low, finding ways to lower incidence should be a major focus of our attention. Prevention of incidence—versus prevention of death, which is the primary goal of treatment—is also preferable, insofar as it is not just cancer that is thereby prevented but also the suffering from the pain and side effects of treatment, not to mention medical costs and the ever-present specter of recurrence subsequent to the five-year fiction used to define "survival."

Incidence rates, however, are not always easy to obtain, and are often less reliable than death rates. In the absence of a comprehensive tumor registry, many tumors will go unrecorded or will be improperly diagnosed.[19] Recording cause of death is a legal obligation in nearly every industrial nation, and though underreporting can occur (a person with advanced colon cancer may finally succumb to a heart attack, for example), death rates are the more common focus of attention among cancer researchers. Those who argue that the war against cancer has failed point to the fact that overall death rates have steadily increased in recent decades. Bailar and Smith thus concluded, in 1986, that "the ugly fact remains that overall cancer mortality is rising. . . . This cannot be explained away as a statistical artifact obscured by the clear evidence of progress here and there, or submerged by rosy rhetoric about research results still in the pipeline."[20] At a September 22, 1993, meeting of President Bill Clinton's Cancer Panel, Bailar repeated his claim that "our decades of war against cancer have been a qualified failure."[21]

Bailar's gloomy conclusions have been challenged by critics who point to hopeful signs in several European nations that mortality rates have finally peaked or begun to decline. Lung cancer rates have in fact begun to fall, most likely due to the decline in smoking that began in the late 1960s.[22] German women's cancer mortality rates have fallen steadily since the 1950s—perhaps partly due to the fact that Nazi social policies discouraged smoking while encouraging physical fitness and early childbearing, but perhaps also to the fact that the protracted war and postwar poverty resulted in a low consumption of fat.[23] In a 1990 review, Richard Doll pointed out that while cancer mortality rates for men in Hungary, Italy, and France are well above what they were in the 1950s, the increase has stopped or even reversed in Germany, Ireland, England, and Wales. Among women, mortality rates have increased in Italy, Hungary, England, and Wales, but have dramatically declined in France and Germany. His conclusion: "we are, for the most part, winning the fight against cancer."[24]

Central in the dispute over whether the war is being won or lost has been the question of how one should regard the apparent increase in cancer among the elderly. In their 1981 *Causes of Cancer,* Doll and Peto ignored persons sixty-five and older on the grounds that medical records are often unreliable for this group. The Oxford team continues to argue that generalizations about long-term trends for the elderly are difficult to make in an unbiased manner, given the improvement of diagnostic techniques (notably CT scans). The fact that more and more people are visiting their physicians gives statisticians yet another chance to record a given cancer, and that, they say, is precisely what happened after the enactment of Medicare in the mid-1960s. Doll and Peto point out that if cancers were in fact becoming more common among the elderly, this would presumably be the result of some environmental agent—but why would we not also see an increase among the young? Cancers require fifteen or twenty years to begin to show up, but not more: "There's no example of any cancer that requires longer than 15 or 20 years to reach a fairly high incidence as a result of exposure to a carcinogen. You're not going to have an epidemic occurring 30 years after first exposure."[25] Doll concedes that a few cancers not associated with smoking are growing—such as melanoma, multiple myeloma, and non-Hodgkin's lymphoma. But these, he suggests, are minor blips in the general trend—lung cancer excepted—of a steady if slight decline in overall cancer death rates.

Devra Lee Davis, senior adviser to the assistant secretary of Health and Human Services, counters that it is wrong to ignore the elderly, especially given that more than half of all cancers occur among the sixty-five and older group. (These are also people with longer histories of exposures to carcinogens.) The diagnostic bias is furthermore not, according to her, as serious as Doll and Peto suggest, since death rates from several of the "unspecified" causes that might be masking cancer deaths have also risen. Brain cancer remains her preferred piece of evidence: she notes that brain cancer rates are five times higher in the United States than in Japan, leading one to suspect environmental factors as responsible for both the geographical disparity and the rising rates over time.[26] Davis has been supported by colleagues who confirm her view that diagnostic shifts alone cannot explain all of the increase. Anthony Polednak of Yale has shown that even if one excludes cases confirmed by radiography alone, one still finds disturbing increases in brain and central nervous system cancer. Drawing from the data of the Connecticut Tumor Registry, the oldest such registry in the United States, Polednak has shown that among people aged seventy to seventy-four, brain cancer incidence has grown from about 10 per 100,000 from 1965 to 1969 to about 22 per 100,000 from 1985 to 1988.[27] The East German Cancer Registry, one of the most comprehensive in the world (now absorbed by the Federal Republic's Deutsches Krebsforschungszentrum), shows a similar rate of increase over the period 1977 to 1987:

from 4.4 to 5.4 per 100,000 for males and from 3.2 to 4.1 per 100,000 for females.[28]

The question of long-term trends remains controversial within the cancer establishment (see figure). A central question, broached time and again, is how much of the overall increase can be attributed to smoking. Peto estimates that roughly one in five persons alive today in the wealthier parts of the world will die from a tobacco-associated illness. Davis counters that smoking (along with sunshine) is being used to explain away all cancer increases when it is not yet clear that occupational exposures and environmental pollutants are negligible. She points out that even if smoking is associated with as many as 87 percent of all lung cancers, the remaining 13 percent still exact a considerable toll. Thirteen percent of a very large number (there are 150,000 annual lung cancer deaths in the United States) is still a large number, large enough to make lung cancer among nonsmokers the fourth largest category of U.S. cancer deaths. Apart from smoking-associated lung cancer, only cancers of the breast and colon in women and prostate and colon cancer in men claim more victims.[29] Davis also notes that for a given level of smoking, men appear to have a much higher chance of developing lung cancer than women, leading one to suspect that men may encounter risks—workplace pollutants, for example—to a greater extent than women. Doll and Peto worry that such speculations distract from the fundamental issue; Alan D. Lopez argues similarly that Davis's work is "dangerous in the sense that it draws the principal public health concern away from cigarettes."[30]

Breast cancer, interestingly, is an area where the politics of trends is more subtle. The Connecticut Tumor Registry indicates a steady rise of about 1 percent per year in age-adjusted breast cancer incidence since the 1940s. The American Cancer Society shows a comparable 21 percent increase in U.S. breast cancer incidence from 1973 to 1989, giving rise to concerns that we are in the midst of a breast cancer epidemic. (Mortality rates have held pretty much steady, despite the increasing incidence.) Angered by the snail's pace of progress against the disease, activists in the early 1990s launched the National Breast Cancer Coalition, uniting 180 advocacy groups across the United States. Using grassroots organizing techniques pioneered by AIDS activists, breast cancer activists managed to persuade the U.S. Congress to increase breast cancer funding by nearly $300 million, including an unprecedented $210 million pledge from the U.S. Army.[31] No one is really sure what lies behind the rise in incidence, but there is disagreement even over whether it is wise to press the argument for increase. Ruth Hubbard of Harvard, for example, a long-standing women's health advocate, worries that the specter of an epidemic is being used by medical authorities to frighten women into massive and often unnecessary screening. She concedes that breast cancer has been neglected, but she also worries that the hype surrounding the disease is being used (by diagnostic firms promoting mammograms, for example) to further medicalize the female body, encouraging screening but not prevention. This view is supported by the fact that Zeneca, a UK-based pharmaceutical firm, is both the manufacturer of the world's best-selling cancer drug (Nolvadex, or tamoxifen citrate, with sales of $470 millon per year) and the founding sponsor of National Breast Cancer Awareness month, with its annual media barrage urging women to "get a mammogram *now.*" *Mother Jones* recently pointed out

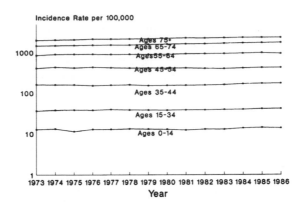

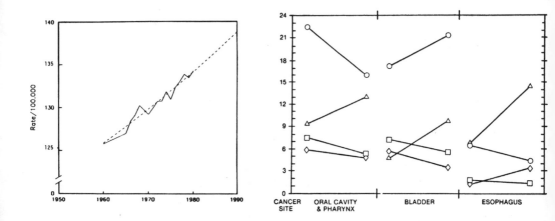

Three views of cancer trends representing, counterclockwise from the top, typical chemical industry, environmental, and tobacco industry views of cancer trends. Note the logarithmic scale in the topmost (chemical industry) figure: the American Industrial Health Council suggests that this "yields a truer picture of time changes than do arithmetic scales." The bottom (tobacco industry) figure compares the rates of three kinds of cancer for the periods 1947–49 and 1969–71.

Source: Reprinted with permission from the American Industrial Health Council's Epidemiology Subcommittee, "How Are Cancer Rates Changing?" *Consumers' Research* (April 1993): 24; Marvin A. Schneiderman, "Trends in Cancer Mortality," in *Statistics in Medical Research,* ed. Valerie Miké and Kenneth E. Stanley (New York: John Wiley & Sons, 1982), p. 92; and a Tobacco Institute publication from the early 1980s reprinted in Elizabeth M. Whelan, *A Smoking Gun: How the Tobacco Industry Gets Away with Murder* (Philadelphia: George F. Stickly, 1984). All figures are derived from U.S. government statistics.

that GE and Du Pont, rivals for the lead in Superfund toxic sites, sell more than $100 million worth of mammography machines every year (GE) and much of the film used in those machines (Du Pont). Zeneca itself does a $300 million annual business in the carcinogenic herbicide acetochlor. Companies such as these profit, so to speak, from both ends of the cancer cycle.[32]

Whether or not overall cancer incidence rates are still are on the rise, they are surely high enough to be cause for alarm. Cancer may not be technically epidemic (rapidly increasing), but it is indisputably endemic (common in the population). The American Cancer Society likes to point out that several of the twenty-five-odd cancers monitored by the society are declining, and one gets the impression that this good news is supposed to balance the bad. The occasional piece of good news is encouraging, but any increase in cancer incidence or mortality should be viewed as a black mark on the society bearing that increase. Even a holding pattern is a shame, given the heights to which rates already have soared.

The war, in short, is not going well. Why not?

COMPETING CONCEPTS OF CAUSE

Causation is a notoriously slippery concept;[33] in our case the problem is magnified by the powerful professional, economic, and activist interest groups competing for a say at every turn. In a toxicological research environment, for example, the question of interest may be whether a substance causes cancer at a specified dose when inhaled or ingested. Causality is also a legal concept, however, where the question of interest is typically whether a particular individual or class of individuals has been injured by a given work environment or toxic effluent.[34] Legal distinctions are sometimes drawn between natural and man-made carcinogens for the purposes of assigning blame and organizing regulation, and such distinctions can end up shaping what gets medical attention and what does not. This multiplicity of meanings can lead to very different avenues of inquiry and imperatives of action.

The questions of interest in molecular biology are often questions that abstract from particular causal agents. The questions are typically: What makes a cancer cell begin to grow? What genetic loci are altered in the process, and what does this do to cellular function? Why does the body's immune system not reject a tumor as foreign, and how might that immunity be reversed? The molecular approach generally looks to processes underlying a broad spectrum of cancers—and not just to causes of specific kinds of tumors (asbestos-induced mesothelioma, for instance). Molecular biologists may question the entire notion of cause, as when Harold Varmus of the National Institutes of Health cautioned that "You can't do experiments to see what causes cancer. It's not an accessible problem, and it's not the sort of thing scientists can afford to do."[35]

The epidemiological approach, by contrast, looks to variations in the practices of specific populations to identify agents or circumstances that might cause cancer. The interest is in why Japanese women have low breast cancer rates, or whether people living near high-voltage power lines develop excess leukemias. The explanations for a particular pattern may be couched in terms of dietary or occupational practices, or in terms of the age distribution or socioeconomic status of the population. The interest is often in knowledge that might lead to policy: epidemiological knowledge is, generally speaking, more immediately useful for prevention than, say, molecular or biochemical approaches, which tend to spin off in the direction of new potential therapies or diagnostics. The distinction is not hard and fast: molecular evidence of the anticancer action of a vitamin, for example, might lead to prevention-relevant knowledge, and not all epidemiological conclusions are of use in prevention. The new field of molecular epidemiology seeks to combine the virtues of both approaches; the hope is that molecular techniques will circumvent some of the inherent limitations of animal tests, giving us better early insights into a population's likelihood of developing cancer.[36]

But how far into the causal chain does one look for the cause of a particular kind of cancer? And what is the relevant expertise? Should the NCI be spending money on the history or philosophy of science, or on cross-cultural studies of anticancer programs? When is a nation's health care or environmental policy (or the absence thereof) the cause of cancer? Historians of medicine teach us that disease can be looked at not just as a consequence of genes or germs or chemical agents, but also of bad habits, bad government, bad business, and perhaps even bad science and bad medicine. Distribution of health and disease follows variations in distribution of wealth and poverty as closely as it follows the distribution of genes, germs, or chemical carcinogens; these associations are effaced when we focus exclusively on, say, the immune system or genetic infirmities or specific chemical or radioactive carcinogens. Disease is not just a result of "bad habits" or the failure to follow good advice; disease is also a social process with social causes and (potentially) social solutions. Two historical examples may help to shed light on what such a broader notion of cause might mean.

In 1846–48, Ireland suffered a potato blight "caused" by the fungus *phytophthora infestans,* which ended only after the starvation of more than a million people and the emigration of another 3 million from the Emerald Isle. What was the cause of the famine?[37] Physiologically speaking, starvation involves a broad range of systemic failures: many of those who died succumbed to infectious diseases such as pneumonia; others died from various and sundry forms of general system collapse. Each of these processes could be described in terms of biochemical deficiencies; one might even find that starvation ran in families and was confined to certain geographic regions. The fungus might be found to have affected potatoes grown in one fashion to a greater extent than potatoes grown in another, and the part played by unusually heavy rainfall in the spread of the blight might be (and has in fact been) studied.

Such an analysis of the "causes" of the Irish potato famine, however, would miss the essence of the disaster. The important question is not the mechanism of starvation or the biochemistry of the offending fungus or even the vagaries of the weather, but rather the social circumstances that gave rise to such vast numbers of people living on the border of starvation in the first place. To discover these circumstances, we would have to look to the social conditions that generated Irish poverty. We would have to look to British policies of the Manchester era, during which affairs of state and industry were governed according to a belief that a competitive struggle of each against all would guarantee the survival of the fittest. We would have to understand what led Thomas H. Huxley to characterize nature as "red in tooth and claw," and what led Herbert Spencer to embrace the dog-eat-dog ethic of the "struggle for existence." We would have to fathom how hunger can exist amid a plenty of food, and why Ireland continued to export grain during the years of famine. Taking this larger view, the election of the Tory government in 1846 or the triumph of Manchester economics or the historical subordination of Ireland to English rule might all be regarded as causes contributing to the famine.

The deadly London fog of December 4–8, 1952, might be similarly regarded. For four days and nights, Britain's largest city suffered a severe atmospheric inversion that left 3,000 to 4,000 people dead and thousands more suffering from bronchitis and pulmonary inflammation.[38] Meteorologically, London's most infamous "fog" was the product of an air inversion that trapped polluted air near the ground, depleting much of the oxygen. But on the thousands of death certificates issued on those days, there was never a mention of air pollution. The cause of death was usually listed as pneumonia, bronchitis, emphysema, or perhaps even heart failure or "old age." Again, as with the Irish famine, a purely physiological understanding of the cause of death would miss much of the point. The more appropriate causes are of a sociopolitical or economic nature. The London fog resulted from a combination of circumstances that included a failure to enact and enforce air pollution laws. Like the dog that did not bark and thereby woke the town, the failure to enact and enforce such laws must be considered one of the primary causes of several thousand people's deaths.

The London fog and the Irish potato famine raise general questions about the nature of disease causation. In each case, one has to look beyond the immediate physiological or biochemical events to the social processes that generate the problem in the first place. Richard Lewontin puts the matter provocatively:

It is undoubtedly true that pollutants and industrial wastes are the immediate physiological causes of cancer, miners' black lung, textile workers' brown lung, and a host of other disorders. Moreover, it is undoubtedly true that there are trace amounts of cancer-causing substances even in the best of our food and water unpolluted by pesticides and herbicides that make farm workers sick. But to say that pesticides cause the death of farm workers is to make a fetish out of inanimate objects. We must distinguish between *agents* and *causes*. Asbestos fibers and pesticides are the agents of disease and disability, but it is illusory to suppose that if we eliminate these particular irritants that the diseases will go away, for other similar irritants will take their place. So long as efficiency, the maximization of profit from production, or the filling of centrally planned norms of production without reference to the means remain the motivating forces of productive enterprises the world over, so long as people are trapped by economic need or state regulation into production and consumption of certain things, then one pollutant will replace another. Regulatory agencies or central planning departments will calculate cost and benefit ratios where human misery is costed out at a dollar value. Asbestos and cotton lint fibers are not the causes of cancer. They are the agents of social causes, of social formations that determine the nature of our productive and consumptive lives, and in the end it is only through changes in those social forces that we can get to the root of problems of health. The transfer of causal power from social relations into inanimate agents that then seem to have a power and life of their own is one of the major mystifications of science and its ideologies.[39]

The distinction between agents and causes is arguably not as sharp as Lewontin makes it out to· be, and one mustn't, of course, allow a social reductionism to dissolve personal responsibilities. The value of the social perspective, however, is that it allows us to broaden our understanding of where one might intervene in the process of carcinogenesis. The natural history of cancer is not irrelevant here. A typical cancer originates in middle age, in one's thirties or forties. Twenty or thirty years later, symptoms begin to appear and a diagnosis becomes possible (breast cancer tumors typically take about ten years to reach the size of a pea). Orthodox forms of treatment are normally administered after the onset of symptoms—in other words, after the cancer has grown large enough to be diagnosed. This is often too late. Cancers are often diagnosed after they have spread to other parts of the body, when therapeutics are least effective. Even locally confined cancers may fail to respond to treatment. For many cancers, the hallowed triad of chemotherapy, surgery, and radiation—"slash, burn and poison," in the words of Susan M. Love, director of UCLA's Breast Center[40]—does little to lengthen the life of cancer victims and can often diminish its quality.

Early detection offers some hope in this regard. For women over fifty, regular mammography clearly lessens the chances of dying from breast cancer. Elderly women stand most to gain, since they are more likely to contract the disease and less likely to be harmed by the procedure (mammograms, one should not forget, are breast X rays). Pap smears, rectal exams, and other routine screenings can dramatically reduce death rates for a number of important cancers. Early detection has improved in recent years, and blood tests are in the process of being developed for several kinds of cancer. This will make it possible to detect tumors earlier than can be done by digital rectal exams or X-ray mammography, to name just two examples.[41]

Early detection is often hailed as the key to survivability, but even this is not the panacea it is sometimes made out to be. It can mean the difference between life and death—this is true for cervical cancer,[42] and for the more common skin cancers. It is, generally speaking, better to have one's cancer detected earlier rather than later: studies have shown that people with little or no access to health care are much more likely to die from easily curable cancers than people with proper health care. For many cancers, however, early diagnosis does little to improve the outcome. A 1993 study of 2,000 men at the University of Chicago showed that many diagnosed with late-stage prostate cancer had passed the standard digital exam within the previous year.[43] Early diagnosis is of little use for lung cancer: screening helps to find the disease earlier, but doesn't seem to improve long-term survival. In some cases screening can even do harm. Irwin D. J. Bross, director of biostatistics at Buffalo's Roswell Park Memorial Institute, predicted in 1977 that the joint ACS-NCI breast cancer screening project could result in "the worst iatrogenic [physician-induced] epidemic of breast cancer in history."[44] The X-ray levels used in standard mammography have since declined, due to stricter supervision and the use of higher-speed photographic film, but criticisms are still raised that too many younger women and not enough older women are being screened.[45] In 1993 the National Cancer Institute dropped its recommendation for routine mammograms for women in their forties, but the American Cancer Society continues to urge the practice. The problem, again, is that detection is often possible only after the damage has been done. By the time breast cancer is detectable by mammography, for example, it has usually been growing for about eight years. Lumps are physically palpable only after about ten years of growth inside the body.[46]

Considerations such as these have led public health activists to ask, What can be done earlier, prior even to the onset of carcinogenesis? This is the question also asked by those who advocate cancer prevention. If we are not soon going to cure cancer, then perhaps we should concentrate more on what can be done to prevent it. Denis Burkitt, an early champion of dietary fiber to prevent colon cancer (and other modern ills, such as diabetes), suggests that we have a leaky faucet, an overflowing sink, and many experts busily mopping the floor. But why, he asks, are there so few trying to turn off the tap?

TWO KINDS OF CONSERVATISM

In Anglo-American law, a person is innocent until proven guilty. The reasoning is that it is worse to convict a person wrongly than to convict no one at all. This legal presumption of innocence has an interesting parallel in classical statistical inference, where the supposition is that a suspected agent is *not* the cause of a given effect. Evidence is adduced to refute the "null hypothesis" (the presumption of no effect), and only if a sufficiently strong case can be made is the null hypothesis rejected. The scientist in this case is the courtroom equivalent of the prosecutor, presenting evidence and arguments to overthrow the initial presumption of innocence. The conservatism of such an approach is thought by some to be necessary in order to preserve the body of reliable knowledge.[47]

In medicine or public health, however, the logic is arguably different. A chemical up for regulation, unlike a person accused of a crime, is not necessarily innocent until proven guilty, harmless until proven hazardous. The burden of proof should be on the defenders, not on the accusers. Conservatism implies a measure of caution in exposing ourselves to a hazard, even if there is uncertainty about the magnitude of that hazard. Prudence often means acting in the midst of uncertainty: in the words of William Drayton, Jr., former assistant administrator of the EPA, regulators responsible for protecting human health and safety "are charged by our society to act before all the evidence is in . . . before every issue is nailed down and ten articles have been published in the rigorous academic literature."[48] To do otherwise, to wait and see, to delay in the face of good but partial evidence, is tantamount to experimenting on humans.[49]

People do not always see eye to eye, of course, on what exactly is to be done in the name of prudence. There is, in fact, a basic conflict of interest between two different kinds of conservatism in cancer research, leading to very different senses of what it means to err on the side of caution. To estimate a hazard "conservatively" in the scientific sense means taking care not to overestimate a hazard: what is conserved in such cases is the body of secure empirical knowledge and perhaps also the scientific reputation of the person making the estimate. Conservatism in the public health sense, by contrast, means taking care not to underestimate or overlook a hazard. These are often contradictory goals: conservatism in the scientific sense can lead to recklessness in the sphere of public health, but conservatism in the public health sense can lead to regulatory excess and damage to the reputations of scientists accused of exaggerating.[50]

The contrast between scientific and public health conservatism usually translates into different perceptions of what is most in need of protection. Most scientists adhere to an implicit empiricism, requiring that they be cautious about making statements of fact. Risks are often expressed as minima and, for that reason, scientific conservatism is sometimes regarded as (and sometimes used as) a ploy to stonewall efforts to limit risks. Scientists who worry first and foremost about the integrity of science are, in fact, more likely to underestimate than to overestimate the hazards of a particular substance or situation. By contrast, administrative agencies such as the EPA are entrusted with protection of public health and safety, and "conservatism" for them generally means erring on the side of caution. Public health conservatism of the kind reflected in the EPA's preference for "worst-case" scenarios is designed to minimize public harm. One can also understand, though, why some scientists may find this irksome or even irresponsible.

Public health conservatism is usually championed by political liberals;[51] scientific conservatism is more often championed by political conservatives. The distinction is also expressed in different branches of government. The U.S. Department of Labor has traditionally served as an advocate for labor, leading it to err on the side of caution in matters of environmental and occupational health. The U.S. Public Health Service, by contrast, has tended to regard itself as a neutral scientific body; its tendency has been to err on the side of minimizing hazards. This epistemological divide was clearly expressed in the original establishment of the National Institute for Occupational Safety and Health not within the Labor Department, along with OSHA, but rather within the Public Health Service. The goal was apparently to create separate spheres for research and advocacy, leaving the former in the hands of the more conservative PHS and the latter in the hands of labor advocates.[52]

Under the Bush administration, the public health conservatism of the EPA and OSHA came under attack from Republican officials. In 1990, the Office of Management and Budget (OMB) criticized traditional risk assessment for its construction of worst-case scenarios involving large margins of safety: "Conservatism in risk assessment distorts the regulatory priorities of the Federal Government, directing societal resources to reduce what are often trivial carcinogenic risks while failing to address more substantial threats to life and health." (Household radon measurements being taken in basements was singled out as a case in point.) The OMB criticized such practices for illegitimately intermingling the science and policy aspects of risk assessment; it also complained that "the continued reliance on conservative or worst-case assumptions distorts risk assessment, yielding estimates that may over-state likely risks by several orders of magnitude." The danger was that attention would be diverted onto minute or trivial risks. The report urged that responsibility for establishing margins of safety be put into the hands of risk managers rather than be preempted by biased risk assessors.[53] Political conservatives have often leveled similar charges.[54]

At stake in such challenges is not just what should be done but *how*. There are several different methods cancer researchers use to establish the gravity of a hazard; whether a hazard is considered real can depend on what level of proof is considered satisfactory before action should be taken. Epidemiological studies, one should keep in mind, are notoriously insensitive: it is virtually impossible to detect, from epidemiology alone, the effect of a carcinogen that increases human cancer risk by less than about 30 percent. As one can well imagine, though, there are many situations that expose us to carcinogens that increase our cancer risk by, say, 2 or 5 or 10 percent. Saccharin, for example, was estimated by the FDA to increase the risk of bladder cancer by about 3 percent. Though epidemiologically invisible, a hazard of this magnitude would cause some 1,200 extra cases of bladder cancers per year in the United States.[55] Epidemiological studies are of no use in revealing hazards such as these; that is one reason so much effort has been put into animal studies and bacterial bioassays. A similar problem is encountered in trying to understand the impact of low-level hazards. If a hazard is sufficiently widely dispersed—asphalt road dust or pesticides in foods or microwave radiation, for instance—it may be difficult to find unexposed controls. Many such cancers are presumably preventable, though the magnitude of the hazard may be difficult or even impossible to ascertain.

Epidemiology, in other words, is good at revealing certain kinds of hazards and bad at revealing others. Strong carcinogens affecting large numbers of people are fairly easy to identify—these often turn out to be "lifestyle" factors, like smoking, or clear-cut occupational hazards, like asbestos and uranium mining. Low-grade occupational or toxic hazards may be more elusive, since exposures tend to vary over time and space.[56]

Such difficulties have legal implications. The key question when a plaintiff is seeking damages for injury is usually: Did this person's work cause his or her injury? The question is deceptively complex, and one's answer may have a lot to do with how one goes about the investigation—through clinical or statistical means, for example. David Ozonoff of Boston University has suggested that requiring an epidemiologic standard of proof in some such cases "will essentially foreclose on the ability of many truly injured workers to recover any damages." Epidemiology can often help one find out whether a particular substance causes cancer and to what degree, but it is often of little help in deciding whether a particular individual was injured by his or her work: "A rare disease or idiosyncratic reaction following a rare exposure, such as aplastic anemia after chlordane exposure, may be obvious from case reports but may not be demonstrable by epidemiological study."[57] The numbers involved may simply be too small to warrant epidemiological conclusions.

There is yet another kind of bias in the use of statistical tests to establish the veracity of hazards. Statistical significance tests conventionally require that there be less than a one-in-twenty possibility that the results could have come about by chance. What this means is that, among twenty studies reporting significance at this level ($p \leq .05$), one, on average, will have produced a positive result simply by chance. Scientists arbitrarily designate this 95 percent "confidence rule" as a conventional index of significance; statistical results that display higher levels of confidence (99 percent, for example, where $p \leq .01$) are sometimes referred to as "very significant."

Ozonoff and Boden point out that it is important to distinguish—though many public health officials do not—between statistical significance and public health significance. A town near a nuclear plant may have a 200 percent increased incidence of leukemia, but if the numbers involved are small the result may not be statistically significant. There might be, say, a 10 percent probability that the results could have occurred by chance: this would be true, for example, if for a neighborhood of 1,000 persons there were 3 leukemias when only 1 was expected. Excess leukemias on this order would surely be of concern to residents, however, if they were viewed as indicative of a potentially real, though not yet proven, hazard. Exclusive reliance on the 95 percent confidence rule to define the reality of a hazard produces a bias against taking certain kinds of hazards seriously—living near toxic waste dumps, for example, where exposures may be heterogeneous, health effects diverse, and the populations affected not very large.[58]

In court, where push comes to shove, debates over statistical significance can become heated. The question often boils down to what kind of error one is willing to tolerate, which in turn depends on one's conception of prudence. Decision theorists distinguish two types of error: type I involves asserting that a proposition is true (that a hazard is real, for example) when in fact it is false; type II involves asserting that a proposition is false when in reality it is true. In matters of public health, attitudes toward these two types of errors reflect a fundamental divide between the two types of conservatism already mentioned. The question reduces to whether it is better to minimize the possibility of wrongly asserting that a hazard *does exist* or to minimize the possibility of wrongly asserting that a hazard *does not exist.* Scientific conservatives stress the former, avoiding so-called "false positives," while public health conservatives stress the latter (avoiding "false negatives").[59]

Scientists are never, of course, fortunate enough to have direct access to reality. Probabilistic judgments are inevitable, especially in fields such as epidemiology where population effects are the objects of interest. How one regards such judgments, and especially whether one chooses to emphasize scientific or public health conservatism, is ultimately a political choice. The customary acceptance of the 95 percent confidence rule is designed to preserve the credibility of research, but it is not necessarily the best guide for deciding whether the water we give to our children is safe. Deciding whether to stress the avoidance of either type I or type II error is a value decision with political consequences. Experimental epidemiology tends to follow the tradition of scientific conservatism in stressing the avoidance of false positives. As Ozonoff and Boden put it: "epidemiological techniques have evolved from analogy with the laboratory sciences where acceptance of a new hypothesis is avoided unless there is great confidence that it reflects the true state of nature."[60] Such a strategy makes sense when there is no human cost to waiting for confirmation. But in cases where human lives may be at stake, other factors must be considered. Otherwise, scientific conservatism can lead to public health recklessness.

THE POVERTY OF PREVENTION

Cancer is a largely preventable disease; virtually no one today denies this. Prevention, though, has languished as a relatively minor part of the American cancer program. Richard Nixon's much celebrated war on cancer emphasized research to perfect chemotherapies, giving little attention to the expertise and political action required for prevention. The majority of funds distributed by cancer research bodies has gone either to efforts to improve treatment (especially surgery, radiation, and chemotherapy) or to efforts to understand the biological mechanisms involved in carcinogenesis. The emphasis, in other words, has been on the pragmatics of cures and the theoretical aspects of biochemical mechanisms, rather than on research into what causes cancer and (most important) what might be done to prevent it.[61] The NCI as late as 1987 spent only $32 million on its Smoking, Tobacco, and Cancer Program, an office that until 1980 had been devoted almost entirely to the development of "safe cigarettes."[62] Even today, the NCI spends less than 3 percent of its budget on antitobacco efforts, despite the broad consensus that cigarettes are responsible for as many as 30 percent of all human cancers.[63] The State of California spends more to combat smoking than the National Cancer Institute, the Centers for Disease Control, and all other federal government agencies combined.

Why is the bulk of cancer research dedicated to the improvement of treatment, when the more sensible approach would be to encourage prevention?

Scholars who have examined the history of medicine in America point to the extraordinary success of the biomedical establishment in convincing Congress, the president, and the American people that science would win the war on cancer. From the outset, the war was geared toward finding a cure or at least a vaccine. The focus on treatment reflected the immediate needs of patients, since once a person has cancer it is usually not of much therapeutic use to know what caused it (lung cancer caused by asbestos inhalation is treated pretty much the same as lung cancer caused by smoking). The curative focus also reflected the unbridled optimism of the era: scientists had tamed the atom and banished polio; surely an all-out assault on cancer could solve that problem, too. Nixon's 1971 declaration of war on cancer reflected a similar optimism, this time modeled on the Apollo moon shot.[64] The nation that had landed a man on the moon could surely conquer cancer. The oft-expressed hope that a cure was "just around the corner" occasionally led to recklessness, as when Benjamin F. Byrd, a former president of the American Cancer Society, defended the society's aggressive breast cancer–screening program on the grounds that, though X-ray mammography might increase a woman's risk of cancer in the future, "there's also an excellent chance that by that time science will have learned to control the disease."[65]

Some like to argue that the root cause of the neglect of prevention is that cancer research, like cancer treatment, is big business with close ties to pharmaceutical and manufacturing interests. The alternative-therapy advocate Ralph Moss, in his 1980 *The Cancer Syndrome* (updated in 1991 as *The Cancer Industry*), showed that many cancer research institutions, like the American Cancer Society and the Memorial Sloan-Kettering Cancer Center, are not free of such ties. Sloan-Kettering's neglect of occupational carcinogens and the possibilities of prevention was overdetermined, in Moss's view, by the fact that most of those sitting on its board of overseers were bankers and industrialists. Leo Wade had a long career as medical director at Standard Oil of New Jersey before coming to Sloan-Kettering first as vice president and eventually as director. Wade was a former member of the Manufacturing Chemists Association, the American Petroleum Institute, and the National Association of Manufacturers; it comes as no surprise, then, to hear Wade ridiculing efforts to control chemicals as "both futile and suspect."[66] Peter Chowka likewise notes that many of the carcinogens slighted by the ACS "are by-products of profitable industries in which its directors have financial interests."[67]

Samuel Epstein points to similar conflicts of interest on the National Cancer Advisory Panel, a three-member group with direct access to the president. As of 1990, Armand Hammer was chairman of both the Advisory Panel and the Occidental Petroleum Company, "a major polluting industry and manufacturer of carcinogenic chemicals." When Hammer announced a fund-raising drive to add $1 billion to the NCI budget, the goal was "to find a cure for cancer in the next ten years." None of the proposed funding was earmarked for prevention; nothing was to be spent to alleviate the cancer costs of Occidental's own carcinogenic activities.[68]

One needn't, of course, be a conspiracy theorist to recognize that the poverty of prevention—by contrast with the wealth of basic research—stems partly from the fact that effective prevention requires changes not just in research priorities but also in deeply ingrained personal habits and the logic of business enterprise. This latter aspect was made clear in the 1970s, when a leading cancer researcher conceded that while the removal of carcinogens from the environment was the most effective way to conquer cancer, "it may require such a rearrangement of the environment that society cannot or will not allow this to be done except slowly over decades." Basic research, by contrast, was politically safe: "A knowledge of the steps in the carcinogenic process will almost certainly lead to ways to interrupt the process in the continuing presence of the carcinogen."[69] Such a logic would have us keep the carcinogens while searching for a cancer cure. No one's boat is rocked if our goal is to cure cancer once it has begun rather than to prevent it before it starts. Governments, for example, have been reluctant to curtail tobacco sales because the tax base generated is enormous: $14 billion every year in the United States alone. Cynics point out that cancer affects the elderly more than the young, relieving governments of the cost of Social Security. It's hard to bite the hand that's keeping you well fed.[70]

The arguments for prevention won't go away, however, and recent years have seen some changes. In 1983, the NCI's Division of Cancer Control and Rehabilitation was renamed the Division of Cancer Prevention and Control. Between 1976 and 1987 the fraction of the NCI's overall budget officially devoted to prevention rose from 27 to 29 percent. In 1992, the Centers for Disease Control was rechristened the Centers for Disease Control and Prevention. But how much of what goes under that name is genuine? The NCI claims to spend more than a quarter of its budget on prevention, but much of what counts as prevention would more accurately be classed as early detection or treatment. Chemoprevention is a case in point. Chemoprevention involves the search for drugs that might render a healthy person less susceptible to cancer. The most celebrated recent case is the NCI's Breast Cancer Prevention Trial, a five-year study to see whether 8,000 high-risk women administered the new drug tamoxifen develop fewer cancers. Critics charge that the $60 million program is not really prevention at all, but another form of chemotherapy—and not one without risk. Tamoxifen has been shown to reduce the odds of recurrence of breast cancer in women who have already had the disease (by 30–40 percent), but it is also known to increase the risk of eye damage, liver failure, and a particularly aggressive form of endometrial cancer.[71] Similar objections may be raised against the NCI's exploration of the drug finasteride as a chemopreventive for prostate cancer.

The effort to redefine treatment as simply another form of prevention leads one to wonder whether the cancer research bureaucracy is trying to do with words what it is unwilling or unable to do with deeds. Cancer researchers now talk in terms of "primary," "secondary," and "tertiary" prevention, the former being closest to the traditional definition of the term. (Tertiary prevention is apparently just a new and fashionable way to say "treatment.") Therapeutics remains the dominant federal research emphasis, followed closely by basic biology, especially molecular genetics. Epstein faults the NCI for being "hostile to prevention," but the NCI responds that the institute's mission precludes a broader focus. NCI's director, Samuel Broder, complains that Epstein "in effect wants us to be the Environmental Protection Agency or the National Institute of Environmental Health Sciences. We follow the advice of peer review, but our primary mission is to generate knowledge."[72]

Basic research is perennially invoked as that which "alone holds out the hope of real progress,"[73] though it is hard to show that basic research has done very much to halt the progress of the disease. John Cairns told a *Science* reporter in 1985 that cancer trends were so discouraging that it was difficult to discuss them in public: "That's why this dispute has been carried on in a gentlemanly way. People do get cancer and they have to be given encouragement. Research has to go on."[74] Samuel Epstein is more cynical: basic cancer research, in his view, "is an excellent slush fund for molecular biologists, but it won't have any impact on cancer."[75] Monies from the public treasury are arguably better spent on cancer research than on stealth bombers or tobacco and sugar subsidies, but one should not be misled into thinking that basic scientific knowledge is the main thing that is lacking in the fight against cancer. Basic cancer research has produced some remarkable biological insights, but surprisingly little in the way of successful treatments and even less that is of relevance to prevention.

The fruitful question is not basic versus applied research, but research for what end? The sad truth is that cancer prevention is low prestige. Prevention is impoverished in an age of heroic medicine, where the reward structure is heavily biased in favor of last-ditch, quick-burst, high-tech interventions and high-profile, Nobel

Prize–potential basic science. In the field of research, this means exorbitant funding for therapies and molecular genetics and a more penurious approach to epidemiology, nutrition, health education, occupational health and safety, and behavioral and social science research—none of which will ever generate a Nobel Prize. (Richard Doll came close: he and several others were nominated for showing the environmental origins of cancer, but nothing further came of this.) In clinical practice, it means that surgeons and radiologists earn hundreds of thousands of dollars while preventive medicine languishes, grossly underfunded. In terms of medical education, it means the relative neglect of things like diet, sexuality, psychology, history, public health, environmental policy, and industrial medicine.[76] The consequence is a lack of interest and therefore lack of competence in these areas. Wilhelm Hueper recognized more than thirty years ago that "the goal of curing the victims of cancer is more exciting, more tangible, more glamorous and rewarding than prevention." Rachel Carson cited these words in *Silent Spring,* and countless public health advocates have communicated a similar message. A cancer cured is tangible, a marvelous success, but a cancer prevented is invisible, a statistical abstraction.[77] Prizes abound for cancer cures,[78] but where are the awards for cancer prevention? Who is there to testify on talk shows or before Congress to being alive and well in consequence of a well-planned program of preventive medicine?

I don't want to leave the impression that steps are not being taken in the direction of prevention. The most powerful moves have come as a result of political action: bans on smoking on commercial aircraft and in other indoor spaces, regulation of asbestos in the workplace, bans on certain pesticides and food additives, and so forth. Technological developments—like refrigeration and transport networks increasing the availability of fresh fruits and vegetables—have unwittingly eased certain cancers (notably, stomach). Food labeling policies have made a difference: the grading of meat by fat content, for example, has allowed consumers to make more intelligent choices about what they eat. Many of the gains made in smoking prevention can be traced to the efforts of groups like John Banzhaf's Action on Smoking and Health, which forced network television stations to run prime-time antismoking ads to counter smoking commercials in the 1960s and early 1970s.[79] A 1994 study showed that California's five-year effort to curb smoking in restaurants and in the workplace (Proposition 99) had reduced cigarette consumption by 26 percent.[80] President Clinton's proposed ban on smoking in the workplace—both military and civilian—could turn out to be the most powerful anticancer measure in history.

There is much, though, that remains to be done. First and foremost, we need stiffer taxes on tobacco sales and a halt to tobacco subsidies. We need financial support for tobacco counteradvertising and support for personal-injury litigation against the tobacco industry.[81] Senator Edward Kennedy of Massachusetts has proposed a bill to authorize the federal government to spend $110 million to combat tobacco via a new Center for Tobacco Products; Congressman Henry Waxman of California has proposed restricting tobacco-industry sponsorship of sporting events, eliminating cigarette vending machines, and banning the use of models and color images in tobacco ads.[82] We need to learn from creative efforts abroad, such as Sweden's policy of printing stories on cigarette cartons of smokers who died from cancer and Hong

Kong's 1986 ban on smokeless tobacco. Canada and New Zealand have had powerful antismoking campaigns: Canada's $3.72 tax per pack cut cigarette consumption by nearly 50 percent, though smuggling from the United States prompted the Canadian government to roll back the tax in February 1994. A $2 federal excise tax on U.S. cigarettes would reduce the number of U.S. smokers by roughly 8 million, preventing nearly 2 million smoking-related deaths. It would also facilitate the resumption of the Canadian program.[83] Prevention must mean research into why young people take up smoking and how they can be encouraged to stop once they've begun. Marvin Schneiderman at the NCI proposed a stronger focus on behavioral questions such as these in the mid-1970s, only to be told that the NCI was not in the business of supporting "soft science." This led him to remark that soft science was hard (to do as well as to get support for).[84]

Prevention should mean stiffer supervision of pesticides and federal support for integrated pest management and other alternatives to petrochemical agriculture. Prevention might mean financial support for farmers wanting to shift from tobacco to less deadly crops[85] and perhaps even incentives to use genetically altered strains of tobacco plants to produce pharmaceuticals or other, more beneficial products. Prevention could mean research designed to ease the transition from gasoline to electric automobiles and increased support for international limits on chlorofluorocarbons and other ozone-depleting compounds. Prevention could mean stiffer regulation of tanning salons and closure of the loopholes that allow nonfood pesticides to be used on crops that wind up in human foods (such as cottonseed oil). Prevention could mean paving dirt roads in areas (such as California) where serpentine, a natural asbestos, forms a major component of the surface rock. It should surely mean empowering OSHA to deal with indoor pollutants like radon and secondhand smoke. It also means developing the tools to identify high-radon regions, so that attention can be drawn to places most in need of remediation.[86] And it must mean granting the FDA authority to limit the substances added to tobacco in the course of its manufacture.

Prevention also means modifying trade practices at home that encourage cancer abroad. The United States is the largest exporter of carcinogens in the world, judging from the brisk pace of the tobacco export business and the ongoing practice of dumping pharmaceuticals, pesticides, and food additives abroad that are banned in this country. Tobacco companies have expanded their foreign sales enormously, assisted by trade policies that treat tobacco as an export commodity like any other. (The United States today exports more than three times as many cigarettes as any other country in the world.)[87] In 1990, the American Medical Association's Council on Scientific Affairs attacked U.S. policy in this regard, pointing out that since tobacco worldwide accounts for nearly 5 percent of all human deaths, it was perverse for the United States to promote smoking abroad. The United States Cigarette Export Association claims that since tobacco is legal there should be no restrictions on its sale; C. Everett Koop, the former surgeon general, has more aptly recognized the export of tobacco as "a moral outrage."[88] Americans worry a lot about the importing of cocaine and other drugs into the United States, but little is done to halt the far more deadly export of tobacco into Third World nations.

Asbestos export is another area that should be curtailed. U.S. production has dropped to levels not seen for more than eighty years, but several other nations continue to mine and export the mineral. Canada, Russia, Brazil, Zimbabwe, and South Africa are among the world's leading suppliers of asbestos, much of which ends up in the cement piping, roofing, and sheeting of Third World nations. Millions of people gather their drinking water from asbestos-roofing rainwater runoff containing hundreds of millions of fibers per liter. The local manufacture of asbestos products often involves a great deal of dust inhalation (from sawing asbestos sheeting, for example); this is particularly troubling given that local substitutes (sisal, hemp, jute) are often available. (In Costa Rica, locally available pine pulp is commonly used.) The World Bank and the U.S. Agency for International Development could do more to require that projects supported by those agencies use building materials that are safe from a cancer point of view.[89]

Finally, prevention should mean a more broad-minded approach to the kinds of things one might consider as causes of cancer. The molecular approach, inspired by the search for proximal causes that underlie all cancers, tends to efface the diversity of more distant causes—things like obstacles to quitting smoking, apathy in the face of radon hazards, body machismo, tobacco advertising, social forces shaping dietary habits, and the failure to enact and enforce effective anticancer policies. If cigarettes are responsible for 30 percent of all cancer, then why hasn't more been done to reduce smoking? If natural carcinogens are contributors to disease, then why have plants bred for insect resistance not come under closer scrutiny?[90] If dietary fat presents a danger, why don't agricultural policies encourage a shift away from meat and dairy-centered production? It is not hard to expand such lists: what is needed in each case is a larger understanding of "the causes of causes" (we could use a comparative anthropology of cancer), combined with the political will to translate knowledge into policy.

THE MYTH OF
IGNORANCE

The cancer research establishment likes us to believe that a shortage of research funds is the primary problem. Ignorance, in this view, is the basic cause of cancer. The key to cancer control is therefore knowledge—the great scientific breakthrough to wash away our stupor. But we already know a lot about cancer. We know that cigarettes and asbestos cause cancer, and that it is not very healthy to eat high-fat, low-fiber, high-salt foods. We know that it is dangerous to burn our skin in the sun or to bathe our foods in pesticides. We know that dust in the lungs is bad, whether from the floors we clean or the hobbies we enjoy or the materials we are consigned to work with. We know that unregulated industry can cause cancer, and that cancer is the product of bad habits, bad government, bad business, and perhaps even bad science. Knowledge about cancer is not in very short supply. What is needed are thoughtful and confident steps to reorient cancer policy.

Knowledge of causes, in other words, is only part of what is needed. It is often true that if you don't ask you don't know, and if you don't know you can't act.[91] But the important question is not just "what causes cancer" but also "what can be done to prevent cancer?" Answering the former is not always necessary for answering the latter.[92] We don't need to know the precise mechanisms by which asbestos triggers mesothelioma to know that breathing the fiber is bad. Duplicitous armies of PR men would have us endlessly chase the red herrings of mechanisms, to make us lose the forest for the trees. The marvelous fact that two questions are raised for every one answered is the bread and butter of science, but we mustn't lose sight of the simple measures we already know will lower cancer incidence. We must recognize that things are not always what they seem: that ignorance may be manufactured, that ideological gaps may leave us sightless, that good news is often partial, that causes may be cultural, that writing checks to scientists is only one of several ways to combat cancer. Activists who push for "more research" must ask: What kind of knowledge, knowledge for what end? It is important to ask why we know what we know, and why we don't know what we don't know. We need to appreciate not just how ignorance can invite knowledge but also how knowledge can abide ignorance, despite all our efforts to clear a path from the one to the other.

Notes

INTRODUCTION
WHAT DO WE KNOW?

1. Roswell Park, "A Further Inquiry into the Frequency and Nature of Cancer," *The Practitioner* 62 (1899): 385.
2. American Cancer Society, *Cancer Facts & Figures—1994* (New York, 1994).
3. Takashi Sugimura, "Multistep Carcinogenesis: A 1992 Perspective," *Science* 258 (1992): 603–607.
4. In 1991, the American Hospital Association predicted that cancer would surpass heart disease as the leading cause of death in the United States by the year 2000. The association also predicted that cancer would account for 20 percent of the nation's health care expenditures in the 1990s; see Tim Friend, "By 2000, Cancer Predicted to Be the No. 1 Killer," *USA Today,* May 13, 1991. The National Cancer Institute calculates that cancer costs the nation more than $100 billion every year, including medical treatment and loss of labor power and earnings; see American Cancer Society, *Cancer Facts & Figures—1994,* p. 26. The cost of treating AIDS, by contrast, was about $6 billion in 1991, projected to rise to about $10 billion by 1994; see Fred J. Hellinger, "Forecasting the Medical Care Costs of the HIV Epidemic: 1991–1994," *Inquiry* 28 (1991): 213–225.
5. D. M. Parkin et al., "Estimates of the Worldwide Frequency of Sixteen Major Cancers in 1980," *International Journal of Cancer* 41 (1988): 184–197.
6. Howe, *Global Geocancerology,* pp. 3–42.
7. Thomas G. Benedek and Kenneth F. Kiple, "Concepts of Cancer," *The Cambridge World History of Human Disease* (Cambridge, England, 1993), p. 107.
8. American Cancer Society, *Cancer Facts and Figures for Minority Americans—1991* (New York, 1991), p. 21.
9. T. J. Mason, *Atlas of Cancer Mortality for U.S. Counties: 1950–1969* (Washington, D.C., 1975); also the publication of New Jersey's Department of Environmental Protection, *Cancer and the Environment* (Trenton, 1976). New Jersey acquired the nicknames "cancer alley" and "cancer capital of the country" after publication of the atlas.
10. Ann Gibbons, "Does War on Cancer Equal War on Poverty?" *Science* 253 (1991): 260. In 1992, American blacks had a 27 percent higher age-adjusted cancer mortality rate than American whites. Blacks also had a much lower five-year survival rate: 38 percent compared to 53 percent for whites. This is particularly disturbing, given that overall cancer incidence rates are actually *lower* for blacks than for whites. Blacks who suffer from cancer are presumably receiving inferior medical treatment and are therefore much more likely to have their cancers diagnosed after they have spread to other parts of the body. There are also cancers for which the black death rate exceeds the white death rate, even though the incidence rate is lower in blacks than in whites; see Catherine C. Boring et al., "Cancer Statistics for African Americans," *CA* 42 (1992): 7–17; Harold P. Freeman, "A Summary of the American Cancer Society Report to the Nation: Cancer in the Poor," *CA* 39 (1989): 263–265, also his "Cancer in the Socioeconomically Disadvantaged," *CA* 39 (1989): 266–288.
11. Eliot Marshall, "Experts Clash over Cancer Data," *Science* 250 (1990): 900.
12. "Cancer Consequences," *Centre Daily Times* (Knight-Ridder), March 28, 1991.
13. See D. P. Burkitt and H. C. Trowell, eds., *Refined Carbohydrate Foods and Disease: Some Impli-*

cations of Dietary Fibre (London, 1975); also their *Western Diseases: Their Emergence and Prevention* (Cambridge, Mass., 1981).

14. One of the most celebrated examples of cancer prevention unfolded after a Chinese study linked cancer of the esophagus to certain pickled foods consumed by the local population. Researchers were able to demonstrate that chickens raised on a diet similar to that of their cancer-prone owners developed cancers similar to those of their owners; a program to eliminate the culprit reduced esophageal cancers significantly; see *The Cancer Detectives of Lin Xian* (1982), a documentary produced by WGBH Boston for Nova and distributed by Time-Life Films, Inc.

15. Henri Picheral, "France," in Howe, *Global Geocancerology*, p. 151.

16. Toshio Oiso, "Incidence of Stomach Cancer and Its Relation to Dietary Habits and Nutrition in Japan between 1900 and 1975," *Cancer Research* 35 (1975): 3254-3258.

17. "Lung Cancer Tied to Vapor from Oil in Stir Frying," *New York Times,* November 1, 1987; Environmental Protection Agency, *Respiratory Health Effects,* pp. 5-53 to 5-54.

18. "Small Study Implicates PCBs in Breast Cancer," *Journal of the National Cancer Institute* 84 (1992): 834–835. A recent study disputing evidence of a fat–breast cancer link looked at 89,494 nurses; see Walter C. Willett et al., "Dietary Fat and Fiber in Relation to Risk of Breast Cancer," *JAMA* 268 (1992): 2037–2044. *USA Today* reported the story as front-page news, in an article that began: "After nearly 10 years of warning women that too much fat in the diet may cause breast cancer, doctors are now saying never mind"; see Tim Friend, "Breast Cancer, Fat: No Link," *USA Today,* October 21, 1992.

19. Love, *Love's Breast Book,* p. 146; Rose E. Frisch et al., "Lower Prevalence of Breast Cancer and Cancers of the Reproductive System among Former College Athletes," *British Journal of Cancer* 52 (1985): 885–891.

20. Doll and Peto, *Causes of Cancer,* p. 1259.

21. "Criticized Panel Backs E.P.A. Smoking Study," *New York Times,* December 6, 1990; "EPA Defines Hazards of Tobacco Smoke," *Science* 257 (1992): 471; Environmental Protection Agency, *Respiratory Health Effects,* p. 1-1.

22. Stanton A. Glantz and William W. Parmley, "Passive Smoking and Heart Disease," *Circulation* 83 (1991): 1–13.

23. Between 1911 and 1935, U.S. deaths from stomach cancer outnumbered deaths from lung cancer by more than a factor of ten; in the 1990s, lung cancer mortality rates are about ten times higher than stomach cancer rates; see Dublin and Lotka, *Twenty-Five Years of Health Progress,* p. 167; American Cancer Society, *Cancer Facts & Figures—1994,* p. 5.

24. American Cancer Society, *Cancer Facts & Figures—1994,* p. 11.

25. Lewis, *Biology of the Negro,* pp. 336–355.

26. W. J. Hunter et al., "Mortality of Professional Chemists in England and Wales, 1965–1989," *American Journal of Industrial Medicine* 23 (1993): 615–627.

27. Roush et al., *Cancer Risk and Incidence Trends,* p. 205.

28. B. A. Miller et al., *SEER Cancer Statistics Review 1973–1990* (Bethesda, Md., 1993).

29. Cairns, "Treatment of Diseases" p. 56.

30. Gastrointestinal Tumor Study Group, "Adjuvant Therapy of Colon Cancer—Results of a Prospectively Randomized Trial," *New England Journal of Medicine* 310 (1984): 737–743.

31. Bailar and Smith, "Progress Against Cancer?" and the correspondence in the *New England Journal of Medicine* 315 (1986): 967–968.

32. Chowka, "National Cancer Institute," p. 23.

33. Epstein, *Politics of Cancer,* p. 511; Sigerist, "Historical Development," p. 644.

34. "Cancer and the Mind: How Are They Connected?" *Science* 200 (1978): 1363–1365. In 1994, Gary D. Friedman found little support for the hypothesis that depression predisposes to cancer occurrence; see his "Psychiatrically-Diagnosed Depression and Subsequent Cancer," *Cancer Epidemiology, Biomarkers & Prevention* (January–February 1994): 11–13.

35. See Brodeur, "Cancer at Slater School"; contrast also the more skeptical review by Kenneth R. Foster, "Weak Magnetic Fields: A Cancer Connection?" in *Phantom Risk: Scientific Inference and the Law,* edited by Kenneth R. Foster et al. (Cambridge, Mass., 1993), pp. 47–86. The radar guns used by police to catch speeders have been implicated in cancer causation; see "Radar Gun Hazards?" *Science* 254 (1991): 1724–1725. Police in several states have filed suits against radar gun manufacturers, claiming they contracted cancer of the eyelid or testicle from the guns.

36. Natalie Angier, "Cellular Phone Scare Discounted," *New York Times,* February 2, 1993. A 1988 study suggested that ham radio operators, exposed to much higher microwave radiation than people using cellular phones, did have higher rates of acute myeloid leukemia, but the study has been criticized for failing to look for possible confounding variables; see Samuel Milham, "Increased Mortality in Amateur Radio Operators Due to Lymphatic and Hematopoietic Malignancies," *American Journal of Epidemiology* 127 (1988): 50–54.

37. In 1993 Abortion Industry Monitor, an anti-abortion group based in Purcellville, Virginia, published *Before You Choose: The Link Between Abortion and Breast Cancer,* a propaganda tract branded "epidemiologically illiterate" by Nancy Krieger of the Kaiser Foundation; see Michael McCarthy, "Abortion Foes' New Claim," *Lancet* 342 (1993): 1290; Troy Parkins, "Does Abortion Increase Breast Cancer Risk?" *Journal of the National Cancer Institute* 85 (1993): 1987–1988.

38. Polly A. Newcomb et al., "Lactation and a Reduced Risk of Premenopausal Breast Cancer," *New England Journal of Medicine* 330 (1994): 81–87.

39. Robert D. Morris et al., "Chlorination, Chlorination By-products, and Cancer: A Meta-analysis," *American Journal of Public Health* 82 (1992): 955–963; Earl V. Anderson, "Chlorine Producers Fight Back Against Call for Chemical's Phaseout," *Chemical & Engineering News* (May 10, 1993): 11–12.

40. Vicki Monks, "See No Evil," *American Journalism Review* (June 1993): 18–25.

41. In 1988, the Department of Energy estimated the cost of cleaning up the nation's twenty-eight nuclear bomb plants and testing facilities at between $40 billion and $110 billion (*New York Times,* July 2, 1988). The General Accounting Office reported shortly thereafter that costs could actually be as high as $175 billion (*New York Times,* July 14, 1988).

42. See Bruce Stutz, "Science for Sale: Ecologists Call Colleagues 'Biostitutes,'" *The Scientist* (November 28, 1988): 1, 24–25.

43. Henderson et al., "Toward the Primary Prevention of Cancer," p. 1137.

44. "Women in NIH Research," *Science* 251 (1991): 159.

45. For particularly brutal examples of people having science done to them, see James H. Jones, *Bad Blood: The Tuskegee Syphilis Experiment* (New York, 1981); also my *Racial Hygiene.*

46. For a spirited critique of constructivism, see Langdon Winner, "Upon Opening the Black Box and Finding It Empty: Social Constructivism and the Philosophy of Technology," *Science, Technology, & Human Values* 18 (1993): 362–378; compare also my *Value-Free Science? Purity and Power in Modern Knowledge,* pp. 224–231.

47. In 1908, V. Ellerman and O. Bang in Denmark showed that certain chicken leukemias were transmissible by viral agents; three years later Peyton Rous demonstrated that viruses could cause sarcoma tumors in chickens, a discovery for which he was awarded the Nobel Prize in 1966; see R. E. Shope, "Evolutionary Episodes in the Concept of Viral Oncogenesis," *Perspectives in Biology and Medicine* 9 (1966): 258–274. In 1913, Danish pathologist Johannes Fibiger claimed to have induced stomach cancer in rats by feeding them nematode-infested cockroaches; the work earned him the 1926 Nobel Prize, the first for cancer research. Subsequent scientists were unable to reproduce his results, however; see Shimkin and Triolo, "History of Chemical Carcinogenesis," pp. 15–16.

48. Alexis Carrel and Montrose T. Burrows, "Cultures de sarcome en dehors de l'organisme," *Comtes rendus hebdomadaires des séances et mémoires de la société de biologie* 69 (1910): 332–334.

49. Harald zur Hausen of the Deutsches Krebsforschungszentrum suggests that viral infections may be involved in as many as 15 percent of all cancers worldwide and that viruses are therefore "the second most important risk factor for cancer development in humans, exceeded only by tobacco"; see his "Viruses in Human Cancers," *Science* 254 (1991): 1167–1173.

50. Alpha-Tocopherol, Beta Carotene Cancer Prevention Study Group, "The Effect of Vitamin E and Beta Carotene on the Incidence of Lung Cancer and Other Cancers in Male Smokers," *New England Journal of Medicine* 330 (1994): 1029–1035.

51. See, for example, Stephen Hilgartner, *Who Speaks for Science? Cultural Authority in the Diet-Cancer Debates* (Berkeley, forthcoming) also Walter C. Willett, "Diet and Health: What Should We Eat?" *Science* 264 (1994): 532–537; and A. B. Miller et al., "Diet in the Aetiology of Cancer," *European Journal of Cancer* 30A (1994): 207–220.

CHAPTER 1
A DISEASE OF CIVILIZATION?

1. A paper lending support to the idea of the rarity of ancient tumors is Michael R. Zimmerman, "An Experimental Study of Mummification Pertinent to the Antiquity of Cancer," *Cancer* 40 (1977): 1358–1362. On the question of tumors in nonhuman species, including plants, an early source is also one of the best: see Julian Huxley, "Cancer Biology: Comparative and Genetic," in *Biological Reviews of the Cambridge Philosophical Society,* edited by H. Munro Fox (Cambridge, England, 1956), pp. 474–514.

2. The bibliography of the history of cancer research is vast. A useful introduction is Michael B. Shimkin, "History of Cancer Research: A Starter Reading List and Guide," *Cancer Research* 34

(1974): 1519–1520; compare also the bibliographic essay in Patterson, *Dread Disease*, pp. 313–322. The most comprehensive early bibliography of cancer, containing nearly 5,000 entries, though many with flaws, is Robert Behla, *Die Carcinomlitteratur. Eine Zusammenstellung der in- und ausländischen Krebsschriften bis 1900* (Berlin, 1901). A useful guide to more recent periodical literature is Olson, *The History of Cancer: An Annotated Bibliography*. A listing of 422 English-language cancer journals can be found in Pauline M. Vaillancourt, *Cancer Journals and Serials: An Analytical Guide* (New York, 1988). The best overview of the history of cancer statistics is Clemmesen, *Statistical Studies in the Aetiology of Malignant Neoplasms*, vol. 1, *Review and Results*.

3. A good discussion of these early ambiguities can be found in Rather, *Genesis of Cancer*, pp. 8–19.

4. Erwin Franck, "Das Carcinom im Hause Napoleon Bonaparte," *Die medicinische Woche* 5 (1904): 110–134. Hume is sometimes said to have died of "chronic ulcerative colitis, following an acute bacillary dysentery"; see Ernest C. Mossner, *The Life of David Hume*, 2nd ed. (Oxford, 1980), p. 596.

5. William Rosenzweig, "Disease in Art: A Case for Carcinoma of the Breast in Michelangelo's 'La Notte,'" *Paleopathology Newsletter* (March 1983): 8–11.

6. Malcolm Weller, "Tutankhamun: An Adrenal Tumour?" *Lancet* 2 (1972): 1312. These and other cases are abstracted in Olson, *History of Cancer*, pp. 85–96.

7. One alternative might be to construct retrospective cancer mortality statistics from biographies of famous personalities; such data could be age-standardized, and though diagnostic ambiguities would certainly be a problem, the results might nevertheless be revealing.

8. On early ideas of probability, see Ian Hacking, *The Taming of Chance* (Cambridge, England, 1990).

9. W. Newman, "Notes from Registers of Market Deeping (1711–1723)," *British Medical Journal* 1 (1896): 915–916.

10. Hoffman, *Mortality from Cancer*, pp. 23, 604. John Hunter (1728–1793) assumed that cancer affected about 1 in 1,000 persons in his 1787 *Lectures*, p. 377. Hunter also suggested that cancer was rare in the West Indies and among the prize-fighting women of the Friendly Islands (p. 381).

11. Williams, *Natural History*, p. 51.

12. Doris Panzer and Gustav P. Wildner, "Zur Geschichte der Krebsstatistik in Deutschland bis 1949," *Zeitschrift für ärztliche Fortbildung* 79 (1985): 41–44. In the United States, it was not until 1914 that lung cancer was listed as a separate cause of death; see U.S. Bureau of the Census, *Mortality Statistics 1914* (Washington, D.C., 1916), pp. 29–30.

13. Joseph J. Lister's development of lenses correcting for spherical and chromatic aberration was a key development in the history of medical microscopy. Lister had designed a microscope correcting for spherical and chromatic aberration in 1826, although large-scale manufacture of the microscopes did not begin until the 1840s; see Brian Bracegirdle, "J. J. Lister and the Establishment of Histology," *Medical History* 21 (1977): 187–191. Microscopy ultimately became the *sine qua non* of tumor classification; it also led to a more general understanding of metastasis and better statistical reporting on cancer incidence; see Stanley J. Reiser, *Medicine and the Reign of Technology* (Cambridge, England, 1978), pp. 77–79.

14. Lung cancer was so rare in 1869 that a physician at Charing Cross Hospital, upon discovering a case during an autopsy, showed it to his students and remarked that they might never see another case; see F. B. Smith, *The People's Health 1830–1910* (London, 1979), p. 329.

15. Jerry Adler et al., "The Killer We Don't Discuss," *Time* (December 27, 1993): 40. In the weeks following Nancy Reagan's breast cancer surgery in October 1987, the number of American women seeking mammograms nearly doubled; see "More Women Take X-Rays of Breast," *New York Times*, November 1, 1987.

16. The idea that city life led to the corruption of morals and therefore a higher statistical death rate was already expressed in Johann P. Süssmilch's *Die göttliche Ordnung in den Veränderungen des menschlichen Geschlechts* (Berlin, 1761–1762), the first work on vital statistics. No mention is made of cancer, however.

17. Tanchou, "Recherches," p. 313.

18. Stanislas Tanchou, "Sur la fréquence croissante des cancers," *Comptes rendues hebdomadaires des séances de l'Académie des Sciences* 18 (1844): 878–882; compare Stefansson, *Cancer*, pp. 24–31. The idea of diseases of civilization predates even Rousseau; see Carl Haffter, "Die Entstehung des Begriffs der Zivilisationskrankheiten," *Gesnerus* 36 (1979): 228–237, also Roy Porter's excellent "Diseases of Civilization," in the *Companion Encyclopedia of the History of Medicine*, edited by W. F. Bynum and Roy Porter, vol. 1 (London, 1993), pp. 585–599.

19. Scotto and Bailar, "Rigoni-Stern."

20. Domenico Rigoni-Stern, "Nota sulle richerche del dottor Tanchou intorno la frequenza del cancro," *Annali Universali di Medicina* (1844), pp. 110ff; Scotto and Bailar, "Rigoni-Stern," p. 66.

21. Le Conte, "On Carcinoma"; also his article by the same title in *Transactions of the Society of Alumni of College of Physicians and Surgeons of the State of New York* 1 (1842): 9–16.

22. Le Conte, "Statistical Researches on Cancer," pp. 273–274.

23. Stefansson, *Cancer,* p. 28.

24. John Le Conte, "Vital Statistics, and the True Coefficient of Mortality, Illustrated by Cancer," *Tenth Biennial Report of the State Board of Health of California* (Sacramento, 1888), pp. 1–19. Le Conte age-adjusts Tanchou's data, using the adjusted figures to correct the popular misconception that cancer was most common between the ages of 35 and 50. He shows instead that cancer incidence increases steadily with age. Compare also his "Vital Statistics: Illustrated by the Laws of Mortality from Cancer," *Western Lancet* (March 1872): 176–189.

25. Le Conte, "On Carcinoma," p. 301.

26. Williams, *Natural History,* pp. 12–49; Hoffman, *Cancer and Diet,* pp. 18–67; Hirsch, *Handbook,* vol. 3, pp. 505–508. J. Lyman Bulkley spent 12 years as a physician travelling among several different tribes of Alaskan natives, during which time he purportedly "never discovered among them a single true case of carcinosis"; see his "Cancer Among Primitive Tribes," *Cancer* 4 (1927): 289–295.

27. Albert Schweitzer, preface to Alexander Berglas, *Cancer: Nature, Cause and Cure* (Paris, 1957).

28. This was not for want of looking: Stefansson tells the remarkable story of how an amateur surgeon and sea captain by the name of Charles Leavitt set out in 1884 to determine whether cancer existed among the Eskimos. Leavitt found after nearly twenty years of travel throughout Alaska that while cancer was occasionally found among peoples in contact with settlers, it was virtually nonexistent where traditional Indian culture still flourished. Leavitt became convinced that western dietary patterns were to blame for cancer, as for scurvy and tooth decay; see Stefansson, *Cancer,* pp. 17–23, 64–69. John Higginson credits David Livingstone, the nineteenth century explorer, with having stated that cancer is a "disease of civilization"; see Maugh, "Cancer and Environment."

29. Park, "Further Inquiry," pp. 385–399. Park believed that infectious agents were responsible for the transmission of cancer from one person to another and from one part of the body to another. He recalls the experiments of one unnamed physician, published by Victor Cornil in 1891, where a woman had cancerous cells from one breast removed and transplanted into her healthy breast, apparently as part of an experiment to see if cancer could be induced in this manner. As hoped, the operation produced a cancerous nodule in the second breast within two months; see M. V. Cornil, "Sur les greffes et inoculations de cancer," *Bulletin de l'Académie de médecine* 25 (1891): 906–909.

30. Newsholme, "Statistics of Cancer," p. 371.

31. Scotto and Bailar, "Rigoni-Stern," p. 66.

32. Walshe, *Nature and Treatment of Cancer,* p. 141. Walshe also suggested that cancer was rare in Hobart Town and Calcutta, as among the natives of Egypt, Algiers, Senegal, Arabia, and the tropical parts of the Americas. Compare his statement that "wherever the disease is particularly rare, it may be remarked that a low state of civilization prevails; wherever social organization is of a highly perfect kind, there cancer flourishes" (pp. 161–170). Walshe illustrated this thesis with tables showing how "gentlemen and ladies" had higher cancer rates than servants, tavern keepers and sailors' wives.

33. King and Newsholme, "On the Alleged Increase of Cancer."

34. "Cancer has often been called a disease of civilisation, but that doctrine finds no support in the information before us"; see Hirsch, *Handbook,* pp. 505–508. Hirsch reported that cancer was rare in Greenland, Iceland, Turkey, Greece, Syria, Persia, Arabia, Egypt, Tunis, Algiers, Abyssinia, Senegambia, and the Faroe Islands. In the 1862–1864 edition of his book, Hirsch had expressed skepticism about cancer as a disease of civilization, but made no reference to the apparent increase of cancer; see his *Handbuch der historisch-geographischen Pathologie,* vol. 2 (Erlangen, 1862–1864), pp. 377–381.

35. Newsholme, "Statistics of Cancer," pp. 382–383. Ernest F. Bashford and James A. Murray of the Imperial Cancer Research Fund in Britain challenged the idea that cancer was on the increase in *The Statistical Investigation of Cancer* (London, 1905).

36. Williams, *Natural History,* pp. 52–55. Williams had earlier criticized King and Newsholme for considering only cancers of the uterus, breast, vagina, and tongue under the rubric of "accessible cancers," leaving all others—skin, mouth, lips, and external genitalia—classed as "inaccessible." This was an especially serious error, he claimed, given that the latter accounted for about 40 percent of all cancers in males; see his "The Question of the Increase of Cancer," *British Medical Journal* (December 30, 1893): 1450–1451.

37. Hoffman *Mortality from Cancer,* pp. 30–33, vii, 218–219. Hoffman was a prominent antiphilanthropist, racial Darwinist, and natural diet enthusiast; see John S. Haller, *Outcasts from Evolution* (Urbana, Ill., 1971), pp. 60–68. Hoffman was also the author of, among other things, *The Indian as a Life Insurance Risk* (Newark, N.J., 1928).

38. Hoffman, *Mortality from Cancer,* pp. vii, 27; for later data see Major Greenwood, "A Review of Recent Statistical Studies of Cancer Problems," *The Cancer Review* 3 (1928): 97–107; Patterson, *Dread Disease,* pp. 78–81, 338 n.74.

39. Patterson, *Dread Disease,* pp. 88–89.

40. Walter F. Willcox, "On the Alleged Increase of Cancer," *Publications of the American Statistical Association* 15 (1916–1917): 701–782.

41. Dublin and Lotka, *Twenty-Five Years of Health Progress,* pp. 181–185. Dublin and Lotka's conclusions were based on an analysis of 262,046 cancer deaths among the industrial policyholders of the Metropolitan Life Insurance Company. Compare also Dublin et al.'s earlier *Mortality Statistics of Insured Wage-Earners and Their Families* (New York, 1919).

42. Haviland, "Medical Geography," p. 401.

43. Sibley, "Contribution," p. 114; compare also Williams, *Natural History,* pp. 312–316. In 1862, W. M. Baker showed that were it not for cancer of the breast, male cancers would greatly outnumber female; see his "Contribution to the Statistics of Cancer," *Medico-Chirurgical Transactions* 45 (1862): 389–406.

44. Le Conte, "Vital Statistics," pp. 11–14. Herbert Snow of the London Cancer Hospital believed that "mental distress" was a leading cause of cancer, explaining why women, "the more neurotic and emotional sex," were its principal victims. He also noted that laundresses appeared to furnish a disproportionate number of cases; see his "Increase of Cancer: Its Probable Cause," *The Nineteenth Century* (July 1890): 84–85.

45. Jens Paulsen, "Konstitution und Krebs," *Zeitschrift für Krebsforschung* 21 (1924): 126. In 1932 Antoine Lacassagne reported the first experimental induction of malignant tumors by a sex hormone; see his "Apparition de cancers de la mamelle chez le souris mâle, soumise à des injections de folliculine," *Comptes rendues de l'Académie des Sciences, Paris* 195 (1932): 630–32. William S. Murray had earlier shown that castrated male mice into which ovaries had been grafted developed high rates of breast cancer ("Ovarian Secretion and Tumor Incidence," *Journal of Cancer Research* 12 [1928]: 18–25). Lacassagne later showed that diethylstilbestrol (DES), a synthetic estrogen, could induce mammary tumors in male mice of susceptible strains; see his "Apparition d'adénocarcinomes mammaires chez des souris mâles," *Comptes rendues de la Société Biologique* 129 (1938): 641–43. DES was banned in the United States in 1971, after several daughters of women who had taken this drug (used since the 1950s to prevent miscarriage) developed vaginal cancer; see Alan I. Marcus, *Cancer from Beef: DES, Federal Food Regulation, and Consumer Confidence* (Baltimore, Md., 1994).

46. Jean L. Marx, "Estrogen Use Linked to Breast Cancer," *Science* 245 (1989): 593.

47. King and Newsholme, "On the Alleged Increase."

48. Dublin and Lotka, *Twenty-Five Years of Health Progress,* pp. 167–170.

49. Wilhelm Weinberg and K. Gastpar, "Die bösartigen Neubildungen in Stuttgart von 1873 bis 1902," *Zeitschrift für Krebsforschung* 2 (1904): 195–260.

50. Harold F. Dorn, "Variations in Cancer Incidence in the United States," in Clemmesen, *Symposium,* pp. 19–20.

51. Devra Lee Davis et al., "Medical Hypothesis: Xenoestrogens as Preventable Causes of Breast Cancer," *Environmental Health Perspectives* 101 (1993): 372–377; Beardsley, "War Not Won," p. 137; Janet Raloff, "EcoCancers: Do Environmental Factors Underlie a Breast Cancer Epidemic?" *Science News* (July 3, 1993): 10–13. The idea that petrochemicals may cause cancer by mimicking estrogens dates back at least to the 1930s: see Adolf Butenandt, *Neue Beiträge der biologischen Chemie zum Krebsproblem* (Berlin, 1940), pp. 27–38. Nancy Krieger has presented evidence questioning the thesis that women with breast cancer have higher levels of DDT derivatives in their tumors; see Nancy Krieger et al., "Breast Cancer and Serum Organochlorines: A Prospective Study Among White, Black, and Asian Women," *Journal of the National Cancer Institute* 86 (1994): 589–599.

52. See Higginson, "Present Trends."

53. Roush et al., *Cancer Risk and Incidence Trends.*

54. "Cancer Facilities and Services; A Report from the National Advisory Cancer Council," *Journal of the National Cancer Institute* 6 (1946): 239–302.

55. John Higginson, "Population Studies in Cancer," *Acta Unio Internationalis Contra Cancrum* 16 (1960): 1667–1670. The age-adjusted incidence of cancer in the United States was about twice that of the Bantu.

56. Ernst L. Wynder, Frank R. Lemon, and Irwin J. Bross, "Cancer and Coronary Artery Disease Among Seventh-Day Adventists," *Cancer* 12 (1959): 1016–1028.

57. Clemmesen, *Symposium.*

58. World Health Organization, *Annual Epidemiological and Vital Statistics* (Geneva, 1955), p. 300.

59. Richard Doll, Peter Payne, and John Waterhouse, eds., *Cancer Incidence in Five Continents,* vols.

1-6 (Berlin, 1966–1992); World Health Organization, *Mortality from Malignant Neoplasms, 1955–1965,* vols. 1, 2 (Geneva, 1970); Howe, *Global Geocancerology,* pp. 3–42.

60. Higginson, "Population Studies," p. 1669.

61. See, for example, "Are 90% of Cancers Preventable?" *Lancet* 1 (1977): 685–687.

62. Bruno Latour, *The Pasteurization of France* (Cambridge, Mass., 1988).

63. George Bell, *Thoughts on the Cancer of the Breast* (Birmingham, England, 1788), pp. 6–10.

64. Richard J. Behan, *Relation of Trauma to New Growths* (Baltimore, Md., 1939), p. 7.

65. John Hill, *Cautions Against the Immoderate Use of Snuff* (London, 1761), pp. 27–38; Samuel T. Soemmerring, *De morbis vasorum absorbentium corporis humani* (Frankfurt, 1795), p. 109. One of the earliest Europeans apparently to die from tobacco-induced cancer was the English naturalist, Thomas Harriot (1560–1621). Harriot brought pipe smoking back to England, where he continued to smoke and eventually died of lip cancer, although no links were drawn at that time between the habit and the disease. Jean Nicot (1530–1600), for whom nicotine is named, believed that tobacco could actually cure certain kinds of tumors; see Juraj Körbler, "Der Tabak in der Krebslehre zu Anfang des 19. Jahrhunderts," *Proceedings of the 21st International Congress of the History of Medicine,* vol. 2 (London, 1969), pp. 1179–1183.

66. Sigerist, "Historical Development," p. 651; Charles Oberling, *The Riddle of Cancer* (New Haven, 1952), p. 37.

67. Pott, *Chirurgical Observations.* For background on Pott and a reprint of his pamphlet, see Potter, "Percivall Pott's Contribution."

68. Historian Donald Hunter notes that the children employed in this trade were often forced by their masters "to climb by means of a fire of straw lighted beneath them. Submerged in soot in narrow or horizontal flues, many were suffocated. They usually slept in cellars with a bag of soot for a bed and another for a coverlet"; see his *Diseases of Occupations,* pp. 142–146; compare also Walter Besant, *London in the Eighteenth Century* (London, 1902), pp. 386–387. Percivall Pott originally reported that he had never seen scrotal cancer in a chimney sweep under the age of puberty, but his son-in-law and editor of his collected works, Sir James Earle, later brought to his attention a case in a child not yet eight; see Pott's *Chirurgical Works,* vol. 3 (London, 1790), pp. 257–258. The miseries of English sweeps are well documented in *Improving the Lot of the Chimney Sweeps: One Book and Nine Pamphlets, 1785–1840* (New York, 1972).

69. Astley Cooper, *Observations on the Structure and Diseases of the Testis* (London, 1841), pp. 325–330; T. B. Curling, *A Practical Treatise on the Diseases of the Testis, and of the Spermatic Cord and Scrotum,* 2nd ed. (London, 1856), p. 507. Pott's discovery inspired the Danish Chimney Sweeps Association in 1778 to recommend daily bathing to all its members.

70. Potter, "Percivall Pott's Contribution," p. 3. No one knows why chimney soot causes cancer almost exclusively on the scrotum, but some radiation biologists suggest that soot contains radioactive alpha emitters that penetrate the relatively thin layer of dead skin on the scrotum, inducing mutations in the living tissues beneath (personal communication, Geoffrey Sea).

71. Thomas Oliver, ed., *Dangerous Trades* (New York, 1902), p. 237. A line drawing of the gardener's cancerous hand can be found in Paget, *Lectures,* p. 599.

72. Ramazzini, *De morbis artificum.*

73. Ibid.

74. Ibid., p. 191.

75. In 1906, H. Hearsey, principal medical officer in British Central Africa, suggested that the growing tendency to wean infants at a very early age might explain the high incidence of breast cancer among "civilized communities"; see his "The Rarity of Cancer Among the Aborigines of British Central Africa," *British Medical Journal* 2 (1906): 1562–1563.

76. Paget, *Lectures,* p. 680; Samuel J. Kowal, "Emotions as a Cause of Cancer: 18th and 19th Century Contributions," *Psychoanalytic Review* 42 (1955): 217–227.

77. Hugh P. Dunn, "An Inquiry Into the Causes of the Increase of Cancer," *British Medical Journal* (April 14, 1883): 708–761.

78. A comprehensive and rather tedious account of early theories of cancer can be found in Wolff, *The Science of Cancerous Disease.*

79. Willard Parker, *Cancer: A Study of Three Hundred and Ninety-Seven Cases of Cancer of the Female Breast* (New York, 1885), p. 38.

80. Friedrich W. Beneke, *Constitution und constitutionelles Kranksein des Menschen* (Marburg, 1881).

81. Williams, *Natural History,* p. 352.

82. Lung cancer was still a rarity in the early years of the twentieth century. In his famous monograph, *Primary Malignant Growths of the Lungs and Bronchi* (New York, 1912), I. Adler noted that lung malignancies were "among the rarest forms" of cancer. Adler believed that while overall incidence rates were probably not rising, lung cancer incidence seemed to be showing "a decided increase"

(pp. 3, 7). He also recognized, following David von Hansemann's comparisons of autopsies and diagnoses, that at least 20 percent of all cancers were never diagnosed (p. 5).

83. Jonathan Hutchinson, "Arsenic Cancer," *British Medical Journal* 2 (1887): 1080–1081; James Whorton, *Before Silent Spring: Pesticides & Public Health in Pre-DDT America* (Princeton, N.J., 1974), pp. 52–53, 262 n.36. In 1898, physicians discovered a high rate of tumors among the inhabitants of Reichenstein, a Silesian town whose spring water became contaminated with arsenic as it filtered through mine tailings; see Hueper, *Occupational Tumors*, p. 40.

84. James N. Hyde, "On the Influence of Light in the Production of Cancer of the Skin," *American Journal of the Medical Sciences* 131 (January 1906): 13–14.

85. William J. Elmslie, "Etiology of Epithelioma Among the Kashmiris," *Indian Medical Gazette* 1 (1866): 324–326; Ernest F. Neve, "Kangri-burn Cancer," *British Medical Journal* 2 (1923): 1255–1256. Elmslie notes that similar practices were not unknown in Europe: children employed in the straw plait districts in England, for example, were said to have used such pots to warm themselves, and Elmslie himself observed the inhabitants of Florence using such devices. Neve originally attributed kangri cancer solely to the irritation of heat; his 1923 paper adds that "the volatile substances resulting from the combustion of the wood" may play a secondary part. A similar disease ("kairo cancer") was observed in the 1930s among Japanese women who carried coals in a wooden container to keep warm. Wilhelm Hueper classed both kangri and kairo cancers as soot or tar cancers; see his *Occupational Tumors*, pp. 84–85, 293.

86. Reginald Harrison, "Specimens of Bilharzia Affecting the Urinary Organs," *Lancet* 2 (1889): 163. The best history of bilharzia is John Farley, *Bilharzia: A History of Imperial Tropical Medicine* (Cambridge, England, 1991), a book that unfortunately does not even mention the cancer hazard posed by the parasite.

87. W. J. Niblock, "Cancer in India," *Indian Medical Gazette,* 37 (1902): 161–163; I. M. Orr, "Oral Cancer in Betel-Nut Chewers in Travancore," *Lancet* 2 (1933): 575–580. In the 1940s, V. R. Khanolkar of Bombay reported three new types of cancer in India: a cancer of the lower trunk and groin associated with the Deccani Hindu custom of wearing—even while bathing—a tight-fitting form of trousers known as a *dhoti;* a cancer of the hard palate linked to the practice, common among the women of Vizagapatam and the outlying districts of Andhra Province, but also among Venezuelan women of African descent, of smoking handmade cigars (*chutta*) in which the burnt end is placed in the mouth; and a cancer of the lower lip found among Bihar men who chew *khaini,* a mixture of powdered tobacco and lime; see V. R. Khanolkar and B. Suryabai, "Cancer in Relation to Usages: Three New Types in India," *Archives of Pathology* 40 (1945): 351–361.

88. John A. Paris, *Pharmacologia,* 5th ed. (London, 1822), p. 208. Paris's observations are disputed; see Ernest L. Kennaway, "A Contribution to the Mythology of Cancer Research," *Lancet* 243 (1942): 769–772; also Wilhelm C. Hueper, "Mythology of Cancer Research: A Reply to Professor Kennaway," *Lancet* 244 (1943): 538–540.

89. The first known case of a skin cancer caused by exposure to tar distillates was reported in 1876; the first cancer of the skin caused by paraffin was reported in 1875. Both were in the German State of Saxony, and both were associated with the lignite coal distillation industry. These and many other examples of environmental carcinogenesis are reviewed in Hueper's *Occupational Tumors;* compare Haagensen, "Occupational Neoplastic Disease"; also his "An Exhibit of Important Books." Another good review is Shimkin and Triolo, "History of Chemical Carcinogenesis," pp. 1–20.

90. C. G. Santesson, "Chronische Vergiftungen mit Steinkohlentheerbenzin: Vier Todesfälle," *Archiv für Hygiene und Bakteriologie* 31 (1897): 336–376.

91. Ludwig Rehn, "Blasengeschwülste bei Fuchsin-Arbeitern," *Archiv für klinische Chirurgie* 50 (1895): 588–600.

92. S. G. Leuenberger, "Die unter dem Einfluss der synthetischen Farbenindustrie beobachtete Geschwulstentwicklung," *Beiträge zur klinischen Chirurgie* 80 (1912): 208–316.

93. Hueper, *Occupational Tumors,* pp. 470–471. Otto Teutschlaender fashioned a similar chain of reasoning in his "Arbeit und Geschwulstbildung," *Monatsschrift für Krebsbekämpfung* 1 (1933): 106.

94. Butlin, "Three Lectures," esp. 2 (1892): 71.

95. In *Cancerous and Cancroid Growths* (Edinburgh, 1849), J. H. Bennet wrote, "If a tendency to fat be an antidote to tubercle, as I believe it is, spareness may possibly be considered opposed to cancer" (p. 251).

96. Friedrich Beneke maintained that cancer was "extremely rare" among herbivores; see his *Constitution und constitutionelles Kranksein des Menschen* (Marburg, 1881), p. 75. Beneke also maintained that certain individuals were predisposed to cancer by virtue of their "carcinomatöse Constitution" (p. 74).

97. W. Roger Williams, "The Continued Increase of Cancer, with Remarks as to Its Causation," *Medical Chronicle* (Manchester), n.s. 5 (1896–1897): 325.

98. "The Month," *The Practitioner* 62 (1899): 369.

99. Henry T. Butlin, "Report on Inquiry No. XIII, Cancer (of the Breast Only)," *British Medical Journal* (February 26, 1887): 439.

100. Ibid., pp. 439–440. Geological theories were common in this period. In 1868, Alfred Haviland pondered why cancer rates in England and Wales appeared highest where the predominant rock was limestone and lowest in regions of flooded clays ("Medical Geography," p. 416). German scholars published similar speculations; see Karl Kolb, *Der Einfluss von Boden und Haus auf die Häufigkeit des Krebses* (Munich, 1904).

101. "The Month," *The Practitioner* 62 (1899): 369. The skeptic is cited in Hoffman, *Cancer and Diet,* p. 470.

102. Patterson, *Dread Disease,* pp. 59–60.

103. Hoffman, *Cancer and Diet,* pp. 167–168, 652–662.

104. Carlo Moreschi, "Beziehungen zwischen Ernährung und Tumorwachstum," *Zeitschrift für Immunitätsforschung* 2 (1909): 651–685.

105. Peyton Rous, "The Rate of Tumor Growth in Underfed Hosts," *Proceedings of the Society for Experimental Biology and Medicine* 8 (1910–1911): 128–130; also his "The Influence of Diet on Transplanted and Spontaneous Mouse Tumors," *Journal of Experimental Medicine* 20 (1914): 433–451.

106. See Michael W. Pariza and Roswell K. Boutwell, "Historical Perspective: Calories and Energy Expenditure in Carcinogenesis," *American Journal of Clinical Nutrition* 45 (1987): 151–156.

107. Le Conte, "Vital Statistics," p. 17.

108. Moore, "The Antecedent Conditions of Cancer," p. 209.

109. James Sawyer, "Note on the Causes of Cancer," *Lancet* (March 24, 1900): 849.

110. Cope, *Cancer,* p. 137. Cope claimed that "masculinism" was a major cause of cancer among emancipated women, as was the unnatural irritation of physical contraceptives (pp. 146–168).

111. In his extraordinary *Krebsverbreitung, Krebsbekämpfung, Krebsverhütung,* Erwin Liek, the patron saint of Nazi medicine, argued that the "unnatural lifestyle" of the *Kulturländer* (use of nitrates to preserve foods, bleaches to whiten bread, metal salts and aniline dyes to color foods, and so forth) was responsible for their elevated cancer rates (pp. 161–208). Kurt Blome, a leading Nazi health official, claimed that since cancer deaths were underreported by as much as 30–50 percent, cancer was probably the leading cause of death in Germany. He, too, blamed "the denaturing of our food through civilization"; see his "Krebsforschung und Krebsbekämpfung," *Die Gesundheitsführung (Ziel und Weg)* 10 (1940): 406–412. Clarence Cook Little, the fearless Spencerian geneticist and tobacco apologist, reversed the usual metaphor in his *Civilization Against Cancer* (New York, 1939), characterizing cancer cells as disobedient, uncontrolled, and "anarchist." Little did worry that Americans were on the verge of trading their pioneer hardihood for a "pottage of mechanized indolence," but he also stressed that science-based civilization was "the last word in social effectiveness" and the best hope for a cancer cure (pp. 30, 31, 37, 142).

CHAPTER 2
THE ENVIRONMENTALIST THESIS

1. Several examples of early transplantation research are reprinted in Shimkin, *Some Classics of Experimental Oncology,* pp. 58–126.

2. Clunet, *Recherches;* Katsusaburo Yamagiwa and Koichi Ichikawa, "Experimentelle Studie über die Pathogenese der Epithelialgeschwülste," *Mitteilungen aus der medizinischen Fakultät der kaiserlichen Universität zu Tokyo* 15 (1916): 295–344; A. E. C. Lathrop and Leo Loeb, "Further Investigations on the Origin of Tumors in Mice," *Journal of Cancer Research* 1 (1916): 1–19.

3. Ernest Kennaway, "The Identification of a Carcinogenic Compound in Coal-Tar," *British Medical Journal* 2 (1955): 749–752.

4. Harold F. Blum, "General History and Outline of the Concept," *National Cancer Institute Monograph* 50 (1978): 211–212. Paul G. Unna is usually credited as the first to associate skin cancer with sunlight; see his *Die Histopathologie der Hautkrankheiten* (Berlin, 1894).

5. J. C. Bridge and S. A. Henry, "Industrial Cancers," *Report of the International Conference on Cancer* (London, 1928), pp. 258–268; Hueper, "Causal and Preventive Aspects."

6. A. H. Southam and S. R. Wilson, "Cancer of the Scrotum," *British Medical Journal* 2 (1922): 971; Haagensen, "An Exhibit of Important Books," p. 119.

7. William H. Woglom, "Experimental Tar Cancer," *Archives of Pathology* 2 (1926): 533–576, 709–754; M. G. Seelig and Zola K. Cooper, "A Review of the Recent Literature of Tar Cancer (1927–1931 Inclusive)," *American Journal of Cancer* 17 (1933): 589–667.

8. Hoffman, *Mortality from Cancer,* pp. 48–76, 305–315.

9. Patterson, *Dread Disease,* p. 74.
10. Wilhelm C. Hueper, "Leukemoid and Leukemic Conditions in White Mice with Spontaneous Mammary Carcinoma," *Folia Haematologica* 52 (1934): 167–178.
11. World Health Organization, *Prevention of Cancer* (Geneva, 1964), p. 4.
12. Peter Hirtle, personal communication; Hueper, "Cigarette Theory."
13. Hueper, *Occupational Tumors,* pp. 403–406; also his "Silicosis, Asbestosis, and Cancer of the Lung, *American Journal of Clinical Pathology* 25 (1955): 1388–1390. See also chapter 5.
14. Lester Breslow, *A History of Cancer Control in the United States 1946–1971,* book 1 (Washington, D.C., 1979), pp. 132–144.
15. Wilhelm C. Hueper, "Über die histologischen Veränderungen im menschlichen Gewebe nach Injektion von Paraffin," *Frankfurter Zeitschrift für Pathologie* 29 (1923): 268–286.
16. Wilhelm C. Hueper, "Primary Gelatinous Cylindrical Cell Carcinoma of the Lung," *American Journal of Pathology* 2 (1926): 81–91; also his "Carcinoma of the Lung," *Illinois Medical Journal* 52 (1929): 296–304.
17. Hueper, "Cigarette Theory"; Waldemar Kaempffert, "Smoking and Cancer," *New York Times,* August 1, 1954; Waldemar Kaempffert, "Smoking Factor Doubted," *New York Times,* December 8, 1954. Morris Fishbein, editor of the *Journal of the American Medical Association* (1924–1949), the statistician-geneticist Ronald A. Fisher, and Joseph Berkson of the Mayo Clinic all believed that the role of tobacco in cancer causation had been exaggerated. On Fisher's views, see Stephen Jay Gould, "The Smoking Gun of Eugenics," *Natural History* (December 1991): 8–17.
18. By the fall of 1932 the company had at least 20 workers with tumors; see Hounshell and Smith, *Science and Corporate Strategy,* p. 559.
19. Hueper, "Adventures," pp. 139–141; also Michaels, "Waiting for the Body Count."
20. Hueper, "Adventures," p. 143.
21. Hounshell and Smith, *Science and Corporate Strategy,* pp. 560–563; Hueper, "Adventures," p. 141.
22. Hueper, "Adventures," pp. 154–155.
23. Hueper published four articles on bladder cancer in 1938; see, for example, his paper with F. H. Wiley and H. D. Wolfe, "Experimental Production of Bladder Tumors in Dogs by Administration of Beta-Naphthylamine," *Journal of Industrial Hygiene and Toxicology* 20 (1938): 46–84.
24. Hueper, "Adventures," p. 152.
25. Michaels, "Waiting for the Body Count," p. 218.
26. Hueper, "Adventures," pp. 156–157.
27. Ibid., p. 159.
28. Incredibly, Du Pont executives in the mid-1980s would cite Hueper's record with the company as proof of their longstanding concern for worker health and safety; see Bruce W. Karrh, "An Illustration of Voluntary Actions to Reduce Carcinogenic Risks in the Workplace," in Deisler, *Reducing the Carcinogenic Risks.*
29. Hueper, *Occupational Tumors,* pp. 4–6. Hueper had originally intended to dedicate his book to the memory of those who had made "better chemicals for a better living of others"; shortly before publication he rededicated it to "the memory of those of our fellow men who have died from occupational diseases contracted while making better things for an improved living for others." The rewording represented a twist on Du Pont's slogan, "Better things for better living through chemistry"; see Hueper, "Adventures," p. 163; also Agran, *Cancer Connection.*
30. Hueper, "Adventures," p. 46.
31. Hueper, *Occupational Tumors,* pp. 154, 162, 196–198.
32. See Arthur Newsholme's extraordinary "Statistics of Cancer," p. 379. Newsholme's article presents an impressive discussion of the importance of age-adjustment, random cancer clusters ("cancer houses"), differential cancer morality rates for England and Ireland and for different occupations, and the various kinds of artifacts that can distort statistical analyses. The notion of "cancer houses" dates back at least to T. Law Webb, "The Etiology of Cancer," *Birmingham Medical Review* (December 1892): 342. Alfred Haviland spoke of "cancer-fields" and "cancer haunts" ("Medical Geography," pp. 400–417), while Levin spoke of "cancerous fraternities" (*Zeitschrift für Krebsforschung* 9 [1912]: 5).
33. Hueper, *Occupational Tumors,* pp. 194–195.
34. Ibid., pp. 42, 59–60. Hueper argued that engineers should design and outfit factories with carcinogenic hazards in mind; see his "Industrial Management and Occupational Cancer," *JAMA* 131 (1946): 738.
35. *Archives of Pathology* 34 (1942): 610. This anonymous reviewer called the book "one of the most valuable reference and source books which has appeared in recent years"; the book was "destined to become widely used . . . because of the store of information which it contains."
36. Hueper, "Adventures," p. 269. Hueper was probably the author of "Environmental Cancer," *JAMA*

126 (1944): 836, and of "Asbestosis and Cancer of the Lung," *JAMA* 140 (1949): 1219–1220.

37. Starr, *Social Transformation*, pp. 341–343.

38. Ibid., pp. 343–344.

39. Thomas F. Mancuso, "Occupational Cancer and Other Health Hazards in a Chromate Plant: A Medical Appraisal," *Industrial Medicine and Surgery* 20 (1951): 393–407. Mancuso elsewhere cited this paper with Hueper as coauthor: see Paul F. Urone and Thomas F. Mancuso, "A Spectrophotometric and Chemical Study of Chromium in Human Blood," *Industrial Medicine and Surgery* 20 (1951): 440. A. J. Lanza met with Surgeon General Leonard A. Scheele and several other PHS representatives on July 24, 1951, to call for the suppression of the Hueper–Mancuso paper. George A. Benington, president of the Mutual Chemical Co. of America, the chromate producer in question, thanked Lanza for his "outstanding job" defending the industry against Hueper (Lanza to George A. Benington et al., July 31, 1951; Benington to John Sargent, August 2, 1951). I'd like to thank Barry Castleman for bringing these letters to my attention.

40. Hueper, "Adventures," p. 203.

41. Ibid., pp. 174–179, 264–265. Former interior secretary Stewart L. Udall confirmed Hueper's account in a 1979 interview with Warren; see his *Myths of August* (New York, 1994), p. 192.

42. Telephone interview with Victor E. Archer, April 10, 1992.

43. Hueper, "Adventures," pp. 245–247.

44. Hueper's June 8, 1959, memo to John Heller, director of the NCI, titled "A 'Dark' and 'Somewhat Shady' History of the Environmental Cancer Section at the National Cancer Institute," was published in the *Drug Research Reports* ("The Blue Sheets") (September 13, 1961): 354-S to 360-S.

45. Ibid.

46. Hueper, "Adventures," p. 238.

47. Richard Pearson, "Dr. Wilhelm Hueper, Retired Official of National Cancer Institute, Dies at 84," *Washington Post*, January 1, 1979.

48. Hueper, "Adventures," p. 300.

49. Ibid., pp. 221–229.

50. Ibid., pp. 226–236, 300–301.

51. Harold L. Stewart, "Spoken on the Occasion of the National Institutes of Health Director's Award to Dr. Wilhelm C. Hueper on Friday, September 1, 1978," *Journal of the National Cancer Institute* 62 (1979): 719.

52. Hueper, "Adventures," p. 214. Hueper derided the viral theory of cancer as "a supposition based on circumstantial evidence, but not a scientific fact" (p. 215). To those (Egon Lorenz, for example) who suggested that the cancers among Europe's uranium workers might be due to inbreeding, he pointed out that both Schneeberg and Joachimsthal had railway stations, "a fact which should make inbreeding rather difficult" (p. 205).

53. See, for example, the essays in Rosner and Markowitz, *Dying for Work.*

54. J. T. Arlidge *Hygiene, Diseases and Mortality of Occupations* (London, 1892); Thomas Oliver, ed. *Dangerous Trades: The Historical, Social, and Legal Aspects of Industrial Occupations as Affecting Health* (New York, 1902); and Alice Hamilton, *Industrial Poisons in the United States* (New York, 1929), pp. 209–211, 412–416. Oliver's *Diseases of Occupations* (London, 1908) devotes very little attention to cancer (see pp. 24–25); far greater attention is devoted to accidents, and when he does discuss cancer it is primarily in reference to cancers caused by accidents (conceived as "trauma," as in a blow to the neck).

55. George Ordish, "150 Years of Crop Pest Control," *World Review of Pest Control* 7 (1968): 204–213. The socialist Murray Bookchin published a similar polemic six months prior to the appearance of Carson's *Silent Spring*, drawing also from Hueper, but the book received none of the attention garnered by Carson's; see his *Our Synthetic Environment* (New York, 1962), published under the pseudonym "Lewis Herber."

56. Carson, *Silent Spring.*

57. Ibid., pp. 25, 47, 179.

58. Ibid., pp. 31–32; compare also Herber, *Our Synthetic Environment*, pp. 141–142.

59. Carson, *Silent Spring*, pp. 221, 226–227.

60. Ibid., p. 238.

61. Ibid., pp. 50–51, 239.

62. Robert Merideth, *The Environmentalist's Bookshelf: A Reader's Guide to the Best* (New York, 1992).

63. William J. Darby, "Silence, Miss Carson," *Chemical and Engineering News* (October 1, 1962): 60–62. Darby, a former chairman of the National Academy of Science's Food Protection Committee, ridiculed Carson's book for ignoring "the essentiality of use of agricultural chemicals."

64. Graham, *Since Silent Spring*, pp. 60, 65; Brooks, *House of Life*, pp. 294–307.

65. Graham, *Since Silent Spring*, pp. 59, 65–67. Robert White Stevens of American Cyanamid gave at

least 28 speeches against the book; see the 1993 film *Rachel Carson* produced by Neil Goodwyn for WGBH and *The American Experience*.

66. Carson's new physician was George Crile, author of *Cancer and Common Sense* (New York, 1955). Carson had admired his book when first published and sought medical advice from him when she found out she had breast cancer; see Brooks, *House of Life*, p. 265.
67. Sandra Steingraber, "'If I Live to Be 90 Still Wanting to Say Something': My Search for Rachel Carson," in Stocker, *Confronting Cancer*, pp. 190–191.
68. Graham, *Since Silent Spring*, p. 94.
69. Carson's chapter on cancer in *Silent Spring* does not even mention the tobacco hazard. An earlier chapter draws attention to the increasing arsenic content of cigarettes (pp. 58–59), but nowhere is there even a hint that smoking by itself might cause cancer.

CHAPTER 3
THE PERCENTAGES GAME

1. Thomas McKeown, *The Role of Medicine: Dream, Mirage, or Nemesis?* (Princeton, 1979), also his *The Modern Rise of Population* (New York, 1976); Sontag, *Illness as Metaphor;* Chowka, "National Cancer Institute," pp. 22–27; Aaron Wildavsky, "Doing Better and Feeling Worse: The Political Pathology of Health Policy," *Daedalus* (Winter 1977): 105; Ivan Illich, *Medical Nemesis* (New York, 1975).
2. Griffin, *World Without Cancer,* pp. 61, 391, 522–526. Griffin recounted the story of the Hunza, the Himalayan mountain people "world renowned for their amazing longevity" among whom "there never has been a case of cancer." The key to this people's long life, he argued, was their ingestion of large amounts of apricot pits containing vitamin B_{17}. Milton Friedman's advice appears in his "Frustrating Drug Advancement," *Newsweek* (January 8, 1973): 49.
3. Vicente Navarro, "The Underdevelopment of Health of Working America: Cause, Consequences and Possible Solutions," *American Journal of Public Health* 66 (1976): 538–547; Ozonoff, "Political Economy," pp. 13–14.
4. John Higginson and A. G. Oettlé, "Cancer Incidence in the Bantu and 'Cape Colored' Races of South Africa," *Journal of the National Cancer Institute* 24 (1960): 589–671; World Health Organization, *Prevention*, p. 4.
5. Higginson, "Present Trends," pp. 40–75; Cairns, "Cancer Problem," p. 64.
6. Greenberg and Randal, "Waging the Wrong War," p. C1.
7. In William J. Broad, "New Strength in the Diet-Disease Link?" *Science* 206 (1979): 666–668.
8. "Are 90% of Cancers Preventable?" *Lancet* 1 (1977): 685–687.
9. Samuel S. Epstein, "Environmental Determinants of Human Cancer," *Cancer Research* 34 (1974): 2425–2435; compare also Gertrude Barna-Lloyd, "'Environmentally' Caused Cancers," *Science* 202 (1978): 469.
10. Rachel Carson Trust, "The Bicentennial of Preventable Cancer" (n.p., 1975), p. 1.
11. Cristine Russell, "The Environment as the Major Cause of Cancer," *Washington Star,* January 12, 1976.
12. Efron, *Apocalyptics,* pp. 387–488.
13. Shimkin and Triolo, "History of Chemical Carcinogenesis," p. 7.
14. Maugh, "Cancer and Environment," p. 1363.
15. Ibid., p. 1364; Higginson and Muir, "Environmental Carcinogenesis," p. 1293.
16. Herbert Black, "Cancer Study Shows Pollutants Possible Cause," *Boston Globe,* March 30, 1969.
17. William Lijinsky and Samuel S. Epstein, "Nitrosamines as Environmental Carcinogens," *Nature* 225 (1970): 21–23.
18. Epstein, "Control of Chemical Pollutants," p. 816.
19. Epstein, *Politics of Cancer,* pp. 13–17.
20. Ibid., pp. 422–434; also his "Polluted Data."
21. Jack Mabley, "Politics Hampers Cancer Prevention," *Chicago Tribune,* November 6, 1978. The Velsicol Chemical Company continued producing chlordane and heptachlor, sold under the brand name "Termide," until 1987, when the product was finally withdrawn. By this time, more than 24 million homes had been treated; see "Maker to Stop Sale of 2 Cancer-Tied Pesticides," *New York Times,* August 12, 1987. For an update on the Velsicol story, see Samuel S. Epstein, "Corporate Crime: Can We Trust Industry-Derived Safety Studies?" *The Ecologist* 19 (1989): 23–30.
22. Epstein, *Politics of Cancer,* p. 327.
23. Samuel Epstein, "Epidemic! The Cancer-Producing Society," *Science for the People* (July 1976): 4–11.

24. Samuel Epstein, "Billions for Cures, Barely a Cent to Prevent," *Environmental Action* (November–December 1984): 10–11.
25. Epstein, *Politics of Cancer,* pp. 472–509.
26. *San Francisco Examiner,* October 30, 1978, p. 22.
27. Doyal and Epstein, *Cancer in Britain.*
28. *Chemical and Engineering News* (January 28, 1980): 47–48; *New York Magazine* (November 13, 1978): 13.
29. *Washington Report,* Oct. 30, 1978, p. 3; *Oil, Chemical, and Atomic Workers Union News* (December 1978–January 1979): 3; William Tucker, "The Environment of Disease," *Washington Post Book World* (December 24, 1978).
30. Michael Halberstam, review of Epstein's *Politics of Cancer,* in *New York Magazine* (November 13, 1978): 131; William Tucker, "The Environment of Disease," *Washington Post Book World* (December 24, 1978).
31. Richard Peto, "Distorting the Epidemiology of Cancer: The Need for a More Balanced Overview," *Nature* 284 (1980): 297–300.
32. Ibid., pp. 298–300.
33. Samuel S. Epstein, "Theories of Cancer" (letter), *Nature* 289 (1981): 115–116.
34. Epstein and Swartz, "Fallacies." In their *Causes of Cancer,* Doll and Peto disputed Epstein's assertion that 20 percent of the annual victims of lung cancer (100,000 in the late 1970s) were nonsmokers. They estimated instead that, if no American had ever smoked, there would be about 12,000 American deaths per annum from lung cancer (p. 1222).
35. "Caution: Industry Is Hazardous to Your Health," *The Capital* Times (Madison), September 24, 1979, pp. 21–22.
36. John Mathews, "Cancer: The Politics of the 'New Plague'" *Time Out* (June 29, 1979).
37. Ruth Hall, "London Diary," *New Statesman* (July 6, 1979); John Mathews, "The Politics of Cancer," *New Statesman* (August 3, 1979): 164; David Basnett, letter to *New Statesman* (July 17, 1979) (Epstein files).
38. *Yorkshire Post,* July 11, 1979, p. 14.
39. Richard Peto to the GMWU, October 22, 1979. The Association of Scientific, Technical and Managerial Staffs issued its own 92-page policy document on *The Prevention of Occupational Cancer* in March 1980, coming to many of the same conclusions as the GMWU. The document called for a complete ban, or "no human contact" condition, on any chemical shown to be a carcinogen by animal tests; it also endorsed the U.S. government's claim that 20–40 percent of all cancers were caused by chemicals used in the workplace. The British Chemical Industries Association prepared a response, citing World Health Organization figures endorsed by the Royal Society of London that at most 1–5 percent of all cancers were occupationally derived. *Chemistry in Britain,* a leading trade journal, reported the controversy along with Richard Doll's suggestion that the union should withdraw the document and prepare a more moderate version. R.D.S., "Cancer: Union Speaks Out," *Chemistry in Britain* (May 1980): 235. For Sheila McKechnie's discussion of the ASTMS document, see Tony Ades, "Danger! Cancer at Work," *Science for People* (Winter 1981–82): 8–11.
40. "Reasonable projections of the future consequences of past exposure to established carcinogens suggest that . . . occupationally related cancers may comprise as much as 20% or more of total cancer mortality in forthcoming decades. Asbestos alone will probably contribute up to 13–18%, and the data {relating to five other carcinogens} suggest at least 10–20% more." See "Estimates of the Fraction of Cancer in the United States," p. 710. Contributors to the Estimates Paper included Kenneth Bridbord, Pierre Decoufle, Joseph F. Fraumeni, David G. Hoel, Robert N. Hoover, David P. Rall, Umberto Saffiotti, Marvin Schneiderman, and Arthur C. Upton. For an overview of the claims and the critical response, see Efron, *Apocalyptics,* pp. 437–448.
41. The 1978 Estimates Paper has been variously called the "Califano report," "Bridbord et al., 1978," the "OSHA paper," the "HEW paper," and the "NCI document." The meaning should be clear in any case.
42. Califano, "Occupational Safety and Health."
43. Center for Disease Control, *Recommendations for a National Strategy for Disease Prevention* (Atlanta, 1978).
44. Maugh, "Industry Council Challenges HEW," pp. 602–603. To calculate the risk from chromium dusts, for example, the Estimates Paper's authors used a 1930s study indicating that workers in chromate-producing plants were five to nine times more likely to develop lung cancer than unexposed populations. The NOHS study indicated that 1.5 million workers had been exposed to chromium in recent decades; the Estimates Paper authors calculated that one could expect 7,900 to 16,000 excess respiratory tumors as a result of chromium exposures. Similar calculations were made for tumors from exposure to arsenic, leukemias from exposure to benzene, and lung cancers from nickel compounds and petroleum fractions.

45. Ibid., p. 602.
46. "What Proportion of Cancers Are Related to Occupation?" *Lancet* 2 (1978): 1238–1240; compare also "A Trade Union Looks at Cancer," *Lancet* 1 (1980): 636–637.
47. "Cancer and the Workplace: 'A Disaster,'" *Science News* 114 (1978): 229.
48. Maugh, "Industry Council Challenges HEW"; American Industrial Health Council, *A Reply to 'Estimates of the Fraction of Cancers in the United States Attributable to Occupational Factors'* (New York, 1978).
49. Reuel A. Stallones and Thomas Downs, *A Critical Review of: Estimates of the Fraction of Cancer in the United States Related to Occupational Factors* (n.d., late 1978).
50. American Industrial Health Council, *Cancer, Pollution and the Workplace* (Washington, D.C., 1983).
51. Higginson, "Present Trends," pp. 40–75; Higginson and Muir, "The Role of Epidemiology in Elucidating the Importance of Environmental Factors in Human Cancer."
52. Ernst L. Wynder and Gio B. Gori, "Guest Editorial: Contribution of the Environment to Cancer Incidence: An Epidemiologic Exercise," *Journal of the National Cancer Institute* 58 (1977): 825–832; Doll, "Strategy for Detection of Cancer Hazards to Man."
53. Philip Cole, "Cancer and Occupation: Status and Needs of Epidemiologic Research," *Cancer* 39 (1977): 1788–1791, plus the discussion on pp. 1807–1808.
54. Schwartz, "1978 Paper," in Napier, *Issues*, p. 26. Edward Beattie, medical director of the Memorial-Sloan Kettering Cancer Center, endorsed a figure of 3 percent about this time, and Margaret Heckler, Reagan's secretary of health and human services, eventually put forward a figure of about 8 percent (Love, "Indecent Exposure," p. 58). In 1980, John Higginson estimated that "only 6% of the variation in occurrence of cancer according to occupation can be traced to exposure in the workplace, while 88% is due to other lifestyle factors"; see his "Proportion of Cancers Due to Occupation," *Preventive Medicine* 9 (1980): 180–188.
55. Abelson, "Cancer—Opportunism and Opportunity." The appendix to the Estimates Paper was the only part of the paper ever published; see Day and Brown, "Multistage Models."
56. Linda Rosenbaum, "Does Progress Cause Cancer?" *Canadian Business Review* (April 1979): 43–47, 132–137.
57. Greenberg, "Severest Critic."
58. Office of Technology Assessment, *Technologies for Determining Cancer Risks from the Environment.* The OTA document concluded diplomatically that, despite the debates surrounding the question of workplace associated cancers, "almost every estimate fits comfortably in the range of 10 plus or minus five percent" (p. 88).
59. Doll and Peto, *Causes of Cancer,* p. 1241.
60. For a review, see Stephen Hilgartner, "The Dominant View of Popularization: Conceptual Problems, Political Uses," *Social Studies of Science* 20 (1990): 519–539.
61. Philip M. Boffey, "Cancer Experts Lean Toward Steady Vigilance, but Less Alarm, on Environment," *New York Times,* March 2, 1982.
62. Michael Gough, "Estimating Cancer Mortality," *Environmental Science and Technology* 23 (1989): 926.
63. See, for example, Brookes, "Wasteful Pursuit," where he cites Ronald Hart, director of the National Center for Toxicological Research in Jefferson, Arkansas, asserting that "we know that diet accounts for about 35% of all cancer deaths, or about 178,000 a year, and maybe a lot more" (p. 166). Peter Greenwald, head of the NCI's Division of Cancer Prevention and Control, recently placed the dietary contribution in the 20–60 percent range; see Beardsley, "War Not Won," p. 136. Doll and Peto had characterized their 35 percent estimate as "uncertain in the extreme" (*Causes of Cancer,* p. 1235): 35 percent was simply an approximate midpoint of the range they considered plausible (10–70 percent), although these and other caveats are rarely mentioned in references to the document.
64. Samuel S. Epstein and Joel B. Swartz, "Cancer and Diet" (letter), *Science* 224 (1984): 660; Davis et al., "Cancer Prevention."
65. Ruth B. Schwartz, "1978 Paper Grossly Overestimates Incidence of Occupationally Induced Cancer" (1980), reprinted in Napier, *Issues,* pp. 24–27. Even Epstein concedes that the Estimates Paper overestimated carcinogenic exposures; see his "Losing the War Against Cancer," p. 55. Joseph Fraumeni, listed as a coauthor, claims never even to have seen the document prior to its release: "By the time I got to look at it, it was already in the newspapers" (personal communication, March 10, 1994).
66. Irving J. Selikoff, "Constraints in Estimating Occupational Contributions to Current Cancer Mortality in the United States," in Peto and Schneiderman, *Banbury Report 9,* pp. 3–13; Irving Selikoff, E. C. Hammond, and J. Churg, "Asbestos Exposure, Smoking, and Neoplasia," *JAMA* 204 (1968): 106–112.

67. In his textbook, *Modern Epidemiology,* Kenneth J. Rothman characterized as "naive" the presumption that the contributory causes to a given disease must add up to 100 percent. Referring explicitly to efforts to allocate cancer causation among diet, smoking, and so forth, he states, "There is, in fact, no upper limit to the sum that was being constructed; the total of the proportion of disease attributable to various causes is not 100 percent but infinity" (p. 14). The Estimates Paper itself noted that "any percentage accounting of contributing causes to cancer well exceeds 100%" (p. 709).

68. Doll and Peto discuss some of these issues in *Causes of Cancer,* pp. 1219–1220; compare also Schneiderman, "Sources, Resources and Tsouris," pp. 451–466.

69. Compare the distinction between proximal and underlying causes in Rose, *The Strategy of Preventive Medicine,* pp. 98–99.

70. *Occupational Health and Safety Newsletter* (November 22, 1978).

71. See, for example, the letters in the August 2, 1979, issue of *New Scientist,* pp. 389–390.

72. N. E. J. Wells, "Cancer in Modern Mortality," Paper prepared for the Chemical Industries Association (Great Britain [1980]), p. 11.

73. "Occupational risks may be no higher than 1 percent of the total, and may possibly be closer to zero"; see Whelan, "Cancer and the Politics of Fear," p. 82.

74. New York State Assembly Standing Committee on Environmental Conservation, "Joint Public Hearing on the N.Y.P.I.G. Report on Toxics in the Niagara River," December 10, 1981, pp. 172–179. The chairman of the Assembly Committee, Maurice D. Hinchey, was so incensed by the CMA paper he had his staff prepare a long rebuttal; see "A Critique of Dr. Geraldine Cox's Testimony," February 10, 1987. In 1985, Cox fielded more than 100 media interviews for the CMA in response to the Union Carbide disaster at Bhopal, in which some 3,000 Indians died; see W. David Gibson, "After Bhopal: First Aid for Chemicals' Public Image," *Chemical Week* (February 20, 1985): 22.

75. William J. Nicholson, "Quantitative Estimates of Cancer in the Workplace," *American Journal of Industrial Medicine* 5 (1984): 341–342. Compare also Rodolfo Saracci et al., "Re: 'Occupational Cancer: Where Now and Where Next?'" *Scandinavian Journal of Work and Environmental Health* 12 (1986): 75–77; and Davis et al., "Cancer Prevention," p. 117.

76. "Japanese, as a general rule, bathe daily; the workmen in these industries [mule spinning, along with gas, tar, and tar by-products manufacture] habitually wear gloves while working, they frequently change their clothes which are often washed, and they daily take a very hot bath"; see Mataro Nagayo and Riojun Kinosita, "Some Significant Features of Cancer Incidence in Japan," *Yale Journal of Biology and Medicine* 12 (1940): 301–303. In 1892, Henry T. Butlin showed that continental chimney sweeps suffered less scrotal cancer than their English counterparts because of their regular bathing; see his "Three Lectures," pp. 5–6, 66.

77. Virginia L. Ernster, "Women and Smoking," *American Journal of Public Health* 83 (1993): 1202–1203; Ernst L. Wynder, "Toward a Smoke-Free Society: Opportunities and Obstacles," *American Journal of Public Health* 83 (1993): 1204–1205; Brandt, "The Cigarette, Risk, and American Culture."

78. Merril Eisenbud has noted that during sunspot activity, which follows an 11-year cycle, solar flares may produce cosmic radiation on the order of 10 or even 100 mrem/hr at altitudes as low as 35,000 feet; see his *Environmental Radioactivity,* 2nd ed. (New York, 1973), p. 193. No airline, so far as I'm aware, has ever cautioned its pregnant flight attendants from working during such flares; nor is the public generally aware of this potential hazard.

79. "Preventive Healthcare Strategies," *Medical World News* (January 1992): 58.

80. Devra Lee Davis, Aaron Blair, and David G. Hoel, "Agricultural Exposures and Cancer Trends in Developed Countries," *Environmental Health Perspectives* 100 (1992): 39–44; Beardsley, "War Not Won," pp. 136–137; Aaron Blair and Shelia Hoar Zahm, "Cancer Among Farmers," *Occupational Medicine* 6 (1991): 335–354.

CHAPTER 4
THE REAGAN EFFECT

1. Tom Alexander, "It's Roundup Time for Runaway Regulators," *Fortune* (December 3, 1979): 126–132; Peter Nulty, "A Brave Experiment in Pollution Control," *Fortune* (February 12, 1979): 120–123.

2. Capital investment in environmental management by the 106 member companies of the Chemical Manufacturers Association grew from 4.4 percent in 1971 to nearly 12 percent in 1979; see Chemical Manufacturers Association, *Protecting the Environment: What We're Doing About It* [1978].

3. Marvin Schneiderman, foreword to Peto and Schneiderman, *Banbury Report 9*, p. xi.
4. This statement prompted one protestor to post a sign on a tree reading "Chop Me Down Before I Kill Again"; see Lou Cannon, *Reagan* (New York, 1982), p. 289.
5. Dick Kirschten, "The New War on Cancer—Carter Team Seeks Causes, Not Cures," *National Journal* (August 6, 1977): 1220–1225.
6. Mark Green, foreword to Bollier and Claybrook, *Freedom from Harm*, p. v.
7. Cannon, *Reagan*, pp. 287–288.
8. Auchter's job was complicated by the fact that there were movements afoot to abolish the agency: Senator Orrin Hatch of Utah met privately with Auchter shortly before his confirmation and told him he was ready to introduce legislation to this effect. Auchter persuaded Hatch to wait a year to see what he could do to introduce a "less adversarial" relationship between business and labor; see Judson MacLaury, "The Occupational Safety and Health Administration: A History of Its First Thirteen Years: 1971–1984," unpublished manuscript, p. 46. Compare Thomas O. McGarity and Sidney A. Shapiro, *Workers at Risk: The Failed Promise of the Occupational Safety and Health Administration* (Westport, Conn., 1993).
9. Crawford, "'Death by a Thousand Cuts,'" p. 207.
10. In September 1983, the *Washington Post* disclosed the existence of an OSHA "Dormant Standards Project" designed to halt efforts to set exposure standards for 116 hazardous substances—including twelve known to cause cancer in laboratory animals. Sheldon Samuels of the AFL-CIO denounced the project, accusing OSHA of having decided it was "cheaper to let workers die"; see MacLaury, "Occupational Safety," p. 58.
11. Bergin and Grandon, *How to Survive*, p. xvi.
12. Crawford, "'Death by a Thousand Cuts,'" pp. 205–208.
13. Sibbison, "Censorship at the New E.P.A.," p. 208.
14. Bob Drogin, "The E.P.A.'s Stall on Toxic Waste," *The Nation* (November 27, 1982): 551–552.
15. James Nathan Miller, "What *Really* Happened at EPA," *Reader's Digest* (July 1983): 59–64. Compare also Russ Bellant, *The Coors Connection* (Boston, 1991).
16. James Ridgeway, "Pollution Is Our Most Important Product," *The Nation* (November 7, 1981): 473–474.
17. "Where the Cuts Will Come From," *Boston Globe*, January 18, 1982.

Support for Federal Agencies: Fiscal Years 1980–1989

Agency	1980	1981	1982	1983	1984	1985	1986	1987	1988	1989
				(in millions of dollars)						
EPA	5602	5232	5004	4299	4057	4511	4868	4903	4872	4906
NIH	3322	3604	3665	3750	4157	4670	5115	5222	6334	6992
OSHA	186	209	195	207	213	220	209	226	235	248
NIOSH	80	68	62	57	66	65	64	70	70	70
FDA	326	337	343	364	390	418	417	422	463	510
CPSC	41	42	32	34	35	36	34	35	33	34
				(in billions of dollars)						
DOD[a]	136	159	186	205	240	264	286	295	304	318

Note: Fiscal years begin in October of the previous calendar year; 1982 was thus the first fiscal year over which Ronald Reagan had full control. Figures are not adjusted for inflation.

[a]Figures include both DOD military and DOD civilian funding.

Sources: U.S. Treasury Department, Bureau of Government Operations; NIOSH; National Archives Civil Records Branch; CPSC; Judson MacLaury, Labor Department Historian.

18. Sidel, "Health Care," p. 38.
19. Miller, "What *Really* Happened," pp. 62–64.
20. Daniel S. Greenberg, "A Sober Anniversary of the 'War on Cancer,'" *Lancet* 338 (1991): 1583.
21. Viscusi, *Risk by Choice*, pp. 16–19. Viscusi's book (one chapter of which is titled "How to Set Standards If You Must") displays a cavalier disregard for worker health and safety, as on the very first page where he trivializes a worker's death on the job as an "unexpected rendezvous with destiny."

22. Rogene A. Buchholz, *Public Policy Issues for Management* (Englewood Cliffs, N.J., 1992), p. 191. Three of the Labor Department's four professional historians were fired within the first couple of years of the Reagan administration (Judson MacLaury, personal communication).

23. Sidel, "Health Care," p. 39.

24. Karen Ball, "Study Finds California Has Best Worker-Safety Laws," *Centre Daily Times* (State College), January 1, 1992. The study in question was done by the National Safe Workplace Institute, based in Chicago.

25. Sibbison, "Censorship," pp. 208–209.

26. Claybrook, *Retreat*, pp. 124–128.

27. Jonathan King, "There Are Good Guys in Science," *Mother Jones* (June 1982): 41; MacLaury, "Occupational Safety," p. 60.

28. Sibbison, "Censorship," pp. 208–209.

29. Ridgeway, "Pollution," pp. 474–475.

30. Miller, "What *Really* Happened," p. 63; compare also Peter Behr, "An Interview with A. Alan Hill," *Environment* (July–August 1981): 38–42.

31. Sidel, "Health Care," p. 40.

32. Marjorie Sun, "Reagan Reforms Create Upheaval at NIOSH," *Science* 214 (1981): 166–168.

33. Epstein, "Losing the War," pp. 59–60.

34. Love, "Indecent Exposure."

35. "Bill Clinton, Environmentalist?" *New York Times,* January 5, 1993.

36. Michael Wines, "Scandals at EPA May Have Done in Reagan's Move to Ease Cancer Controls," *National Journal* (June 18, 1983): 1264; U.S. House of Representatives, Democratic Study Group, *Reagan's Toxic Pollution Record: A Public Health Hazard* (Washington, D.C., 1984). Compare also Anthony Robbins, "Can Reagan Be Indicted for Betraying Public Health?" *American Journal of Public Health* 73 (1983): 12–13.

37. Claybrook, *Retreat,* p. 121.

38. Bergin and Grandon, *How to Survive,* pp. xvi–xvii. Ruckelshaus came under criticism for failing to act more decisively when EDB was discovered in Florida's groundwater in the summer of 1983: the new EPA chief barred the fumigant's use on soil but allowed its continued use on grain and citrus. Ruckelshaus eventually announced an emergency ban on the use of EDB on grain on February 3, 1984. Florida health officials by this time had ordered dozens of products found to contain EDB residues removed from supermarket shelves; see Love, "Indecent Exposure," pp. 58–59.

39. For a critique of risk assessment, especially its tendency to imply that "the chance of harm in question is accepted willingly in the expectation of gain," see Winner, "Risk"; Iain Boal's equally incisive "Rhetoric of Risk"; and Robert Ginsburg, "Quantitative Risk Assessment and the Illusion of Safety," *New Solutions* (Winter 1993): 8–15.

40. Love, "Indecent Exposure," p. 61.

41. Luther J. Carter, "Costs of Environmental Regulation Draw Criticism, Formal Assessment," *Science* 201 (1978): 140–144.

42. EPA, *Environmental Radiation Protection Requirements for Normal Operations of Activities in the Uranium Fuel Cycle,* vol. 1 (Washington, D.C., 1976).

43. PSAC, *Chemicals and Health, Report of the Panel on Chemicals and Health, Science and Technology Policy Office, NSF* (Washington, D.C., 1973).

44. For an overview and critique, see David F. Noble, "Cost-Benefit Analysis," *Health/PAC Bulletin* (August 1980): 1–40.

45. The American Enterprise Institute for Public Policy Research, publisher of *Regulation* magazine, was founded in 1943 as the "American Enterprise Association" by the chairman of the Johns-Manville Corporation, Lewis H. Brown, to provide Congress with scholarly information on national economic problems. By 1980 the AEI had a staff of 140 and an annual budget approaching $10 million. Milton Friedman once served on the AEI's Council of Academic Advisors; other fellows have included former President Gerald R. Ford, Ben Wattenberg, and Jeanne Kirkpatrick; see Joseph C. Kiger, ed., *Research Institutions and Learned Societies* (Westport, Conn., 1982), pp. 56–59.

46. Ludlam, *Undermining Public Protections,* p. 2.

47. MacLennan, "From Accident to Crash."

48. Ronald Brownstein, "Making the Worker Safe for the Workplace," *The Nation* (June 6, 1981): 692–694.

49. Kazis and Grossman, *Fear at Work,* pp. 127–129. In 1981, Mark Green of Congress Watch pointed out that 25–30 percent of Weidenbaum's $100 billion was paperwork required by the IRS; Green characterized Weidenbaum's calculation as "Chicken Little economics" designed "to scare the government and the public into thinking that government is too expensive"; see John C. Daly et al., *Health, Safety, and Environmental Regulation: How Effective?* (Washington, D.C., 1981), p. 8. An

excellent critique of Weidenbaum's proposal is Mark Green and Norman Waitzman, "Cost, Bene-fit, and Class," *Working Papers* (May–June 1980): 39–51.

50. Noble, "Chemistry of Risk," p. 26.
51. Sidel, "Health Care," p. 42.
52. Ludlam, *Undermining Public Protections,* pp. 22–23.
53. Truax, "Surviving the 1980s," pp. 14–15.
54. Albert L. Nichols, "Comparing Risk Standards," *Regulation* (Fall 1991): 91–93.
55. Truax, "Surviving the 1980s," p. 13.
56. Ibid.
57. Ridgeway, "Pollution," p. 475.
58. See Robert H. Harris et al., "Reducing Environmental Risks," *Society* (March–April 1981): 18; compare Kazis and Grossman, *Fear at Work,* pp. 123–138.
59. Harris et al., "Reducing Environmental Risks," p. 18; Bergin and Grandon, *How to Survive,* pp. 38–49.
60. Bryan G. Norton, *Why Preserve Natural Variety?* (Princeton, N.J., 1987).
61. Sylvia N. Tesh, *Hidden Arguments: Political Ideology and Disease Prevention* (New Brunswick, N.J., 1988); Boal, "Rhetoric of Risk."
62. Efron, *Apocalyptics,* p. 389.
63. Readings of this sort have been gathered together in Lehr, *Rational Readings.* Other recent exam-ples include Social Affairs Unit/Manhattan Institute, *Health, Lifestyle & Environment: Countering the Panic* (New York, 1991); J. R. Johnstone and C. Ulyatt, *Health Scare: The Misuse of Science in Public Health Policy* (Perth, 1991); Michael J. Bennett, *The Asbestos Racket: An Environmental Parable* (Bellevue, Wash., 1991); Ronald E. Gots, *Toxic Risks: Science, Regulation, and Percep-tion* (Boca Raton, Fla., 1993); Ronald Bailey, *Eco-Scam: The False Prophets of Environmental Doom* (New York, 1993). Until his death in 1991, Warren T. Brookes of the *Boston Herald-Ameri-can* and *Detroit News* was a vocal champion of anti-environmentalist causes. See also the refer-ences in note 96 below.
64. Efron first wrote on environmental cancer in response to Dan Rather's October 15, 1975, CBS News special on "The American Way of Death," in which the young anchorman-to-be asserted that, according to the National Cancer Institute, "if you're living in America your chances of get-ting cancer are higher than anywhere in the world." Efron wrote a scathing critique for *TV Guide,* arguing that there was "no substance on earth which when ingested in varying amounts by human beings will not cause problems for some of them" ("Biased 'Science' Reporting Scares TV View-ers" [January 10, 1976]: A-7 to A-8). It was Efron's disgust with Rather's interviewing "sobbing wives" in an acute spasm of "moral elephantiasis" (Irving Kristol's phrase, appropriated by Efron) that prompted her to write a book-length critique. Efron today describes herself as a "feminist, atheist, neo-conservative" who "fell in love with the industrial revolution" after living in poverty-stricken Haiti in the 1940s and 1950s, where she was married to a Haitian businessman. She un-derwent another epiphany in the late 1950s, when she befriended Ayn Rand shortly after the publication of *Atlas Shrugged* (personal communication, January 19, 1994).
65. Efron, *Apocalyptics,* p. 232.
66. Ibid., pp. 421–423.
67. Ibid., pp. 348, 383, 76, 372, 418, 472.
68. Ibid., pp. 12–13.
69. Love, "Indecent Exposure," p. 61. Neo-conservatives and libertarians returned the favor: *Barron's* praised Efron's book, and excerpts appeared in *Reason* and *The American Spectator.* Other en-dorsers included Herman Kraybill ("a masterpiece"), James Lovelock ("powerfully written and ac-curate"), John Weisburger ("a landmark book"), and Norman Borlaug. Reviews are excerpted in the University of Rochester's *Annual Report* for June 1987.
70. Edith Efron, *The News Twisters* (Los Angeles, 1971), pp. x–xi.
71. Ibid., pp. xi, 3, 42–43.
72. Ames, "Mother Nature." Efron, interestingly, provides no hint that environmental cancer policies might have changed during the period of her writing (1977–1982); there is no reference to the fact that Reagan had taken power or that cancer policy was beginning to look more like what Efron might have had in mind.
73. Fredrick J. Stare founded Harvard's Department of Nutrition in 1942, and did not retire until 1981. Stare headed the department for 34 years, during which time he was also editor of *Nutrition Re-views,* a food industry publication funded by the Nutrition Foundation. Critics accused him of being a paid mouthpiece for industry: his Harvard department received gifts from dozens of food companies in the 1970s. The Kellog Company, for example, donated $2 million six months after he testified before Congress that cereals containing as much as 70 percent sugar were more nourish-ing than an old-fashioned breakfast. See Hess, "Harvard's Sugar-Pushing Nutritionist."

74. See, for example, her *Preventing Cancer: What You Can Do to Cut Your Risks by up to 50 Percent* (New York, 1978). Whelan's other books include *Making Sense out of Sex*, with S. T. Whelan (New York, 1975); *Panic in the Pantry: Food Facts, Fads and Fallacies*, with Fredrick J. Stare (New York, 1975); *Eat OK—Feel OK! Food Facts and Your Health*, with Fredrick J. Stare (North Quincy, Mass., 1978); and *The 100% Natural, Purely Organic, Cholesterol-free, Megavitamin, Low Carbohydrate Nutrition Hoax*, with Fredrick J. Stare (New York, 1983).

75. Hess, "Harvard's Sugar-Pushing Nutritionist," p. 12.

76. Mix, "The Truth Comes Out." Compare Bruce Ames's effusive confidence that "the future of the planet has never been brighter" ("Science and the Environment," *The Freeman* [September 1993]: 343).

77. Elizabeth M. Whelan, *Toxic Terror* (Ottawa, Ill., 1985), pp. xvi–xvii.

78. Mix, "The Truth Comes Out," p. 52.

79. *Food Chemical News* announced the formation of the ACSH with the goal of providing information for scientists "who are alarmed that chemical and cancer-phobia has damaged this country" (July 24, 1978, p. 38). The ACSH has published hundreds of pamphlets since its founding; in the 1980s it also published the ACSH *Newsletter*. For background, see Kurtz, "Dr. Whelan's Media Operation."

80. ACSH, *Introducing American Council on Science and Health, Inc.* (New York, n.d.). The council's advisors included, as of 1992, Bernard L. Cohen, Dixy Lee Ray, Kenneth J. Rothman, Julian L. Simon, and Aaron Wildavsky, just to name some of the better known; see the ACSH *Directory* (New York, 1992).

81. Gail Bronson, "A Health Group Gets a Beating from the Press," *Wall Street Journal*, February 12, 1979; ACSH, "Court Dismisses Libel Suit Against Nutrition Scientists," press release, June 26, 1980. The columnist Ann Landers and the AMA's director of nutrition education, Philip White, were named as co-conspirators in the suit.

82. Harnik, *Voodoo Science.*

83. ACSH supporters have included Coca-Cola, the National Soft Drink Association, the Pepsico Foundation, International Flavors and Fragrances, Inc., several major timber and paper companies, Bethlehem Steel, Chevron, Consolidated Edison, General Motors, the US Steel Foundation, American Cyanamid, the Amoco Foundation, Dow Chemical of Canada, Hooker Chemical and Plastics, the Mobil Foundation, the Monsanto Fund, the Shell Companies Foundation, Tenneco, and dozens of others; see Harnik, *Voodoo Science*, pp. 3–4. A 1992 internal council memo by Whelan bemoans the loss of funding from the Shell Oil Co. Foundation: "When one of the largest international petrochemical companies will not support ACSH, the great defender of petrochemical companies, one wonders who will." See "The ACSH: Forefront of Science, or Just a Front?" *Consumer Reports* (May 1994): 319.

84. *Does Nature Know Best? Natural Carcinogens in American Food*, ACSH publication distributed by the National Agricultural Chemicals Association.

85. Mix, "The Truth Comes Out," pp. 51–52.

86. Boal, "Rhetoric of Risk," pp. 7–8. Wildavsky defined democratic capitalism as "the decentralized search for safety"; see his "Thanks for the Commentary," *Society* (November–December 1989): 28. Elsewhere, Wildavsky praised Reagan for having done "more for capitalism than any president since FDR, possibly since Grover Cleveland" ("The Triumph of Ronald Reagan," *The National Interest* [Winter 1988–89]: 3). The Berkeley political scientist died on September 4, 1993, of lung cancer.

87. Aaron Wildavsky, "Pollution as Moral Coercion: Culture, Risk Perception, and Libertarian Values," *Cato Journal* 2 (1982): 305–307.

88. Ibid., pp. 307, 319.

89. Ibid., p. 321.

90. Interview with Bruce Ames, March 16, 1992. I was in Ames's office when he received a call from his collaborator, Lois Gold, whom I was about to interview. Apparently not recognizing that I could hear the call, projected into the room by speaker phone, Gold asked Ames whether I was "an environmentalist." Ames, recognizing that I had heard the question, replied "we are all environmentalists."

91. Rush H. Limbaugh III, *The Way Things Ought to Be* (New York, 1992), p. 164.

92. Ray, *Trashing the Planet*, pp. 6–7, 82, 163.

93. Ray, *Environmental Overkill.* Ray lived long enough to see the AEC accused of subjecting thousands of people—including prisoners, cancer patients, and the mentally retarded—to inhumane radiation experiments. Ray dismissed as "alarmist" the public's outrage over the experiments: "Everybody is exposed to radiation. A little bit more or a little bit less is of no consequence"; see Don Duncan, Mark Matassa, and Jim Simon, "Dixy Lee Ray: Unpolitical, Unique, Uncompromising," *Seattle Times*, January 3, 1994.

94. Louis R. Guzzo, *Is It True What They Say About Dixy?* (Mercer Island, Wash., 1980), pp. 101–148.

95. "Dixy Rocks the Northwest," *Time* (December 12, 1977): 27.

96. Ben Bolch and Harold Lyons, *Apocalypse Not: Science, Economics, and Environmentalism* (Washington, D.C., 1993), esp. pp. vii, 39–62; Fumento, *Science Under Siege,* pp. 46, 51; Andrea Arnold, *Fear of Food: Environmentalist Scams, Media Mendacity, and the Law of Disparagement* (Bellevue, Wash., 1990), p. 75; M. Alice Ottoboni, *"The Dose Makes the Poison": A Plain-Language Guide to Toxicology,* 2nd ed. (New York, 1991). Ottoboni is a long-standing science adviser for the ACSH; her book claims to be an "objective," "plain-language" guide to toxicology," though the APHA *Section Newsletter* for Occupational Safety and Health criticized it for mixing "ideological polemic line by line with material alleged to introduce the basic scientific principles of toxicology" (January 1994).

97. Accuracy in Media (AIM) was a key player in organizing a response to the controversial *60 Minutes* report of February 26, 1989, pronouncing Alar, the apple-ripening agent, a dangerous carcinogen; see the *AIM Report* following this period. AIM's director of media analysis, Joseph Goulden, describes himself as an East Texan "stock-car racing nut" with an interest in mercenary warfare and low-intensity conflict; see Lorne Manly, "Without Bias?" *Folio* (October 15, 1993): 57–58.

98. The words are those of John W. Robbins, cited on the back cover of *The Freeman,* published by the Foundation for Economic Education in New York. Compare Dixy Lee Ray's assertion that "the strategy of the prime movers behind Earth Summit was to use environmentalism as a Trojan horse to advance the socialist agenda" ("Dear Concerned American," AIM mailing, January 1994).

99. Wildavsky, "Topic of Cancer," p. 74.

100. Ray, *Trashing the Planet,* p. 6.

101. Ray, *Environmental Overkill,* p. 157.

102. Compare Efron's insistence that "asbestos is a natural mineral . . . everything is chemicals; Epstein is chemicals" (*Apocalyptics,* p. 132).

103. Duncan et al., "Dixy Lee Ray."

104. Aaron Wildavsky, "The Secret of Safety Lies in Danger," *Society* (November–December 1989): 4–5; Paul Johnson, "This Side of Apocalypse," *National Review* (June 29, 1984): 44. Compare Johnson's assertion that "the Industrial Revolution itself was a gigantic risk" that, thankfully, was able to proceed without the pesky disturbances of "the ecological lobby" (p. 44). Iain Boal points out that the *International Encyclopedia of the Social Sciences* lists "profit" as the only cross-indexed entry under "risk" ("Rhetoric of Risk.")

105. Ray traces environmentalism back not just to the Luddites and Saboteurs but to older religious warnings of "punishment in the sulfurous fires of hell" (*Trashing the Planet,* pp. 169, 5). Paul Johnson characterizes environmentalism as "a quasi-mystical vision of total purity" combining a hatred of capitalism, "the itch to interfere," and "the eternal nanny principle" ("The Perils of Risk Avoidance," *Regulation* (May–June 1980): pp. 15–16). Compare also William C. Clark, "Witches, Floods, and Wonder Drugs: Historical Perspectives on Risk Management," in *Societal Risk Assessment,* edited by R. C. Schwing and W. A. Albers (New York, 1980), p. 290; also *Forbes* magazine's critique of OSHA regulators for their imputation of a "satanic quality" to cancer-causing substances ("Diseased Regulation" [February 19, 1979]: 34.)

106. Ray, *Environmental Overkill,* p. 78; Ray, *Trashing the Planet,* p. 118.

107. Dixy Lee Ray to "Concerned American."

108. Ruth S. Cowan, *More Work for Mother: The Ironies of Household Technology from the Open Hearth to the Microwave* (New York, 1983).

109. Ray, *Trashing the Planet,* pp. 20–21.

110. Ibid., p. 8.

111. Christine Triano and Nancy Watzman, "Quayle's Hush-Hush Council," *New York Times,* November 20, 1991.

112. William D. Ruckelshaus, "Who Will Regulate the Regulators?" *New York Times,* February 3, 1993.

113. David Kotelchuck, "Bush Declares Open Season on OSHA," *Health/PAC Bulletin* (Spring 1992): 35–36.

114. Samuel S. Epstein, "Editorial Misconduct in *Science,*" *International Journal of Health Services* 20 (1990): 349–352.

115. Barbara Presley Noble, "Breathing New Life Into OSHA," *New York Times,* January 23, 1994.

116. "NRC Report Spurs Changes in Policies on Pesticides in Foods," *The Nation's Health* (August 1993): 24.

117. Gwen Ifill, "Middle-of-Roaders Leave Liberals a Cold Shoulder," *New York Times,* January 27, 1994.

118. "Tight Budget Caps Pinch Most Public Health Programs in FY 1995," *The Nation's Health* (March 1994): 9. One should not overestimate the Democratic Party's willingness or ability to reorient funding priorities: President Clinton's 1995 budget requested a higher percentage increase for bal-

listic missile defense (14 percent) than for either the National Institutes of Health (5 percent) or EPA research and development (8 percent). Star Wars lives! For data, see Jeffrey Mervis, "Cutting the President's Target Is Mixed Blessing for Agencies," *Science* 266 (1994): 211.

119. In Reagan's first budget, issued in February 1981, the Surgeon General's Office on Smoking and Health was zeroed out, although it was later restored after Senator Lowell Schweiker defended it to OMB director David Stockman.

120. In 1983, Michael Jacobson of the CSPI noted that the USDA under Reagan was "being managed by former meat industry executives who have done everything possible to scuttle the nutrition policies and education programs which encourage the reduced consumption of fatty foods, including meat"; see Claybrook, *Retreat,* p. 30.

121. "Statement of George H. R. Taylor," U.S. Congress, Committee on Government Operations, Office of Management and Budget, Control of OSHA Rulemaking, 1982, p. 5. In December 1982, Congressman Albert Gore told *Science* reporter Eliot Marshall that the Reagan administration's abandonment of efforts to prevent cancer "will probably result in hundreds of thousands of additional deaths attributable to cancer"; see Eliot Marshall, "EPA's High-Risk Carcinogen Policy," *Science* 218 (1982): 975.

122. "Delay on Aspirin Warning Label Cost Children's Lives, Study Says," *New York Times,* October 23, 1992; Claybrook, *Retreat,* pp. 41-45.

CHAPTER 5
"DOUBT IS OUR PRODUCT": TRADE ASSOCIATION SCIENCE

1. Notable exceptions are Nelkin, *Selling Science,* pp. 53-69, 144-153; and Hilgartner, "The Political Language of Risk." Compare also Warner, "Tobacco Industry Scientific Advisors," and the lively exposés in Cavanaugh, "The Corruptions of Science"; Noble's "Chemistry of Risk"; and "Public-Interest Pretenders," *Consumer Reports* (May 1994): 316-320.

2. See, for example, Martin Gardner, *Fads and Fallacies in the Name of Science* (New York, 1957); also his *Science—Good, Bad and Bogus* (Buffalo, N.Y., 1989).

3. In 1900, there were only about 100 trade associations in the United States; today there are approximately 40,000; see Shiv K. Gupta and David R. Brubaker, "The Concept of Corporate Social Responsibility Applied to Trade Associations," *Socio-Economic Planning Science* 24 (1990): 261-271. A good reference manual on the staffing and budgets of trade associations is the *Encyclopedia of Associations,* edited by Deborah M. Burek, 28th ed. (Detroit, Mich., 1994). The American Society of Association Executives suggests that more than half a million Americans work for associations (not just trade associations), a work force comparable in size to the airline and computer industries; see Helen F. Bensimon and Patricia A. Walker, "Associations Gain Prestige and Visibility by Serving as Expert Resources for Media," *Public Relations Journal* (February 1992): 14-16. Many trade associations boast budgets in the millions of dollars: the American Hospital Association, for example, based on Lake Shore Drive in Chicago, has a staff of 884 and an annual budget of about $79 million.

4. Edward L. Aduss and Mathew C. Ross, "Integrating Efforts of Advertising Agencies and Public Information Organizations," in Heath, *Strategic Issues Management,* p. 289; "Issues Management: Preparing for Social Change," *Chemical Week* (October 28, 1981): 46-51. The expression "issues management" was apparently coined in the mid-1970s by W. Howard Chase of the American Can Company, who subsequently became founding chairman of the Issues Management Association. The term "advocacy advertising" was first used about the same time; see Robert L. Heath and Richard A. Nelson, *Issues Management* (Beverly Hills, Calif., 1986), pp. 12-13.

5. Victor R. Hirsch, "Industry Performance and Public Opinion," in *The Cosmetic Industry: Scientific and Regulatory Foundations,* edited by Norman F. Estrin (New York, 1993), p. 275.

6. "Bush Asks Cautious Response," *New York Times,* February 6, 1990.

7. See Peter Passell, "The Garbage Problem: It May Be Politics, Not Nature," *New York Times,* February 26, 1991. Critics had pointed out that 18 billion diapers make their way to landfills every year, enough to stretch to the moon and back seven times. On October 23, 1992, the *New York Times* reported that disposables had again become popular in consequence of new evidence showing that disposables were environmentally safe ("Among the Earth Baby Set, Disposable Diapers are Back"). Several weeks later, two letters to the editor pointed out that the *Times* had failed to mention that the data were drawn from studies financed by Procter & Gamble; see Nancy Skinner, Dave Kershner, and Ethan Seidman, "Too Soon to Hail Throwaway Diaper's Victory?" *New York Times,* November 9, 1992.

8. Al Ries and Jack Trout, *Positioning: The Battle for Your Mind* (New York, 1981), pp. 167-170.

9. Details of the Chemical Industry Communications Action Program are presented in S. Prakash Sethi, *Handbook of Advocacy Advertising: Concepts, Strategies and Applications* (Cambridge, Mass., 1987), pp. 317–336.

10. Epstein, *Politics of Cancer,* pp. 107–108.

11. Falguni Sen and William G. Egelhoff, "Six Years and Counting: Learning from Crisis Management at Bhopal," *Public Relations Review* 17 (1991): 69–83.

12. Marconi, *Crisis Marketing,* p. ix.

13. *Hill and Knowlton: The Power of Communication* (New York, 1992), p. 2.

14. Patterson, *Dread Disease,* p. 212. An excellent history of tobacco industry public relations in the 1950s is Pollay, "Propaganda, Puffing and the Public Interest."

15. Matthew M. Swetonic, "Death of the Asbestos Industry," in Gottschalk, *Crisis Response,* pp. 299–300.

16. Heath, *Strategic Issues Management,* p. 41.

17. Sethi, *Handbook,* p. 322.

18. J. A. Jacobs, "Supreme Court Increases Antitrust Risks," *Association Trends* (July 8, 1988): 1, 4.

19. There are actually several hundred association management firms in the United States, the largest being Smith, Bucklin, and Associates, Inc., in Chicago, with a staff of about 450 and a clientele of some 190 professional and trade associations. There is even an overarching Institute of Association Management Companies, also based in Chicago, to which most of the nation's association management firms belong.

20. Society of the Plastics Industry, Inc., *SPI at Your Service* (Washington, D.C., 1992); SOCMA, *SOCMA—Annual Report, 1992–1993* (Washington, D.C., 1993). SOCMA was founded in 1921 and now boasts some 225 member companies.

21. Commoner, *Making Peace with the Planet,* p. 129.

22. Keith Schneider, "Science Academy Says Chemicals Do Not Necessarily Increase Crops," *New York Times,* September 8, 1989. The academy document in question was the National Research Council's *Alternative Agriculture* (Washington, D.C., 1989).

23. Joseph Palca, "Get-the-Lead-Out Guru Challenged," *Science* 253 (1991): 842–844.

24. "Delay on Aspirin Warning Label Cost Children's Lives."

25. See Hilgartner, *Who Speaks for Science?* The coalition included the National Milk Producers Federation, National Pork Producers Council, National Livestock and Meat Board, National Broiler Council, National Turkey Federation, United Egg Producers, National Cattlemen's Association, and Egg Institute of America.

26. Brian MacMahon et al., "Coffee and Cancer of the Pancreas," *New England Journal of Medicine* 304 (1981): 630–633; National Coffee Association, *Coffee Consumption and Pancreatic Cancer: A Scientific Update* (New York, January 1, 1987).

27. Sheila Kaplan, "The Ugly Face of the Cosmetics Lobby," *Ms.* (January–February 1994): 88–89.

28. Centers for Disease Control, "Smoking-Attributable Mortality and Years of Potential Life Lost—United States, 1988," *Morbidity and Mortality Weekly Report* 40 (1991): 62–71.

29. "Deaths from Smoking Fall in U.S.," *International Herald Tribune,* August 28–29, 1993. The 434,000 figure is for 1988, the 419,000 figure for 1990.

30. Hoffman, *Mortality from Cancer,* p. 185. See Chapter 1 for references to Soemmerring's and Hill's writings.

31. Herbert L. Lombard and Carl R. Doering, "Cancer Studies in Massachusetts," *New England Journal of Medicine* 198 (1928): 481–487; Fritz Lickint, "Tabak und Tabakrauch als ätiologischer Factor des Carcinoms," *Zeitschrift für Krebsforschung* 30 (1929): 349–365. Lickint's *Tabak und Organismus* (Berlin, 1939) is the most comprehensive medical indictment of tobacco of the prewar era.

32. Ernst L. Wynder and Evarts A. Graham, "Tobacco Smoking as a Possible Etiologic Factor in Bronchiogenic Carcinoma," *JAMA* 143 (1950): 329–336; Richard Doll, "On the Aetiology of Cancer of the Lung," in Clemmesen, *Symposium,* p. 47; Richard Doll and A. Bradford Hill, "A Study of the Aetiology of Carcinoma of the Lung," *British Medical Journal* 2 (1952): 1271–1286.

33. Patterson, *Dread Disease,* p. 209; Edward C. Hammond, "Smoking in Relation to the Death Rates of One Million Men and Women," *National Cancer Institute Monographs* 19 (1966): 127–204.

34. Doll and Peto, *Causes of Cancer,* pp. 1220–1224; R. T. Ravenholt, "Tobacco's Impact on Twentieth-Century US Mortality Patterns," *American Journal of Preventive Medicine* 1 (1985): 4–17.

35. Taylor, *Smoke Ring,* p. xiv.

36. If one assumes that annual smoking-induced deaths in the United States grew linearly from 0 to 400,000 between 1940 and 1990, this means that tobacco killed about 10 million Americans over this period. If Americans have smoked one-tenth of all the cigarettes ever smoked, this would put the world death toll for this period at about 100 million. (Americans are presently about 5 percent of the world's population but have consumed a disproportionate total of the world's cigarettes.) In

1990, R. T. Ravenholt estimated an annual world death toll from "tobaccosis" at 3 million persons, and a total world death toll this century at 50 million, approaching 100 million by the year 2000; see his "Tobacco's Global Death March"; compare also Nigel J. Gray, "Tobacco and Smoking: Evidence and Overview of Global Tobacco Problem," *ICCR International Conference on Cancer Prevention* (Washington, D.C., 1992). China is the world's major consumer of cigarettes: 80 percent of Chinese males were smokers in 1990; see Anne Simon Moffat, "China: A Living Lab for Epidemiology," *Science* 248 (1990): 554. Early in the twenty-first century one can expect Chinese deaths from smoking to be in the neighborhood of 2 million per year. China produces about 1.9 trillion cigarettes per year, more than any other country; see "Can China Quit the Habit?" *Nova* (April 12, 1994).

37. Tobacco Institute, *In the Public Interest: Three Decades of Initiatives by a Responsible Cigarette Industry* (Washington, D.C., 1986); Pollay, "Propaganda," pp. 43–50. The Council for Tobacco Research is funded by five of the six largest manufacturers of tobacco products—all but Liggett & Myers. The immediate cause for its formation was the public relations "emergency" caused by Ernst L. Wynder's widely reported demonstration that mice painted with tars condensed from cigarette smoke developed skin cancer. See the litigation documents filed at the U.S. District Court of Southern California on April 29, 1994, by plaintiffs in the class action suit of Allman et al. v. Philip Morris et al., available from the Tobacco Products Liability Project in Boston. The British equivalent of the CTR is the Tobacco Research Council, which spent £6 million on research between its founding in 1956 and 1969. In the mid-1970s, the TRC had its own laboratory with over 250 scientists and staff; see Friedman, *Public Policy and the Smoking-Health Controversy*, p. 133.

38. *Report of the Council for Tobacco Research-U.S.A., Inc., 1992* (New York, n.d.), p. 5; Annette Piccirelli, ed., *New Research Centers* (Detroit, 1993), p. 2535. Many CTR-funded projects have been jointly sponsored by the NCI, the ACS, and other leading sponsors of cancer research. The CTR, according to its chairman and chief executive officer, James F. Glenn, has sponsored "pioneering work in identifying familial cancers, the role of genetic factors in cancer formation, and the identification of oncogenes. The Council was instrumental in supporting early work in the role of free radicals in the etiology of disease and in opening up the new field of growth factors research." See his testimony before the Subcommittee on Health and the Environment, U.S. House of Representatives, May 26, 1994. In 1988, Philip Morris, R.J. Reynolds, Lorillard, and Svenska Tobaks AB established a Center for Indoor Air Research (CIAR) in Linthicum, Maryland; the center has become one of the leading supporters of indoor air pollution research.

39. Tobacco Institute, *The Cigarette Controversy: Why More Research Is Needed* (Washington, D.C., n.d.).

40. Obscured in such images are the results of independent surveys indicating that smokers tend to be poor, uneducated, and overweight by comparison with nonsmokers; see "Why the Turk Can't Get It Up," *Mother Jones* (January 1979): 37.

41. Clarence C. Little, writing for the Tobacco Industry Research Committee, *Report of the Scientific Director, July 1, 1956–June 30, 1957;* Ronald A. Fisher, "Dangers of Cigarette-Smoking," *British Medical Journal* 2 (1957): 297–298. In 1956, the American Cancer Society's scientific and medical director, Charles Cameron, suggested that there might be a gene that leads one both to smoke and to contract cancer; see his *The Truth About Cancer* (Englewood Cliffs, N.J., 1956), pp. 54–66. Samuel Epstein aptly distilled this effort into the slogan "Cancer causes smoking"; see his *Politics of Cancer*, p. 168.

42. R. L. Phillips et al., "Mortality Among California Seventh-Day Adventists for Selected Cancer Sites," *Journal of the National Cancer Institute* 65 (1980): 1097–1107; T. Hirayama, "Nonsmoking Wives of Heavy Smokers Have a Higher Risk of Lung Cancer: A Study from Japan," *British Medical Journal* 282 (1981): 183–185. Suspicions of a cancer hazard from secondhand smoke date back at least to the 1930s; see Boehncke, "Krebs und Tabak," *Fortschritte der Medizin* 53 (1935): 631. For more recent reviews, see David M. Burns, "Environmental Tobacco Smoke: The Price of Scientific Certainty," *Journal of the National Cancer Institute* 84 (1992): 138–139; Dimitrios Trichopoulos et al., "Active and Passive Smoking," *JAMA* 268 (1992): 1697–1701; and the EPA study, *Respiratory Health Effects.*

43. Tobacco Institute, *Smokers' Rights in the Workplace: An Employee Guide* (Washington, D.C., 1990), p. 7; also their *Smoking Restrictions: The Hidden Threat to Public Health* (1987) and *Indoor Air Pollution: Is Your Workplace Making You Sick?* (1988). A full-page ad published by the R.J. Reynolds Tobacco Company in the *New York Times* ("Passive Smoking: An Active Controversy," February 11, 1985) recognized, perhaps unwittingly, that "it is rhetoric, more than research, which makes passive smoking an active controversy."

44. Marshall, "Tobacco Science Wars." Arthur C. Upton, director of the NCI under President Carter, was visited by half a dozen tobacco industry representatives in late 1977 or early 1978, shortly after Califano declared smoking "public health enemy number 1"; the tobacco group threatened to

have its friends in Congress cut off NCI funds unless the NCI director eased up on the smoking issue. Upton's family about this time received several threatening telephone calls, which he still today attributes to the industry's effort to intimidate him (personal communication, October 27, 1994).

45. Taylor, *Smoke Ring,* pp. 190–207; Michael P. Traynor, Michael E. Begay, and Stanton A. Glantz, "New Tobacco Industry Strategy to Prevent Local Tobacco Control," *JAMA* 270 (1993): 479–486. Efforts to ban smoking go back a long time. By 1909, for example, 17 U.S. states had adopted prohibitions on the sale of tobacco. Antitobacco forces were dealt a crippling blow in World War I, when the U.S. government distributed tobacco along with K-rations; see the historical essays in Robert L. Rabin and Stephen D. Sugarman, eds., *Smoking Policy: Law, Politics, and Culture* (New York, 1993).

46. Warren E. Leary, "U.S. Ties Secondhand Smoke to Cancer," *New York Times,* January 8, 1993.

47. K. Michael Cummings et al., "What Scientists Funded by the Tobacco Industry Believe About the Hazards of Cigarette Smoking," *American Journal of Public Health* 81 (1991): 894–896.

48. Andrew Tobias, "The Dividends for Quitters," *Time* (October 12, 1992): 76.

49. "Editorial: The Cigarette Message," *New York Times,* August 22, 1967.

50. Among 253 published papers acknowledging CTR support in 1988, only 4 were classed as epidemiology; see the *Report of the Council for Tobacco Research-U.S.A., Inc., 1988* (New York, 1988).

51. Cummings et al., "What Scientists Funded by the Tobacco Industry Believe."

52. Patterson, *Dread Disease,* p. 213; compare also Howard Wolinsky and Tom Brune, *The Serpent on the Staff: The Unhealthy Politics of the American Medical Association* (New York, 1994), pp. 144–173. In 1989, the public interest group Common Cause pointed out that the Tobacco Institute had paid more speaking fees to congressmen than had any other industry group; see Charles R. Babcock, "Tobacco Industry Led '88 Hill Honoraria Givers," *Washington Post,* July 11, 1989.

53. Taylor, *Smoke Ring,* p. 17; Robert Pear, "A.M.A. Found to Donate More to Foes of Its Views," *New York Times,* January 6, 1994.

54. *Smoking and Health: Report of the Advisory Committee to the Surgeon General of the Public Health Service* (Washington, D.C., 1964). Compare also the earlier report by Britain's Royal College of Physicians, *Smoking and Health* (London 1962).

55. National Cancer Institute, *National Cancer Program: 1975 Annual Plan for FY 1977–1981* (Bethesda, Md., 1975), pp. I-1 to I-15. In 1978, the American Cancer Society recommended that the Defense Department stop selling tax-free cigarettes at PXs, where a pack could sell for as little as 7–10 cents. The society's annual budget for that year included only $200,000 for smoking-related topics out of a total budget of over $100 million; see Deborah Shapley, "ACS Group Urges Anti-Smoking Drive," *Science* 199 (1978): 753.

56. Hueper, "Cigarette Theory."

57. "Caution: Industry Is Hazardous to Your Health," *The Capital Times* (Madison), September 24, 1979, pp. 21–22; Susan Q. Stranahan, "How Life-Saving Legislation Sits in Congress," *Philadelphia Inquirer,* March 15, 1976.

58. "Focus on Smoking Said to Divert Scientific Attention," *The Tobacco Observer* (December 1980): 1.

59. Whelan, *Smoking Gun,* pp. 20–23, 120–125; Ravenholt, "Tobacco's Global Death March," p. 225; Joseph A. Califano, Jr., *Governing America* (New York, 1981), pp. 181–197, 432; U.S. Congress, Committee on Interstate and Foreign Commerce, "Hearing on Antismoking Initiatives of the Department of Health, Education, and Welfare," February 15, 1978 (Washington, D.C., 1978), pp. 5–184. Jimmy Carter in 1978 at a North Carolina political rally defended his support for tobacco subsidies; the Georgia native also argued that the NCI should continue to try to make smoking "even more safe than it is today" ("The Tobacco Auctioneer," *Washington Post,* August 9, 1978).

60. Freedman and Cohen, "Smoke and Mirrors"; compare also Richard Harris's update for National Public Radio's *All Things Considered,* June 17, 1994.

61. Interview with Klaus Hueper, June 20, 1994. Hueper may have been offered the job that Clarence Cook Little eventually assumed—namely, Chairman of the Science Advisory Board of the Tobacco Industry Research Council.

62. Marshall, "Tobacco Science Wars," p. 251.

63. David Margolick, "Federal Judge in Tobacco Cases Ousted for Bias Against Industry," *New York Times,* September 8, 1992. For the views of several leading tobacco executives on the question of cigarettes and cancer, see Michael Janofsky, "On Cigarettes, Health and Lawyers," *New York Times,* December 6, 1993.

64. Freedman and Cohen, "Smoke and Mirrors."

65. R. H. Jones, *Asbestos and Asbestic* (London, 1897), pp. 1–2. Jones's passage and some of what follows is derived from David Ozonoff's excellent "Failed Warnings."

66. Ozonoff, "Failed Warnings," pp. 148–151.

67. D. S. Wright, "Man Made Mineral Fibres: A Historical Note," *Journal of the Society of Occupational Medicine* 30 (1980): 138–140.
68. Murray, "Asbestos."
69. Michael J. Bennett, *The Asbestos Racket* (Bellevue, Wash., 1991), pp. 65–87. Malcolm Ross of the U.S. Geological Survey reportedly told columnist Warren Brookes of the *Detroit News* that there was "no doubt" in his mind that the Challenger disaster was caused by "asbestos paranoia"; see "The Real Asbestos Horror Story," *AIM Report* (September 1990).
70. Ozonoff, "Failed Warnings," p. 156.
71. Frederick Hoffman, "Mortality from Respiratory Diseases in Dusty Trades (Inorganic Dusts)," *Bulletin of the U.S. Bureau of Labor Statistics* 231 (1918): 176–180.
72. W. E. Cooke, "Fibrosis of the Lungs Due to Inhalation of Asbestos Dust," *British Medical Journal* 2 (1924): 147.
73. Martin Cherniack, *The Hawk's Nest Incident: America's Worst Industrial Disaster* (New Haven, Conn., 1986); Rosner and Markowitz, *Deadly Dust,* pp. 4, 38–41.
74. Anthony J. Lanza et al., "Effects of the Inhalation of Asbestos Dust on the Lungs of Asbestos Workers," *U.S. Public Health Reports* 50 (1935): 1–12; Ozonoff, "Failed Warnings," pp. 170–171.
75. W. C. Dreessen et al., *A Study of Asbestosis in the Asbestos Industry* (Washington, D.C., 1938); Ozonoff, "Failed Warnings," pp. 188–190.
76. In the mid-1960s, workers at a Tyler, Texas, asbestos insulation plant owned by the Pittsburgh Corning Corporation were exposed to up to ten times the 5 mppcf standard. For years, workers and union officials at the plant were not informed of the hazard, despite being the objects of a Public Health Service study; see Brodeur, *Expendable Americans,* pp. 11–20. Today, OSHA allows only 1 asbestos fiber (not particle) per 10 milliliters of air, less than one-thousandth the level allowed in the 1950s.
77. S. R. Gloyne, "The Morbid Anatomy and Histology of Asbestos," *Tubercle* 14 (1933): 550–558; also his "Two Cases of Squamous Carcinoma of the Lung Occurring in Asbestosis," *Tubercle* 17 (1935): 4–10.
78. E. L. Middleton, "Industrial Pulmonary Disease Due to the Inhalation of Dust with Special Reference to Silicosis," *Lancet* 2 (1936): 59–64; Martin Nordmann, "Der Berufskrebs der Asbestarbeiter," *Zeitschrift für Krebsforschung* 47 (1938): 288–302. Franz Koelsch concluded that asbestos workers must be counted as "among those groups most exposed to carcinoma of the lungs"; see his "Lungenkrebs und Beruf," *Acta Unio Internationalis Contra Cancrum* 3 (1938): 243–251; also Ozonoff, "Failed Warnings," pp. 199–201.
79. Hueper, *Occupational Tumors,* pp. 399–405; Barry I. Castleman, "Asbestos and Cancer: History and Public Policy" (letter), *British Journal of Industrial Medicine* 48 (1991): 427–430.
80. Arthur J. Vorwald and John W. Karr, "Pneumonoconiosis and Pulmonary Carcinoma," *American Journal of Pathology* 14 (1938): 49.
81. Ozonoff, "Failed Warnings," pp. 202–203; Lilienfeld, "The Silence," pp. 794–795. The abbreviated paper is A. J. Vorwald et al., "Experimental Studies of Asbestosis," *Archives of Industrial Hygiene* 3 (1951): 1–43.
82. Castleman, *Asbestos,* pp. 91–93.
83. K. Smith to Lindell, January 28, 1950, cited in Lilienfeld, "The Silence," p. 795.
84. Daniel C. Braun, *An Epidemiological Study of Lung Cancer in Asbestos Miners* (Pittsburgh, 1957), p. 76, cited in Ozonoff, "Failed Warnings," p. 205.
85. Daniel C. Braun and T. David Truan, "An Epidemiological Study of Lung Cancer in Asbestos Miners," *Archives of Industrial Health* 17 (1958): 634–653.
86. Cited in Castleman, *Asbestos,* p. 96.
87. Lilienfeld, "The Silence," p. 797.
88. Hueper, "Adventures," p. 259; Hueper to Michael Shimkin, May 9, 1961, reprinted in *Drug Reporter* (January 11, 1961).
89. Richard Doll, "Mortality from Lung Cancer in Asbestos Workers," *British Journal of Industrial Medicine* 12 (1955): 81–86. Doll had originally wanted to coauthor this paper with J. F. Knox, the medical advisor to a Turner Brothers asbestos plant in Rochdale, who had gathered the cancer data. Knox withdrew his name when the company told him the data were "confidential"; the company's managing director also threatened Doll with legal action if he pursued his efforts to publish. The relevant correspondence can be found in the documents produced for the trial of *Chase* v. *Turner & Newall,* prepared in 1993.
90. R. Kiviluoto, "Pleural Calcification as Roentgenologic Sign of Non-Occupational Endemic Anthophyllite-Asbestosis," *Acta Radiologica, Suppl.* 194 (1960): 1–67.
91. J. C. Wagner, C. A. Sleggs, and P. Marchand, "Diffuse Pleural Mesothelioma and Asbestos Exposure in the North Western Cape Province," *British Journal of Industrial Medicine* 17 (1960): 260–271. Gerrit W. H. Schepers, a member of the Pneumonoconiosis Bureau of South Africa from

1944 to 1954, found high levels of mesothelioma at the Penge amosite mine as early as 1949; see his "Asbestos-Related Disease," unpublished documentation produced in the discovery process for the trial of *Chase Manhattan* v. *Turner & Newall.*

92. Thomas F. Mancuso and Elizabeth J. Coulter, "Methodology in Industrial Health Studies," *Archives of Environmental Health* 6 (1963): 210–226.
93. Selikoff, Churg, and Hammond, "Asbestos Exposure and Neoplasia."
94. Brodeur, *Expendable Americans.*
95. Ibid., p. 249.
96. Califano, "Occupational Safety and Health," p. 739.
97. Brodeur, *Outrageous Misconduct,* pp. 3–5. For an update on the bankruptcy proceedings showing how much of the trust fund set up to compensate victims has gone to lawyers and a Manville insurance plan, see Stephen Labaton, "The Bitter Fight Over the Manville Trust," *New York Times,* July 8, 1990.
98. Heath, *Strategic Issues Management,* p. 8.
99. Irving J. Selikoff and J. Churg, eds., *Biological Effects of Asbestos* (New York, 1965).
100. *Job Safety and Health Report* (November 7, 1978): 180; also Epstein, *Politics of Cancer,* pp. 89–101.
101. G. Hobart to Vandiver Brown, December 15, 1934, cited in Lilienfeld, "The Silence," p. 794.
102. Ozonoff, "Failed Warnings," pp. 140; compare also Barry I. Castleman, *The Development of Knowledge About Asbestos Disease* (Washington, D.C., 1977).
103. Memo from Linke to McKinney, January 12, 1979, cited in Lilienfeld, "The Silence," p. 797.
104. Barry I. Castleman, "Preventing Catastrophe," *Society* (March–April 1981): 9.
105. James A. Talcott et al., "Asbestos-Associated Diseases in a Cohort of Cigarette-Filter Workers," *New England Journal of Medicine* 321 (1989): 1220–1222; Myron Levin, "Smoking's Asbestos Episode," *Los Angeles Times,* October 17, 1991.
106. Peters and Peters, *Sourcebook,* p. B17.
107. B. T. Commins, "The Significance of Asbestos and Other Mineral Fibres in Environmental Ambient Air," *Scientific and Technical Report* (June 1985): 67.
108. Debbie Socolar, "Breath Taken: Exposing the Ongoing Tragedy of Asbestos," *Health/PAC Bulletin* (Winter 1990): 9. Asbestos is also present in drinking water: a 1983 survey of 538 U.S. water supplies found that 8 percent had more than 10 million asbestos fibers per liter. Some had more than 1 billion fibers per liter; see James R. Millette et al., "Asbestos in Water Supplies of the United States, *Environmental Health Perspectives* 53 (1983): 45–48. In 1989, the EPA recommended no more than 7 million asbestos fibers exceeding 10 microns in length per liter of drinking water. The Asbestos Information Association maintains that there should be no limit whatsoever, because animal studies have thus far demonstrated only "benign" tumors caused by drinking contaminated water; see its "New EPA Regulatory Initiative," *News and Notes* (May 31, 1989): 1–2. Asbestos enters drinking water through several sources, the primary one being asbestos-reinforced cement water pipes, 400,000 miles of which have been laid across the United States. Asbestos also enters water supplies through natural serpentine rock, as on the San Francisco Peninsula.
109. Asbestos manufacturers (including the AIA and the Asbestos Institute) protested the ban, and in March 1992 a federal appeals court in New Orleans struck it down on the grounds that the EPA had failed to give industry a chance to challenge the health benefits of the ban; see "Asbestos Regs to be Re-examined," *Science* 255 (1992): 1639. In 1994, U.S. asbestos production was only about 4 percent of what it was at its peak in the mid-1970s.
110. Philip J. Landrigan, Homayoun Kazemi, and Irving J. Selikoff, eds., *The Third Wave of Asbestos Disease: Exposure to Asbestos in Place* (New York, 1991).
111. Fumento, "Asbestos Rip-Off."
112. Ibid.
113. Susan Wood, "Firms Still Busy in Post-Panic Asbestos Industry," *Dallas Business Journal* (March 15, 1991): 16.
114. Wilhelm Hueper to Ward M. O'Donnell, March 11, 1964, Box 17, Stewart Papers, National Library of Medicine.
115. Mossman and Gee, "Asbestos-Related Diseases," pp. 1722–1723. Animal evidence appears to indicate that long, invisibly fine fibers (you need an electron microscope to see them) are the most potent mesothelioma-causing agents. Most asbestos standards require only that long fibers be counted, but this is primarily due to the fact that most monitoring systems simply cannot count short fibers and it is easier to design protective gear to guard against long fibers. The visibility of the longer fibers seems to have focused attention on these as the hazardous elements. As one sourcebook on the topic puts it: "It is difficult to count what you can't see"; see Peters and Peters, *Sourcebook,* pp. B11–B24; Brodeur, *Expendable Americans,* pp. 30–32.
116. A 1990 report in *Forbes* magazine endorses the "Harvard symposium"/"two-fiber" hypothesis and

calls it "amazing" that Congress and the EPA still regulate asbestos fibers according to a single standard; see Gary Slutsker, "Paratoxicology," *Forbes* (January 8, 1990): 303. Dixy Lee Ray exploits the two-fiber theory in her assertion that amphibole varieties are "very dangerous," while chrysotile forms are "generally benign"; see her *Trashing the Planet*, pp. 85–86, and her *Environmental Overkill*, pp. 151–156.

117. Mossman and Gee, "Asbestos-Related Diseases," pp. 1727–1728; Brooke T. Mossman et al., "Asbestos: Scientific Developments and Implications for Public Policy," *Science* 247 (1990): 294–301; Philip H. Abelson, "The Asbestos Removal Fiasco," *Science* 247 (1990): 1017; also the correspondence in *Science* 248 (1990): 795–802.

118. J. Corbett McDonald and Alison D. McDonald, "Asbestos and Carcinogenicity" (letter), *Science* 249 (1990): 844.

119. Gary Slutsker, "Paratoxicology," *Forbes* (January 8, 1990): 303.

120. Richard Stone, "No Meeting of the Minds on Asbestos," *Science* 254 (1991), 928–931; compare also David Kotelchuck, "The Third-Wave Asbestos Conference: High Drama in Science," *Health/PAC Bulletin* (Winter 1990): 13–17.

121. Health Effects Institute-Asbestos Research, *Asbestos in Public and Commercial Buildings: A Literature Review and Synthesis of Current Knowledge* (Cambridge, Mass., 1991).

122. Stone, "No Meeting of the Minds on Asbestos," p. 929.

123. Doll and Peto, *Causes of Cancer*, pp. 1307–1308.

124. William J. Nicholson, George Perkel, and Irving J. Selikoff, "Occupational Exposure to Asbestos: Population at Risk and Projected Mortality—1980–2030," *American Journal of Industrial Medicine* 3 (1982): 259–311.

125. David Lilienfeld et al., "Projection of Asbestos-Related Disease in the United States, 1985–2009," *British Journal of Industrial Medicine* 45 (1988): 283–291; Philip J. Landrigan, "Environmental Disease—A Preventable Epidemic," *American Journal of Public Health* 82 (1992): 941. Even the neoconservative Michael Fumento concedes that asbestos causes 3,300–12,000 deaths per year; see his "Asbestos Rip-Off," pp. 21–26.

126. Barry I. Castleman, "Asbestos and Cancer: History and Public Policy" (letter), *British Journal of Industrial Medicine* 48 (1991): 429.

127. Ozonoff, "Failed Warnings," p. 208.

128. Suzanne L. Oliver and Leslie Spencer, "Who Will the Monster Devour Next?" *Forbes* (February 18, 1991): 76; David Kotelchuck, "Asbestos—Science for Sale," *Science for the People* (September 1975): 13.

129. An official history has been published as *The Manufacturing Chemists' Association, 1872–1972: A Centennial History* (Washington, D.C., 1972). The CMA publishes a monthly magazine, *Chem-Ecology*, the *CMA News*, and a newsletter. The CMA's 51-member board of directors for 1991–1992 includes not a single female—an extraordinary situation for an association that prides itself on "balance" and "neutrality"; see the CMA's *1991–1992 Annual Report* (Washington, D.C., n.d.), pp. 36–37.

130. Chemical Manufacturers Association, *Who We Are, What We Do* (Washington, D.C., 1990), p. 33.

131. Chemical Manufacturers Association, *CMA—What It Is, What It Does* (Washington, D.C., 1981), p. 2.

132. CMA, *Chemstar: A Report on the Year 1990–1991*, pp. 1–2, 33.

133. AIHC, "The American Industrial Health Council's Comment and Proposed Procedures for Implementing Section 112 of the Clean Air Act," February 21, 1980. A brief history of the AIHC can be found in William J. Murphy, "The American Industrial Health Council," published by the Harvard Business School's Case Services (Cambridge, Mass., 1982). The AIHC was originally created as an ad hoc committee within SOCMA, a member organization of the CMA. By 1978, the council had raised an estimated $30 million for lobbying and media campaigns against OSHA's new standards. Apart from its corporate members, the AIHC has some 70 association members, ranging from the Adhesives and Sealants Council to the Truck Trailer Manufacturers Association.

134. "Carcinogen Crackdown Proposed," *Chemical Week* (October 12, 1977): 18–19.

135. Peter Behr, "Controlling Chemical Hazards," *Environment* (July–August 1978): 26; "Carcinogen Crackdown Proposed," p. 18. For a defense of the case-by-case argument, see the Interdisciplinary Panel on Carcinogenicity's "Criteria for Evidence of Chemical Carcinogenicity," *Science* 225 (1984): 682–687.

136. See the report prepared by Booz, Allen & Hamilton, Inc., for the AIHC: *Executive Summary of Preliminary Estimates of Direct Compliance Costs and Other Economic Effects of OSHA's Generic Carcinogen Proposal on Substance Producing and Using Industries* (Scarsdale, N.Y., 1978).

137. James G. Robinson and Dalton G. Paxman, "OSHA's Four Inconsistent Carcinogen Policies," *American Journal of Public Health* 81 (1991): 775–780.

138. AIHC, *Report to the Membership, 1982* (Washington, D.C., 1982), pp. 7, 19. The AIHC had spent

$1.275 million to combat the OSHA proposal by the summer of 1978; see Ronald A. Lang to Designated Contacts, "AIHC Membership, Finances," May 4, 1978.

139. William D. Rowe, *Food and Drug Association OM Hearings,* September 4, 1979, FDA Docket no. 77 N-0026, p. 11.

140. Peter F. Infante and Gwen K. Pohl characterized "we need more study" as "the grandfather of all arguments for taking no action"; see their "Living in a Chemical World: Actions and Reactions to Industrial Carcinogens," *Teratogenesis, Carcinogenesis, and Mutagenesis* 8 (1988): 244.

141. Epstein, *Politics of Cancer,* pp. 422–434.

142. Chemical Manufacturers Association, *Food Additives: What They Are/How They Are Used* (Washington, D.C., 1971).

143. AIHC, *Cancer,* p. 5.

144. Ray Sentes, "Poisonous Pits," *Canadian Dimension* (April–May 1990): 43.

145. Cavanaugh, "Corruptions," p. 195.

146. Cited in Epstein, *Politics of Cancer,* p. 299.

147. Tobacco Institute, *Tobacco Smoke and the Nonsmoker: Scientific Integrity at the Crossroads* (Washington, D.C., 1986), p. 1.

148. John O'Toole, "Is Advertising Second-Class Speech?" *Gannett Center Journal* (1987): 105.

149. Green, "Scientific McCarthyism," p. 402.

150. Boal, "Rhetoric of Risk," p. 10; Winner, "Risk," p. 7.

151. The term "cancerphobia" has been used since at least the early decades of the twentieth century. In his *Mortality from Cancer,* for example, Hoffman used the term to designate the public's fears of either the heritability (and hence inevitability) of cancer or its possible contagious nature (pp. 207–208).

152. Michael Thoryn, "Chemicals and Plastics: The Catalysts of Living," *Nation's Business* (March 1979): 70.

153. Elizabeth Whelan, "Chemicals and Cancerphobia," *Society* (March–April 1981): 7; also her "Cancer and the Politics of Fear," her "Politics of Cancer" (p. 46), and her "The Era of Rotten Apples," *Successful Farming* (mid-February 1990). The fear of pesticide ranking is from Mix, "Truth Comes Out," p. 51.

154. Hilgartner, "Political Language," pp. 30–31.

155. "Nosophobia causes unnecessary anxiety and blurred vision, which negates our ability to distinguish between real and hypothetical risks"; see Whelan, "The Facts Behind the Health Scares," pp. 19–22. Compare also "The Real Asbestos Horror Story," *AIM Report* (September 1990); and Abelson, "Cancer Phobia."

156. Caufield, *Multiple Exposures,* p. 177; Weart, *Nuclear Fear;* Arthur J. Barsky, *Worried Sick: Our Troubled Quest for Wellness* (New York, 1988); Robert L. DuPont, *Nuclear Phobia* (Washington, D.C., 1980).

157. Commoner, *Making Peace,* p. 121.

158. Richard S. Brown and Paul R. Lees-Haley, "Fear of Future Illness, Chemical AIDS, and Cancerphobia: A Review," *Psychological Reports* 71 (1992): 187–207.

159. Green, "Scientific McCarthyism," p. 406. In 1906, a Dr. Römer of Stuttgart suggested that fear of cancer might well drive a person insane; see his "Über Krebsangst," *Zeitschrift für Krebsforschung* 4 (1906): 75–82. In 1955, George Crile speculated that "cancer phobia" might well cause "more suffering than cancer itself"; see his *Cancer and Common Sense* (New York, 1955), p. 8.

160. Maureen C. Hatch et al., "Cancer Rates After the Three Mile Island Nuclear Accident and Proximity of Residence to the Plant," *American Journal of Public Health* 81 (1991): 719–724; Susan Fitzgerald, "Stress, Not Radiation, Caused Cancers Near Three Mile Island," *Centre Daily Times* (State College, Penn.), May 27, 1991.

161. Robert Crawford, "Cancer and Corporations," *Society* (March–April 1981): 26.

162. Brookes, "Wasteful Pursuit," p. 162.

163. Tobacco Institute, *The Cigarette Controversy: Why More Research Is Needed* (Washington, D.C., n.d.), p. 2.

164. In a 1972 memo to Tobacco Institute president Horace R. Kornegay, Fred Panzer, a vice president of the institute, recommended sticking with the "brilliantly conceived" strategy of "'creating doubt about the health charge without actually denying it, advocating the public's right to smoke without actually urging them to take up the practice, and encouraging objective scientific research as the only way to resolve the question of health hazard"; see Geoffrey Cowley, "Science and the Cigarette," *Newsweek* (April 11, 1988): 67. The memo is reproduced in *Harper's Magazine* (June 1988): 25–26.

165. Green, "Scientific McCarthyism," p. 404.

166. For one recent case, see Gershon Fishbein, "Alar PR: A Media Victory," *Chemtech* (May 1990): 264–267.

167. Anthony Ramirez, "Market Place," *New York Times,* February 3, 1993.

CHAPTER 6
NATURAL CARCINOGENS AND THE MYTH OF TOXIC HAZARDS

1. Bruce N. Ames, "Dietary Carcinogens and Anticarcinogens," *Science* 221 (1983): 1256-1264; compare also Philip H. Abelson, "Dietary Carcinogens," *Science* 221 (1983): 1249. In March 1991, the director of the National Cancer Institute's Division of Cancer Etiology announced that if cooking at a high temperature were regulated by regulatory agencies "it would have been banned yesterday" (Knight Ridder Newspapers, March 30, 1991). A good recent update on Ames can be found in Jane E. Brody, "Strong Views on Origins of Cancer," *New York Times,* July 5, 1994.

2. Ames, "Dietary Carcinogens," pp. 1257-1259.

3. Ibid., pp. 1259-1260.

4. In his 1976 autobiography, Hueper suggested that however many natural carcinogens were in foods we should not allow food producers to add more: "the money-hungry food producers apparently feel that they have a moral right to imitate nature and to add their share of carcinogenicity to that present and often avoidable in the food contained in the daily food basket." Food producers were not to be granted the "immoral right to kill"; no one should be allowed to introduce "potentially lethal ingredients into the daily food for enriching himself." See his "Adventures," pp. 294-295.

5. M. C. Lancaster et al., "Toxicity Associated with Certain Samples of Groundnuts," *Nature* 192 (1961): 1095-1096; S. J. Van Rensburg et al., "Primary Liver Cancer Rate and Aflatoxin Intake in High Cancer Area," *South African Medical Journal* 48 (1974): 2508a-d.

6. National Research Council, *Toxicants Occurring Naturally in Foods* (Washington, D.C., 1966). E. Boyland devoted a whole section to natural carcinogens in "The Correlation of Experimental Carcinogenesis and Cancer in Man," *Progress in Experimental Tumor Research* 11 (1969): 231; compare also Doll, *Prevention of Cancer,* p. 92.

7. For references, see Efron, *Apocalyptics,* pp. 494-516, and National Research Council, *Diet, Nutrition, and Cancer* (Washington, D.C., 1982), pp. 234-276.

8. Cairns, *Cancer,* p. 164.

9. Gail Bronson, "New Group to Study Health Issues, Give Nader Competition," *New York Times,* July 18, 1978; compare also Whelan's article, "The Politics of Cancer," for the Heritage Foundation's *Policy Review,* esp. p. 40. Philip Abelson, in his 1979 critique of the Califano thesis, pointed out that Takashi Sugimura, director of Tokyo's National Cancer Center, had used the Ames test to detect mutagens in raw foods; see his "Cancer," p. 11.

10. Bruce N. Ames, "Identifying Environmental Chemicals Causing Mutations and Cancer," *Science* 204 (1979): 587-593; compare also Elizabeth K. Weisburger, "Not All Carcinogens Are Created by Chemists," *Sciquest* (October 1979): 12-15.

11. "Monsanto Speaks Up About Chemicals" [1977], p. 7. The brochure grimly warned that, under the 1958 Delaney amendment, "all green vegetables, if they were man-made chemicals, would have to be banned." In point of fact, the amendment bars only the *addition* of *synthetic* carcinogens to foods—and is very imperfectly enforced.

12. Ames, "Dietary Carcinogens," p. 1258. Compare also his "Cancer and Diet" (letter), *Science* 224 (1984): 757, where he states, "We are eating more than 10,000 times more of nature's pesticides than of man-made pesticides."

13. Ames, "Dietary Carcinogens," pp. 1257-1258.

14. Bruce N. Ames, "Natural Carcinogens: They're Found in Many Foods," *Health and Environment Digest* (February 1990): 5.

15. Ames, "Dietary Carcinogens," p. 1259.

16. Bruce N. Ames, "The Detection of Chemical Mutagens with Enteric Bacteria," in *Chemical Mutagens: Principles and Methods for Their Detection,* edited by Alexander Hollaender (New York, 1971), p. 271.

17. Bruce N. Ames et al., "Carcinogens Are Mutagens: A Simple Test System Combining Liver Homogenates for Activation and Bacteria for Detection," *Proceedings of the National Academy of Sciences* 70 (1973): 2281-2285.

18. Blum and Ames, "Flame-Retardant Additives." For a review of Ames's early career, see Tierney, "Not to Worry." When Ames first introduced his bacterial bioassay in the mid-1970s, more than 85 percent of chemicals known (from animal and human epidemiological studies) to be carcinogens showed up as carcinogens on the new test. As more chemicals were tested, however, its reliability came under question. Long-term studies by the National Cancer Institute and the National Toxicology Program showed that the Ames test was an imperfect indicator of human carcinogenic potential and that there was only about a 50–70 percent correlation between proven carcinogenicity and bacterial mutagenicity. Several known carcinogens (hydrazine and heavily chlorinated hydrocarbons, for example) showed little mutagenicity, and several of the most notorious—DDT, chloro-

form, and heavy metals such as cadmium, for example—failed to show any mutagenic activity at all. See R. W. Tennant et al., "Prediction of Chemical Carcinogenicity from *in vitro* Genetic Toxicity Assays," *Science* 236 (1987): 933–941; Errol Zeiger, "Carcinogenicity of Mutagens," *Cancer Research* 47 (1987): 1287–1296; also Efron, *Apocalyptics*, pp. 204–212.

19. Bruce N. Ames, "Six Common Errors Relating to Environmental Pollution," *Regulatory Toxicology and Pharmacology* 7 (1987): 380.

20. Ames, Magaw, and Gold, "Ranking Possible Carcinogenic Hazards." Ames had presented an early version of his HERP database in R. Peto et al. (including Ames), "The TD_{50}: A Proposed General Convention for the Numerical Description of the Carcinogenic Potency of Chemicals in Chronic-Exposure Animal Experiments," *Environmental Health Perspectives* 58 (1984): 1–8.

21. Ames et al., "Ranking Possible Carcinogenic Hazards," p. 277; Jane E. Brody, "Putting the Risk of Cancer in Perspective," *New York Times,* May 6, 1987.

22. Bruce N. Ames, "Cancer Scares over Trivia: Natural Carcinogens in Food Outweigh Traces in Our Water," *Los Angeles Times,* May 15, 1986; also his "Peanut Butter, Parsley, Pepper and Other Carcinogens," *Wall Street Journal,* February 14, 1984.

23. Tierney, "Not to Worry," pp. 30–32.

24. Bruce N. Ames et al., "Risk Assessment" (letter), *Science* 237 (1987): 1400.

25. Jane E. Brody, "New Index Finds Some Cancer Dangers Are Overrated and Others Ignored," *New York Times,* April 17, 1987.

26. Brody, "Putting the Risk of Cancer in Perspective."

27. ABC News, *20/20,* show no. 810, "Much Ado About Nothing," March 18, 1988.

28. Samuel S. Epstein and Joel B. Swartz, et al., "Cancer and Diet" (letter), *Science* 224 (1984): 660–667. The letter was cosigned by sixteen public health scientists and activists, including Eula Bingham, director of OSHA under Carter, John Gofman of the University of California, Robert Harris of Princeton, and Ruth Hubbard and George Wald of Harvard. Epstein later lampooned Ames with a vitriolic essay on "The Dangers of a Glass of Wine: Scientific Mythologies of Bruce Ames," in *Environmental Health Monthly* (Newsletter of the Citizens Clearinghouse Against Hazardous Waste) (April 1989): 1–3.

29. David Bollier, "Leading Scientist Laughs at DDT, Worries About Peanut Butter, Believe It or Not!" *Public Citizen* (September–October 1988): 12–20; "Too Much Fuss About Pesticides?" *Consumer Reports* (October 1989): 655–658.

30. Terry Gips to Samuel Epstein, March 23, 1988. Gips, a Yale-trained agricultural economist, was author of *Breaking the Pesticide Habit: Alternatives to 12 Hazardous Pesticides* (Minneapolis, Minn., 1987).

31. Terry Gips to Pesticide Action Network, Sustainable Agriculture Policy Group, and Organic Agriculture Groups, March 31, 1988.

32. Ibid. Many of these charges are contained in Terry Gips's March 21, 1988, letter to Meredith White, senior editor of *20/20.*

33. Bruce Ames, "Testimony Before the State of California Department of Industrial Relations, Occupational Safety and Health, Dibromochloropropane Inquiry," October 18, 1977, p. 7.

34. Ibid., p. 18. Compare his statement, "I'm getting a little nervous what all this is going to mean having all these chlorinated compounds in our body fat" (p. 27).

35. Ibid., pp. 23–27; compare also Bruce N. Ames, "Identifying Environmental Chemicals Causing Mutations and Cancer," *Science* 204 (1979): 587–593. In 1977, Ames warned that "thousands of chemicals" had been introduced into the environment without adequate testing and that "a steep increase" in human cancer rates from exposure to such chemicals might soon occur as the twenty- to thirty-year lag time for chemical carcinogenesis in humans was "almost over." See Blum and Ames, "Flame-Retardant Additives," pp. 17, 22.

36. The Bruce Ames political "flip-flop" is sketched in "Baffled by Bruce," *Everyone's Backyard* (Spring 1989): 7; compare also Piller, "Scientist Questions Modern Cancer Hazards," pp. 1, 8, and his "Experts Refute Ames' Theories."

37. Tierney, "Not to Worry," p. 29.

38. Efron, *Apocalyptics,* pp. 419–420.

39. Ames, "Mother Nature."

40. Lester, "Mothers," p. 6.

41. Terry Gips to Bruce Ames Action Committee, "Meeting Notes," April 14, 1988, p. 2.

42. Leslie Spencer, "Ban All Plants—They Pollute," *Forbes* (October 25, 1993): 104.

43. Ames, personal communication, March 28, 1994.

44. Epstein and Swartz et al., "Cancer and Diet," p. 666.

45. Marvin Schneiderman to Samuel Epstein, September 9, 1987.

46. Bruce N. Ames, "Science and the Spontaneous Order," *Lectures in Social Philosophy and Policy* (Bowling Green, Ohio, 1989).

47. Postrel, "Of Mice and Men," p. 22. Compare Ames's assertion that "the EPA kills people" by over-regulating (Spencer, "Ban All Plants," p. 104).

48. Postrel, "Of Mice and Men," p. 19.

49. Ames, personal communication, March 28, 1994.

50. Samuel S. Epstein and Joel B. Swartz et al., "Carcinogenic Risk Estimation," *Science* 240 (1988): 1043–1045. The letter is cosigned by 15 health activist-scholars, including John Bailar, David Ozonoff, Beverly Paigen, William Nicholson, and Philip Landrigan.

51. William Lijinsky, statement before the Committee on Labor and Human Resources, U.S. Senate, June 6, 1989. T. Colin Campbell's study of health and eating habits in China has come up with similar conclusions; see Gail Vines, "China's Long March to Longevity," *New Scientist* (December 8, 1990): 37–41.

52. Devra Lee Davis, "Paleolithic Diet, Evolution, and Carcinogens," *Science* 238 (1987): 1633–1634; also her "Natural Anticarcinogens, Carcinogens, and Changing Patterns in Cancer: Some Speculation," *Environmental Research* 50 (1989): 322–340.

53. Thomas W. Culliney, David Pimentel, and Marcia H. Pimentel, "Pesticides and Natural Toxicants in Foods," *Agriculture, Ecosystems and Environment* 41 (1992): 297–320. Frederica Perera and Paolo Boffetta of Columbia University's School of Public Health pointed out that several of the substances Ames had classed as "natural pesticides and dietary toxins" were known to result from improper harvesting, manufacturing, or cooking; see their "Perspectives." The authors construct an interesting HERP table of their own, excluding some of the rarer exposures ranked by Ames and correcting some of its omissions (pp. 1286–1287).

54. Edward Groth to Bruce Ames, December 7, 1989, pp. 18–19.

55. David Pimentel et al., "Environmental and Economic Costs of Pesticide Use," *BioScience* 42 (1992): 750–760. Pimentel points out that sugar beets treated with 2,4-D can have more than 20-fold increased levels of potassium nitrate; see his letter to Giorgio Celli, March 29, 1989; also his *Ecological Effects of Pesticides on Non-Target Species* (Washington, D.C., 1971).

56. Leonard Stoloff, "Carcinogenicity of Aflatoxins" (letter), *Science* 237 (1987): 1283; compare also Ames's "Response" (letter), *Science* 237 (1987): 1283–1284.

57. In 1962, Rachel Carson had suggested that although natural cancer-causing agents were a factor in producing malignancy, these were few in number and "belong to that ancient array of forces to which life has been accustomed from the beginning" (*Silent Spring*, p. 219).

58. Davis, "Paleolithic Diet, Evolution, and Carcinogens," p. 1633.

59. Peter Weiner to Samuel Epstein, February 9, 1984. I. Bernard Weinstein has made a similar argument; see Marx, "Animal Carcinogen Testing Challenged," p. 745.

60. Marvin Schneiderman to Samuel Epstein, September 9, 1987.

61. Bruce N. Ames and Lois S. Gold, "Paleolithic Diet, Evolution, and Carcinogens" (letter), *Science* 238 (1987): 1634.

62. MacKenzie, Bartecchi, and Schrier, "Human Costs of Tobacco Use."

63. Lester, "Mothers," p. 8; Perera and Boffetta, "Perspectives," p. 1285.

64. Edward Groth to Bruce Ames, December 7, 1989, p. 3.

65. Daniel Wartenberg and Michael A. Gallo, "The Fallacy in Ranking Carcinogens Using $TD_{50}s$," *Risk Analysis* 10 (1990): 609–613.

66. World Health Organization, *Public Health Impact of Pesticides Used in Agriculture* (Geneva, 1990), pp. 85–89.

67. David Pimentel et al., "Environmental and Economic Costs of Pesticide Use," *BioScience* 42 (1992): 750–760; David Pimentel to Giorgio Celli, March 29, 1989.

68. Postrel, "Of Mice and Men," p. 21.

69. Efron, *Apocalyptics*, p. 419.

70. National Agricultural Chemicals Association, "Update on Agrichemical Issues," April 1, 1988.

71. "Too Much Fuss About Pesticides?" *Consumer Reports* (October 1989): 656. Videotape copies of "Big Fears, Little Risks" and "Grape Boycott: More Smoke and Mirrors" are available through the California Table Grape Commission.

72. *SIRC Review* (October 1990): 69–71.

73. Jeffrey B. Kaplan, "Food Pesticide Ban Won't Reduce Cancer" (letter), *New York Times*, March 22, 1993.

74. Bruce N. Ames, "Misconceptions About Pollution and Cancer," *National Review* (December 3, 1990): 34–35; "Tests on Trial," *The Economist* (September 21, 1991): 103–104; "The Topic of Cancer," *American Spectator* (June 1993): 38–39.

75. Philip H. Abelson, "California's Proposition 65," *Science* 237 (1987): 1553.

76. Piller, "Scientist Questions Modern Cancer Hazards," p. 8.

77. Ames and Gold, "Pesticides, Risk, and Applesauce."

78. Robert J. Scheuplein, director of the FDA's Office of Toxicological Sciences, is an Ames supporter: in February 1990, Scheuplein made headlines by claiming that no one "will ever die of pesticide residues from any food, anywhere" and that a day's consumption of protein caused more genetic damage than smoking five cigarettes. See Lisa Y. Lefferts, "Carcinogens au Naturel?" *Nutrition Action Health Letter* (July–August 1990): 1, 5–7.

79. Office of Management and Budget, "Regulatory Program of the United States Government," reprinted in *The SIRC Review* (October 1990): 9–21.

80. Philip H. Abelson, "Incorporation of New Science Into Risk Assessment," *Science* 250 (1990): 1497.

CHAPTER 7

THE POLITICAL MORPHOLOGY OF DOSE-RESPONSE CURVES

1. Maugh, "Chemical Carcinogens," p. 37.

2. Peto, "Distorting the Epidemiology," p. 297.

3. Richard A. Kerr, "Indoor Radon: The Deadliest Pollutant," *Science* 240 (1988): 608.

4. In the nineteenth century, psychologists sometimes spoke of a *threshold of consciousness* designating "a limit below which our several sensibilities are unable to discriminate"; the *Oxford English Dictionary* traces the notion back to Johann Herbart's 1824 *Psychologie als Wissenschaft*.

5. For the history of petrochemical and mineral threshold values, see Castleman's *Asbestos,* pp. 223–311; on radiation thresholds, see Whittemore, "National Committee"; for early German concepts, see Henschler, "Exposure Limits."

6. K. B. Lehmann, 'Experimentelle Studien über den Einfluss technisch und hygienisch wichtiger Gase und Dämpfe auf den Organismus," *Archiv für Hygiene* 5 (1886): 1–126; Castleman, *Asbestos,* pp. 224–225. Lehmann spoke in terms of *Giftigkeitsgrenze* and *zulässige Grenzwerthe* for specific substances (pp. 5–7).

7. Henschler, "Exposure Limits," pp. 79–92. In her autobiography, Alice Hamilton recalled how, in Würzburg in 1933, Lehmann waxed poetic over the "symbolic" beauty of Nazi book burning. She quotes him: "I stood for two hours in the square watching the leaping flames and the crowd of silent worshipers. To all of us it was a symbolic act, a renunciation of the religion of class hatred and conflict, of sexual looseness and of scoffing at all that was old and revered, a freeing of the people from all the decadence that followed the war. And to me it was also a purification of the spot where, in the fall of 1918, I witnessed a Communist attack, a mob of the most degraded people I ever saw, who swooped down on the city, nobody knew from where, and took possession of the public buildings. As I watched the flames that night I felt that at last that crime was wiped out." See her *Exploring the Dangerous Trades* (Boston, 1943), p. 377.

8. Paull, "Origin and Basis."

9. Ferdinand Flury, "Über Kampfgasvergiftungen," *Zeitschrift für die gesamte experimentelle Medizin* 13 (1921): 1–15.

10. Henschler, "Exposure Limits," p. 82.

11. Hounshell and Smith, *Science and Corporate Strategy,* p. 557.

12. Paull, "Origin and Basis," p. 229; compare also Ludwig Teleky, *History of Factory and Mine Hygiene* (New York, 1948), pp. 285–317.

13. Hounshell and Smith, *Science and Corporate Strategy,* p. 557.

14. Alan Derickson, "'On the Dump Heap': Employee Medical Screening in the Tri-State Zinc-Lead Industry, 1924–1932," *Business History Review* 62 (1988): 656–677.

15. Robert A. Kehoe et al., "On the Normal Absorption and Excretion of Lead," *Journal of Industrial Hygiene* 15 (1933): 257–289; William Graebner, "Hegemony Through Science: Information Engineering and Lead Toxicology, 1925–1965," in Rosner and Markowitz, *Dying for Work,* p. 143.

16. David Rosner and Gerald Markowitz, "'A Gift of God'? The Public Health Controversy over Leaded Gasoline During the 1920s," in Rosner and Markowitz, *Dying for Work,* p. 133.

17. In 1943, J. H. Sterner defined "maximum allowable concentration" as "the upper limit of concentration of an atmospheric contaminant which will not cause injury to an individual exposed continuously during his working day and for indefinite periods of time"; see Paull, "Origin and Basis," pp. 229–230.

18. Paull, "Origin and Basis," pp. 232–233; American Conference of Governmental Industrial Hygienists, *Threshold Limit Values . . . for 1984–1985* (Cincinnati, 1984).

19. Barry I. Castleman and Grace E. Ziem, "Corporate Influence on Threshold Limit Values," *American Journal of Industrial Medicine* 13 (1988): 531–559. Chairman of the committee at this time was Herbert Stokinger. By the mid-1970s, the ACGIH had established TLVs for more than 400

suspect substances; TLVs developed by the ACGIH were adopted or influential in Japan, West Germany, Italy, Holland, the United Kingdom, most other European nations (France excepted), and many developing nations. See V. E. Rose, "Standards for the Control of Carcinogens in the Workplace," *Journal of Occupational Medicine* 18 (1976): 81–84.

20. S. A. Roach and S. M. Rappaport, "But They Are Not Thresholds: A Critical Analysis of the Documentation of Threshold Limit Values," *American Journal of Industrial Medicine* 17 (1990): 727–753.

21. Hounshell and Smith, *Science and Corporate Strategy,* p. 563.

22. Manfred Bowditch et al., "Code for Safe Concentrations of Certain Common Toxic Substances Used in Industry," *Journal of Industrial Hygiene and Toxicology* 22 (1940): 251.

23. Paull, "Origin and Basis," pp. 234–235. Several substances are mentioned as suspected carcinogens (beryllium and selenium, for example) in the ACGIH's 1966 *Documentation of Threshold Limit Values,* but there is no mention of cancer in the discussion of asbestos, benzene, uranium, or vinyl chloride, and arsenic is expressly dismissed as a possible cause of cancer.

24. Edward W. Schroder and Paul H. Black, "Retinoids: Tumor Preventers or Tumor Enhancers?" *Journal of the National Cancer Institute* 65 (1980): 671–674.

25. Hermann J. Muller, "The Manner of Production of Mutations by Radiation," in *Radiation Biology,* vol. 1: *High-Energy Radiation,* edited by A. Hollaender (New York, 1954), pp. 475–626.

26. Robert Bierich, "Über den Einfluss genetischer Faktoren auf Entstehung und Ausbildung der Krebsanlage," *Zeitschrift für Krebsforschung* 48 (1938): 87.

27. Alice M. Stewart et al., "Preliminary Communication: Malignant Disease in Childhood and Diagnostic Irradiation in Utero," *Lancet* 2 (1956): 447; Alice M. Stewart and G. W. Kneale, "Radiation Dose Effects in Relation to Obstetric X-rays and Childhood Cancers," *Lancet* 1 (1970): 1185–1188.

28. Whittemore, "National Committee," pp. 172–174; Hacker, *Dragon's Tail,* pp. 38–39. The first "tolerance dose" for radiation was established in 1925 by the radiologist Arthur Mutscheller.

29. Caufield, *Multiple Exposures,* pp. 73, 120.

30. William L. Russell et al., "Radiation Dose Rate and Mutation Frequency," *Science* 128 (1958): 1550.

31. William L. Russell, "The Genetic Effects of Radiation," *Peaceful Uses of Atomic Energy* 13 (1972): 496–499.

32. Whittemore, "National Committee," pp. 34, 121–127.

33. Holaday et al., *Control of Radon,* pp. 1, 8.

34. Doll, *Prevention,* pp. 99, 123.

35. Mantel, "The Concept of Threshold," p. 104.

36. Office of Technology Assessment, *Identifying and Regulating Carcinogens* (Washington, D.C., 1987). "Linearity" is often used to mean "no-threshold," and I've adopted that use here. The most commonly used linear "one-hit" model is actually an exponential curve that is nearly linear at low doses. The key fact, for our purposes, is that like several other standard models (Weibull, multihit, and the EPA's linearized multistage model), the linear model presumes no threshold. See Lauren Zeise, Richard Wilson, and Edmund A. C. Crouch, "Dose-Response Relationships for Carcinogens: A Review," *Environmental Health Perspectives* 73 (1987): 259–308; contrast also Marvin A. Schneiderman et al., "Thresholds for Environmental Cancer: Biologic and Statistical Considerations," *Annals of the New York Academy of Sciences* 329 (1979): 92–130; and Cranor's *Regulating Toxic Substances,* pp. 12–48.

37. Epstein, Brown, and Pope, *Hazardous Waste,* p. 36.

38. Peto, "Need for Ignorance," pp. 129–130.

39. The Conservation Foundation, *Risk Assessment and Risk Control* (Washington, D.C., 1985), p. 31.

40. Arthur C. Upton, "Are There Thresholds for Carcinogenesis?" in *Living in a Chemical World: Occupational and Environmental Significance of Industrial Carcinogens,* edited by Cesare Maltoni and Irving J. Selikoff (New York, 1988), pp. 863–884; also his "Historical Perspectives on Radiation Carcinogenesis," in *Radiation Carcinogenesis,* edited by Arthur C. Upton et al. (New York, 1986), pp. 1–10.

41. Lappé, *Chemical Deception,* pp. 96–124. Contrast Efron's *Apocalyptics,* pp. 340–343.

42. The metaphor of threshold as point of overflow is also discussed in Mantel's "Concept of Threshold," p. 105.

43. Ames, "Testimony," p. 28.

44. Leonard A. Sagan, "On Radiation, Paradigms, and Hormesis," *Science* 245 (1989): 574, 621.

45. Edward J. Calabrese, "Biological Effects of Low Level Exposures," *AIHC Journal* (Spring 1994): 7–15.

46. Bruce N. Ames et al., "DNA Lesions, Inducible DNA Repair, and Cell Division: Three Key Fac-

tors in Mutagenesis and Carcinogenesis," *Environmental Health Perspectives* 101 (suppl. 5, 1993): 35–44.

47. Latarjet, "Radiation Carcinogenesis." The most comprehensive defense of hormesis is T. Don Luckey's *Radiation Hormesis* (Boca Raton, Fla., 1991); compare also his earlier *Hormesis with Ionizing Radiation.*

48. Académie des Sciences, *Risques des rayonnements ionisants et normes de radioprotection* (Paris, 1989); Ray, *Trashing the Planet,* pp. 109–112. Contrast the suggestion by W. H. Koppenol and P. L. Bounds of Louisiana State University that the concept of hormesis "smacks of homeopathy"; see *Science* 246 (1989): 311.

49. Linda E. Ketchum, "First Conference on Radiation Hormesis Explores Nonhazardous Effects of Exposure," *Journal of Nuclear Medicine* 26 (1985): 1363–1364. Papers from the conference, held at Oakland, August 14–16, 1985, were published in *Health Physics* 52 (1987): 517–680. Leonard A. Sagan, a senior staff scientist at EPRI and a longtime champion of hormesis, noted in his opening remarks that the purpose of the conference was to challenge "the conventional radiation paradigm" according to which low-level exposures are harmful; compare also his "A Brief History and Critique of the Low Dose Effects Paradigm," *BELLE Newsletter* (December 1993): 1–7.

50. Sheldon Wolff, "Are Radiation-Induced Effects Hormetic?" *Science* 245 (1989): 575, 621.

51. Carson, *Silent Spring,* p. 232. Carson also argued that the FDA's certification of contaminated foods below a certain level (the "tolerance") as safe was meaningless (pp. 181–184).

52. Gofman, *Radiation-Induced Cancer,* figure 13-C and chapter 14.

53. Gofman's *Radiation-Induced Cancer* documents several dozen assertions of "safe levels of radiation," including three that do so on the basis of radiation repair and five that assert a hormetic effect (p. 34-1). His analysis has been influential within the antinuclear community—the Oakland-based Atomic Reclamation and Conversion Project, for example, and several radiation victims alliances.

54. Gofman, *Radiation and Human Health,* pp. 225–229.

55. Efron, *Apocalyptics,* p. 285; compare also pp. 246–251.

56. Love, "Indecent Exposure," p. 61.

57. Bruce N. Ames, "Six Common Errors Relating to Environmental Pollution," *Regulatory Toxicology and Pharmacology* 7 (1987): 381.

58. Marx, "Animal Carcinogen Testing," p. 745.

59. Wilhelm Hueper, "The Potential Role of Non-Nutritive Food Additives and Contaminants as Environmental Carcinogens," *A.M.A. Archives of Pathology* 62 (1956): 226. As early as 1938, Hueper had shown that betanaphthylamine was not a carcinogen in dogs until it was metabolized in the bladder.

60. Stannard, *Radioactivity and Health,* pp. 410–412; personal communication, Elizabeth Hanson.

61. Eileen Welsome, "The Plutonium Experiment," *Albuquerque Tribune* November 15–17, 1993; Geoffrey Sea, "The Radiation Story No One Would Touch," *Columbia Journalism Review* (March–April 1994): 37–40. Sea's paper, prior to publication, was titled "Waiting for the *New York Times*: How the Radiation Story Became 'News Fit to Print'" (personal communication).

62. Sigismund Peller, *Cancer in Man* (New York, 1952), p. 114.

63. Herber [Bookchin], *Our Synthetic Environment,* p. 135.

64. For a review, see Efron, *Apocalyptics,* pp. 246–248.

65. H. F. Kraybill, "Conceptual Approaches to the Assessment of Nonoccupational Environmental Cancer," in *Environmental Cancer,* edited by H. F. Kraybill and Myron A. Mehlman (Washington, D.C., 1977), pp. 35–37; also his "By Appropriate Methods: The Delaney Clause," in *Regulatory Aspects of Carcinogenesis and Food Additives: The Delaney Clause,* edited by Frederick Coulston (New York, 1979), pp. 70–73.

66. Peto, "Distorting the Epidemiology," p. 300; compare also Cairns, *Cancer,* pp. 94–97.

67. James D. Wilson, "Interpreting Cancer Tests," *Science* 251 (1991): 257–258.

68. BELLE Advisory Committee members include Max Eisenberg, head of the tobacco industry's Center for Indoor Air Research, and Leonard Sagan, EPRI's hormesis theorist. The committee has published the *BELLE Newsletter* since 1990; see also their flagship publication, *Biological Effects of Low Level Exposures to Chemicals and Radiation* (Chelsea, Mich., 1992).

69. David G. Hoel et al., "The Impact of Toxicity on Carcinogenicity Studies: Implications for Risk Assessment," *Carcinogenesis* 9 (1988): 2045–2052.

70. Bailar et al., "One-Hit Models."

71. Ibid., p. 485.

72. In 1994, the American Industrial Health Council estimated that the total amount spent over the past twenty-five years by both government and industry to conduct animal tests of human cancer hazards was on the order of $1 billion to $3 billion—not a very substantial amount when one consid-

ers the diversity of chemicals to which we are exposed. See "Need for Re-Examination of Risk," *AIHC Science Commentary* (April 1994): 9.

73. Clive Jenkins, "Risk Assessment and Control of Toxic Substances in the Workplace: Who Counts the Cost?" *The Ecologist* 13 (1983): 230.

74. See my *Racial Hygiene*, pp. 223–250.

CHAPTER 8
NUCLEAR NEMESIS

1. Suggestions of a possible radioactive hazard from tobacco smoke emerged in the 1960s, when it was discovered that tobacco leaves concentrate radioactive polonium 210 and lead 210 from the phosphate fertilizers commonly used to grow tobacco. See E. P. Radford and V. R. Hunt, "Polonium-210: A Volatile Radioelement in Cigarettes," *Science* 143 (1964): 247–249; Edward A. Martell, "Radioactivity of Tobacco Trichomes and Insoluble Cigarette Smoke Particles," *Nature* 249 (1974): 215–217. Thomas H. Winters and Joseph R. Di Franza of the University of Massachusetts Medical Center resurrected the issue in 1981, provoking a flood of supportive letters in the *New England Journal of Medicine*. C. R. Hill of England's Institute of Cancer Research suggested that it was not phosphate fertilizer but rather natural fallout (radon 222) that accounted for the radioactivity of tobacco smoke, and R. T. Ravenholt of the Centers for Disease Control hypothesized that the radioactive elements in tobacco smoke might pass through the lungs and into the bloodstream, causing cancer in other parts of the body. See Thomas H. Winters and Joseph R. Di Franza, "Radioactivity in Cigarette Smoke," *New England Journal of Medicine* 306 (1982): 364–365; also the letters in "Radioactivity in Cigarette Smoke," *New England Journal of Medicine* 307 (1982): 309–312. Ravenholt believes that the American public is exposed to "far more radiation from the smoking of tobacco than they are from any other source"; Joseph Di Franza believes that the radiation inhaled with tobacco smoke could account for "about half of all lung cancers in smokers." A Tobacco Institute spokesperson has predictably dismissed the entire affair. See "Smokers Said to Risk Cancers Beyond Lungs," *New York Times*, July 29, 1982; Lowell Ponte, "Radioactivity: The New-Found Danger in Cigarettes," *Reader's Digest* (March 1986): 123–127.

2. Henri Becquerel, "Sur diverses propriétés des rayons uraniques," *Comptes rendus des séances de l'Académie des Sciences* 123 (1896): 855–858.

3. Herber, *Our Synthetic Environment*, pp. 166–172. In Germany in the 1930s, women with tuberculosis were often advised to have their ovaries X-rayed to "moderate" the course of the disease and to relieve them of the "onerous" effects of menstruation; see Felix von Mikulicz-Radecki, "Über Misserfolge der Röntgenkastration und -sterilisierung wegen Lungentuberkulose," *Strahlentherapie* 50 (1934): 658–663.

4. Ruth Brecher and Edward Brecher, *The Rays—A History of Radiology in the United States and Canada* (Baltimore, Md., 1969), p. 146. Emil H. Grubbe, a Chicago physician, provided therapeutic X-rays to a woman with inoperable breast cancer as early as January 1896; see his "Priority in the Therapeutic Use of X-rays," *Radiology* 21 (1933): 156–162. Radium pastes were first used to treat skin cancer in Paris in 1901. A good history of early radiation carcinogenesis is Antoine Lacassagne, *Les cancers produits par les rayonnements électromagnétiques* (Paris, 1945); compare also Jacob Furth and Egon Lorenz, "Carcinogenesis by Ionizing Radiation," in *Radiation Biology*, vol. 1 (New York, 1954), pp. 1145–1201.

5. Caufield, *Multiple Exposures*, pp. 6–12.

6. Carlson, *Genes*, pp. 336–337.

7. Hueper, *Occupational Tumors*, pp. 245–246.

8. Caufield, *Multiple Exposures*, pp. 9–11.

9. Albert Frieben, "Cancroid des rechten Handrückens," *Deutsche medicinische Wochenschrift, Vereins-Beilage* 28 (1902): 335.

10. K. Sick, "Karzinom der Haut," *Münchener medizinische Wochenschrift* 50 (1903): 1445; P. G. Unna, "Die chronische Röntgendermatitis der Radiologen," *Fortschritte auf dem Gebiete der Röntgenstrahlen* 8 (1904): 67–91.

11. Caufield, *Multiple Exposures*, p. 13. In 1908, Jean Clunet of Paris became the first person to induce a cancer experimentally by means of an externally applied agent: Clunet bombarded four white rats with X rays and found that one of the two survivors developed a sarcoma precisely where it had been exposed. In 1911 two German reviews of the human medical evidence documented fifty-four separate cases of radiation-caused cancers (mostly among physicians and their assistants), and a French review three years later listed more than a hundred cases. Clunet, *Recherches;* Piero Mustacchi and Michael B. Shimkin, "Radiation Cancer and Jean Clunet," *Can-*

cer 9 (1956): 1072–1074; Otto Hesse, *Symptomologie, Pathogenese und Therapie des Röntgenkarzinoms* (Leipzig, 1911); Paul Krause, "Zur Kenntnis der Schädigung der menschlichen Haut durch Röntgenstrahlen," *Zeitschrift für Röntgenkunde und Radiumforschung* 13 (1911): 256–264; Sophie Feygin, *Du cancer radiologique* (Paris, 1914).

12. Brenner, *Radon*, p. 85.

13. Helmuth Ulrich, "The Incidence of Leukemia in Radiologists," *New England Journal of Medicine* 234 (1946): 45–46. Two years earlier, Paul S. Henshaw et al. had shown that physicians as a whole were 1.7 times more likely to contract leukemia than other white males. See their "Incidence of Leukemia in Physicians," *Journal of the National Cancer Institute* 4 (1944): 339–346; also Herman C. March, "Leukemia in Radiologists," *Radiology* 43 (1944): 275–278.

14. Caufield, *Multiple Exposures*, pp. 141, 153.

15. Mazur, *Dynamics of Technical Controversy* pp. 2–9.

16. Brenner, *Radon*, p. 76.

17. Two of the best accounts of the radium-dial painters are Angela Nugent's "The Power to Define a New Disease: Epidemiological Politics and Radium Poisoning," in Rosner and Markowitz, *Dying for Work*, pp. 177–191, and William D. Sharpe, "The New Jersey Radium Dial Painters: A Classic in Occupational Carcinogenesis," *Bulletin of the History of Medicine* 52 (1979): 560–570. See also Daniel P. Serwer, "The Rise of Radiation Protection," Ph.D. thesis, Princeton University, 1977.

18. Harrison S. Martland and Robert E. Humphries, "Osteogenic Sarcoma in Dial Painters Using Luminous Paint," *Archives of Pathology* 7 (1929): 406–417.

19. Frederick L. Hoffman, "Radium (Mesothorium) Necrosis," *JAMA* 85 (1925): 963–965.

20. Eisenbud, *Environmental Odyssey*, p. 34.

21. Caufield, *Multiple Exposures*, pp. 29–37.

22. Bill Richards, "The Dial Painters: The Live and the Dead Still Raise Questions About Job Radiation," *Wall Street Journal*, September 19, 1983.

23. Ibid.

24. Cole, *Element of Risk*, pp. 112–123.

25. Caufield, *Multiple Exposures*, pp. 47–50. The best in-depth history of radiation safety in the Manhattan Project is Hacker's *Dragon's Tail*. The best history of radiation safety prior to the 1940s is Whittemore's "National Committee."

26. Hacker, *Dragon's Tail*, pp. 62–69; compare also Jim Lerager, *In the Shadow of the Cloud* (Golden, Colo., 1988), and Caufield, *Multiple Exposures*, pp. 53–55.

27. Caufield, *Multiple Exposures*, pp. 62–63.

28. Lindee, *Suffering Made Real;* John Beatty, "Genetics in the Atomic Age: The Atomic Bomb Casualty Commission, 1947–1956," in *The Expansion of American Biology*, edited by Keith Benson et al. (New Brunswick, N.J., 1991).

29. A summary of these calculations can be found in Brenner, *Radon*, pp. 79–87; compare also Upton et al., *Health Effects*, pp. 4–6, which provides cancer risk estimates 3 times higher for solid cancers and 4 times higher for leukemia than the previous, much disputed, 1980 BEIR III report.

30. Alice Stewart, "Low Level Radiation: The Cancer Controversy," *Bulletin of the Atomic Scientists* (September 1990): 15–18. Stewart's hypothesis has been disputed by the National Academy of Science's Committee on the Biological Effects of Ionizing Radiations; see Upton et al., *Health Effects* (BEIR V), p. 184.

31. Mancuso, Stewart, and Kneale, "Radiation Exposures."

32. Ethel A. Gilbert, G. R. Petersen, and J. A. Buchanan, "Mortality of Workers at the Hanford Site, 1945–1981," *Health Physics* 56 (1989): 11–25.

33. Christopher Anderson, "NAS to Redo Atomic Studies Found to be Flawed," *Nature* 359 (1992): 354.

34. Keith Schneider, "Radiation Records of 44,000 Released," *New York Times*, July 18, 1990.

35. Physicians for Social Responsibility, *Dead Reckoning: A Critical Review of the Department of Energy's Epidemiologic Research* (Cambridge, Mass., 1992).

36. Broad, "New Study."

37. Ibid.

38. Roberts, "British Radiation Study"; compare also the update by Sharon Kingman, "New Sellafield Study Poses a Puzzle," *Science* 262 (1993): 648. The cluster artifact argument was used in the 1960s to dispute the existence of familial cancer clusters: see C. M. Woolf and E. A. Isaacson, "An Analysis of 5 'Stomach Cancer Families' in the State of Utah," *Cancer* 14 (1961): 1005–1016.

39. Jon Payne, "Lies and Statistics," *Nuclear Safety* (March 1989): 29.

40. Roberts, "British Radiation Study," p. 25. Gardner's views are consistent with the earlier views of F. E. Lundin et al., *Radon Daughter Exposure* (Springfield, Va., 1971).

41. Warren E. Leary, "British and U.S. Researchers Find a New Form of Radiation Injury," *New York Times*, February 21, 1992; M. A. Kadhim et al., "Transmission of Chromosomal Instability After Plutonium Alpha-particle Irradiation," *Nature* 355 (1992): 738–740. For a critique of Gardner's

hypothesis, see Louise Parker et al., "Geographical Distribution of Preconceptional Radiation Doses to Fathers Employed at the Sellafield Nuclear Installation, West Cumbria," *British Medical Journal* 307 (1993): 966–971.

42. Robert J. Roscoe et al., "Lung Cancer Mortality Among Nonsmoking Uranium Miners Exposed to Radon Daughters," *JAMA* 262 (1989): 629–633. Contrast the assertion of Olav Axelson, a physician at University Hospital in Linköping, Sweden, that smoking "actually seems to offer some (relative) protection against lung cancer" by stimulating the secretion of a protective layer of mucus that shields lung tissues from inhaled alpha emitters ("Reevaluated Risk for Radon-Cancer Link," *Science News* [April 14, 1979]: 247).

43. F. H. Härting and W. Hesse, "Der Lungenkrebs, die Bergkrankheit in den Schneeberger Gruben," *Vierteljahrsschrift für gerichtliche Medizin* 30 (1879): 300, and 31 (1879): 109–112, 325.

44. Alfred Arnstein, "Sozialhygienische Untersuchungen über die Bergleute in den Schneeberger Kobaltgruben," *Wiener Arbeiten auf dem Gebiete der sozialen Medizin, Beihefte* 5 (1913): 64–83; also his "Über den sogenannten 'Schneeberger Lungenkrebs,'" *Verhandlungen der deutschen pathologischen Gesellschaft* 16 (1913): 332–342. The mines of this region are dug into the mountains along the German-Czech border. The original 1879 diagnosis was made for miners on the German side; the disease was not identified among Czech miners until 1929, fifty years later. David Brenner attributes this to an extraordinary lack of communication between the mines, separated by only a few miles of mountains but a great cultural distance (*Radon*, p. 86). The original Czech paper is by Julius Löwy ("Über die Joachimsthaler Bergkrankheit," *Medizinische Klinik* 25 [1929]: 141–142). A good review of early Czech literature is F. Běhounek and M. Fořt, "Joachimsthaler Bergmannskrankheit," *Strahlentherapie* 70 (1941): 487–498.

45. Hueper, *Occupational Tumors,* pp. 435–459.

46. Kurt Lange, "Krebserkrankungen und geologische Verhältnisse im Erzgebirge," *Zeitschrift für Krebsforschung* 49 (1935): 306–310.

47. A. Pirchan and H. Šikl, "Cancer of the Lung in the Miners of Jáchymov (Joachimsthal)," *American Journal of Cancer* 16 (1932): 681–722.

48. On the inbreeding hypothesis, see M.-S. Vesin, "Cancer pulmonaire provoqué par les émanations radioactives," *Archives des maladies professionnelles* 9 (1948): 280–283; Lorenz, "Radioactivity and Lung Cancer," p. 13.

49. Uhlig, "Über den Schneeberger Lungenkrebs," pp. 86–87.

50. Ludewig and Lorenser, "Untersuchung der Grubenluft," pp. 178–185. H. Mache and S. Meyer in 1905 were the first to detect radium emanation in the mines of Joachimsthal; see Werner Schüttmann, "Aus den Anfängen der Radontherapie," *Zeitschrift für die gesamte innere Medizin* 41 (1986): 451–456.

51. Hueper, *Occupational Tumors,* p. 441.

52. The "death mine" referred to in these reports were the "Siebenschlehen" shafts. In the late 1930s, after the mine was abandoned (for economic reasons), radon in the stagnant air of the now flooded mine was found to be on the order of 54,000 picocuries per liter. The other mines still operating in the region showed radon levels between 2,000 and 3,000 picocuries per liter. See Boris Rajewsky, "Bericht über die Schneeberger Untersuchungen," *Zeitschrift für Krebsforschung* 49 (1939): 315–340. In 1938, A. Brandt of Dresden reported excess tumors among experimental rats and mice raised in the mines; see his "Bericht über die im Schneeberger Gebiet auf Veranlassung des Reichsausschusses für Krebsbekämpfung durchgeführten Untersuchungen," *Zeitschrift für Krebsforschung* 47 (1938): 108–111.

53. Viorst and Reistrup, "Radon Daughters and the Federal Government."

54. E. S. London, "Über die physiologischen Wirkungen der Emanation des Radiums," *Zentralblatt für Physiologie* 18 (1904): 185–188; John Read and J. C. Mottram, "The 'Tolerance Concentration' of Radon in the Atmosphere," *British Journal of Radiology* 12 (1939): 54–60.

55. Hueper, *Occupational Tumors,* pp. 435–459.

56. Lorenz, "Radioactivity and Lung Cancer," pp. 5, 13.

57. Stewart, "Occupational and Post-Traumatic Cancer," p. 146. A good history of uranium mine epidemiology can be found in Stannard, *Radioactivity and Health*, pp. 113–194.

58. Eisenbud, *Environmental Odyssey,* pp. 60–62. The Joint Committee on Atomic Energy, the congressional group established to oversee the AEC, was equally negligent; see Ball, *Cancer Factories,* p. 20.

59. Carlson, *Genes,* pp. 6, 356–367.

60. Schneider, "Uranium Miners."

61. Holaday et al., *Control of Radon,* pp. ix–x, 2–4. The highest value apparently ever recorded in a U.S. mine was 120,000 picocuries per liter; see Ball, *Cancer Factories,* pp. 26–27.

62. Holaday et al., *Control of Radon,* pp. ix–x, 2–4. Holaday had first proposed a "working level" equal to 100 picocuries per liter in his *Interim Report: Health Study of the Uranium Mines and*

Mills (Washington, D.C., 1952), coauthored with W. David and H. N. Doyle. The working level proposed in 1957 was defined as any combination of short-lived radon decay products in one liter of air that will result in the emission of 1.3×10^5 MeV of potential alpha energy. In a confined space like an unventilated mine, where radon and its decay products are in radioactive equilibrium, 1 WL = 100 pCi/l. In homes and ventilated mines, however, where the gas is allowed to dissipate into open air, a radioactive equilibrium is not established—primarily because long-lived isotopes travel much farther than short-lived isotopes. In such circumstances, 1 WL is roughly equal to 300 pCi/l. One Working Level Month (WLM) was later defined as the cumulative radiation received by a miner working 170 hours in a mine where radon levels average 100 picocuries per liter. Miners in the Public Health Service study received 882 WLM of radiation on average; see Richard W. Hornung and Theodore J. Meinhardt, "Quantitative Risk Assessment of Lung Cancer in U.S. Uranium Miners," *Health Physics* 52 (1987): 417–430.

63. Holaday, "History of the Exposure of Miners to Radon," p. 551.
64. Archer et al., "Hazards to Health." This paper analyzed deaths through 1959.
65. Joseph K. Wagoner et al., "Unusual Cancer Mortality Among a Group of Underground Metal Miners," *New England Journal of Medicine* 269 (1963): 284–289; Victor E. Archer, Joseph K. Wagoner, and Frank E. Lundin, "Lung Cancer Among Uranium Miners in the United States," *Health Physics* 25 (1973): 351–371.
66. Holaday et al., *Control of Radon*, p. 4.
67. Holaday had concluded as early as 1951 that a human health tragedy was unfolding in consequence of the failure of the AEC and the mining companies to ventilate the mines; see Ball, *Cancer Factories*, pp. 49–51.
68. Holaday is cited in Ringholz, *Uranium Frenzy*, p. 148; Archer is cited in Ball, *Cancer Factories*, pp. 46 and 59–60. Ball's book is the best history of the efforts of uranium miners to obtain compensation for lung cancers caused by mining for the AEC.
69. Ball, *Cancer Factories*, pp. 11–12, 49.
70. Ivins, "Uranium Mines"; compare also Ringholz, *Uranium Frenzy*, pp. 166–222.
71. Ball, *Cancer Factories*, p. 49.
72. Viorst and Reistrup, "Radon Daughters and the Federal Government," pp. 26–27. In 1971, the newly created Environmental Protection Agency reduced the maximum allowable level of ambient radon to one-third of a working level, or 4 working level months (WLMs) per year. Exposures at the mines dropped significantly, but in 1975 a monitoring blitz by the Mine Safety and Health Administration found that miners were still being exposed to 4.6 WLMs per year on average, slightly above the EPA limit and several times higher than what was being claimed by the companies. See Richard W. Hornung and Theodore J. Meinhardt, "Quantitative Risk Assessment of Lung Cancer in U.S. Uranium Miners," *Health Physics* 52 (1987): 417–430.
73. Caufield, *Multiple Exposures*, pp. 84–86; Schneider, "Uranium Miners."
74. Rosalie Bertell, *No Immediate Danger* (London, 1985), pp. 83–88; Ball, *Cancer Factories*, p. 50. In 1979, Joseph Wagoner suggested a possible racial bias in the government's dealings with Navajos: "In 1978, the government started notifying all shipyard workers and others who had been exposed to asbestos, but nothing like that has been done for the miners. Is it because they are Westerners with little power in their Congressional representation, because they are Indians so no one needs to care?" (Ivins, "Uranium Mines").
75. Ball, *Cancer Factories*, p. 12.
76. Schneider, "Uranium Miners."
77. Sandra D. Atchison, "'These People Were Used as Guinea Pigs,'" *Business Week* (October 15, 1990): 98. Compare also the statement of a Colorado health inspector concerning the 1950s: "anybody that said a thing against uranium mining was suspected of being a communist"; cited in Wasserman and Solomon, *Killing Our Own*, p. 149.
78. Uhlig, "Über den Schneeberger Lungenkrebs," p. 83; Lorenz, "Radioactivity and Lung Cancer," p. 1.
79. A pamphlet published by the Soviet-German Arbeitsgemeinschaft Wismut just prior to German reunification asserts that the uranium mines pose "no immediate danger," "no danger to the general public," and "negligible risk." See *Uranerzbergbau contra Umwelt?* (Karl-Marx-Stadt, 1990), preface by Dr. S. Richter; also Reimar Paul, *Das Wismut Erbe* (Göttingen, 1991).
80. Kahn, "Grisly Archive"; "The Legacy of Schneeberg," *Nuclear Engineering* (February 1991): 7. Schneeberg's mines were Soviet-run until 1954; they first used forced labor, then recruited local residents by offering special privileges.
81. Kahn, "Grisly Archive," pp. 448–451. In February 1991, a member of a West German delegation inspecting the East German records characterized the mood among epidemiologists as reminiscent of "prospectors in San Francisco during the gold rush." See Steven Dickman, "Gold Mine in East Germany?" *Nature* 349 (1991): 728.
82. Dickman, "Gold Mine," p. 728.

83. Kahn, "Grisly Archive," p. 448; Taryn Toro, "Uranium Mines Leave Heaps of Trouble for Germany," *New Scientist* (February 1991): 29.

84. Kahn, "Grisly Archive," pp. 448–451; compare Werner Schüttmann, "Deutsche Opfer für Moskaus Atombombe," *Der Tagesspiegel,* January 13, 1991; also Schüttmann and Karl Aurand, *Die Geschichte der Aussenstelle Oberschlema des Kaiser-Wilhelm-Instituts für Biophysik, Frankfurt am Main* (Salzgitter, 1991). Schüttmann notes that from the defeat of the Nazi armies until June 1945, Americans occupied most of the uranium mine deposits of the Erzgebirge but soon thereafter turned these regions over to the Soviets, apparently not recognizing their military value. See also Norman M. Naimark, *The Russians in Germany: A History of the Soviet Zone of Occupation* (Cambridge, Mass., 1995).

85. See my letter, "The Oberrothenbach Catastrophe," *Science* 260 (1993): 1676–1677.

86. Interview with Vladimír Řeřicha, May 27, 1992.

87. The idea of using prisoners in the mines arose with the Nazis: on November 19, 1938, SS Oberführer and Regierungspräsident Hans Krebs wrote to Himmler, asking him to install KZ prisoners in Joachimsthal to free "our poor Sudetendeutsche miners" from work that was killing them in their forties. Himmler's office agreed to the proposal on December 15, 1938, on the condition that the prisoners be released after two to three years if they showed good conduct. See T-175 #87, folder 193, Captured German Documents, National Archives. It is not yet clear how many (if any) prisoners were forced into the mines under the Nazis.

88. Avraham Shifrin, *The First Guidebook to Prisons and Concentration Camps of the Soviet Union* (Uhldingen, 1980). Shifrin describes six camps east of Tashkent in which 10,000 prisoners were required to work in uranium mines without protective clothing; these include mines at Fergana (2,500 prisoners), Andizhan (2,500 prisoners), Zeravshan, Leninsk, Margilan, and Almalyk. He also mentions: a "death camp" outside Frunze, in Kirghizia, where 2,500 uranium miners worked without protection; 10,000 uranium mine prisoners in four camps near Ol'ga Bay on the Pacific Ocean; 10,000 prisoners in seven uranium death camps northeast of Verkhnekamsk and Omutninsk, east of Kirov; 10,000 prisoners in uranium mines and enrichment facilities at Sovetabad, Zeravshan, Bekabad, Asht, and Sotsgorod near Leninabad in Tadzhikistan; 5,000 prisoners at the open-pit mines near Cherepovets, west of Vologda; 2,500 prisoners at Kyshtym; 500 at Bashkir; and an estimated 14,000 to 17,000 prisoners in various uranium mine camps of Kazakhstan (Aksu, Tselinograd, and Karagaily, for example). Shifrin also reports uranium mine death camps with an unspecified number of prisoners at Novaya Borovaya, Cholovka, Rakhov, and Zheltye Vody, in Ukraine; at Kokand in Uzbekistan; at Oimyakon in Yakutiya; on Vaigach Island in the Arctic; in Lermontov, near Stavropol; on the Mangyshlak Peninsula; at Groznyi, in the Chechen-Ingush region; at Cape Medvezhii in Novaya Zemlya; and at Achinsk, in the Krasnoyarsk Territory. See pp. 32–35, 178, 228–231, 266, 270, 276, 308, 340, 350. He also mentions a number of other camps where uranium was enriched or where other hazardous radioactive military work was undertaken.

89. See, for example, J. Ševc, E. Kunz, and V. Plaček, "Lung Cancer in Uranium Miners and Long-Term Exposure to Radon Daughter Products," *Health Physics* 30 (1976): 433–437. Ševc et al.'s report, widely cited in American scientific literature, does not even mention how many workers were involved; it simply notes that "a somewhat larger group" was involved in the Czech study than in Lundin et al.'s Public Health Service study of 3,366 white workers (p. 433).

90. J. Ševc et al., "A Survey of the Czechoslovak Follow-Up of Lung Cancer Mortality in Uranium Miners," *Health Physics* 64 (1993): 355–369.

91. U.S. Congress, Senate Committee on Labor and Human Resources, *Radiation Exposure Compensation Act,* 101st Congress, 2d session (1990), p. 2. On the "downwinders," see Howard Ball, *Justice Downwind: America's Atomic Testing Program in the 1950s* (New York, 1988); Carole Gallagher, *American Ground Zero: The Secret Nuclear War* (Cambridge, Mass., 1993).

92. Keith Schneider, "Valley of Death: Late Rewards for Navajo Miners," *New York Times,* May 3, 1993.

CHAPTER 9
RADON'S DEADLY DAUGHTERS

1. Environmental Protection Agency, *A Citizen's Guide to Radon,* 2nd ed. (Washington, D.C., 1992), p. 2. The EPA's 1986 *Citizen's Guide* estimated 5,000 to 20,000 deaths; compare also the figures produced by Fabrikant et al., *Health Risks of Radon.*

2. The EPA estimates that some 200 lung cancer deaths are caused every year by water-released indoor radon; the highest levels are typically found in homes using well water. For a review and critique, see Richard Stone, "EPA Analysis of Radon in Water Is Hard to Swallow," *Science* 261

(1993): 1514–1516; also Nancy Chiu et al., "Radon Risk Estimates" (letter), *Science* 264 (1994): 1239–1240.

3. Environmental Protection Agency, *National Residential Radon Survey: Summary Report* (Washington, D.C., 1992). The EPA here concludes, on the basis of 6,000 measurements nationwide, that radon in U.S. homes averages about 1.25 picocuries per liter. The report also estimates that a lifetime exposure at 1 picocurie per liter (eighteen hours per day for seventy years) will result in about 3 extra lung cancer deaths per 1,000 people exposed and that residents of homes with annual radon levels greater than 10 picocuries per liter will have about a 1 in 25 chance of contracting lung cancer due to radon (p. 3).

4. According to Anthony Nero, the Environmental Defense Fund in the early 1980s refused to publicize this issue because they were dealing with radon from uranium mine tailings (personal communication). The EDF later reversed this policy and published *Radon: The Citizens' Guide*, written by E. K. Silbergeld et al. (New York, 1987).

5. Teresa Opheim, "How Serious Is the Radon Risk?" *Utne Reader* (November–December 1988): 12–13.

6. E. Dorn, "Versuche über Sekundärstrahlen und Radiumstrahlen," *Abhandlungen der naturforschenden Gesellschaft für Halle* 22 (1900): 155ff.

7. See Brenner, *Radon*, p. 72.

8. Dawson Turner, "An Address on the Nature and Physiological Action of Radium Emanations and Rays," *British Medical Journal* 2 (1903): 1523–1524.

9. Ekkehard Schmid, "Messungen des Radium-Emanationsgehaltes von Kellerluft," *Mitteilungen aus dem physikalischen Institut der Universität Graz* 82 (1932): 233–242. Schmid expressed no apparent concern about the health effects of such exposures. In a previous paper, he had sought to elucidate the dependence of ambient radiation on factors such as temperature, barometric pressure, humidity, wind velocity, and several other meteorological conditions; see his "Der Gehalt der Freiluft an Radiumemanation," *Sitzungsberichte der Akademie der Wissenschaften, Wien* 140 (1931): 27–48. The first indication of natural radiation emanation in homes was published by the physicist Heinrich W. Schmidt in 1907; see Werner Schüttmann, "Das Radon bedeutet keine neue Strahlengefahr—ein historischer Rückblick auf das Radonproblem in Wohnungen, *Forum Städte-Hygiene* (Berlin-Hannover) 41 (1990): 250–257.

10. For the Austrian case, see Friedrich Steinhäusler, "Long-Term Measurements of ^{222} Rn, ^{220} Rn, ^{214} Pb and ^{212} Pb Concentrations in the Air of Private and Public Buildings," *Health Physics* 29 (1975): 705–713.

11. Harley, "Radon Is Out."

12. Nancy Wood, "America's Most Radioactive City," *McCall's* (September 1970): 46–50, 122.

13. Metzger, "'Dear Sir.'"

14. Ibid., pp. 14, 62.

15. Krimsky and Plough, *Environmental Hazards,* p. 143; Environmental Protection Agency, "Standards for Remedial Actions at Inactive Uranium Processing Sites, *Federal Register* 48 (January 5, 1983): 591–592. It is in fact difficult to reduce high radon homes below this level. Silvio O. Funtowicz and Jerome R. Ravetz have suggested that the EPA's guideline of 4 picocuries per liter was set "so as to have 90% of homes deemed 'safe,' and also incidentally to save the Federal Government millions of dollars in decontamination costs for homes in the West that had been affected by uranium tailings"; see their *Uncertainty and Quality in Science for Policy* (Dordrecht, 1990), p. 26.

16. Cole, *Element of Risk*, pp. 10–11.

17. The AEC began its project on "Indoor Radon Daughters and Radiation Measurements in East Tennessee and Central Florida" in 1967 to refute charges that radon levels in Colorado homes were abnormally high as a result of construction using uranium tailings. The AEC paper was classified "for internal use only" and the results were apparently never made public. See Metzger, "'Dear Sir,'" p. 62.

18. Office of Technology Assessment, *Residential Energy Conservation* (Washington, D.C., 1979). The OTA figure ignored the $2 million being spent at the DOE's Lawrence Berkeley Laboratory.

19. U.S. Council on Environmental Quality, *Environmental Quality–1980* (Washington, D.C., 1980), p. 397.

20. Bengt Hultqvist, "Studies on Naturally Occurring Ionizing Radiations," *Kungliga Svenska Vetenskapsakademiens Handlingar,* 4th ser., 6 (1956), no. 3.

21. Gun Astri Swedjemark, *Radon in Dwellings* (in Swedish; Stockholm, 1974); Wendy Barnaby, "Very High Radiation Levels Found in Swedish Homes," *Nature* 281 (1979): 6.

22. Gun Astri Swedjemark, "Radon in Dwellings in Sweden," in *Natural Radiation Environment III,* edited by Thomas F. Gesell and Wayne M. Lowder, vol. 1 (Springfield, Va., 1980), pp. 1237–1259.

23. Harold Glauberman and A. J. Breslin, *Environmental Radon Concentrations: An Interim Report* (New York, 1957); interview with Naomi Harley, February 10, 1993.

24. A. C. George and A. J. Breslin, "The Distribution of Ambient Radon and Radon Daughters in Res-

idential Buildings in the New Jersey–New York Area," in *Natural Radiation Environment III*, edited by Thomas F. Gesell and Wayne M. Lowder, vol. 2 (Springfield, Va., 1980), pp. 1272–1292.

25. R. J. Budnitz et al. (including Nero), "Human Disease from Radon Exposures: The Impact of Energy Conservation in Residential Buildings," *Energy and Buildings* 2 (1979): 209–215. This paper originally appeared as an in-house publication of the LBL in August 1978.

26. J. Rundo et al., "Observation of High Concentrations of Radon in Certain Houses," *Health Physics* 36 (1979): 729–730.

27. In 1976, J. A. Auxier, director of the Health Physics Division at Oak Ridge National Laboratory, published an article indicating high levels of radon in buildings; see his "Respiratory Exposure in Buildings Due to Radon Progeny," *Health Physics* 31 (1976): 119–125. The Pennsylvania Power and Light Company had also found high levels of radon in the homes of several of its employees prior to TMI; see Thomas M. Gerusky, "The Pennsylvania Radon Story," *Journal of Environmental Health* (January–February 1987): 197–201.

28. Nero, "Indoor Radon," p. 30.

29. Richard J. Guimond and Samuel T. Windham, *Radioactivity Distribution in Phosphate Products, By-Products, Effluents, and Wastes* (Washington, D.C., 1975).

30. A Mr. Homer Hooks, executive director of the Florida Phosphate Council, responded to charges of radioactive contamination from phosphate strip mining by suggesting that "our senators and representatives walking in the halls of Congress are probably exposed to more radiation than you get from phosphate. Granite is radioactive to some degree, too." Mr. Hooks defended strip mining on the grounds that "God put the phosphate here and we have to mine it here." See Wayne King, "Florida Phosphate Pollution Stirs Alarm," *New York Times*, July 24, 1976.

31. Galen, "Nowhere to Run."

32. Lisa B. Belkin, "Warning: Home Energy Conservation May Be Dangerous to Your Health," *National Journal* (August 2, 1980): 1274. The EPA produced its own estimates of 5,000 to 20,000 killed by radon about this same time; see Mazur's *Dynamics of Technical Controversy*.

33. General Accounting Office, *Indoor Air Pollution: An Emerging Health Problem* (Washington, D.C., 1980), p. iii.

34. Council on Environmental Quality, *Environmental Quality–1980* (Washington, D.C., 1980), p. 184; Walter Sullivan, "Radiation Danger Seen in Seepage of Radon in Homes," *New York Times*, October 7, 1980.

35. Henry Hurwitz, "The Indoor Radiological Problem in Perspective," *Risk Analysis* 3 (1983): 63–77. Hurwitz also complained that, while the Nuclear Regulatory Commission required modifications of reactor design whenever a means could be found to reduce public exposure to radiation at a cost of less than $1,000 per person-rem, the same logic had not been implemented for energy conservation: savings of $1,000 gained through energy conservation were generally accompanied by increased radiation exposures far in excess of 1 person-rem.

36. Cohen was one of the first to rank the dangers of nuclear power on a par with deaths from meteor strikes, fires, and other rare or trivial hazards—the stock and trade of much subsequent risk assessment rhetoric. In 1974, for example, he introduced a session of the American Physical Society by suggesting that the danger to the average citizen from a nuclear accident was "equivalent to that of being 1/4 ounce overweight, or of smoking one cigarette every four months, or of a farmer's spending four hours per year in a city"; see his "Refining a Statement," *Nuclear News* (May 1974): 29). Cohen later compared the genetic risk of nuclear power to the risk of "men wearing pants an extra 8 hours per year, or 1.5 minutes every day," based on the fact that trousers warm the gonads and thereby increase the rate of spontaneous mutation. See his "Basic Facts About Radiation," in Napier, *Issues*, p. 64.

37. Cohen, "Health Effects of Radon."

38. Bernard L. Cohen, "Radon: Our Worst Radiation Hazard," *Consumers' Research* (April 1986): 12–13. Dixy Lee Ray repeated Cohen's warning in 1990: "Radon has become a national health problem because of our well-meant but stupid insistence on sealing up our homes and buildings to conserve energy, without considering the possible ill effects" (*Trashing the Planet*, p. 6).

39. Interview with Anthony Nero, March 17, 1992.

40. Belkin, "Warning," p. 1274.

41. General Accounting Office, *Indoor Air Pollution*, pp. 22–23.

42. Measurements at the Watras home showed dramatic fluctuations according to weather and barometric pressure; the highest value recorded there was 4,400 picocuries per liter. The highest persistent household radon recording as of 1993 was in the Whispering Hills development in northwest New Jersey, where 3,500 picocuries per liter were measured in the basement of a home in 1988. Basement vents were installed at a cost of about $1,200 to put the house within EPA safety guidelines. See Egginton, "Menace of Whispering Hills." The highest level of radon ever recorded for household water is 2 million picocuries per liter, according to C. Tom Hess of the University of

Maine in Orono. See Malcolm W. Brown, "Scientist Says Low Radon Levels May Be Harmless," *New York Times,* September 28, 1988.

43. R. C. Smith et al., "Radon: A Profound Case," *Pennsylvania Geology* (April 1987): 2. The radon remediation industry's *Radon Directory* for 1990–1991 (2nd ed., Chevy Chase, Md., 1990) states that living in the Watras home was equivalent to smoking 238 packs of cigarettes per day (p. v).

44. Philip Shabecoff, "Radioactive Gas in Soil Raises Concern in Three-State Area," *New York Times,* May 24, 1985; "Radon Gas Tied to Cancer Deaths," *Chicago Tribune,* August 6, 1985; "Radon in Homes Could Kill 30,000 Yearly," *Atlanta Constitution,* July 15, 1985; "Radon Gas: A Deadly Threat—A Natural Hazard Is Seeping Into 8 Million Homes," *Newsweek* (August 18, 1986): 60–61.

45. McCally, "What the Fight Is All About," p. 13. Compare also Jamie Murphy, "The Colorless, Odorless Killer," *Time* (July 22, 1985): 72. In 1984, the NCRP released a report advising an indoor radon standard of 8 pCi/l = 2 WLM/yr; see their *Exposures* from *the Uranium Series.*

46. Nero, "Indoor Radon Story," p. 31; Shepherd, "Indoor Radon," p. 45.

47. The Watras family was evacuated from their home and Stanley's employer, the Philadelphia Electric Company, owner of the Limerick Nuclear Power Plant, paid $32,000 to have the family room and cellar dug up and vented. Stanley Watras had originally requested funds from the EPA's Superfund, but federal money was not available for natural radioactive contamination. The family was able to move back in after six months; the ventilation system reduced radioactivity in the home to about 4 picocuries per liter. As of 1993, the Watras family was alive and well, still living in the now infamous house, and apparently cancer-free.

48. Jason Gaertner, "Radon: The Pennsylvania Response," *The Pennsylvanian* (August 1987): 4.

49. Kerr, "Indoor Radon," p. 607.

50. Cole, *Element of Risk,* p. 83.

51. Harvey Sachs, "Was There a Radon Coverup in Pennsylvania?" in *Radon and the Environment,* edited by William Makofske and Michael Edelstein (Mahwah, N.J., 1987), pp. 376–378; Mazur, "Putting Radon," p. 89.

52. Anthony V. Nero et al., *Indoor Radon and Decay Products: Concentrations, Causes, and Control Strategies* (Washington, D.C., 1990), p. 108.

53. Shepherd, "Indoor Radon," p. 45.

54. Cole, *Element of Risk,* pp. 15–16.

55. "Lautenberg Accuses EPA, OMB of 'Cover-Up' of Agency Report on Health Risks of Radon," *Environment Reporter* BNA 16 (1985): 1046. Lautenberg was particularly sensitive to radon issues because his hometown of Montclair, New Jersey, had been—and still is—the site of a protracted dispute over what to do with radioactive wastes dumped by the US Radium Corporation in the 1920s. For a history of the fiasco, see Cole, *Element of Risk,* pp. 112–123, and Mazur, "Putting Radon," pp. 91–92.

56. U.S. Congress, Committee on Energy and Commerce, *Hearing on the Radon Pollution Control Act of 1987,* April 23, 1987 (Washington, D.C., 1987).

57. "Radon: The Problem No One Wants to Face," *Consumer Reports* (October 1989): 623–625.

58. "Public Apathy Said Barrier to Control of Indoor Air Contamination by Radon Gas," *Environment Reporter* (BNA), 17 (1987): 1793; Viveca Novak, "Profits in the Air," *Venture* (December 1988): 75.

59. Jonathan Hartwell, *A Survey of Compounds That Have Been Tested for Carcinogenic Activity* (Washington, D.C., 1941).

60. Epstein, *Politics of Cancer,* p. xvii. In an interview, Epstein told me that he had ignored radiation because he thought John Gofman had already done a good job of this.

61. See Kathryn Harrison, "Out-of-Sight—Out-of-Mind: The Absence of Indoor Air Pollution from the Regulatory Agenda," *Technology and Society* 8 (1986): 177–186; Galen, "Nowhere to Run," pp. 180–182.

62. Indoor air pollution had become a priority of research in the final years of the Carter administration. Apart from the GAO's already mentioned report on *Indoor Air Pollution,* there is the National Research Council's *Indoor Pollutants* (Washington, D.C., 1981), and James L. Repace and Alfred H. Lowrey's "Indoor Air Pollution," *Science* 208 (1980): 464–472. Isaac Turiel's *Indoor Air Quality and Human Health* (Stanford, Calif., 1985), claimed that contaminants from indoor heating, cooking, and smoking, together with microbes and allergens, radon gas, and various contaminants from treated woods, constitute a far greater threat to human health than outdoor or workplace pollutants.

63. "Cancer Risk from Domestic Radon," *Lancet* 1 (1989): 93.

64. Edward Groth, personal communication.

65. H. Ward Alter and Richard A. Oswald, "Nationwide Distribution of Indoor Radon Measurements," *Journal of Air Pollution Control Association* 37 (1987): 227–231; Anthony Nero et al., "Distribu-

tion of Airborne Radon-222 Concentrations in U.S. Homes," *Science* 234 (1986): 992–997; and Nero, personal communication.

66. A 1991 study by Jonathan Samet of the University of New Mexico argued that since miners tend to breathe more heavily at work, and since mines are dusty, a given level of radon in a home is roughly 30 percent less hazardous than that same level in a mine. See David P. Hamilton, "Indoor Radon: A Little Less to Worry About," *Science* 251 (1991): 1019. EPA analysts have argued, by contrast, that extrapolations from miner data are as likely to underestimate as to overestimate household risks. A 1979 review by the EPA's Office of Radiation Programs noted that since alpha particles do not penetrate deeply into tissues, and since the bronchial epithelium appears thicker in miners than in the general population, miners actually receive lower doses (by a factor of about 2) than one might imagine; see Richard Guimond et al., *Indoor Radiation Exposure Due to Radium-226 in Florida Phosphate Lands* (Washington, D.C., 1979).

67. Samet and Nero, "Indoor Radon and Lung Cancer." In 1989, Jay Lubin of the National Cancer Institute suggested that household radon would induce 13,000 lung cancers per year but that 11,000 of these would be among smokers. See "Cancer Risk from Domestic Radon," *Lancet* 1 (1989): 93.

Anthony Nero has voiced a concern that the EPA is going about its publicity efforts all wrong. Instead of cautioning everyone to test their homes, it would be better, he suggests, to identify and remediate the 100,000 houses whose occupants face a clear and very high risk. Instead, the EPA has established an action guideline that encompasses millions of homes but does little to assist the much smaller number of homes at very high risk. Nero's recommendations are in his "National Strategy"; another good review is William W. Nazaroff and Kevin Teichman, "Indoor Radon," *Environmental Science & Technology* 24 (1990): 774–781.

68. Edward A. Martell, "Critique of Current Lung Dosimetry Models for Radon Progeny Exposure," in *Radon and Its Decay Products,* edited by P. Hopke (New York, 1987), pp. 444–461. Consider the following three scenarios for the life-time risk of breathing indoor air with a radon content of 4 pCi/1. The National Council on Radiation Protection (NCRP) predicts an extra 7.4 lung cancer deaths per 100,000 nonsmokers and a similar figure for smokers. The International Commission on Radiological Protection (ICRP) predicts 23 extra deaths per 100,000 nonsmokers and 280 extra deaths per 100,000 smokers. The National Research Council predicts 9.6 and 110 extra deaths per 100,000 for the two groups, respectively. The NCRP, in other words, assumes zero synergy; the ICRP predicts a synergetic factor of more than tenfold. (These data—for males only—are summarized in Samet and Nero, "Indoor Radon and Lung Cancer," p. 592.) It is difficult to sort out which of these estimates is correct because—as Philip Abelson has pointed out on the editorial page of *Science*—the funding has simply not been made available to do the appropriate epidemiological studies (Philip H. Abelson, "Uncertainties About Health Effects of Radon," *Science* 250 [1990]: 353).

69. Jackson, Geraci, and Bodansky, "Observations of Lung Cancer," p. 105.

70. Christer Edling et al., "Radon in Homes—A Possible Cause of Lung Cancer," *Scandinavian Journal of Work, Environment, and Health* 10 (1984): 25–34.

71. Judith B. Klotz et al., *Mortality Experience of Residents Exposed to Elevated Indoor Levels of Radon from an Industrial Source* (Trenton, N.J., 1988); Janet Schoenberg and Judith Klotz, *A Case-Control Study of Radon and Lung Cancer Among New Jersey Women* (Trenton, N.J., 1989), pp. iii–iv.

72. High Background Radiation Research Group, China, "Health Survey in High Background Radiation Areas in China," *Science* 209 (1980): 877–880.

73. William H. Blot et al., "Indoor Radon and Lung Cancer in China," *Journal of the National Cancer Institute* 82 (1990): 1025–1030.

74. Göran Pershagen et al., "Residential Radon Exposure and Lung Cancer in Sweden," *New England Journal of Medicine* 330 (1994): 159–164.

75. Richard Stone, "New Radon Study: No Smoking Gun," *Science* 263 (1994): 465.

76. Bernard L. Cohen, "Expected Indoor ^{222}Rn Levels in Counties with Very High and Very Low Lung Cancer Rates," *Health Physics* 57 (1989): 897–907; compare also "Radon: Is a Little Good for You?" *Science News* 134 (1988): 254. Arthur B. Robinson summarizes Cohen's arguments in the pronuclear *Access to Energy* (December 1993), arguing that radon "may soon be listed as a vitamin," defined as "substances that are beneficial to health in small amounts and that are under attack by federal bureaucrats" (p.4).

77. Philip H. Abelson, "Uncertainties About Health Effects of Radon," *Science* 250 (1990): 353; also his "Mineral Dusts and Radon in Uranium Mines," *Science* 254 (1991): 777. Compare the response by Margo T. Oge and William H. Farland, "Radon Risk in the Home," *Science* 255 (1992): 1194.

78. Malcolm W. Brown, "Scientist Says Low Radon Levels May Be Harmless," *New York Times,* September 28, 1988.

79. Interview with Bernard Cohen, August 5, 1992. Cohen recalled his dismay following the Arab oil

crisis of 1973, when scientists failed to unite "in a World War II–style effort" to tackle the energy problem. Cohen is not sure of the fate of nuclear power over the next few decades, but he is convinced that a hundred or a thousand years from now we will be using nuclear power if we have power at all.

80. R. H. Johnson, "Radon: A Health Problem and a Communication Problem," in Cross, *Indoor Radon,* p. 1107.

81. Richard Stone, "Radon Risks Up in the Air," *Science* 261 (1993): 1515.

82. Mead, "Riddle of Radon," p. 66. *East West* editors don't seem to know where to come down on the issue: a 1987 report featured an attack on Reagan officials for neglecting the problem; see Kirk Johnson, "A Radon Primer," *East West* (November 1987): 59. The *Alternative Press Index* for April–June 1985 lists no articles on radon, although it has nine on the topic of nuclear radiation and twelve on nuclear waste.

83. Viveca Novak, "Profits in the Air," *Venture* (December 1988): 72–75.

84. The American Society of Heating, Refrigerating, and Air-Conditioning Engineers, for example, whose members design or install remediation fans and vents, has advised an action level of 2 pCi/l rather than the EPA's 4 pCi/l or the NCRP's 8 pCi/l. Michael Lafavore, *Radon: The Invisible Threat* (Emmaus, Pa., 1987), p. 89.

85. Committee on Environment and Public Works, *Hearing on Radon Contamination: How Federal Agencies Deal with It,* May 18, 1988 (Washington, D.C.), pp. 135–136.

86. Radon is present in many commercial caves, threatening spelunkers and occasional visitors, but especially tour guides who stay in the caves for hours every day. Annual exposures of park service personnel range from 2 WLM at Carlsbad Caverns to 3 or even 4 WLM at Mammoth Cave in Kentucky. Winter levels are highest in most cases. See Keith A. Yarborough, "Radon- and Thoron-Produced Radiation in National Park Service Caves," in *Natural Radiation Environment III,* edited by Thomas F. Gesell and Wayne M. Lowder, vol. 2 (Springfield, Va., 1980), 1371–1395.

87. Tenner, "Warning."

CHAPTER 10
GENETIC HOPES

1. Natalie Angier, "A New Gene Therapy to Fight Cholesterol Is Being Prepared," *New York Times,* October 29, 1991.

2. Shannon Brownlee and Joanne Silberner, "The Age of Genes," *U.S. News and World Report* (November 4, 1991): 64–76.

3. Doll, *Prevention,* p. 103.

4. Schimke, "Cancer in Families."

5. Jean Marx, "A New Tumor Suppressor Gene?" *Science* 252 (1991): 1067; also her "Zeroing In on Individual Cancer Risk," *Science* 253 (1991): 612.

6. Paul Broca, *Traité des tumeurs,* vol. 1 (Paris, 1866), pp. 150–154. Compare also Paget, *Lectures,* pp. 187, 460. Cancer was believed to be hereditary in the family of Napoleon I: the emperor's father and sister died of cancer of the stomach, and he himself eventually succumbed to the same disease. See Edmund Andrews, "The Diseases, Death and Autopsy of Napoleon I," *JAMA* 25 (1895): 1081–1085.

7. Henry T. Butlin, "Report on Inquiry No. XIII, Cancer (of the Breast Only)," *British Medical Journal* (February 26, 1887): 436–441; Paget, *Lectures,* pp. 674–675; Henry T. Butlin, "Is Cancer Hereditary?" *British Medical Journal* (May 4, 1895): 1006–1007.

8. Septimus W. Sibley of Middlesex Hospital in England calculated in 1859 that heredity accounted for no more than about 9 percent of all cancers, though he also recorded a remarkable case where a mother and five of her daughers all had cancer of the left breast. Charles H. Moore, also of Middlesex Hospital, conceded the occasional heritability of the disease but suggested that direct inheritance from a parent "does not happen thrice in a hundred cases." Sibley, "Contribution," pp. 132–133; Moore, "Antecedent Conditions," p. 205; W. Roger Williams, "Is Cancer Hereditary?" *British Medical Journal* (May 4, 1895): 1007; Wolff, *Science of Cancerous Disease,* pp. 299–302. See also W. Harrison Cripps, "The Relative Frequency with Which Cancer Is Found in the Direct Offspring of a Cancerous or Non-Cancerous Parent," *Saint Bartholomew's Hospital Reports* 14 (1878): 287–290; Snow, "Is Cancer Hereditary?"

9. The latest in this carnival line is the purported discovery of a "sweet-tooth gene" predisposing nearly half the population to crave sweets. See Ernest P. Noble et al., "D_2 Dopamine Receptor Gene and Obesity," *International Journal of Eating Disorders* 15 (1994): 205–217.

10. Stewart, "Occupational and Post-Traumatic Cancer," p. 146. For a more recent example of the con-

fusion, see Charles E. Perkins, *What Price Civilization? The Causes, Prevention and Cure of Human Cancer* (Washington, D.C., 1946) pp. 74–75.

11. Henry Earle, "On Chimney Sweeps Cancer," *Medico-Chirurgical Transactions* 12 (1823): 300.

12. Early support for the idea of a heritable cancer predisposition (*haereditaria dispositio*) can be found in Friedrich Hoffmann's *Opera Omnia physico-medica* (Geneva, 1761), vol. 3, p. 446; Hoffmann also cautioned that "forceful breast squeezing" during lovemaking could have cancer consequences (p. 446). Hereditary ideas are also put forward in Hunter's *Lectures*, pp. 380–382; T. B. Curling's *A Practical Treatise on the Diseases of the Testis, and of the Spermatic Cord and Scrotum*, 2nd ed. (London, 1856), pp. 506–510; and Paget's *Lectures*, p. 629.

13. Tito Spannocchi, "Contributo alla ereditarietà dei fibromi dell' utero," *Archivio italiano di Ginecologia* 2 (1899): 251–254. For a review of early cancer-among-twins literature, see Heinrich Kranz, "Tumoren bei Zwillingen," *Zeitschrift für induktive Abstammungs- und Vererbungslehre* 62 (1932): 173–181.

14. James N. Hyde, "On the Influence of Light in the Production of Cancer of the Skin," *American Journal of the Medical Sciences* (January 1906): 4–5.

15. Otto Jüngling, "Polyposis intestini. Hereditäre Verhältnisse und Beziehungen zum Carcinom," *Beiträge zur klinischen Chirurgie* 143 (1928): 476–483.

16. Haagensen, "An Exhibit of Important Books," p. 121.

17. Aldred S. Warthin, "Heredity with Reference to Carcinoma: As Shown by the Study of the Cases Examined in the Pathological Laboratory of the University of Michigan, 1895–1913," *Archives of Internal Medicine* 12 (1913): 546–555.

18. Theodor Leber, "Die Geschwulstbildungen der Netzhaut," in *Handbuch der gesamten Augenheilkunde*, edited by A. Elschnig, 2nd ed., vol. 7 (Leipzig, 1916), pp. 1723–1957.

19. Ernest E. Tyzzer, "A Study of Heredity in Relation to the Development of Tumors in Mice," *Journal of Medical Research* 12 (1907): 199–211.

20. Maud Slye, "Cancer and Heredity," *Annals of Internal Medicine* 1 (1928): 951–976.

21. A. Rosenbohm, "Untersuchungen über die Disposition zu Benzpyrentumoren bei Mäusern," *Zeitschrift für Krebsforschung* 52 (1942): 335–340.

22. Patterson, *Dread Disease*, p. 38.

23. J. J. McCoy, *The Cancer Lady: Maud Slye and Her Heredity Studies* (Nashville, Tenn., 1977), p. 92.

24. Ibid., p. 70.

25. Clarence C. Little, "The Inheritance of a Predisposition to Cancer in Man," in *Eugenics, Genetics, and the Family*, vol. 1 (Baltimore, Md., 1923), pp. 186–190; Erwin Baur, Eugen Fischer, and Fritz Lenz, *Menschliche Erblichkeitslehre* (Munich, 1923), pp. 258–264; B. Fischer-Wasels, *Die Vererbung der Krebskrankheit* (Berlin, 1935). Wasels's book was published as part of the series, "Schriften zur Erblehre und Rassenhygiene," edited by Günther Just.

26. Papanicolaou, "New Cancer Diagnosis."

27. Carl V. Weller, "The Inheritance of Retinoblastoma and Its Relationship to Practical Eugenics," *Cancer Research* 1 (1941): 517–535.

28. The official commentary on the Nazi Sterilization Law recommended the sterilization of all persons suffering from retinoblastoma (*Netzhautgliom*); see Arthur Gütt, Ernst Rüdin, and Falk Ruttke, *Gesetz zur Verhütung erbkranken Nachwuchses vom 14. Juli 1933* (Munich, 1934), p. 111. K. A. Reiser argued that only familial cases should be sterilized; see his "Bemerkungen zur Erblichkeitsfrage beim Glioma retinae," *Klinische Monatsblätter für Augenheilkunde* 99 (1937): 350–355.

29. Hyde, "On the Influence of Light," p. 15.

30. Mary Gover, "Trend of Mortality Among Southern Negroes Since 1920," *Journal of Negro Education* 6 (1937): 280–285.

31. Arthur Purdy Stout, "Tumors of the Neuromyo-Arterial Glomus," *American Journal of Cancer* 24 (1935): 255–272.

32. Alfredo Niceforo, "Cancer in Relation to Race in Europe," p. 502, and Eugene Pittard, "Can We Ignore the Race Problem in Connection with Cancer?" pp. 503–507, both in *Report of the International Conference on Cancer* (London, 1928).

33. Lewis, *Biology of the Negro*, pp. 327–355. Lewis suggested that "the reaction to disease is no less a characteristic of race than is head form or skin color" (p. ix).

34. Hans R. Schinz and Franz Buschke, *Krebs und Vererbung* (Leipzig, 1935), pp. 33–34; Johannes Schottky, ed., *Rasse und Krankheit* (Munich, 1937); Fischer-Wasels, *Vererbung*.

35. Otmar von Verschuer and E. Kober, "Die Frage der erblichen Disposition zum Krebs," *Zeitschrift für Krebsforschung* 50 (1940): 5–14.

36. Gehrmann, "The Carcinogenetic Agent," p. 135.

37. "Lung Lab Pays Off Quickly," *Business Week* (July 20, 1946): 54, cited in Rosner, *Deadly Dust*, p. 182.

38. Hueper, "Krankheit und Rasse," pp. 41–55.

39. Hueper, "Causal and Preventive Aspects," pp. 10–11. Eugene Pólya, a surgeon at St. Stephens Hospital in Budapest, advised against "intermarriage between members of cancerous families"; see his "The Prevention of Cancer," *Yale Journal of Biology and Medicine* 13 (1941): 384. Compare also Blome, "Krebsforschung," p. 412.

40. Maurice Sorsby, *Cancer and Race: A Study of the Incidence of Cancer Among Jews* (London, 1931).

41. Ernest L. Kennaway, "Cancer of the Liver in the Negro in Africa and in America," *Cancer Research* 4 (1944): 571–577. As early as 1885, Willard Parker, a New York physician, argued that race could not explain cancer rates since cancer was "not uncommon" among the Negroes of the United States, despite being "scarcely known" among the native populations of Africa. See his *Cancer* (New York, 1885), p. 37.

42. R. L. Smith, "Recorded and Expected Mortality Among Japanese of the United States and Hawaii, with Special Reference to Cancer," *Journal of the National Cancer Institute* 17 (1956): 459–473; William Haenszel and Minoru Kurihara, "Mortality from Cancer and Other Diseases Among Japanese in the United States," *Journal of the National Cancer Institute* 40 (1968): 43–68.

43. Peter C. Nowell and David A. Hungerford, "A Minute Chromosome in Human Chronic Granulocytic Leukemia," *Science* 132 (1960): 1497.

44. K. D. Zang and H. Singer, "Chromosomal Constitution of Meningiomas," *Nature* 216 (1967): 84–85; Kenneth W. Kinzler and Bert Vogelstein, "A Gene for Neurofibromatosis 2," *Nature* 363 (1993): 495.

45. J. E. Cleaver, "Defective Repair Replication of DNA in Xeroderma Pigmentosum," *Nature* 218 (1968): 652–656.

46. Bodmer, "Cancer Genetics," p. 1.

47. John J. Mulvihill, "Host Factors in Human Lung Tumors: An Example of Ecogenetics in Oncology," *National Cancer Institute Monographs* 52 (1979): 115–121.

48. Spatz et al., *Detection of Cancer Predisposition,* p. 5.

49. W. Kalow, *Pharmacogenetics: Heredity and the Response to Drugs* (Philadelphia, 1962); Steven A. Atlas and Daniel W. Nebert, "Pharmacogenetics: A Possible Pragmatic Perspective in Neoplasm Predictability," *Seminars in Oncology* 5 (1978): 89–106; Arno G. Motulsky, "Pharmacogenetics and Ecogenetics in 1991," *Pharmacogenetics* 1 (1991): 2–3.

50. Theodor H. Boveri, *Zur Frage der Entstehung maligner Tumoren* (Jena, 1914), pp. 25, 41.

51. Victor A. McKusick, "Marcella O'Grady Boveri (1865–1950) and the Chromosome Theory of Cancer," *Journal of Medical Genetics* 22 (1985): 436.

52. Knudson et al., "Heredity," pp. 115–116.

53. Ruth Sager, "Genomic Rearrangements and the Origin of Cancer. Rediscovering Boveri: The Problem of Causality," in *Chromosome Mutation and Neoplasia,* edited by J. German (New York, 1983), pp. 333–346. In 1916, E. E. Tyzzer argued that tumors could be regarded as "a modification of the somatic tissue which may be termed *somatic mutation*" but there is no evidence he regarded mutation as an alteration in the chromosomal material. See his "Tumor Immunity," *Journal of Cancer Research* 1 (1916): 125–156, esp. 147, 151.

54. For background, see Elof Axel Carlson, *The Gene: A Critical History* (Philadelphia, 1966); also Oluf Jacobsen's excellent *Heredity in Breast Cancer* (Cophenhagen, 1946), pp. 11–29.

55. Burdette, "Significance of Mutation," p. 204; Leonell C. Strong, "The Induction of Mutations by a Carcinogen," *Hereditas, Supplement* (1949): 486–499; E. L. Tatum, "Chemically Induced Mutations and their Bearing on Carcinogenesis," *Annals of the New York Academy of Sciences* 49 (1947): 87–97.

56. Robert Briggs and Thomas King, "Transplantation of Living Nuclei from Blastula Cells into Enucleated Frogs' Eggs," *Proceedings of the National Academy of Sciences* 38 (1952): 455–463.

57. Karl H. Bauer, *Mutationstheorie der Geschwulst-Entstehung* (Berlin, 1928). In 1936, Bauer and Felix von Mikulicz-Radecki published a widely read textbook on male sterilization, endorsing the racial hygienists' ideas of the need to eliminate genetic inferiors; see *Die Praxis der Sterilisierungsoperationen* (Leipzig, 1936). American occupation authorities named Bauer the first postwar president of the University of Heidelberg. In 1968, in his new position as head of the National Cancer Institute of Heidelberg, the German geneticist faced a sit-in by medical students confronting him with his involvement in the Nazi sterilization program; see Christian Pross, "Nazi Doctors, German Medicine, and Historical Truth," in *The Nazi Doctors and the Nuremberg Code,* edited by George Annas and Michael Grodin (New York, 1992), pp. 41, 49 n.31.

58. Burdette, "Significance of Mutation," pp. 201–226. A good overview of cancer genetics circa 1955 is Charles M. Woolf, *Investigations on Genetic Aspects of Carcinoma of the Stomach and Breast* (Berkeley, 1955); compare also Henry T. Lynch, *Hereditary Factors in Carcinoma* (New York, 1967). Woolf, interestingly, does not discuss the somatic mutation hypothesis and does not even

mention Boveri. Leonell C. Strong resurrects Boveri in his "Genetic Concept for the Origin of Cancer: Historical Review," *Annals of the New York Academy of Sciences* 71 (1958): 810–838. For a contemporary critique, see Austin M. Brues, "Critique of Mutational Theories of Carcinogenesis," *Acta Unio Internationalis Contra Cancrum* 16 (1960): 415–417.

59. Rous, "Nearer Causes," p. 581. Isaac Berenblum and Philippe Shubik are usually credited as the originators of the multistep view of cancer, but Peyton Rous here defends a "two stage process" involving the action of "provocative" and "actuating" carcinogens—benzpyrene or tar acting on a viral papilloma, for example.

60. Isaac Berenblum, "The Cocarcinogenic Action of Croton Resin," *Cancer Research* 1 (1944): 44–48; William F. Friedewald and Peyton Rous, "The Initiating and Promoting Elements in Tumor Production," *Journal of Experimental Medicine* 80 (1944): 101–131.

61. Isaac Berenblum and Philippe Shubik, "The Role of Croton Oil Applications, Associated with a Single Painting of a Carcinogen, in Tumour Induction of the Mouse's Skin," *British Journal of Cancer* 1 (1947): 379–382; also their "A New, Quantitative, Approach to the Study of the Stages of Chemical Carcinogenesis in the Mouse's Skin," *British Journal of Cancer* 1 (1947): 383–391; and for background, Efron, *Apocalyptics,* pp. 198–202.

62. Isaac Berenblum and Philippe Shubik, "An Experimental Study of the Initiating Stage of Carcinogenesis, and a Re-Examination of the Somatic Cell Mutation Theory of Cancer," *British Journal of Cancer* 3 (1949): 109–118.

63. In 1954, Peter Armitage and Richard Doll suggested that carcinogenesis involved "a complex process of perhaps six or seven stages"; see their "The Age Distribution of Cancer and a Multi-Stage Theory of Carcinogenesis," *British Journal of Cancer* 8 (1954): 9; also their "A Two-Stage Theory of Carcinogenesis in Relation to the Age Distribution of Human Cancer," *British Journal of Cancer* 11 (1957): 161–169.

64. Cairns, "The Cancer Problem," p. 67.

65. Alfred G. Knudson, "Mutation and Cancer: Statistical Study of Retinoblastoma," *Proceedings of the National Academy of Sciences* 68 (1971): 820–823; compare also Knudson et al., "Heredity," and Knudson, "Genetic Predisposition to Cancer." Earlier discussions of retinoblastoma genetics had hinted at a two-stage process; see Weller's 1941 "Inheritance of Retinoblastoma," p. 532; also Charlotte Auerbach, "A Possible Case of Delayed Mutation in Man," *Annals of Human Genetics* 20 (1956): 266–275.

66. Alfred G. Knudson et al., "Chromosomal Deletion and Retinoblastoma," *New England Journal of Medicine* 295 (1976): 1120–1123; Bishop and Waldholz, *Genome,* pp. 132–153.

67. Webster K. Cavenee et al., "Expression of Recessive Alleles by Chromosomal Mechanisms in Retinoblastoma," *Nature* 305 (1983): 779–784. In 1985, Cavenee became the first person to diagnose a cancer predisposition gene in a clinical setting. His Cincinnati-based team showed that a Swedish infant suspected of having inherited the retinoblastoma gene was in fact a carrier; Swedish physicians were then able to find and vaporize the early-stage malignant clumps in the infant's eyes. See Bishop and Waldholz, *Genome,* pp. 132–133. Thaddeus P. Dryja and his colleagues at the Massachusetts Eye and Ear Infirmary were the first to clone the Rb gene: see their "Molecular Detection of Deletions Involving Band q14 of Chromosome 13 in Retinoblastoma," *Proceedings of the National Academy of Sciences* 83 (1986): 7391–7394.

68. Alfred G. Knudson, "Model Hereditary Cancers of Man," *Progress in Nucleic Acid Research and Molecular Biology* 29 (1983): 17–25; also his "Hereditary Cancer, Oncogenes, and Antioncogenes," *Cancer Research* 45 (1985): 1437–1443.

69. Stephen H. Friend of the Massachusetts General Hospital Cancer Center in Boston claims to have cloned the first human tumor suppressor gene in 1986, but this is partly a matter of definition. Arnold Levine cloned p53 in 1983, although it was not yet known to be a tumor suppressor. For the Friend claim, and his mockery of the term *anti-oncogenes* ("reminiscent of the separate worlds of matter and anti-matter"), see his "Genetic Models for Studying Cancer Susceptibility," *Science* 259 (1993): 774–775. His 1986 paper, coauthored with Robert Weinberg and others, used the term "recessive oncogenes"; see "A Human DNA Segment with Properties of the Gene that Predisposes to Retinoblastoma and Osteosarcoma," *Nature* 323 (1986): 643–646.

70. Paul D. Robbins et al., "Negative Regulation of Human c-*fos* Expression by the Retinoblastoma Gene Product," *Nature* 346 (1990): 668–671.

71. Thilly, "What Actually Causes Cancer," p. 53.

72. Schneiderman, "Cancer," p. 3.

73. Day and Brown, "Multistage Models," pp. 977–989.

74. Knudson et al., "Heredity," p. 117; Wynbrandt and Ludman, *Encyclopedia,* pp. 30, 44–50.

75. D. Barker et al. (including Ray White and Mark Skolnick), "Gene for von Recklinghausen Neurofibromatosis Is in the Pericentromeric Region of Chromosome 17," *Science* 236 (1987): 1100–1102; Bernd R. Seizinger et al. (including James F. Gusella), "Common Pathogenetic Mech-

anism for Three Tumor Types in Bilateral Acoustic Neurofibromatosis," *Science* 236 (1987): 317–319.

76. Margaret R. Wallace et al. (including Francis S. Collins), "Type 1 Neurofibromatosis Gene," *Science* 249 (1990): 181–186; David Viskochil et al. (including Ray White), "Deletions and a Translocation Interrupt a Cloned Gene at the Neurofibromatosis Type 1 Locus," *Cell* 62 (1990): 187–192; Richard M. Cawthon et al. (including Ray White), "A Major Segment of the Neurofibromatosis Type 1 Gene," *Cell* 62 (1990): 193–201.

77. Kenneth W. Kinzler and Bert Vogelstein, "A Gene for Neurofibromatosis 2," *Nature* 363 (1993): 495.

78. Leslie Roberts, "Down to the Wire for the NF Gene," *Science* 249 (1990): 236–237; also her "NF's Cancer Connection," *Science* 249 (1990): 744.

79. R. Neil Schimke, *Genetics and Cancer in Man* (Edinburgh, 1978), p. 92.

80. Ray White and Peter O'Connell, "Identification and Characterization of the Gene for Neurofibromatosis Type 1," *Current Opinion in Genetics & Development* 1 (1991): 15–19; Isamu Nishisho et al. (including Bert Vogelstein), "Mutations of Chromosome 5q21 Genes in FAP and Colorectal Cancer Patients," *Science* 251 (1991): 665.

81. David Botstein et al. "Construction of a Genetic Linkage Map in Man Using Restriction Fragment Polymorphisms," *American Journal of Human Genetics* 32 (1980): 314–331.

82. Bert Vogelstein et al., "Genetic Alterations During Colorectal-Tumor Development," *New England Journal of Medicine* 319 (1988): 525–532; Cannon-Albright et al., "Common Inheritance." In 1984, Bert Vogelstein and Eric Fearon became one of four research teams independently and almost simultaneously to trace the gene for Wilms' tumor to chromosome 11; see Eric R. Fearon et al. (including Vogelstein), "Somatic Deletion and Duplication of Genes on Chromosome 11 in Wilms' Tumours," *Nature* 309 (1984): 176–178. The other reports are on pp. 170–176.

83. Walter F. Bodmer et al., "Localization of the Gene for Familial Adenomatous Polyposis on Chromosome 5," *Nature* 328 (1987): 614–616; Mark Leppert et al. (including Ray White and Yusuke Nakamura), "The Gene for Familial Polyposis Coli Maps to the Long Arm of Chromosome 5," *Science* 238 (1987): 1411–1413.

84. Kinzler et al., "Identification of a Gene."

85. Kenneth W. Kinzler et al. (including Vogelstein), "Identification of FAP Locus Genes from Chromosome 5q21," *Science* 253 (1991): 661–665; Joanna Groden et al. (including Ray White), "Identification and Characterization of the Familial Adenomatous Polyposis Coli Gene," *Cell* 66 (1991): 589–600.

86. Robert A. Weinberg, "Oncogenes and Tumor Suppressor Genes," in *Unnatural Causes: The Three Leading Killer Diseases in America,* edited by Russell C. Maulitz (New Brunswick, N.J., 1989).

87. Varmus and Weinberg, *Genes,* pp. 67–100. On the discovery of oncogenes, especially Robert Weinberg's contribution, see Angier, *Natural Obsessions.* On the sociology of oncogene laboratory life, see Joan H. Fujimura, "Ecologies of Action: Recombining Genes, Molecularizing Cancer, and Transforming Biology," in *Ecologies of Knowledge: Work and Politics in Science and Technology,* edited by Susan Leigh Star (Albany, 1995).

88. Varmus and Weinberg, *Genes,* pp. 102, 114.

89. Hans Breider, "Über Melanosarkome, Melaninbildung und homologe Zellmechanismen," *Strahlentherapie* 88 (1952): 619–639.

90. Varmus and Weinberg, *Genes,* pp. 101–119; Weinberg, "Tumor Suppressor Genes," p. 1139.

91. Daniel Linzer and Arnold Levine, "Characterization of 54K Dalton Cellular SV40 Tumor Antigen Present in SV40 Transformed Cells and Uninfected Embryonal Carcinoma Cells," *Cell* 17 (1979): 43–52; David P. Lane and L. V. Crawford, "T Antigen Is Bound to a Host Protein in SV40-transformed Cells," *Nature* 278 (1979): 261–263.

92. Moshe Oren and Arnold J. Levine, "Molecular CLoning of a cDNA Specific for the Murine p53 Cellular Tumor Antigen," *Proceedings of the National Academy of Sciences* 80 (1983): 56–59; Samuel Benchimol et al., "Transformation Associated p53 Protein Is Encoded by a Gene on Human Chromosome 17," *Somatic Cell and Molecular Genetics* 11 (1985): 505–509.

93. Cathy A. Finlay, Philip W. Hinds, and Arnold J. Levine, "The p53 Proto-Oncogene Can Act as a Suppressor of Transformation," *Cell* 57 (1989): 1083–1093. Finlay et al. argue here that p53 should be regarded as a "recessive oncogene" (p. 1084).

94. Michael Mowat et al. (including Samuel Benchimol), "Rearrangements of the Cellular p53 Gene in Erythroleukaemic Cells Transformed by Friend Virus," *Nature* 34 (1985): 633–636.

95. Suzanne J. Baker et al. (including Ray White, Eric Fearon, and Bert Vogelstein), "Chromosome 17 Deletions and p53 Gene Mutations in Colorectal Carcinomas," *Science* 244 (1989): 217–221.

96. Steven M. Powell et al. (including Bert Vogelstein and Kenneth Kinzler), "APC Mutations Occur Early During Colorectal Tumorigenesis," *Nature* 359 (1992): 235–237.

97. David Malkin et al. (including Stephen Friend, Frederick Li, Louise C. Strong, and Joseph Frau-

meni), "Germ Line p53 Mutations in a Familial Syndrome of Breast Cancer, Sarcomas, and Other Neoplasms," *Science* 250 (1990): 1233–1238.

98. Monica Hollstein, David Sidransky, Bert Vogelstein, and Curtis C. Harris, "P53 Mutations in Human Cancers," *Science* 253 (1991): 49–53. This paper became the most often cited paper in all of science in the early 1990s.

99. Natalie Angier, "Ultraviolet Radiation Tied to Gene Defect Producing Skin Cancer," *New York Times,* November 19, 1991; Douglas E. Brash, "A Role for Sunlight in Skin Cancer: UV-Induced p53 Mutations in Squamous Cell Carcinoma," *Proceedings of the National Academy of Sciences* 88 (1991): 10124–10128.

100. Ralph H. Hruban et al., "Molecular Biology and the Early Detection of Carcinoma of the Bladder—the Case of Hubert H. Humphrey," *New England Journal of Medicine* 330 (1994): 1276–1278.

101. Weinberg, "Tumor Suppressor Genes," p. 1138.

102. Bishop and Waldholz, *Genome,* p. 174.

103. Jean Marx, "Learning How to Suppress Cancer," *Science* 261 (1993): 1387; "Hottest Scientist of 1993? It's Bert Vogelstein," *ScienceWatch* (December 1993): 1–2.

104. Lawrence A. Donehower et al., "Mice Deficient for p53 Are Developmentally Normal but Susceptible to Spontaneous Tumours," *Nature* 356 (1992): 215–221.

105. Lane, "A Death."

106. Ibid., p. 786.

107. Anthony M. Carr et al., "Checkpoint Policing by p53," *Nature* 359 (1992): 486.

108. Michael B. Kastan et al., (including Bert Vogelstein), "A Mammalian Cell Cycle Checkpoint Pathway Utilizing p53 and *GADD45* Is Defective in Ataxia-Telangiectasia," *Cell* 71 (1992): 587–597; Jean Marx, "How p53 Suppresses Cell Growth," *Science* 262 (1993): 1644–1645; also her "How Cells Cycle Toward Cancer," *Science* 263 (1994): 319–321.

109. Scott W. Lowe et al., "P53 is Required for Radiation-Induced Apoptosis in Mouse Thymocytes," *Nature* 362 (1993): 847; Curtis C. Harris, "P53: At the Crossroads of Molecular Carcinogenesis and Risk Assessment," *Science* 262 (1993): 1980–1981.

110. Daniel E. Koshland, "Molecule of the Year," *Science* 262 (1993): 1953; Elizabeth Culotta and Daniel E. Koshland, "P53 Sweeps Through Cancer Research," *Science* 262 (1993): 1958–1959.

111. Harris, "P53," p. 1980.

112. Roberts, "Zeroing In," p. 625.

113. Perera et al., "Molecular and Genetic Damage."

114. K. H. Vähäkangas et al., "Mutations of p53 and *ras* Genes in Radon-Associated Lung Cancer from Uranium Miners," *Lancet* 339 (1992): 576–580.

115. Frederica P. Perera and I. Bernard Weinstein, "Molecular Epidemiology and Carcinogen-DNA Adduct Detection: New Approaches to Studies of Human Cancer Causation," *Journal of Chronic Disease* 35 (1982): 581–600; Peter G. Shields and Curtis C. Harris, "Molecular Epidemiology and the Genetics of Environmental Cancer," *JAMA* 266 (1991): 681–687.

116. Richard Fishel et al., "The Human Mutator Gene Homolog *MSH2* and Its Association with Hereditary Nonpolyposis Colon Cancer," *Cell* 75 (1993): 1027–1038; Fredrick S. Leach et al., "Mutations of a *mutS* Homolog in Hereditary Nonpolyposis Colorectal Cancer," *Cell* 75 (1993): 1215–1225.

117. Jean Weissenbach at Généthon in Paris and James Weber of the Marshfield Medical Research Foundation in Wisconsin developed the new probes; see Jean Marx, "New Colon Cancer Gene Discovered," *Science* 260 (1993): 751–752.

118. S. N. Thibodeau et al. "Microsatellite Instability in Cancer of the Proximal Colon," *Science* 260 (1993): 816–818; Peltomäki and Aaltonen et al., "Genetic Mapping," pp. 810–812; Lauri A. Aaltonen and Päivi Peltomäki et al. (including de la Chapelle and Vogelstein), "Clues to the Pathogenesis of Familial Colorectal Cancer," *Science* 260 (1993): 812–816; Yurij Ionov et al., "Ubiquitous Somatic Mutations in Simple Repeated Sequences Reveal a New Mechanism for Colonic Carcinogenesis," *Nature* 363 (1993): 558–561.

119. *MSH2* actually appears to account for only about 60 percent of all cases of hereditary nonpolyposis colon cancer. In March 1994, Vogelstein and Kinzler's group at Johns Hopkins and Kolodner's laboratory at Harvard isolated another gene, *MLH1,* which is expected to play a role in about 30 percent of all cases of HNPCC. The new gene, like *MSH2,* appears to operate by repairing mismatched DNA base pairs. See Nicholas Papadopoulos et al. (including Kinzler and Vogelstein), "Mutation of a *mutL* Homolog in Hereditary Colon Cancer," *Science* 263 (1994): 1625–1629; C. Eric Bronner (including Kolodner), "Mutation in the DNA Mismatch Repair Gene Homologue *MLH1,*" *Nature* 368 (1994): 258–261.

120. Natalie Angier, "Scientists Isolate Novel Gene Linked to Colon Cancer," *New York Times,* December 3, 1993.

121. Sibley, "Contribution," pp. 127–135; W. R. Williams and D. E. Anderson, "Genetic Epidemiology

of Breast Cancer," *Genetic Epidemiology* 1 (1984): 1–7; Wright, "Breast Cancer."

122. Claus et al., "Genetic Analysis."
123. Roberts, "Zeroing In," pp. 622–625.
124. Jeff M. Hall et al. (including Mary-Claire King), "Linkage of Early-Onset Familial Breast Cancer to Chromosome 17q21," *Science* 250 (1990): 1684–1689.
125. Geoffrey Cowley, "Family Matters: The Hunt for a Breast Cancer Gene," *Newsweek* (December 6, 1993): 49.
126. Yoshio Miki et al. (including Mark Skolnick), "A Strong Candidate for the Breast and Ovarian Cancer Susceptibility Gene *BRCA1*," *Science* 266 (1994): 66–71; Richard Wooster et al., "Localization of a Breast Cancer Susceptibility Gene, *BRCA2*, to Chromosome 13q12–13," *Science* 266 (1994): 2088–2089.
127. Mary-Claire King et al., "Inherited Breast and Ovarian Cancer: What Are the Risks? What Are the Choices?" *JAMA* 269 (1993): 1976. King et al. state that 1 percent of all breast cancers among women under forty may be due to an inherited p53 mutation; they also note that male breast cancer can be heritable, although apparently not via the *BRCA1* or p53 genes (pp. 1975–1976).
128. Claus et al., "Genetic Analysis."
129. Dorothy Nelkin and Laurence Tancredi, *Dangerous Diagnostics: The Social Power of Biological Information* (New York, 1989).
130. "Screening for BRCA 1 mutations is likely to be the first widespread presymptomatic test that finds its way into general medical practice"; see Barbera B. Biesecker et al., "Genetic Counseling for Families with Inherited Susceptibility to Breast and Ovarian Cancer," *JAMA* 269 (1993): 1970–1974.
131. Oncor, Inc., press release, December 14, 1993; National Advisory Council for Human Genome Research, "Statement on Use of DNA Testing for Presymptomatic Identification of Cancer Risk," *JAMA* 271 (1994): 785; Elyse Tanouye, "Gene Testing for Cancer to Be Widely Available, Raising Thorny Questions," *Wall Street Journal,* December 14, 1993. Founder and head of the Hereditary Cancer Institute is Henry T. Lynch, a geneticist involved in the early characterization of hereditary nonpolyposis of the colon and the popularizer of the term *SBLA* for Li-Fraumeni syndrome.
132. Myriad Genetics of Salt Lake City, Utah, was a major funder of the effort to clone *BRCA1;* the company has filed for a patent, but the NIH, a co-sponsor of the project, also wants part of the action. See Eliot Marshall, "A Showdown over Gene Fragments," *Science* 266 (1994): 208–210; Rachel Nowak, "NIH in Danger of Losing Out on *BRCA1* Patent," *Science* 266 (1994): 209.
133. In 1963, G. K. Tokuhata and A. M. Lilienfeld found that lung cancer was more than two times higher than expected among the nonsmoking relatives of lung cancer victims. The study did not control for factors such as exposure to secondhand smoke, however. See their "Familial Aggregation of Lung Cancer in Humans," *Journal of the National Cancer Institute* 30 (1963): 289–312. Tokuhata later claimed to have shown that the smoking habit itself had a genetic basis, and in a 1992 review John J. Mulvihill of the University of Pittsburgh agreed ("the tendency to smoke may be genetic"). See John J. Mulvihill, "Lung Cancer," in *The Genetic Basis of Common Diseases,* edited by Richard A. King et al. (New York, 1992), p. 692. Conclusions based purely on genealogical correlations are often challenged on the grounds that people raised in a smoking household may simply learn to tolerate or appreciate the smoking habit.
134. The enzymes aryl hydrocarbon hydroxylase, debrisoquine hydroxylase, and glutathione S-transferase have all been shown to assist in the detoxification of polycyclic aromatic hydrocarbons, and there is strong evidence that these capacities are genetically rooted and unevenly distributed in the human population. See Neil Caporaso et al., "Relevance of Metabolic Polymorphisms to Human Carcinogenesis," *Pharmacogenetics* 1 (1991): 4–19; Valle Nazar-Stewart et al. (including Arno Motulsky), "The Glutathione S-Transferase—Polymorphism as a Marker for Susceptibility to Lung Carcinoma," *Cancer Research* 53 (1993): 2313–2318. The first such polymorphism was reported by G. Kellermann et al., "Aryl Hydroxylase Inducibility and Bronchogenetic Carcinoma," *New England Journal of Medicine* 289 (1973): 934–937.
135. Kenneth J. Pienta, Alan W. Partin, and Donald S. Coffey, "Cancer as a Disease of DNA Organization and Dynamic Cell Structure," *Cancer Research* 49 (1989): 2525–2532; compare also Harry Rubin, "Understanding Cancer" (letter), *Science* 219 (1983): 1170–1171.
136. Lewontin, *Biology as Ideology,* pp. 47–57.
137. Bishop and Waldholz, *Genome,* pp. 155–156; Peltomäki and Aaltonen et al., "Genetic Mapping," p. 810.
138. Henry T. Lynch et al., "Familial Breast Cancer and Its Recognition in an Oncology Clinic," *Cancer* 47 (1981): 2730–2739; Wynbrandt and Ludman, *Encyclopedia,* p. 49.
139. Graham A. Colditz et al., "Family History, Age, and Risk of Breast Cancer," *JAMA* 270 (1993): 338–343.
140. Hrafn Tulinius et al., "Epidemiology of Breast Cancer in Families in Iceland," *Journal of Medical*

Genetics 29 (1992): 158–164. Iceland's cancer registry was established in 1954 but has information on "all breast cancer cases diagnosed in the population since 1910." Family trees for almost all Icelanders alive in 1840 or thereafter are contained in the files of the Genetical Committee of the University of Iceland; these are also being used to analyze breast cancer genetics (p. 162).

141. M. Miles Braun et al., "Genetic Component of Lung Cancer: Cohort Study of Twins," *Lancet* 344 (1994): 440–443.

142. Cited in Bishop and Waldholz, *Genome,* pp. 156, 163. Compare Skolnick's statement that "only a fraction of the population is at risk for adenomatous polyps and colorectal cancer," in Cannon-Albright et al., "Common Inheritance," p. 536. Similar generalizations accompanied the November 30, 1990, publication in *Science* implicating the p53 gene in Li-Fraumeni syndrome: commenting on his discovery, David Malkin of Boston's Massachusetts General Hospital Cancer Center noted, "We'll be able to say, 'Yes, you carry this mutation . . . and you are at risk,' or 'No, you don't.'" See R. Weiss, "Genetic Propensity to Common Cancers Found," *Science News* 138 (1990): 342.

143. Claus et al., "Genetic Analysis," is a recent example; compare also Thomas A. Sellers et al., "Evidence for Mendelian Inheritance in the Pathogenesis of Lung Cancer," *Journal of the National Cancer Institute* 82 (1990): 1272–1279.

144. Doll and Peto, *Causes of Cancer,* pp. 1202–1205.

145. Skolnick was not the first to suggest that only some people can get cancer: in 1978, for example, R. Neil Schimke suggested that "the population may actually consist of a large number of individuals with no or perhaps only a minimal risk of cancer and a small group with a large risk." See his *Genetics and Cancer in Man,* p. 2. Compare also Herbert Snow's nineteenth-century complaint that people with cancer tend to delay seeking treatment, believing wrongly that they cannot possibly have the disease since it does not run in their family. Such people, chastised for waiting, reply, "Oh, there has never been any cancer in my family, and I thought cancer was always hereditary" ("Is Cancer Hereditary?" p. 691).

146. Gina Kolata, "L. I. Cancer Found to Be Explainable," *New York Times,* December 19, 1992.

147. Thomas J. Lueck, "New Studies on Breast Cancer Sought by D'Amato and Women," *New York Times,* January 8, 1992.

148. Snow, "Is Cancer Hereditary?" p. 691.

149. A 1993 article in the American Cancer Society journal, *CA,* shows that there is a great deal of geographical variation in the treatment of breast cancer: in Kentucky, Tennessee, Mississippi, and Alabama, for example, only about 21 percent of all women have been treated by partial mastectomy (lumpectomy), compared with 55 percent in New England states. See "Breast Cancer," *CA* 43 (1993): 79. Dr. Sidney Salmon, director of the Arizona Cancer Center, recommends that no more than about 10 percent of all women with breast cancer should be getting mastectomies.

150. Gina Kolata, "Why Do So Many Women Have Breasts Removed Needlessly?" *New York Times,* May 5, 1993.

151. Juliet Wittman, *Breast Cancer Journal: A Century of Petals* (Golden, Colo., 1993).

152. Jane E. Brody, "Why Cancer-Free Women Have Breasts Removed," *New York Times,* May 5, 1993.

153. Leslie Roberts, "Genetic Counseling: A Preview of What's in Store," *Science* 259 (1993): 624.

154. Age-adjusted lung cancer mortality rates for U.S. males rose from 5/100,000 in 1930 to 75/100,000 in 1985—a fifteen-fold increase attributable primarily to smoking. Women's lung cancer death rates rose 420 percent over the last thirty years, and breast cancer rates have also grown, although by a lesser margin. The American Cancer Society's 1991 *Cancer Facts and Figures* (New York, 1991) estimated that American women face a one in nine chance of developing breast cancer over a lifetime–a 10 percent increase from only four years earlier (p. 9). The 1993 *Cancer Facts and Figures* qualified this by suggesting that "one of every nine women will develop breast cancer *by age 85*" (p. 11, emphasis added); the 1994 edition deleted all reference to changing risks.

155. Even Claus and colleagues concede that "the great majority of breast cancers are nongenetic"; see Claus et al., "Genetic Analysis," p. 241.

156. "Poverty Blamed for Blacks' High Cancer Rate," *New York Times,* April 17, 1991.

157. Wexler is cited in Robin Marantz Henig, "High-Tech Fortune Telling," *New York Times Magazine* (December 24, 1989): 20.

158. See Eric S. Lander, "The New Human Genetics: Mapping Inherited Diseases," *Princeton Alumni Weekly* (March 25, 1987): 10–15. Compare Natalie Angier's claim that "the study of oncogenes may not be the best hope of banishing cancer; it may be the only hope." More to the point may be the quip of Richard Rifkind, which Angier herself cites: "You want a cure for cancer? Tell the bastards to quit smoking" (*Natural Obsessions,* pp. 17, 141).

159. It is also possible, of course, that a germ-line cancer gene or genetic susceptibility may have arisen through an environmental insult. This is obviously a much rarer event than the induction of a somatic mutation; even so, a germ-line alteration could have dramatic long-term consequences if it were to be passed from generation to generation.

160. J. Michael Bishop, "Oncogenes," *Scientific American* 246 (1982): 92; Robert A. Weinberg, "Finding the Oncogene," *Scientific American* 258 (1988): 44–51. Parts of the argument presented here follow closely my "Genomics and Eugenics: How Fair Is the Comparison?" in *Gene Mapping: Using Law and Ethics as Guides,* edited by George J. Annas and Sherman Elias (New York, 1992), pp. 57–93.

161. Dulbecco, "Turning Point," p. 11. Compare also his claim that "the genome project holds the key to a full understanding and perhaps also the treatments of cancer," in his "Co-chairman's Remarks: The Human Genome Project and Cancer," *Gene* 135 (1993): 260.

162. Murray Bookchin came to a similar conclusion more than thirty years ago when he suggested, with reference to cancer causation, that "if the 'soil' is prepared by heredity, the 'seed' is planted by the environment." See Herber, *Our Synthetic Environment,"* p. 123.

163. Bishop and Waldholz, *Genome,* p. 155.

164. Sporadic (nonfamilial) cancers are also sometimes characterized as "spontaneous"; see Vogelstein et al., "Genetic Alterations," p. 528. Compare Samuel Epstein's 1970 remark that "many diseases hitherto regarded as spontaneous are caused by environmental pollutants" ("Control of Chemical Pollutants," p. 816).

165. Cited in Bishop and Waldholz, *Genome,* p. 156.

166. Roberts, "Zeroing In," p. 622.

167. The words are Arno Motulsky's in the *American Journal of Human Genetics* 48 (1991): 174.

168. Gina Kolata, "Scientists Pinpoint Genetic Changes that Predict Cancer," *New York Times,* May 16, 1989.

169. "The Telltale Gene," *Consumer Reports* (July 1990): 485. Compare also Ruth Hubbard and Elijah Wald, *Exploding the Gene Myth* (Boston, 1993), pp. 81–92.

170. "Statement of the First Workshop of the Joint Working Group on the Ethical, Legal and Social Issues Related to Mapping and Sequencing the Human Genome," Williamsburg, February 5–6, 1990, in the HHS and DOE booklet, *Understanding Our Genetic Inheritance,* pp. 65–73; Committee for Responsible Genetics, "Position Paper on the Human Genome Initiative," January 10, 1990; compare also Thomas H. Murray, "Warning: Screening Workers for Genetic Risk," *Hastings Center Report* (February 1983): 5–8. Americans, of course, are not the only ones prone to genetic screening. In 1989, seven distinguished Soviet scientists led by Boris A. Katsnelson of the Institute of Industrial Hygiene in Sverdlovsk published a paper arguing that industries involving high exposures to silica dust should find ways to employ only workers having a low susceptibility to dust-induced diseases. The authors were especially concerned about silicosis, a common complaint in the mining, stone-cutting, and brick-making industries. The goal would be to find ways of identifying workers especially susceptible to the disease using distinctive physical or genetic markers. "Biological prophylaxis" of this sort was supposed to supplement traditional efforts to improve workers' health and safety. These scholars rejected the idea that it was wrong "to mold man to the machine," noting that governments often pass laws barring certain kinds of people from certain kinds of work (child labor laws, for example). See Boris A. Katsnelson et al., "Trends and Perspectives of the Biological Prophylaxis of Silicosis," *Environmental Health Perspectives* 82 (1989): 311–321.

171. Lappé, *Genetic Politics,* p. 120.

172. Angier, "Cigarettes Trigger Lung Cancer Gene." In August 1992, nine leading cancer researchers and ethicists urged p53 carriers, among other things, to "pursue a healthier lifestyle and diet, with avoidance of cigarette smoking, excess alcohol use, and exposures to other carcinogens." See Frederick P. Li et al., "Recommendations on Predictive Testing for Germ Line p53 Mutations Among Cancer-Prone Individuals," *Journal of the National Cancer Institute* 84 (1992): 1160.

173. Peter Boyle, "Diet in the Aetiology of Cancer," *European Journal of Cancer* 30A (1994): 133.

174. Pienta et al., "Cancer as a Disease," p. 2525.

175. Genetic tests may soon be used to detect cancers in the blood, urine, or stools; see Jean Marx, "Test Could Yield Improved Colon Cancer Detection," *Science* 256 (1992): 32–33.

CONCLUSION
HOW CAN WE WIN THE WAR?

1. Henderson et al., "Toward the Primary Prevention," p. 1131.

2. Medicine had little to do with the increasing life span of people in the wealthy nations in the nineteenth century. More important were the increased consumption of fresh fruits and vegetables, the construction of sewers, decreasing hours of work, bans on child labor, and new and cleaner clothing habits. In 1959, Rene Dubos suggested that "the introduction of inexpensive cotton undergarments easy to launder and of transparent glass that brought light into the most humble dwelling,

contributed more to the control of infection than did all drugs and medical practices" combined; see his *Mirage of Health* (New York, 1959), p. 20; compare also Herber, *Our Synthetic Environment*, p. 11, and McKeown, *Role of Medicine.*

3. American Cancer Society, *Research Report* (New York, 1984), inside cover. The words are those of ACS past president, Saul B. Gusberg.

4. Several examples of breathless optimism can be found in Ross, *Crusade,* e.g. p. 126.

5. Philip M. Boffey, "Scientists Find Striking Change in Cancer Study," *New York Times,* December 8, 1986.

6. Ross, *Crusade,* p. 126; National Cancer Institute, *Cancer Control: Objectives for the Nation: 1985–2000* (Washington, D.C., 1986). The NCI goal was to be achieved through a combination of "a reduction in tobacco smoking by 50% from 1980 levels, the adoption of a prudent low fat, high fiber diet by all Americans, recommended cancer screening measures, and accelerated and widespread application of gains in state-of-the-art cancer treatment methods" (p. 3). In 1989, James E. Enstrom pointed out that the nonsmoking, nondrinking, high priests of the Mormon Church already had a cancer rate 50 percent below that of the nation as a whole; see his "Health Practices and Cancer Mortality Among Active California Mormons," *Journal of the National Cancer Institute* 81 (1989): 1807–1814.

7. Bailar and Smith, "Progress Against Cancer?"; also the responses to this article in the *New England Journal of Medicine* 315 (1986): 966–968.

8. Becker, Smith, and Wahrendorf, "Time Trends."

9. L. A. G. Ries et al., *Cancer Statistics Review: 1973–1988* (Bethesda, Md., 1991). Compare also the update by B. A. Miller et al., *SEER Cancer Statistics Review: 1973–1990* (Bethesda, Md., 1993).

10. Death rates have risen much faster for African Americans than for European Americans: since 1955, cancer deaths rates have risen about 17 percent for white men and 2 percent for white women, but 66 percent for black men and 10 percent for black women. See the American Cancer Society, *Cancer Facts and Figures for Minority Americans,* pp. 2–3; also Catherine C. Boring et al., "Cancer Statistics for African Americans," *CA* 42 (1992): 7–17.

11. Jonathan M. Liff et al., "Does Increased Detection Account for the Rising Incidence of Breast Cancer?" *American Journal of Public Health* 81 (1991): 462–465. Andrew G. Glass and Robert N. Hoover similarly found that the age-adjusted incidence of breast cancer rose 45 percent in the period 1960–1985, and that increased detection by mammography could explain at most one third of the increase; see their "Rising Incidence of Breast Cancer," *Journal of the National Cancer Institute* 82 (1990): 693–696.

12. Lawrence K. Altman, "Lymphomas Are on the Rise in U.S. and No One Knows Why," *New York Times,* May 24, 1994; Shelia H. Zahm and Aaron Blair, "Pesticides and Non-Hodgkin's Lymphoma," *Cancer Research* (supp.) 52 (1992): 5485s–5488s.

13. Rachel Nowak, "A New Test Gives Early Warning of a Growing Killer," *Science* 264 (1994): 1847–1848.

14. Davis et al., "International Trends." Analysis of data from 1984 indicates that black men and black women had 65 percent and 15 percent more nonsmoking lung cancer, respectively, than white men and white women; see Davis, "Trends," p. 491; compare also Francesco Forastiere et al., "Indirect Estimates of Lung Cancer Death Rates in Italy Not Attributable to Active Smoking," *Epidemiology* 4 (1993): 502–510.

15. Bailar, "Some Recent Trends." et al., "Trends in Cancer Mortality."

16. American Cancer Society, *Cancer Facts and Figures—1994,* p. 1.

17. Beardsley, "War Not Won," p. 135; Bailar, "Some Recent Trends."

18. Gibbons, "Does War on Cancer Equal War on Poverty?" p. 260.

19. Clemmesen, *Statistical Studies,* pp. 1–11.

20. Bailar and Smith, "Progress Against Cancer?" pp. 1226; also their response to letters in *New England Journal of Medicine* 315 (1986): 968.

21. Compare also: National Cancer Institute, *Evaluating the National Cancer Program: An Ongoing Process* (Bethesda, Md., 1994).

22. Alan D. Lopez, "Competing Causes of Death: A Review of Recent Trends in Mortality in Industrialized Countries with Special Reference to Cancer," *Annals of the New York Academy of Sciences* 609 (1990): 58–76.

23. See my "Nazi Cancer Research and Policy," unpublished manuscript.

24. Doll, "Are We Winning the Fight Against Cancer?" p. 508.

25. Wright, "Going by the Numbers."

26. Devra Lee Davis et al., "Is Brain Cancer Mortality Increasing in Industrial Countries?" *Annals of the New York Academy of Sciences* 609 (1990): 191–214. Misclassification—e.g., of Japanese brain cancers as "stroke"—might also be involved.

27. Anthony P. Polednak, "Time Trends in Incidence of Brain and Central Nervous System Cancers in

Connecticut," *Journal of the National Cancer Institute* 83 (1991): 1679–1681.

28. Zentralinstitut für Krebsforschung, *Krebsinzidenz in der DDR 1987* (Berlin, 1991), pp. 34–35.

29. Davis, "Trends," p. 490; also Devra Lee Davis and David G. Hoel, "Tobacco-associated Deaths" (letter), *Lancet* 340 (1992): 666. Marvin Schneiderman notes that the smoking omni-etiology for lung cancer is difficult to square with the racial distribution of cancer. Why, for example, he asks, do U.S. blacks have a higher incidence of lung cancer than whites (70 per 100,000 vs. 52 per 100,000 for the period 1977 to 1983), when black smokers smoke 20–25 percent fewer cigarettes on average and start smoking at slightly older ages? Could it be that blacks, on average, have dirtier jobs and live in more polluted places, and that this might multiply their smoking risks? This question is difficult to answer in the affirmative, since more blacks than whites are smokers. The data for female smokers give a clearer picture, since about the same proportion of black and white women are smokers. Black women smoke about 30 percent fewer cigarettes than white women, and yet black women are no less likely to develop lung cancer. This suggests that something other than smoking may be playing a role (Schneiderman, personal communication). For data, see the ACS's *Cancer Facts and Figures for Minority Americans*, pp. 5, 17–18.

30. Wright, "Going by the Numbers," p. 79.

31. Marshall, "The Politics of Breast Cancer."

32. Monte Paulsen, "The Cancer Business," *Mother Jones* (June 1994): 41.

33. See Alfred S. Evans, "Causation and Disease: The Henle-Koch Postulates Revisited," *Yale Journal of Biology and Medicine* 49 (1976): 175–179.

34. Leslie I. Boden et al., "Science and Persuasion: Environmental Disease in U.S. Courts," *Social Science and Medicine* 27 (1988): 1019–1029; Foster, Bernstein, and Huber, *Phantom Risk;* Huber, *Galileo's Revenge.*

35. Angier, *Natural Obsessions*, p. 13.

36. Perera et al., "Molecular and Genetic Damage"; Harris, "p53," pp. 1980–1981.

37. The expression "potato famine" is somewhat of a misnomer: there was plenty of food on the island, although many people were too poor to purchase it. Ireland exported grain during the entire period in question; see Cecil Woodham Smith, *The Great Hunger* (London, 1962).

38. William Wise, *Killer Smog* (New York, 1970).

39. Lewontin, *Biology as Ideology*, pp. 45–46.

40. Susan Ferraro, "The Anguished Politics of Breast Cancer," *New York Times Magazine* (August 15, 1993): 27.

41. Robert J. Mayer, "Blood Tests for Cancer," *Harvard Health Letter* (August 1993): 1–3.

42. The Pap smear test for cervical cancer, developed in 1928, is probably the most widely used cancer diagnostic test, but as of 1985 randomized controlled clinical trials had never been performed to determine its value. See Gina Kolata, "Is the War on Cancer Being Won?" *Science* 229 (1985): 543; also Jean L. Marx, "The Annual Pap Smear: An Idea Whose Time Has Gone?" *Science* 205 (1979): 177–178.

43. Glenn S. Gerber et al., "Disease-Specific Survival Following Routine Prostate Cancer Screening by Digital Rectal Examination," *JAMA* 269 (1993): 61–64. A 1994 study showed that "watchful waiting" is an appropriate option for elderly men with prostate cancer; see Gerald W. Chodak et al., "Results of Conservative Management of Clinically Localized Prostate Cancer," *New England Journal of Medicine* 330 (1994): 242–248.

44. Greenberg and Randal, "Waging the Wrong War."

45. Gina Kolata, "Mammography Campaigns Draw In the Young and Healthy," *New York Times,* January 10, 1993.

46. Love, *Love's Breast Book,* p. 209. A 1994 NIH report recommended that women not undergo routine screening for ovarian cancer on the grounds that it often leads to unnecessary surgery and does not generally improve one's outcome; see "Routine Screening for Ovarian Cancer Is Called Dangerous," *New York Times,* April 10, 1994.

47. Bolch and Lyons, *Apocalypse Not,* pp. 46–47.

48. Cited in Bollier and Claybrook, *Freedom from Harm,* p. 1. For an early example of this argument, see A. J. Lehman, "Conservatism in Estimating the Hazards of Pesticidal Residues," *Association of Food and Drug Officials of the US, Quarterly Bulletin* 18 (1954): 87–90.

49. David Corn, "Death in the Sandbox," *The Nation* (August 27–September 3, 1990): 194–198.

50. Carl F. Cranor has discussed this in terms of the tension between "clean hands science" and "dirty hands regulation"; see his "Some Moral Issues in Risk Assessment," *Ethics* (October 1990): 123–143; compare also his *Regulating Toxic Substances.*

51. In 1990, Vernon Houk, director of environmental health at the Centers for Disease Control in Atlanta, labeled the EPA's "massive expenditures of money on minuscule risks . . . not conservative but very radical" (Brookes, "Wasteful Pursuit," p. 166).

52. David Rosner and Gerald Markowitz, "Research or Advocacy: Federal Occupational Safety and

Health Policies During the New Deal," *Journal of Social History* 18 (1985): 365–381. The conflict between the EPA and the DOE over how to regulate radon can also be regarded as a conflict between scientific and public health conservatism.

53. Office of Management and Budget, "Regulatory Program of the United States Government," reprinted in *The SIRC Review* (October 1990): 9–21.

54. See, for example, Albert L. Nichols and Richard J. Zeckhauser, "The Perils of Prudence," *Regulation* (November–December 1986): 13–24; also Richard B. Belzer, "The Peril and Promise of Risk Assessment," *Regulation* (Fall 1991): 41.

55. Ozonoff, "Public Relations Cancer," p. 13.

56. Ozonoff and Boden, "Truth and Consequences," p. 72.

57. David Ozonoff, "Medical and Legal Causation," in *Occupational Health in the 1990s,* edited by Philip J. Landrigan and Irving J. Selikoff (New York, 1989), p. 25.

58. Ozonoff and Boden, "Truth and Consequences," p. 74. Beverly Paigen, a biochemist-geneticist who became a consultant for residents of Love Canal, was moved to question conventional canons of statistical significance: "Before Love Canal, I also needed to have 95 percent certainty before I was convinced of a result. But seeing this rigorously applied in a situation where the consequences of an error meant that pregnancies were resulting in miscarriages, stillbirths, and children with medical problems, I realized I was making a value judgment . . . whether to make errors on the side of protecting human health or on the side of conserving state resources." See her "Controversy," p. 32; also Phil Brown's excellent "Popular Epidemiology and Toxic Waste Contamination: Lay and Professional Ways of Knowing," *Journal of Health and Social Behavior* 33 (1992): 267–281.

59. Rothman, *Modern Epidemiology,* pp. 115–125; Paigen, "Controversy," p. 32. Ozonoff and Boden argue that a similar logic is implicit in traditional medical practice, where it is usually considered best to "err on the side of safety" by invoking a regimen of treatment above and beyond what is imagined to be required. Antibiotics, for example, are commonly prescribed for longer than is thought necessary to combat an infection. Ozonoff and Boden point out that we do not generally approach problems of public health this way. One could argue, though, that public health "conservatism" is actually closer to what is usually classed as "aggressive" medical treatment. Treating a disease conservatively often implies a "let's wait and see" trust in the body's ability to heal itself— an attitude opposite to what is normally considered cautious in the realm of environmental health.

60. Ozonoff and Boden, "Truth and Consequences," p. 74.

61. In 1991, the NCI's budget was $1.7 billion, $270 million of which was devoted to environmental carcinogenesis. The proportion has grown from about one-ninth in 1979 to about one-sixth in 1991.

62. National Cancer Institute, *National Cancer Institute Response to Article by Dr. Samuel S. Epstein* (Washington, D.C., n.d), p. 6.

63. Donald R. Shopland, personal communication.

64. Rettig, *Cancer Crusade,* pp. 77–114.

65. Greenberg and Randal, "Waging the Wrong War," p. C1.

66. Moss, *Cancer Industry,* pp. 347–348, 441–450.

67. Chowka, "National Cancer Institute"; also his "Cancer 1988," *East West* (December 1987): 47–51.

68. Epstein, "Losing the War," p. 62; compare also David Brown, "Cancer Research Groups' Efforts Called Misdirected," *Washington Post,* February 5, 1992, reporting on a statement by Epstein and 63 others critical of the NCI and ACS; also the response by Walter Lawrence, "Cancer Cause and Prevention," *Washington Post,* February 12, 1992.

69. Emmanuel Farber, "Chemical Carcinogenesis," in *Current Research in Oncology 1972,* edited by C. B. Anfinsen et al. (New York, 1973), p. 97.

70. MacKenzie et al., "Human Costs of Tobacco," p. 979.

71. Critics argued that too little attention had been given to the potential harms of tamoxifen: women participating in the trial were not initially told of the uterine cancer risk, for example. See Cindy Pearson, "NCI Warns Women on Tamoxifen of Risk of Fatal Uterine Cancer," *Network News* (National Women's Health Network) (March–April 1994): 4. Adriane Fugh-Berman notes that although 62 of the 8,000 women receiving the drug are projected to avoid developing breast cancer, this also means that 7,938 women will risk endometrial cancer, blood clots, and potential eye, liver, and gynecological problems with little or no benefit ("The High Risks of Prevention," *The Nation* [December 21, 1992]: 769–772). In May 1994, Cynthia A. Pearson, director of the National Women's Health Network, told a Senate panel that the trial was "more dangerous than any other drug which has ever been studied for long-term prevention" and should be stopped; see Kathy Sawyer, "Breast Cancer Drug Testing Will Continue," *Washington Post,* May 12, 1994. For a review of clinical evidence pro and con, see Eliot Marshall (with reporting by Lisa Seachrist), "Tamoxifen: Hanging in the Balance," *Science* 264 (1994): 1524–1527.

72. Beardsley, "War Not Won," p. 137. Compare also Epstein's article in the September 9, 1987, *Con-*

gressional Record. The NCI responded to Epstein's critique by noting that "since we are primarily a research institute, our view has always been that basic research should have first priority." The institute's neglect of prevention was purportedly due to the fact that, in the years after the passage of the 1971 National Cancer Act, "much of the world felt there were no leads to pursue in prevention" and "too few people were trained in public health research aimed at preventing cancer." See the preamble to the NCI's *National Cancer Institute Response,* pp. 1–2. No author is listed, though one presumes that then-director Vincent T. DeVita must have played a role.

73. Albert S. Braverman, letter, *New England Journal of Medicine* 315 (1986): 967.
74. Kolata, "Is the War on Cancer Being Won?" pp. 543–544.
75. Alisa Solomon, "The Politics of Breast Cancer," *Village Voice* (May 14, 1991): 22–27.
76. A 1983 survey of U.S. medical schools showed that only 66 percent of responding institutions taught occupational health and only 54 percent included this as a required course. See U.S. Congress, Committee on Government Operations, *Occupational Illness and Injuries* (Washington, D.C., 1989), pp. 24–25.
77. Schneiderman, "Sources," p. 460.
78. See, for example, the $2 million award announced by the Prostate Cancer Cure Foundation Ltd. in *Science* 263 (1994): 1298.
79. The $40 million spent on these "equal time" ads probably did more to make people quit than any single policy before or since: U.S. per capita consumption fell from about 4,200 cigarettes in 1966 to less than 4,000 in 1970—though the industry's concession to stop TV advertising and the consequent removal of the "equal time" antismoking ads slowed the downward trend. See Brandt, "The Cigarette," pp. 166–168. Most U.S. domestic airlines barred in-flight smoking in 1990, and in 1994, McDonald's announced a smoking ban in all of its U.S. eateries.
80. Paul Raeburn, "California Anti-Smoke Campaign Sees Results," *USA Today,* March 21, 1994.
81. MacKenzie et al., "Human Costs of Tobacco," p. 979. Tobacco subsidies are complex and confusing; for a review and critique, see Kenneth E. Warner, "The Tobacco Subsidy: Does It Matter?" *Journal of the National Cancer Institute* 80 (1988): 81–83.
82. Rinker Buck, "The Hallucinatory Logic of Tobacco," *Adweek* (July 30, 1990): 12.
83. Eugene Feingold, "Tobacco Tax Hike Would Pay Wide Dividend, *The Nation's Health,* July 1994, p. 2.
84. Marvin Schneiderman, personal communication.
85. Marvin A. Schneiderman, "Legislative Possibilities to Reduce the Impact of Cancer," *Preventive Medicine* 7 (1978): 424–438.
86. Anthony Nero has proposed requiring people selling their homes to pay into a radon remediation insurance fund at the time of sale; home buyers would then receive a free long-term testing kit and an option to have their homes remediated free of charge, the costs being paid from the insurance fund. Nero also argues that radon remediation should be part of commercial building codes and that steps should be taken to ensure that the designs are actually implemented (personal communication).
87. MacKenzie et al., "Human Costs of Tobacco," p. 977.
88. "A.M.A. Assails Nation's Export Policy on Tobacco," *New York Times,* June 27, 1990. Increased attention also needs to be placed on cancers that do not affect Americans: liver and cervical cancer are the most common forms of cancer worldwide, and more aggressive steps could be taken to develop vaccines to the viruses known to contribute to these diseases. See Philip Cole and Yaw Amoateng-Adjepong, "Cancer Prevention: Accomplishments and Prospects," *American Journal of Public Health* 84 (1994): 8–10.
89. Barry I. Castleman, "Building a Future Without Asbestos," unpublished manuscript (1994).
90. D. B. Clayson, "An Overview of Current and Anticipated Methods for Cancer Prevention," *Cancer Letters* 50 (1990): 3–9.
91. Nancy Krieger, "The Making of Public Health Data: Paradigms, Politics, and Policy," *Journal of Public Health Policy* 13 (1992): 412.
92. Schneiderman, "Tsouris," p. 465.

Bibliography

Abelson, Philip H. "Cancer—Opportunism and Opportunity." *Science* 206 (1979): 11.
———. "Cancer Phobia." *Science* 237 (1987): 473.
Agran, Larry. *The Cancer Connection—And What We Can Do About It.* Boston: Houghton Mifflin, 1977.
American Cancer Society. *Cancer Facts and Figures for Minority Americans—1991.* New York: American Cancer Society, 1991.
———. *Cancer Facts and Figures—1994.* New York: American Cancer Society, 1994.
American Industrial Health Council. *A Reply to 'Estimates of the Fraction of Cancers in the United States Attributable to Occupational Factors'.* New York, 1978.
———. *Cancer, Pollution and the Workplace* (Washington, D.C., 1983).
Ames, Bruce N. Testimony before the State of California Department of Industrial Relations, Occupational Safety and Health, Dibromochloropropane Inquiry, October 18, 1977.
———. "Dietary Carcinogens and Anticarcinogens." *Science* 221 (1983): 1256–1264.
———. "Mother Nature Is Meaner Than You Think," *Science 84* (July–August 1984): 98.
Ames, Bruce N., and Lois S. Gold. "Pesticides, Risks and Applesauce." *Science* 244 (1989): 755–757.
———. "Too Many Rodent Carcinogens: Mitogenesis Increases Mutagenesis." *Science* 249 (1990): 970–971.
———. "Chemical Carcinogenesis: Too Many Rodent Carcinogens." *Proceedings of the National Academy of Sciences* 87 (1990): 7772–7781.
———. "Response to Perera." *Science* 250 (1990): 1645.
Ames, Bruce N., R. Magaw, and L. S. Gold. "Ranking Possible Carcinogenic Hazards." *Science* 236 (1987): 271–280.
Ames, Bruce N., Margie Profet, and Lois S. Gold. "Nature's Chemicals and Synthetic Chemicals: Comparative Toxicology." *Proceedings of the National Academy of Sciences* 87 (1990): 7782–7786.
Anderson, Alan, Jr. "The Politics of Cancer: How Do You Get the Medical Establishment to Listen?" *New York* (July 29, 1974).
Angier, Natalie. *Natural Obsessions: Striving to Unlock the Deepest Secrets of the Cancer Cell.* New York: Warner Books, 1988.
———. "Cigarettes Trigger Lung Cancer Gene, Researchers Find." *New York Times,* August 21, 1990.
Archer, Victor E. "Effects of Low-Level Radiation: A Critical Review." *Nuclear Safety* 21 (1980): 68–82.
———, Harold J. Magnuson, Duncan A. Holaday, and Pope A. Lawrence. "Hazards to Health in Uranium Mining and Milling." *Journal of Occupational Medicine* 4 (1962): 55–60.
Austoker, Joan. *A History of the Imperial Cancer Research Fund 1902–1986.* Oxford: Oxford University Press, 1988.

Bailar, John C., III. "Re-Thinking the War on Cancer." *Issues in Science and Technology* (Fall 1987): 16–21.

———. "Some Recent Trends in Cancer: Cancer Undefeated." Unpublished manuscript (1994).

Bailar, John C. III, and Elaine M. Smith. "Progress Against Cancer?" *New England Journal of Medicine* 314 (1986): 1226–1232.

Bailar, John C. III, et al. "One-Hit Models of Carcinogenesis: Conservative or Not?" *Risk Analysis* 8 (1988): 485–497.

Ball, Howard. *Cancer Factories: America's Tragic Quest for Uranium Self-Sufficiency.* Westport, Conn.: Greenwood Press, 1993.

Beardsley, Tim. "A War Not Won." *Scientific American* (January 1994): 130–138.

Becker, Nikolaus, Elaine M. Smith, and Jürgen Wahrendorf. "Time Trends in Cancer Mortality in the Federal Republic of Germany: Progress Against Cancer?" *International Journal of Cancer* 43 (1989): 245–249.

Bergin, Edward J., and Ronald E. Grandon. *How to Survive in Your Toxic Environment.* New York: Avon Books, 1984.

Berman, Daniel M. *Death on the Job: Occupational Health and Safety Struggles in the United States.* New York: Monthly Review Press, 1978.

Bernstein, Geoffrey. "Cascade: Worker Resistance at the Portsmouth Uranium Plant." Senior thesis, Harvard University, 1981.

Bishop, Jerry E., and Michael Waldholz. *Genome.* New York: Simon and Schuster, 1990.

Blome, Kurt. "Krebsforschung und Krebsbekämpfung." *Die Gesundheitsführung (Ziel und Weg)* 10 (1940): 406–412.

Bloom, A. D., et al., eds. *Genetic Susceptibility to Environmental Mutagens and Carcinogens.* White Plains, N.Y.: March of Dimes, 1989.

Blum, Arlene, and Bruce N. Ames. "Flame-Retardant Additives as Possible Cancer Hazards." *Science* 195 (1977): 17–23.

Boal, Iain. "The Rhetoric of Risk," *PsychoCulture* (Spring 1994): 7–10.

Bodansky, David, Maurice A. Robkin, and David R. Stadler, eds. *Indoor Radon and Its Hazards.* Seattle: University of Washington Press, 1987.

Bodmer, Walter F. "Cancer Genetics." In *Inheritance of Susceptibility to Cancer in Man,* edited by Walter F. Bodmer. Oxford: Oxford University Press, 1982.

Bollier, David. "Leading Scientist Laughs at DDT, Worries About Peanut Butter, Believe It or Not!" *Public Citizen* (September–October 1988): 12–20.

———, and Joan Claybrook. *Freedom from Harm: The Civilizing Influence of Health, Safety and Environmental Regulation.* Washington, D.C.: Public Citizen and Democracy Project, 1986.

Brandt, Allan M. "The Cigarette, Risk, and American Culture." *Daedalus* (Fall 1990): 155–176.

Brenner, David J. *Radon: Risk and Remedy.* New York: W. H. Freeman, 1989.

Breslow, Lester. "From Cancer Research to Cancer Control." In *Advances in Cancer Control: Research and Development,* edited by Paul F. Engstrom et al. New York: Alan R. Liss, 1983.

Broad, William J. "New Study Questions Hiroshima Radiation." *New York Times,* October 13, 1992.

Brodeur, Paul. *Expendable Americans.* New York: Viking Press, 1974.

———. *Outrageous Misconduct: The Asbestos Industry on Trial.* New York: Pantheon Books, 1985.

———. *Currents of Death: Power Lines, Computer Terminals, and the Attempt to Cover Up Their Threat to Your Health.* New York: Simon and Schuster, 1989.

———. "The Cancer at Slater School." *The New Yorker* (December 7, 1992): 86–119.

Brookes, Warren T. "The Wasteful Pursuit of Zero Risk." *Forbes* (April 30, 1990): 160–172.

Brooks, Paul. *The House of Life: Rachel Carson at Work.* New York: Houghton Mifflin, 1972.

Brugge, Joan, et al., eds. *Origins of Human Cancer: A Comprehensive Review.* Cold Spring Harbor, N.Y.: Cold Spring Harbor Laboratory, 1991.

Burdette, W. J. "The Significance of Mutation in Relation to Origin of Tumors." *Cancer Research* 15 (1955): 201–226.

Butlin, Henry T. "Three Lectures on Cancer of the Scrotum in Chimney Sweeps and Others." *British Medical Journal* 1 (1892): 1341–1346, and 2 (1892): 1–6, 66–71.

Cairns, John. "The Cancer Problem." *Scientific American* (November 1975): 64–78.
———. *Cancer: Science and Society.* San Francisco: W. H. Freeman, 1978.
———. "The Treatment of Diseases and the War Against Cancer." *Scientific American* (November 1985): 51–59.
Califano, Joseph A. "Occupational Safety and Health." *Vital Speeches of the Day* (October 1, 1978): 738–741.
Cannon, Lou. *Reagan.* New York: G. P. Putnam's Sons, 1982.
Cannon-Albright, Lisa A., et al. "Common Inheritance of Susceptibility to Colonic Adenomatous Polyps and Associated Colorectal Cancers." *New England Journal of Medicine* 319 (1988): 533–537.
Carlson, Elof A. *Genes, Radiation, and Society: The Life and Work of H. J. Muller.* Ithaca: Cornell University Press, 1981.
Carson, Rachel. *Silent Spring.* Boston: Houghton Mifflin, 1962.
Castleman, Barry I. *Asbestos: Medical and Legal Aspects.* 3rd ed. Englewood Cliffs, N.J.: Prentice-Hall, 1990.
Caufield, Catherine. *Multiple Exposures: Chronicles of the Radiation Age.* New York: Harper & Row, 1989.
Cavanaugh, Gerald. "The Corruptions of Science." *Dissent* (Spring 1981): 195–202.
Chowka, Peter B. "The National Cancer Institute and the Fifty-Year Cover-Up." *East West* (January 1978): 22–27.
Claus, E. B., N. Risch, and W. D. Thompson. "Genetic Analysis of Breast Cancer in the Cancer and Steroid Study." *American Journal of Human Genetics* 48 (1991): 232–242.
Claybrook, Joan. *Retreat from Safety: Reagan's Attack on America's Health.* New York: Pantheon Books, 1984.
Clemmesen, Johannes, ed. *Symposium on Geographical Pathology and Demography of Cancer.* Oxford: Oxford University Press, 1950.
———. *Statistical Studies in the Aetiology of Malignant Neoplasms.* Vol. 1, *Review and Results.* Copenhagen: Munksgaard, 1965.
Clunet, Jean. *Recherches expérimentales sur les tumeurs malignes.* Paris: G. Steinheil, 1910.
Cohen, Bernard L. "Health Effects of Radon from Insulation of Buildings." *Health Physics* 39 (1980): 937–941.
Cole, Leonard A. *Element of Risk: The Politics of Radon.* Washington, D.C.: American Association for the Advancement of Science, 1993.
Commoner, Barry. *Making Peace with the Planet.* New York: Pantheon, 1990.
Cook-Deegan, Robert Mullan. "The Alta Summit, December 1984." *Genomics* 5 (1989): 661–663.
Cope, John. *Cancer: Civilization: Degeneration.* London: H. K. Lewis, 1932.
Cranor, Carl F. *Regulating Toxic Substances: A Philosophy of Science and the Law.* New York: Oxford University Press, 1993.
Crawford, James. "'Death by a Thousand Cuts': The Dismantling of OSHA." *The Nation* (September 12, 1981): 205–208.
Cross, Fredrick T., ed. *Indoor Radon and Lung Cancer: Reality or Myth?* Columbus, Ohio: Battelle Press, 1992.
Cummings, K. Michael, et al. "What Scientists Funded by the Tobacco Industry Believe About the Hazards of Cigarette Smoking." *American Journal of Public Health* 81 (1991): 894–896.
Davis, Devra Lee. "Cancer in the Workplace: The Case for Prevention." *Environment* 23 (1981): 25–37.
———. "Trends in Nonsmoking Lung Cancer." *Epidemiology* 4 (1993): 489–492.
Davis, Devra Lee, and Brian H. Magee. "Cancer and Industrial Chemical Production" (letter). *Science* 206 (1979): 1356–1357.
Davis, Devra Lee, Kenneth Bridbord, and Marvin Schneiderman. "Cancer Prevention: Assessing Causes, Exposures, and Recent Trends in Mortality for U.S. Males 1968–1978." *Teratogenesis, Carcinogenesis, and Mutagenesis* 2 (1982): 105–135.
Davis, Devra Lee, et al. "Increasing Trends in Some Cancers in Older Americans: Fact or Artifact?" *International Journal of Health Services* 18 (1988): 35–68.
———. "International Trends in Cancer Mortality in France, West Germany, Italy, Japan, England and Wales, and the USA." *Lancet* 336 (1990): 474–481.

Day, Nicholas E., and Charles C. Brown. "Multistage Models and Primary Prevention of Cancer." *Journal of the National Cancer Institute* 64 (1980): 977–989.

Deisler, Paul F., ed. *Reducing the Carcinogenic Risks in Industry.* New York: Marcel Dekker, 1984.

Demopoulos, H. B., and M. A. Mehlman, eds. *Cancer and the Environment: An Academic Review of the Environmental Determinants of Cancer Relevant to Prevention.* Park Forest South, Ill.: Pathotox, 1980.

Department of Energy, Office of Health and Environmental Research. *Human Genome: 1989–90 Program Report.* Springfield, Va.: National Technical Information Service, 1990.

"Diseased Regulation." *Forbes* (February 19, 1979): 34.

Doll, Richard. *Prevention of Cancer: Pointers from Epidemiology.* London: Nuffield Provincial Hospitals Trust, 1967.

———. "Strategy for Detection of Cancer Hazards to Man." *Nature* 265 (1977): 589–596.

———. "An Overview of the Epidemiological Evidence Linking Diet and Cancer." *Proceedings of the Nutrition Society* 49 (1990): 119–131.

———. "Are We Winning the Fight Against Cancer? An Epidemiological Assessment." *European Journal of Cancer* 26 (1990): 500–508.

———, and Richard Peto. *The Causes of Cancer.* Oxford: Oxford University Press, 1981.

Doyal, Lesley, and Samuel S. Epstein. *Cancer in Britain: The Politics of Prevention.* London: Pluto Press, 1983.

Dublin, Louis I., and Alfred J. Lotka. *Twenty-Five Years of Health Progress.* New York: Metropolitan Life Insurance, 1937.

Dulbecco, Renato. "A Turning Point in Cancer Research: Sequencing the Human Genome." In *Viruses and Human Cancer,* edited by Robert C. Gallo et al. New York: Alan R. Liss, 1987, pp. 1–14.

Efron, Edith. *The Apocalyptics: Cancer and the Big Lie—How Environmental Politics Controls What We Know About Cancer.* New York: Simon and Schuster, 1984.

Egginton, Joyce. "Menace of Whispering Hills." *Audubon* (January 1989): 28–35.

Eisenbud, Merril. "Environmental Causes of Cancer." *Environment* 20 (October 1978): 6–16.

———. *An Environmental Odyssey.* Seattle: University of Washington Press, 1990.

Environmental Protection Agency. *Respiratory Health Effects of Passive Smoking.* Washington, D.C.: Environmental Protection Agency, 1992.

Epstein, Samuel S. "Control of Chemical Pollutants." *Nature* 228 (1970): 816–819.

———. "Polluted Data." *The Sciences* (July–August 1978): 16–21.

———. *The Politics of Cancer.* New York: Anchor Press, 1979.

———. "Losing the War Against Cancer: Who's to Blame and What to Do About It." *International Journal of Health Services* 20 (1990): 53–71.

Epstein, Samuel S., Lester O. Brown, and Carl Pope. *Hazardous Waste in America.* San Francisco: Sierra Club Books, 1982.

Epstein, Samuel S., and Joel Swartz. "Fallacies of Lifestyle Cancer Theories." *Nature* 289 (1981): 127–130.

Epstein, Samuel S., and Joel Swartz, et al. "Carcinogenic Risk Estimation." *Science* 240 (1988): 1043–1045.

"Estimates of the Fraction of Cancer in the United States Related to Occupational Factors." National Cancer Institute-National Institute of Environmental Health Sciences, National Institute of Occupational Safety and Health. Washington, D.C., September 15, 1978. Reprinted as an appendix in *Banbury Report 9. Quantification of Occupational Cancer,* edited by Richard Peto and Marvin Schneiderman. Cold Spring Harbor, N.Y.: Cold Spring Harbor Laboratory, 1981.

Fabrikant, Jacob I., et al. *Health Risks of Radon and Other Internally Deposited Alpha-Emitters* (BEIR IV). Washington, D.C.: National Academy Press, 1988.

Fischer-Wasels, Bernhard. *Die Vererbung der Krebskrankheit.* Berlin: A. Metzner, 1935.

Foster, Kenneth R., David E. Bernstein, and Peter W. Huber, eds. *Phantom Risk: Scientific Inference and the Law.* Cambridge, Mass.: MIT Press, 1993.

Freedman, Alex M., and Laurie P. Cohen. "Smoke and Mirrors: How Cigarette Makers Keep Health Question 'Open' Year After Year." *Wall Street Journal,* February 11, 1993.

Friedman, Kenneth M. *Public Policy and the Smoking-Health Controversy.* Lexington, Mass.: D. C. Heath, 1975.

Fumento, Michael. "The Asbestos Rip-Off." *American Spectator* (October 1989): 21–26.
———. "The Politics of Cancer Testing." *American Spectator* (August 1990): 18–23.
———. *Science Under Siege: Balancing Technology and the Environment.* New York: William Morrow, 1993.
Galen, Michele. "Nowhere to Run from Radon." *The Nation* (February 14, 1987): 180–182.
Gehrmann, G. H. "The Carcinogenetic Agent—Chemistry and Industrial Aspects." *Journal of Urology* 31 (1934): 126–137.
General Accounting Office. *Indoor Air Pollution: An Emerging Health Problem.* Washington, D.C.: Government Printing Office, 1980.
Gibbons, Ann. "Does War on Cancer Equal War on Poverty?" *Science* 253 (1991): 260.
Gofman, John W. *Radiation and Human Health.* Updated and abridged. New York: Pantheon, 1983.
———. *Radiation-Induced Cancer from Low-Dose Exposure.* San Francisco: Committee for Nuclear Responsibility, 1990.
Gottschalk, Jack, ed. *Crisis Response: Inside Stories on Managing Image Under Siege.* Detroit: Gale Research, 1993.
Gough, Michael. "How Much Cancer Can EPA Regulate Away?" *Risk Analysis* 10 (1990): 1–6.
Graham, Frank, Jr. *Since Silent Spring.* Greenwich, Conn.: Fawcett, 1970.
Green, Milton. "Scientific McCarthyism." *Chemtech* (July 1981): 402–406.
Greenberg, Daniel S. "A Critical Look at Cancer Coverage." *Columbia Journalism Review* (January–February 1975): 40–44.
———. "Severest Critic Gets a Hearing on the War on Cancer." *Science & Government Report* (May 15, 1992): 1–4.
———, and Judith E. Randal. "Waging the Wrong War on Cancer." *Washington Post,* May 1, 1977.
Griffin, G. Edward. *World Without Cancer: The Story of Vitamin B_{17}.* Westlake Village, Calif.: American Media, 1974.
Griffin, Melanie. "Setting the Record Straight" (on the Council for Energy Awareness), *Sierra* (March–April 1986): 23–28.
Haagensen, Cushman D. "Occupational Neoplastic Disease." *American Journal of Cancer* 15, part 1 (1931): 641–703.
———. "An Exhibit of Important Books, Papers, and Memorabilia Illustrating the Evolution of the Knowledge of Cancer." *American Journal of Cancer* 18 (1933): 42–126.
Hacker, Barton C. *The Dragon's Tail: Radiation Safety in the Manhattan Project, 1942–1946.* Berkeley: University of California Press, 1987.
Harley, John H. "Radon Is Out." In *Indoor Radon and Lung Cancer: Reality or Myth?* edited by Fredrick T. Cross. Columbus, Ohio: Battelle Press, 1992, pp. 741–763.
Harnik, Peter. *Voodoo Science, Twisted Consumerism: The Golden Assurances of the American Council on Science and Health.* Washington, D.C.: Center for Science in the Public Interest, 1982.
Harris, Curtis C. "P53: At the Crossroads of Molecular Carcinogenesis and Risk Assessment." *Science* 262 (1993): 1980–1981.
Härting, F. H., and W. Hesse. "Der Lungenkrebs, die Bergkrankheit in den Schneeberger Gruben." *Vierteljahrsschrift für gerichtliche Medizin* 30 (1879): 296–309, and 31 (1879): 102–132, 313–337.
Haviland, Alfred. "The Medical Geography of Cancer in England and Wales." *The Practitioner* 62 (1899): 400–417.
Heath, Robert L., and Associates, eds. *Strategic Issues Management: How Organizations Influence and Respond to Public Interests and Policies.* San Francisco: Jossey-Bass, 1988.
Henderson, Brian E., et al. "Toward the Primary Prevention of Cancer." *Science* 254 (1991): 1131–1138.
Henschler, Dietrich. "Exposure Limits: History, Philosophy, Future Developments." *Annals of Occupational Hygiene* 28 (1984): 79–92.
Herber, Lewis [Murray Bookchin]. *Our Synthetic Environment.* New York: Alfred A. Knopf, 1962.
Hess, John L. "Harvard's Sugar-Pushing Nutritionist." *Saturday Review* (August 1978): 10–14.
Hiatt, H. H., J. D. Watson, and J. A. Winsten, eds. *Origins of Human Cancer.* 3 vols. Cold Spring Harbor, N.Y.: Cold Spring Harbor Laboratory, 1977.

Higginson, John. "Present Trends in Cancer Epidemiology." *Proceedings of the Canadian Cancer Conference* 8 (1969): 40–75.

———, and Calum S. Muir. "The Role of Epidemiology in Elucidating the Importance of Environmental Factors in Human Cancer." *Cancer Detection and Prevention* 1 (1976): 79–105.

———. "Environmental Carcinogenesis: Misconceptions and Limitations to Cancer Control." *Journal of the National Cancer Institute* 63 (1979): 1291.

Hilgartner, Stephen. "The Political Language of Risk: Defining Occupational Health." In *The Language of Risk,* edited by Dorothy Nelkin. Beverly Hills, Calif.: Sage, 1985.

———, et al. *Nukespeak: The Selling of Nuclear Technology in America.* Harmondsworth: Penguin Books, 1983.

Hirsch, August. *Handbook of Geographical and Historical Pathology.* Vol. 3. London: New Sydenham Society, 1886.

Hoel, David G., et al. "Trends in Cancer Mortality in 15 Industrialized Countries, 1969–1986." *Journal of the National Cancer Institute* 84 (1992): 313–320.

Hoffman, Frederick L. *The Mortality from Cancer Throughout the World.* Newark: Prudential Press, 1915.

———. *Cancer and Diet.* Baltimore, Md.: Williams and Wilkins, 1937.

Holaday, Duncan A. "History of the Exposure of Miners to Radon." *Health Physics* 16 (1969): 547–552.

———, et al. *Control of Radon and Daughters in Uranium Mines and Calculations on Biologic Effects.* Washington, D.C.: U.S. Public Health Service, 1957.

Hounshell, David A., and John Kenly Smith. *Science and Corporate Strategy: Du Pont R&D, 1902–1980.* Cambridge: Cambridge University Press, 1988.

Howe, G. Melvyn, ed. *Global Geocancerology: A World Geography of Human Cancer.* Edinburgh: Churchill Livingstone, 1986.

Huber, Peter W. *Galileo's Revenge: Junk Science in the Courtroom* New York: Basic Books, 1991.

Hueper, Wilhelm C. "Krankheit und Rasse." *Rasse* 3 (1936): 41–55.

———. *Occupational Tumors and Allied Diseases.* Springfield, Ill.: Charles C. Thomas, 1942.

———. "Significance of Industrial Cancer in the Problem of Cancer." *Occupational Medicine* 2 (1946): 190–200.

———. "Environmental and Occupational Cancer." *Public Health Reports,* supp. 209 (1948).

———. "Recent Developments in Environmental Cancer." *AMA Archives of Pathology* 58 (1954): 475–523.

———. "The Cigarette Theory of Lung Cancer." *Current Medical Digest* (October 1954): 35–39.

———. "Causal and Preventive Aspects of Environmental Cancer." *Minnesota Medicine* (January 1956): 5–22.

———. "Adventures of a Physician in Occupational Cancer: A Medical Cassandra's Tale" (1976). Unpublished autobiography in the Hueper Papers, National Library of Medicine.

Hunter, Donald. *The Diseases of Occupations.* 6th ed. Boston: Little, Brown, 1978.

Hunter, John. *Lectures on the Principles of Surgery.* Philadelphia: Haswell, Barrington, and Haswell, 1839.

International Labour Organisation. *Prevention of Occupational Cancer–International Symposium.* Geneva: International Labour Office, 1982.

Ivins, Molly. "Uranium Mines in West Leave Deadly Legacy." *New York Times,* May 20, 1979.

Jackowitz, Anne R. "Radon's Radioactive Ramifications: How Federal and State Governments Should Address the Problem." *Boston College Law Review: Environmental Affairs* 16 (1988): 329–381.

Jackson, Kenneth L., Joseph P. Geraci, and David Bodansky. "Observations of Lung Cancer: Evidence Relating Lung Cancer to Radon Exposure." In *Indoor Radon and Its Hazards,* edited by David Bodansky et al. Seattle: University of Washington Press, 1987.

Jacobson, Michael F. "Diet and Cancer" (letter). *Science* 207 (1980): 258–259.

Kahn, Patricia. "A Grisly Archive of Key Cancer Data." *Science* 259 (1993): 448–451.

Kazis, Richard, and Richard L. Grossman. *Fear at Work: Job Blackmail, Labor and the Environment.* New York: Pilgrim Press, 1982.

King, George, and Arthur Newsholme. "On the Alleged Increase of Cancer." *Proceedings of the Royal Society of London* 54 (1893): 209–242.

Kinzler, Kenneth W., et al. "Identification of a Gene Located at Chromosome 5q21 that Is Mutated in Colorectal Cancers." *Science* 251 (1991): 1366–1370.

Knudson, Alfred G. "Genetic Predisposition to Cancer." *Cancer Detection and Prevention* 7 (1984): 1–8.

———, Louise C. Strong, and David E. Anderson. "Heredity and Cancer in Man." *Progress in Medical Genetics* 9 (1973): 113–158.

Kothari, Manu L., and Lopa A. Mehta. *Cancer: Myths and Realities of Cause and Cure.* London: Marion Boyars, 1979.

Krimsky, Sheldon, and Alonzo Plough. *Environmental Hazards: Communicating Risks as a Social Process.* Dover, Mass.: Auburn House, 1988.

Kurtz, Howard. "Dr. Whelan's Media Operation." *Columbia Journalism Review* (March–April 1990): 43–47.

Landrigan, Philip J. "Commentary: Environmental Disease—A Preventable Epidemic." *American Journal of Public Health* 82 (1992): 941–943.

Lane, David P. "A Death In the Life of p53." *Nature* 362 (1993): 786–787.

Lappé, Marc. *Genetic Politics: The Limits of Biological Control.* New York: Simon and Schuster, 1979.

———. *Chemical Deception: The Toxic Threat to Health and the Environment.* San Francisco: Sierra Club Books, 1991.

Latarjet, R. "Radiation Carcinogenesis and Radiation Protection." *The Cancer Journal* 5 (1992): 23–27.

Le Conte, John. "On Carcinoma in General, and Cancer of the Stomach." *New York Lancet* 2 (1842): 284–287, 299–304.

———. "Statistical Researches on Cancer." *Southern Medical and Surgical Journal* 2 (1846): 257–293.

———. "Vital Statistics and the True Coefficient of Mortality, Illustrated by Cancer. *Biennial Report of the State Board of Health of California.* Sacramento: California State Board of Health, 1888, pp. 1–19.

Lehr, Jay H. *Rational Readings on Environmental Concerns.* New York: Van Nostrand Reinhold, 1992.

Leppert, M., et al. "Genetic Analysis of an Inherited Predisposition to Colon Cancer in a Family with a Variable Number of Adenomatous Polyps." *New England Journal of Medicine* 322 (1990): 904–908.

Lester, Stephen U. "Mothers: Throw Away Your Peanut Butter!" *Everyone's Backyard* (Citizen's Clearinghouse for Toxic Wastes, Arlington, Va.) (Spring 1989): 6–8.

Lewis, Julian H. *The Biology of the Negro.* Chicago: University of Chicago Press, 1942.

Lewontin, Richard C. *Biology as Ideology: The Doctrine of DNA.* Concord, Ont.: House of Anansi Press, 1991.

Liek, Erwin. *Krebsverbreitung, Krebsbekämpfung, Krebsverhütung.* Munich: J. F. Lehmann, 1932.

Lilienfeld, David E. "The Silence: The Asbestos Industry and Early Occupational Cancer Research." *American Journal of Public Health* 81 (1991): 791–800.

Lindee, M. Susan. *Suffering Made Real: American Science and the Survivors at Hiroshima.* Chicago: University of Chicago Press, 1994.

Little, Clarence Cook. *Report of the Scientific Director.* New York: Tobacco Industry Research Committee, 1957.

Lorenz, Egon. "Radioactivity and Lung Cancer: A Critical Review of Lung Cancer in the Miners of Schneeberg and Joachimsthal." *Journal of the National Cancer Institute* 5 (1944): 1–15.

Love, Robert. "Indecent Exposure: Reagan Guts the Rules on Cancer Protection." *Rolling Stone* (June 7, 1984): 8–9, 58–61.

Love, Susan M. *Dr. Susan Love's Breast Book.* Reading, Mass.: Addison-Wesley, 1990.

Luckey, T. Don. *Hormesis with Ionizing Radiation.* Boca Raton, Fla.: CRC Press, 1980.

Ludewig, P., and S. Lorenser. "Untersuchung der Grubenluft in den Schneeberger Gruben auf den Gehalt an Radiumemanation." *Zeitschrift für Physik* 22 (1924): 178–185.

Ludlam, Charles E. *Undermining Public Protections: The Reagan Administration Regulatory Program.* Washington, D.C.: Alliance for Justice, 1981.

MacKenzie, Thomas D., Carl E. Bartecchi, and Robert W. Schrier. "The Human Costs of Tobacco Use." *New England Journal of Medicine* 330 (1994): 975–980.

MacLaury, Judson. "The Occupational Safety and Health Administration: A History of Its First Thirteen Years: 1971–1984." Unpublished manuscript available through the U.S. Department of Labor.

MacLennan, Carol A. "From Accident to Crash: The Auto Industry and the Politics of Injury." *Medical Anthropology Quarterly,* n.s. 2 (1988): 233–250.

Majumdar, Shyamal, Robert F. Schmalz, and E. Willard Miller. *Environmental Radon: Occurrence, Control, and Health Hazards.* Easton: Pennsylvania Academy of Science, 1990.

Mancuso, Thomas F., Alice Stewart, and George Kneale. "Radiation Exposures of Hanford Workers Dying from Cancer and Other Causes." *Health Physics* 33 (1977): 369–385.

Mantel, Nathan. "The Concept of Threshold in Carcinogenesis." *Clinical Pharmacology and Therapeutics* 4 (1963): 104–109.

Marconi, Joe. *Crisis Marketing: When Bad Things Happen to Good Companies.* Chicago: American Marketing Association, 1992.

Marshall, Eliot. "Tobacco Science Wars." *Science* 236 (1987): 250–251.

———. "Experts Clash over Cancer Data." *Science* 250 (1990): 900–902.

———. "A Is for Apple, Alar, and . . . Alarmist?" *Science* 254 (1991): 20–22.

———. "The Politics of Breast Cancer." *Science* 259 (1993): 616–617.

Marx, Jean. "Burst of Publicity Follows Cancer Report." *Science* 230 (1985): 1367–1368.

———. "Animal Carcinogen Testing Challenged." *Science* 250 (1990): 743–745.

———. "Possible New Colon Cancer Gene Found." *Science* 251 (1991): 1317.

Mason, T. J. *Atlas of Cancer Mortality for U.S. Counties: 1950–1969.* Washington, D.C.: U.S. Department of Health, Education and Welfare, 1975.

Maugh, Thomas H., II. "Chemical Carcinogens: How Dangerous Are Low Doses?" *Science* 202 (1978): 37–41.

———. "Industry Council Challenges HEW on Cancer in the Workplace." *Science* 202 (1978): 602–604.

———. "Cancer and Environment: Higginson Speaks Out." *Science* 205 (1979): 1363–1366.

———, and Jean L. Marx. *Seeds of Destruction: The* Science *Report on Cancer Research.* New York: Plenum, 1975.

Mazur, Allan. *The Dynamics of Technical Controversy.* Washington, D.C.: Communications Press, 1981.

———. "Putting Radon on the Public's Risk Agenda." *Science, Technology, and Human Values* 12 (1987): 86–93.

McAuliffe, Sharon, and Kathleen McAuliffe. "The Genetic Assault on Cancer." *New York Times Magazine* (October 24, 1982): 38–54.

McCally, Michael. "What the Fight Is All About." *Bulletin of the Atomic Scientists* (September 1990): 11–14.

McKeown, Thomas. *The Role of Medicine: Dream, Mirage, or Nemesis?* Princeton, N.J.: Princeton University Press, 1979.

Mead, Nathaniel. "The Riddle of Radon." *East West* (July 1990): 64–71, 110.

Metzger, H. Peter. "'Dear Sir: Your House Is Built on Radioactive Uranium Waste'." *New York Times Magazine* (October 31, 1971): 14–65.

Michaels, David. "Waiting for the Body Count: Corporate Decision Making and Bladder Cancer in the U.S. Dye Industry." *Medical Anthropology Quarterly,* n.s. 2 (1988): 215–232.

Mix, Jerry. "The Truth Comes Out." *Pest Control* (January 1989): 50–52.

Moore, Charles H. "The Antecedent Conditions of Cancer." *British Medical Journal* (August 26, 1865): 201–210.

Moss, Ralph W. *The Cancer Syndrome* (1980). 2nd. rev. ed. published as *The Cancer Industry.* New York: Paragon House, 1991.

Mossman, Brooke T., and J. Bernard L. Gee. "Asbestos-Related Diseases." *New England Journal of Medicine* 320 (1989): 1721–1730.

Murray, R. "Asbestos: A Chronology of Its Origins and Health Effects." *British Journal of Industrial Medicine* 47 (1990): 361–365.

Nakamura, Y., et al. "Localization of the Genetic Defect in Familial Adenomatous Polyposis Within a Small Region of Chromosome 5." *American Journal of Human Genetics* 43 (1988): 638–644.

Napier, Kristine, ed. *Issues in the Environment.* New York: American Council on Science and Health, 1992.

National Cancer Institute. *National Cancer Institute Response to Article by Dr. Samuel S. Epstein.* Bethesda, Md.: NCI, n.d.

National Council on Radiation Protection and Measurements. *Exposures from the Uranium Series with Emphasis on Radon and Its Daughters.* Bethesda, Md.: NCRP, 1984.

———. *Radon Exposure of the U.S. Population—Status of the Problem.* Bethesda, Md.: NCRP, 1991.

National Research Council. *Indoor Air Pollutants.* Washington, D.C.: National Academy Press, 1981.

———. *Diet, Nutrition, and Cancer.* Washington, D.C.: National Academy Press, 1982.

Nelkin, Dorothy. *Selling Science: How the Press Covers Science and Technology.* New York: W. H. Freeman, 1987.

Nero, Anthony V. "The Indoor Radon Story." *Technology Review* (January 1986): 28–40.

———. "A National Strategy for Indoor Radon." *Issues in Science and Technology* (Fall 1992): 33–40.

Newsholme, Arthur. "The Statistics of Cancer." *The Practitioner* 62 (1899): 371–384.

Noble, Charles. *Liberalism at Work: The Rise and Fall of OSHA.* Philadelphia: Temple University Press, 1986.

Noble, David F. "The Chemistry of Risk." *Seven Days* (June 5, 1979): 23–34.

Office of Technology Assessment (OTA). *Technologies for Determining Cancer Risks from the Environment.* Washington, D.C.: OTA, 1981.

Olson, James S. *The History of Cancer: An Annotated Bibliography.* New York: Greenwood Press, 1989.

Ozonoff, David. "The Political Economy of Cancer Research." *Science and Nature,* no. 2 (1979): 13–18.

———. "Public Relations Cancer." *Society* (March–April 1981): 10–14.

———. "Failed Warnings: Asbestos-Related Disease and Industrial Medicine." In *The Health and Safety of Workers,* edited by Ronald Bayer. New York: Oxford University Press, 1988.

———, and Leslie I. Boden. "Truth and Consequences: Health Agency Responses to Environmental Health Problems." *Science, Technology, and Human Values* 12 (1987): 70–77.

Paget, James. *Lectures on Surgical Pathology.* 3rd American ed. Philadelphia: Lindsay & Blakiston, 1865.

Paigen, Beverly. "Controversy at Love Canal." *Hastings Center Report* (March 1982): 29–37.

Papanicolaou, George N. "New Cancer Diagnosis." In *Third Race Betterment Conference Proceedings.* Battle Creek, Mich.: Race Betterment Foundation, 1928, pp. 528–534.

Patterson, James T. *The Dread Disease: Cancer and Modern American Culture.* Cambridge, Mass.: Harvard University Press, 1987.

Paull, Jeffrey M. "The Origin and Basis of Threshold Limit Values." *American Journal of Industrial Medicine* 5 (1984): 227–238.

Peller, Sigismund. *Cancer Research Since 1900: An Evaluation.* New York: Philosophical Library, 1979.

Peltomäki, Päivi, and Lauri A. Aaltonen et al. (including de la Chapelle and Vogelstein). "Genetic Mapping of Locus Predisposing to Human Colorectal Cancer." *Science* 260 (1993): 810–812.

Perera, Frederica P. "Carcinogens and Human Health: Part 1" (letter). *Science* 250 (1990): 1644–1645.

———, and Paolo Boffetta. "Perspectives on Comparing Risks of Environmental Carcinogens. *Journal of the National Cancer Institute* 80 (1988): 1282–1293.

Perera, Frederica P. "Molecular and Genetic Damage in Humans from Environmental Pollution in Poland." *Nature* 360 (1992): 256–258.

Peters, George A., and Barbara J. Peters. *Sourcebook on Asbestos Diseases: Medical, Legal, and Engineering Aspects.* 3 vols. New York: Garland STPM Press, 1980.

Peto, Richard. "Distorting the Epidemiology of Cancer: The Need for a More Balanced View." *Nature* 284 (1980): 297–300.

———. "The Need for Ignorance in Cancer Research." In *The Encyclopedia of Medical Ignorance,* edited by Ronald Duncan and Miranda Weston-Smith. Oxford: Pergamon Press, 1984.

———, and Marvin Schneiderman, eds. *Banbury Report 9. Quantification of Occupational Cancer.* Cold Spring Harbor, N.Y.: Cold Spring Harbor Laboratory, 1981.

Petrakis, Nicholas L. "Historic Milestones in Cancer Epidemiology." *Seminars in Oncology* 6 (1979): 433–444.

Pienta, Kenneth J., Alan W. Partin, and Donald S. Coffey. "Cancer as a Disease of DNA Organization and Dynamic Cell Structure." *Cancer Research* 49 (1989): 2525–2532.

Piller, Charles. "Scientist Questions Modern Cancer Hazards." *Synapse* (May 21, 1987): 1–8.

———, "Experts Refute Ames' Theories." *Synapse* (May 28, 1987): 1–2, 11.

Pollay, Richard W. "Propaganda, Puffing and the Public Interest." *Public Relations Review* 16 (1990): 39–54.

Pool, Robert. "Is There an EMF-Cancer Connection?" *Science* 249 (1990): 1096–1098.

Postrel, Virginia I. "Of Mice and Men" (Interview with Bruce Ames). *Reason* (December 1991): 18–22.

Pott, Percivall. *Chirurgical Observations Relative to the Cataract, the Polypus of the Nose, the Cancer of the Scrotum, the Different Kinds of Ruptures, and the Mortification of the Toes and Feet.* London: Hawes, Clarke, and Collins, 1775.

Potter, Michael. "Percivall Pott's Contribution to Cancer Research." In *The First International Conference on the Biology of Cutaneous Cancer,* edited by Frederick Urbach. Washington, D.C.: Government Printing Office, 1962, pp. 1–13.

Proctor, Robert N. *Racial Hygiene: Medicine Under the Nazis.* Cambridge, Mass.: Harvard University Press, 1988.

———. *Value-Free Science? Purity and Power in Modern Knowledge.* Cambridge, Mass.: Harvard University Press, 1991.

Ramazzini, Bernardino. *De morbis artificum diatriba.* 2nd ed. 1713. Translated by W. C. Wright as Diseases of Workers. New York: Hafner, 1964.

Rather, L. J. *The Genesis of Cancer: A Study in the History of Ideas.* Baltimore, Md.: Johns Hopkins University Press, 1978.

Ravenholt, R. T. "Tobacco's Global Death March." *Population and Development Review* 16 (1990): 213–240.

Ray, Dixy Lee, with Lou Guzzo. *Trashing the Planet: How Science Can Help Us Deal with Acid Rain, Depletion of the Ozone, and Nuclear Waste.* New York: Harper Perennial, 1990.

———. *Environmental Overkill: Whatever Happened to Common Sense?* Washington, D.C.: Regnery Gateway, 1993.

Rennie, Susan. "Breast Cancer Prevention: Diet vs. Drugs." *Ms.* (May–June 1993): 38–46.

Rettig, Richard A. *Cancer Crusade: The Story of the National Cancer Act of 1971.* Princeton, N.J.: Princeton University Press, 1977.

Richards, Evelleen. "The Politics of Therapeutic Evaluation: The Vitamin C and Cancer Controversy. *Social Studies of Science* 18 (1988): 653–701.

Ringholz, Raye C. *Uranium Frenzy: Boom and Bust on the Colorado Plateau.* Albuquerque: University of New Mexico Press, 1989.

Roberts, Leslie. "British Radiation Study Throws Experts into Tizzy." *Science* 248 (1990): 24–25.

———. "Zeroing In on a Breast Cancer Susceptibility Gene." *Science* 259 (1993): 622–625.

Rose, Geoffrey. *The Strategy of Preventive Medicine.* Oxford: Oxford University Press, 1992.

Rosner, David, and Gerald Markowitz. *Deadly Dust: Silicosis and the Politics of Occupational Disease.* Princeton, N.J.: Princeton University Press, 1991.

———, eds. *Dying for Work: Worker's Safety and Health in Twentieth-Century America.* Bloomington: Indiana University Press, 1989.

Ross, Walter S. *Crusade: The Official History of the American Cancer Society.* New York: Arbor House, 1987.

Rothman, Kenneth J. *Modern Epidemiology.* Boston: Little, Brown, 1986.

Rous, Peyton. "The Nearer Causes of Cancer." *JAMA* 122 (1943): 573–581.

Roush, George C., et al. *Cancer Risk and Incidence Trends: The Connecticut Perspective.* Washington, D.C.: Hemisphere, 1987.

Rubin, Philip, ed. *Clinical Oncology: A Multidisciplinary Approach for Physicians and Students.* 7th ed. Philadelphia: W. B. Saunders, 1993.

Rushefsky, Mark E. *Making Cancer Policy.* Albany: State University of New York Press, 1986.

Samet, Jonathan M., and Anthony Nero. "Indoor Radon and Lung Cancer." *New England Journal of Medicine* 320 (1989): 591–594.

Schimke, R. Neil. "Cancer in Families." In *The Genetic Basis of Common Diseases,* edited by Richard A. King et al. New York: Oxford University Press, 1992, pp. 641–649.

Schneider, Keith. "Uranium Miners Inherit Dispute's Sad Legacy." *New York Times* January 9, 1990.

Schneiderman, Marvin A. "Sources, Resources and Tsouris." In *Persons at High Risk of Cancer,* edited by Joseph F. Fraumeni, Jr. New York: Academic Press, 1975.

———. "Cancer: Scientific Policy, Public Policy, and the Prevention of Disease." *Carolina Environmental Essay Series* 4 (1983).

Schottenfeld, David, and Joanna F. Haas. "Carcinogens in the Workplace." *CA—A Cancer Journal for Clinicians* 29 (1979): 144–168.

Scotto, Joseph, and John C. Bailar III. "Rigoni-Stern and Medical Statistics." *Journal of the History of Medicine* 24 (1969): 65–75.

Selikoff, Irving J., Jacob Churg, and E. Cuyler Hammond. "Asbestos Exposure and Neoplasia." *Journal of the American Medical Association* 188 (1964): 22–26.

Selikoff, Irving J., and Douglas H. K. Lee. *Asbestos and Disease.* New York: Academic Press, 1978.

Sethi, S. Prakash. *Handbook of Advocacy Advertising: Concepts, Strategies and Applications.* Cambridge, Mass.: Ballinger, 1987.

Shepherd, Kevin L. "Indoor Radon Rouses the Commercial Real Estate Industry." *Journal of Real Estate Development* 5 (1989): 45–50.

Shimkin, Michael B. *Science and Cancer.* Washington, D.C.: National Institutes of Health, 1973.

———. *Contrary to Nature.* Washington, D.C.: National Institutes of Health, 1977.

———. *Some Classics of Experimental Oncology.* Washington, D.C.: National Institutes of Health, 1980.

———. "Oncology (Neoplastic Diseases)." In *The Oxford Companion to Medicine,* edited by John Walton et al. Oxford: Oxford University Press, 1986.

———, and V. A. Triolo. "History of Chemical Carcinogenesis: Some Prospective Remarks." *Progress in Experimental Tumor Research* 11 (1969): 1–20.

Sibbison, Jim. "Censorship at the New E.P.A." *The Nation* (September 11, 1982): 208–209.

Sibley, Septimus W. "A Contribution to the Statistics of Cancer." *Medico-Chirurgical Transactions* 24 (1859): 111–152.

Sidel, Victor W. "Health Care." In *What Reagan Is Doing to Us,* edited by Alan Gartner et al. New York: Harper and Row, 1982.

Sigerist, Henry E. "The Historical Development of the Pathology and Therapy of Cancer." *Bulletin of the New York Academy of Medicine* 9 (1932): 642–653.

Snow, Herbert. "Is Cancer Hereditary?" *British Medical Journal* (October 10, 1885): 690–692.

Solomon, Alisa. "The Politics of Breast Cancer." *The Village Voice* (May 14, 1991): 22–27.

Sontag, Susan. *Illness as Metaphor.* New York: Farrar, Straus & Giroux, 1978.

Spencer, Leslie. "Ban All Plants—They Pollute." *Forbes* (October 25, 1993): 104.

Spatz, Lawrence, et al., eds. *Detection of Cancer Predisposition: Laboratory Approaches.* White Plains, N.Y.: March of Dimes, 1990.

Stannard, J. Newell. *Radioactivity and Health: A History.* 3 vols. Springfield, Va.: Battelle Memorial Institute, 1988.

Starr, Paul. *The Social Transformation of American Medicine.* New York: Basic Books, 1982.

Stefansson, Vilhjalmur. *Cancer: Disease of Civilization?* New York: Hill and Wang, 1960.

Stewart, Alice M. "Delayed Effects of A-bomb Radiation." *Journal of Epidemiology and Community Health* 36 (1982): 80–86.

———, and G. W. Kneale. "Non-Cancer Effects of Exposure to A-Bomb Radiation." *Journal of Epidemiology and Community Health* 38 (1984): 108–112.

Stewart, Fred W. "Occupational and Post-Traumatic Cancer." *Bulletin of the New York Academy of Medicine* 23 (1947): 145–162.

Stocker, Midge, ed. *Confronting Cancer, Constructing Change: New Perspectives on Women and Cancer.* Chicago: Third Side Press, 1993.

Stranahan, Susan Q. "How Life-Saving Legislation Sits in Congress." *Philadelphia Inquirer,* March 15, 1976.

Strickland, Stephen P. *Politics, Science and Dread Disease.* Cambridge, Mass.: Harvard University Press, 1972.

Studer, Kenneth E., and Daryl E. Chubin. *The Cancer Mission: Social Contexts of Biomedical Research.* Beverly Hills, Calif.: Sage, 1980.

Tamplin, Arthur R., and John W. Gofman. *'Population Control' Through Nuclear Policy.* Chicago: Nelson-Hall, 1970.

Tanchou, Stanislas. "Recherches sur la fréquence du cancer." *Gazette des hopitaux,* July 6, 1843, p. 313.

Taylor, Peter. *The Smoke Ring: Tobacco, Money, and Multinational Politics.* New York: Pantheon Books, 1984.

Tenner, Edward. "Warning: Nature May Be Hazardous to Your Health." *Harvard Magazine* (September–October 1987): 35–38.

Tesh, Sylvia. "Disease Causality and Politics." *Journal of Health Politics, Policy and Law* 6 (1981): 369–390.

Thilly, William G. "What Actually Causes Cancer." *Technology Review* (May–June 1991): 49–54.

Tierney, John. "Not to Worry." *Hippocrates* (January–February 1988): 29–38.

"Too Much Fuss About Pesticides?" *Consumer Reports* (October 1989): 655–658.

Truax, Hawley. "Surviving the 1980s at EPA." *Environmental Action* (January–February 1989): 12–16.

Uhlig, Margarete. "Über den Schneeberger Lungenkrebs." *Virchows Archiv für pathologische Anatomie* 230 (1921): 76–98.

United Nations Scientific Committee on the Effects of Atomic Radiation. *Sources, Effects and Risks of Ionizing Radiation.* New York: United Nations, 1988.

Upton, Arthur C. "Are There Thresholds for Carcinogenesis? The Thorny Problem of Low Level Exposure." *Annals of the New York Academy of Sciences* 534 (1988): 863–884.

———, et al. *Health Effects of Exposure to Low Levels of Ionizing Radiation* (BEIR V). Washington, D.C.: National Academy Press, 1990.

Vainio, H., M. Sorsa, and K. Hemminki, eds. *Occupational Cancer and Carcinogenesis.* Washington, D.C.: Hemisphere, 1981.

Varmus, Harold, and Robert A. Weinberg. *Genes and the Biology of Cancer.* New York: Scientific American Library, 1993.

Viorst, Milton, and J. V. Reistrup. "Radon Daughters and the Federal Government." *Bulletin of the Atomic Scientists* 23 (October 1967): 25–29.

Viscusi, W. Kip. *Risk by Choice: Regulating Health and Safety in the Workplace.* Cambridge, Mass.: Harvard University Press, 1983.

Wagner, Gustav, and Andrea Mauerberger. *Krebsforschung in Deutschand.* Berlin: Springer, 1989.

Walshe, Walter H. *The Nature and Treatment of Cancer.* London: Taylor and Walton, 1846.

Warner, Kenneth E. "Tobacco Industry Scientific Advisors: Serving Society or Selling Cigarettes?" *American Journal of Public Health* 81 (1991): 839–842.

Wasserman, Harvey, and Norman Solomon. *Killing Our Own: The Disaster of America's Experience with Atomic Radiation.* New York: Dell, 1982.

Weart, Spencer R. *Nuclear Fear: A History of Images.* Cambridge, Mass.: Harvard University Press, 1988.

Weinberg, Robert A. "Tumor Suppressor Genes." *Science* 253 (1991): 1138–1146.

Weinstein, Henry. "Did Industry Suppress Asbestos Data?" *Los Angeles Times,* October 23, 1978.

Whelan, Elizabeth M. "Cancer and the Politics of Fear." *Toxic Substances Journal* 1 (1979): 78–94.

———. "The Politics of Cancer." *Policy Review* (Fall 1979): 33–46.

———. *A Smoking Gun: How the Tobacco Industry Gets Away with Murder.* Philadelphia: George F. Stickly, 1984.

———. *Toxic Terror.* Ottawa, Ill.: Jameson Books, 1985.

———. "The Facts Behind the Health Scares." *Across the Board* (June 1990): 19–24.

White, Ray. "Genetics of Colon Cancer Predisposition." Paper presented to the Human Genome Workshop. Bethesda, Md., January 24–25, 1991.

Whittemore, Gilbert F. "The National Committee on Radiation Protection, 1928–1960." Ph.D. thesis, Harvard University, 1986.

Wildavsky, Aaron. "Pollution as Moral Coercion: Culture, Risk Perception, and Libertarian Values." *Cato Journal* 2 (1982): 305–307.

———. "Topic of Cancer." *Policy Review* (Fall 1984): 73–75.

Willcox, Walter F. "On the Alleged Increase of Cancer." *Publications of the American Statistical Association* 15 (1916–1917): 701–782.

Williams, W. Roger. *The Natural History of Cancer, with Special Reference to Its Causation and Prevention.* New York: William Wood, 1908.

Winner, Langdon. "Risk: Another Name for Danger." *Science for the People* (May–June 1986): 5–15, 27.

Witkowski, J. A. "The Inherited Character of Cancer—An Historical Survey." *Cancer Cells* 2 (1990): 229–257.

Wolff, Jacob. *The Science of Cancerous Disease from Earliest Times to the Present* (1907). New Delhi: Amerind, 1989.

Working Group. "Statement of the First Workshop of the Joint Working Group on the Ethical, Legal and Social Issues Related to Mapping and Sequencing the Human Genome, Williamsburg, Va., February 5–6, 1990." In the HHS and DOE booklet, *Understanding Our Genetic Inheritance. The U.S. Human Genome Project: The First Five Years FY 1991–1995.* Springfield, Va.: National Technical Information Service, 1990, pp. 65–73.

Wright, Karen. "Breast Cancer: Two Steps Closer to Understanding." *Science* 250 (1990): 1659.

———. "Going by the Numbers." *New York Times Magazine* (December 15, 1991): 59–60, 77–78.

Wynbrandt, James, and Mark D. Ludman. *The Encyclopedia of Genetic Disorders and Birth Defects.* New York: Facts on File, 1991.

Index